T0211057

Lecture Notes in Computer Science 10398

More information about this series at http://www.springer.com/series/7407

S. Arumugam · Jay Bagga
Lowell W. Beineke · B.S. Panda (Eds.)

Theoretical Computer Science and Discrete Mathematics

First International Conference, ICTCSDM 2016
Krishnankoil, India, December 19–21, 2016
Revised Selected Papers

Editors
S. Arumugam
Kalasalingam University
Krishnankoil
India

Lowell W. Beineke
Indiana University – Purdue University
Fort Wayne, IN
USA

Jay Bagga
Ball State University
Muncie, IN
USA

B.S. Panda
Indian Institute of Technology
New Delhi
India

ISSN 0302-9743 ISSN 1611-3349 (electronic)
Lecture Notes in Computer Science
ISBN 978-3-319-64418-9 ISBN 978-3-319-64419-6 (eBook)
DOI 10.1007/978-3-319-64419-6

Library of Congress Control Number: 2017948183

LNCS Sublibrary: SL1 – Theoretical Computer Science and General Issues

Printed on acid-free paper

This Springer imprint is published by Springer Nature
The registered company is Springer International Publishing AG
The registered company address is: Gewerbestrasse 11, 6330 Cham, Switzerland

Preface

This volume contains selected papers presented at ICTCSDM 2016, the International Conference on Theoretical Computer Science and Discrete Mathematics. ICTCSDM 2016 was held during December 19–21, 2016, at Kalasalingam University in rural Tamil Nadu, India. The conference was jointly organized by the National Centre for Advanced Research in Discrete Mathematics (n-CARDMATH), Kalasalingam University, India, Department of Computer Science, Ball State University, USA, and Department of Mathematics, Indiana University-Purdue University, Fort Wayne, USA.

The conference was sponsored and financially supported by the Science and Engineering Research Board, Government of India, New Delhi, the National Board for Higher Mathematics, Mumbai, the Council of Scientific and Industrial Research, New Delhi, the Indian National Science Academy, New Delhi, and Kalasalingam University.

The conference attracted around 150 participants from all over the world. The nations represented were USA, Indonesia, Czech Republic, Slovakia, South Africa, Malaysia, Dubai, Iran, and India.

Twelve experts from outside India and 14 experts from leading institutions within India delivered invited talks on various topics such as line graphs and their generalizations, large graphs of given degree and diameter, graphoidal covers, adjacency spectrum, distance spectrum, b-coloring, separation dimension of graphs and hypergraphs, domination in graphs, graph labeling problems, subsequences of words and Parike matrices, λ-design conjecture, graph algorithms and interference model for wireless sensor networks.

The call for papers for ICTCSDM 2016 was distributed around the world, resulting in 210 submission. After the review process a total of 57 papers (27%) were accepted subject to revision for publication in this volume.

We thank the authors for their valuable contributions and the Program Committee members and other referees for their constructive and enlightening comments on the manuscripts. We thank Springer for publishing the proceedings in the *Lecture Notes in Computer Science* series. We thank the Organizing Committee from Kalasalingam University for the excellent arrangements and for its efficient management of the conference-related activities. We thank all the sponsors for their financial support.

June 2017

S. Arumugam
Jay Bagga
L.W. Beineke
B.S. Panda

Organization

Program Committee

S. Arumugam (Chair)	Kalasalingam University, India
Jay Bagga (Co-chair)	Ball Sate University, USA
R. Balakrishnan	Bharathidasan University, India
L.W. Beineke (Co-chair)	Indiana University-Purdue University, USA
S. Francis Raj	Pondicherry University, India
Bruce W. Watson	Stellenbosch University, South Africa
G.L. Chia	Universiti Tunku Abdul Rahman, Malaysia
Daniela Ferrero	Texas State University, USA
Dominique Haughton	Bentley University, USA
T. Karthick	Indian Statistical Institute, Chennai, India
Kiki Ariyanti Sugeng	University of Indonesia, Indonesia
Kiran R. Bhutani	The Catholic University of America, USA
L. Lesniak	Western Michigan University, USA
N.S. Narayanaswamy	Indian Institute of Technology Madras, Chennai, India
B.S. Panda	Indian Institute of Technology Delhi, New Delhi, India
Petr Kovar	VSB, Technical University of Ostrava, Czech Republic
Purnima Gupta	University of Delhi, India
Rinovia Simanjuntak	Institut Teknologi Bandung, Indonesia
Saeid Alikhani	Yazd University, Iran
E. Sampathkumar	University of Mysore, India
S.S. Sane	Indian Institute of Technology Bombay, Mumbai, India
T. Singh	BITS-Pilani K.K. Birla Goa Campus, India
K.G. Subramanian	Madras Christian College, Chennai, India
L. Sunil Chandran	Indian Institute of Science, Bangalore, India
Stanislav Jendrol	Pavol Jozef Safarik University in Kosice, Kosice
Sona Pavlikova	Slovak University of Technology, Slovak Republic
Tao-Ming Wang	Tunghai University, Taichung, Taiwan
P. Titus	University College of Engineering Nagercoil, India
Siba K. Udgata	University of Hyderabad, India
Venkatesh Raman	Institute of Mathematical Sciences, Chennai, India
A. Vijayakumar	Cochin University of Science and Technology, India

Organizing Committee (Kalasalingam University, Tamil Nadu, India)

K. Sridharan (Chancellor, Chief Patron)
S. Shasi Anand (Director, Patron)
S. Arjun Kalasalingam (Director, Patron)
S. Saravanasankar (Vice-chancellor, President)
S. Arumugam (Organizing Chair)
Jay Bagga (Co-chair)
L.W. Beineke (Co-chair)
V. Vasudevan
D. Devaraj
K. Sundar
M. Venkatesulu
C. Sivapragasam
S.R. Srikumar
P. Paulraja
K. Vaikundamoorthy
K. Karuppasamy
A. Thuraiswamy
P. Thangaraj
K. Raja Chandrasekar

Additional Reviewers

Martin Bača
Andrea Feňovčíková
M. Venkatesulu
Joe Ryan
S. Slamin
J. Paulraj Joseph
R. Kala
T. Tamizh Chelvam
K. Raja Chandrasekar
P. Paulraja
A.P. Santhakumaran
S. Ramachandran
Lakshmana Gomathinayagam
T. Rajaretnam
M.G. Karunambigai
K. Suthendran
K. Thulasiraman
A. Somasundaram

Contents

Degree Associated Reconstruction Number of Biregular Bipartite Graphs Whose Degrees Differ by at Least Two

A. Anu and S. Monikandan(✉)

Department of Mathematics, Manonmaniam Sundaranar University,
Tirunelveli 627012, India
esa.anu1188@gmail.com, monikandans@gmail.com

Abstract. A vertex-deleted subgraph of a graph G is called a *card* of G. A card of G with which the degree of the deleted vertex is also given is called a *degree associated card* (or *dacard*) of G. The *degree associated reconstruction number* of a graph G (or $drn(G)$) is the size of the smallest collection of dacards of G that uniquely determines G. It is shown that $drn(G) = 1$ or 2 for all biregular bipartite graphs with degrees d and $d + k$, $k \geq 2$ except the bistar $B_{2,2}$ on 6 vertices and that $drn(B_{2,2}) = 3$.

Keywords: Reconstruction · Reconstruction number · Isomorphism

1 Introduction

All graphs considered in this paper are finite, simple and undirected. We shall mostly follow the graph theoretic terminology of [7]. A *vertex-deleted subgraph* or *card* $G - v$ of a graph (digraph) G is the unlabeled graph (digraph) obtained from G by deleting the vertex v and all edges (arcs) incident with v. The *deck* of a graph (digraph) G is its collection of cards. Following the formulation in [2], a graph (digraph) G is *reconstructible* if it can be uniquely determined from its deck. The well-known Reconstruction Conjecture (RC) due Kelly [10] and Ulam [16] asserts that every graph with at least three vertices is reconstructible. The conjecture has been proved for many special classes, and many properties of G may be deduced from its deck. Nevertheless, the full conjecture remains open [6]. Harary and Plantholt [9] defined the reconstruction number of a graph G, denoted by $rn(G)$, to be the minimum number of cards which can only belong to the deck of G and not to the deck of any other graph H, $H \ncong G$, these cards thus uniquely identifying G. Reconstruction numbers are known for only few classes of graphs [4].

An extension of the RC to digraphs is the *Digraph Reconstruction Conjecture* (DRC) proposed by Harary [8]. It was disproved by Stockmeyer [15] by exhibiting several infinite families of counter-examples and this made people doubt the

S. Monikandan—Research is supported by the SERB, Govt. of India, Grant no. EMR/2016/000157.

S. Arumugam et al. (Eds.): ICTCSDM 2016, LNCS 10398, pp. 1–9, 2017.
DOI: 10.1007/978-3-319-64419-6_1

RC itself. To overcome this, Ramachandran [12] introduced degree associated reconstruction for digraphs and proposed a new conjecture in 1981. It was proved [12] that the digraphs in all these counterexamples to the DRC obey the new conjecture, thereby protecting the RC from the threat posed by these digraph counterexamples.

The ordered triple (a, b, c) where a, b and c are respectively the number of unpaired outarcs, unpaired inarcs and symmetric pair of arcs incident with v in a digraph D is called the *degree triple of* v. The *degree associated card* or *dacard* of a digraph (graph) is a pair (d, C) consisting of a card C and the degree triple (degree) d of the deleted vertex. The *dadeck* of a digraph is the multiset of all its dacards. A digraph is said to be *N-reconstructible* if it can be uniquely determined from its dadeck. The *new digraph reconstruction conjecture* [12] (NDRC) asserts that all digraphs are N-reconstructible. Ramachandran [13] then studied the degree associated reconstruction number of graphs and digraphs in 2000. The *degree* (*degree triple*) *associated reconstruction number* of a graph (digraph) D is the size of the smallest collection of dacards of D that uniquely determines D. Articles [1–3,5,11,14] are recent papers on this parameter.

A graph G is *bipartite* if its vertex set is the union of two disjoint independent sets, called *partite sets* of G. A graph whose vertices all have one of two possible degrees is called a *biregular* graph. We show that if G is a biregular bipartite graph, other than the bistar $B_{2,2}$ on 6 vertices, with degrees d and $d+k$, $k \geq 2$, then $drn(G) = 1$ or 2 and that $drn(B_{2,2}) = 3$.

2 *Drn* of Biregular Bipartite Graphs

The degree of a vertex v in G is denoted by $deg_G\ v$ or simply $deg\ v$. A vertex of degree m is called an *m-vertex*. The *neighbourhood* of a vertex v in G, written $N_G(v)$ or simply $N(v)$, is the set of all vertices adjacent to v in G.

Notation. By x, x' with or without subscripts, we mean respectively a d-vertex and a $(d+k)$-vertex in the partition X. Such vertices but in the partition Y are denoted by y and y' respectively.

An *extension* of a dacard $(d(x), G-x)$ of G is a graph obtained from the dacard by adding a new vertex v and joining it to $d(x)$ vertices of the dacard and it is denoted by $H(d(x), G-x)$ (or simply by H). Throughout this paper, H and v are used in the sense of this definition. For a graph G, to prove $drn(G) = k$, we show that every extension (other than G) of at least one dacard has at most $k-1$ dacards in common with those of G (thus $drn(G) \leq k$), and that at least one extension has precisely $k-1$ dacards in common with those of G (thus $drn(G) \geq k$).

In their paper [5], Barrus and West have characterized (Theorem A) graphs G with $drn(G) = 1$.

Theorem A. The dacard $(d, G-v)$ belongs to the dadeck of only one graph (up to isomorphism) if and only if one of the following holds:

(i) $d = 0$ or $d = |V(G-v)|$;
(ii) $d = 1$ or $d = |V(G-v)| - 1$, and $G-v$ is vertex-transitive; or
(iii) $G-v$ is complete or edgeless.

Ramachandran [13] has verified that $drn(G) = 1$, 2 or 3 for all graphs G on at most 6 vertices, and in particular, $drn(G) = 1$ or 2 for all biregular bipartite graphs G on at most 6 vertices with degrees d and $d+k$ such that $k \geq 2$, except the bistar $B_{2,2}$ and that $drn(B_{2,2}) = 3$. So, *we assume that all biregular bipartite graphs G considered here onwards have order at least 7 and no dacard of G satisfies the conditions of Theorem A and so* $drn(G) \geq 2$.

Theorem 1. *If G is a biregular bipartite graph with a vertex adjacent to all the vertices in the other partite set, then* $drn(G) = 2$.

Proof. The graph G is clearly connected. Let z be a vertex adjacent to all the vertices in the other partite set of G. If $deg\ z = d$, then G is a complete bipartite graph and $drn(G) = 2$, since it is known [13] that $drn(K_{m,n}) = 2$ for $2 \leq m < n$. So, let $deg\ z = d+k,\ k \geq 2$.

Suppose that z is adjacent to a d-vertex. Consider the two dacards $(d+k, G-z)$ and $(d, G-w)$, where w is a d-vertex in $N(z)$. It is clear that the dacard $G-w$ is connected and so it has exactly one partite set such that every vertex in the partite set has degree d or $d+k$. To get an extension $H(d+k, G-z)$, add a new vertex v to the dacard $G-z$ and join it with precisely $d+k$ vertices. Clearly $G-z$ contains exactly one partite set (say Z_1) having a $(d-1)$-vertex and a $(d+k-1)$-vertex. If v were joined to all the vertices in Z_1, then the resulting extension H would be isomorphic to G. Otherwise, in every extension $H(d+k, G-z)$, the newly added vertex v is joined to at least one vertex not in Z. But then any d-vertex deleted dacard of $H(d+k, G-z)$ contains a vertex of degree different from d and $d+k$ from each partite set and so it is not isomorphic to $G-w$.

Suppose that no d-vertex is adjacent to z. Now consider the two dacards $(d+k, G-z)$ and $(d+k, G-w)$, where $w \in N(z)$. In $G-w$, exactly one partite set is $(d+k)$-regular. In the extension $H(d+k, G-z)$, if the newly added vertex v were joined only to the $(d+k-1)$-vertices, then H would be isomorphic to G. Otherwise, any $(d+k)$-vertex deleted dacard of $H(d+k, G-z)$ must contain irregular partite sets or a partite set is $(d+k-1)$-regular and hence it is not isomorphic to $(d+k, G-w)$. Thus no graph other than G contains both the two dacards $(d+k, G-z)$ and $(d+k, G-w)$ in its dadeck and hence $drn(G) = 2$.

Theorem 2. *Let G be a biregular bipartite graph. If G has a vertex adjacent to no d-vertices and has a vertex adjacent to no* $(d+k)$*-vertices, then* $drn(G) = 2$.

Proof. Let z be a vertex adjacent to no $(d+k)$-vertices and let z' be that adjacent to no d-vertices. Here we use the two dacards $(d(z), G-z)$ and $(d(z'), G-z')$. Clearly $\delta(G-z') > d-1$. In the extension of $(d(z), G-z)$, if the newly added vertex v were joined to all the $(d-1)$-vertices, then the resulting extension H would be isomorphic to G. Thus, in every extension of $(d(z), G-z)$, vertex v is not joined to a $(d-1)$-vertex of $(d(z), G-z)$. But then each dacard of $H(d(z), G-z)$ contains at least one vertex of degree at most $d-1$ and so it is not isomorphic to $(d(z'), G-z')$. Thus no graph other than G has both the dacards $(d(z), G-z)$ and $(d(z'), G-z')$ in its dadeck and $drn(G) = 2$.

Theorem 3. *If G is a biregular bipartite graph such that all the vertices in a partite set have the same degree in G, then $drn(G) = 2$.*

Proof. Let Y be such a partite set. Let G have exactly a vertices of degree d, and b vertices of degree $d+k$, where $a, b \geq 1$ and $k \geq 2$. We consider the two dacards $(d, G-x)$ (or $(d+k, G-x')$) and $(d+k, G-y')$ (or $(d, G-y)$). In $G-y'$ (or $G-y$), exactly one partite set is $(d+k)$-regular (or d-regular). In the extension of $(d, G-x)$ (or $(d+k, G-x')$), if the newly added vertex v were joined to all $(d+k-1)$-vertices (or $(d-1)$-vertices), then the resulting extension H would be isomorphic to G. Thus, in every extension of $(d, G-x)$ (or $(d+k, G-x')$), vertex v is not joined to at least one $(d+k-1)$-vertex (or $(d-1)$-vertex). But then any $(d+k)$-vertex (or d-vertex) deleted dacard would contain either at least one vertex of different degree in each partite set or exactly one d-regular (or $(d+k)$-regular) partite set. Hence no dacard of H is isomorphic to $G-y'$ (or $G-y$). Thus no graph other than G has both the dacards $(d, G-x)$ (or $(d+k, G-x')$) and $(d+k, G-y')$ (or $(d, G-y)$) in its dadeck and $drn(G) = 2$.

Theorem 4. *If G is a biregular bipartite graph such that all but one vertex in a partite set have degree $d+k$, $(k \geq 2)$ then $drn(G) = 2$.*

Proof. Let Y be such a partite set and let y be the unique d-vertex in Y. If the other partite set X contains at least two d-vertices, then we choose the two dacards $(d+k, G-x')$ and $(d, G-y)$, where $x' \in N(y)$. In $G-y$, exactly one partite set is $(d+k)$-regular. Now we consider the extension of $(d+k, G-x')$. If the newly added vertex v were not joined to any vertex of degree d or $d+k$, then the resulting extension H would be isomorphic to G. Otherwise, any d-vertex deleted dacard of H contains at least one $(d+k+1)$- vertex, at least one vertex of different degree in both the partite sets, or a $(d+1)$-regular partite set. Hence such a dacard is not isomorphic to $G-y$. Therefore no graph other than G contains both these two dacards in its dadeck, we have $drn(G) = 2$. So, we assume that X contains a unique d-vertex.

Let us first consider the case that G is disconnected. Suppose that the two d-vertices of G are belonging to the same component of G. Then consider the two dacards $(d+k, G-x')$ and $(d, G-x)$, where x' and x are belonging to different components. Clearly $\Delta(G-x) \leq d+k$. Consider the extension of $(d+k, G-x')$. If the newly added vertex v were joined to all the $(d+k-1)$-vertices, then H would be isomorphic to G. Otherwise, any d-vertex deleted dacard of H contains at least one vertex of degree $d+k+1$ and so it is not isomorphic to $G-x$. Suppose that the two d-vertices of G are belonging to different components of G. Then consider the two dacards $(d+k, G-x')$ and $(d, G-x)$ where x' and x are in different components and $y \in N(x')$. Clearly $\Delta(G-x) \leq d+k$. In the extension $H(d+k, G-x')$, if the newly added vertex v were not joined to any d-vertex and to any $(d+k)$-vertex, then H would be isomorphic to G. Otherwise, any d-vertex deleted dacard of H contains at least one vertex of degree $d+k+1$ and so it is not isomorphic to $G-x$, we have $drn(G) = 2$.

Assume now that G is connected and if the two d-vertices are adjacent, then consider the dacards $(d, G-x)$ and $(d, G-y)$. In $G-y$, exactly one partite set is $(d+k)$-regular. In $H(d, G-x)$, if v were not joined to any $(d+k)$-vertex, then H would be isomorphic to G. Otherwise, any d-vertex deleted dacard of H contains a $(d+k+1)$-vertex and so it is not isomorphic to $G-y$. Hence $drn(G) = 2$. So, *let us assume that the two d-vertices are nonadjacent in G*. Let X have m $(d+k)$-vertices. Suppose that $|Y| \neq m+1$. Consider the two dacards $(d, G-y)$ and $(d, G-x)$. In $G-x$, exactly one partite set is $(d+k)$-regular and it has size m. In $H(d, G-y)$, if v were joined to all the $(d+k-1)$-vertices, then H would be isomorphic to G. Otherwise, any d-vertex deleted dacard of H contains at least one $(d+k+1)$-vertex, at least one vertex of different degree in both the partite sets, or there is a unique $(d+k)$-regular partite set of size not equal to m. Therefore H has no dacard isomorphic to $G-x$. Suppose that $|Y| = m+1$ and $m = d+k$. Consider the two dacards $(d, G-x)$ and $(d+k, G-x')$, where $x' \notin N(y)$. In $G-x'$, exactly one d-vertex in each partite set. In $H(d, G-x)$, if v were joined to all the $(d+k-1)$-vertices, then H would be isomorphic to G. Otherwise, any d-vertex deleted dacard of H has at least one vertex of degree at most $d+k-1$, or at least two d-vertices in the same partite set. Therefore H has no dacard isomorphic to $G-x'$.

Finally, assume that $m \geq d+k+1$. Consider the two dacards $(d+k, G-x')$ and $(d, G-y)$, where $y \notin N(x')$. Exactly one partite set of $G-x'$, say Y_1 contains a $(d+k-1)$-vertex. Similarly, exactly one partite set of $G-y$, say X_1 contains a $(d+k-1)$-vertex. Now we construct two new dacards of a supergraph, say G_1 obtained from G by adding a new vertex z to the partite set X and joining it with all the vertices in the partite set Y and therefore the new vertex attains the degree at least $d+k+1$. By adding a new vertex w_1 and joining it to all the vertices in Y_1 of $G-x'$ gives a new dacard $(d+k, G_1-x')$. Similarly, by adding a new vertex w_2 and joining it to the neighbours of every vertex in X_1 of $G-y$ gives a new dacard $(d+1, G_1-y)$. Clearly, $deg_{G_1-x'}\ w_1 = m+1$ and $deg_{G_1-y}\ w_2 = m$.

In the extension $H_1(d+1, G_1-y)$, if v were not joined to the pair of vertices of degrees d and $d+k$, then H_1 would be isomorphic to G_1. Otherwise, any $(d+k)$-vertex deleted dacard of the extension H_1 is connected and it contains a unique $(m+1)$-vertex and a $(d+k-1)$-vertex (or $(d+1)$-vertex) in the same partite set. Hence such a dacard of H_1 is not isomorphic to $(d+k, G_1-x')$. Therefore $drn(G_1) = 2$. This means that G_1 can be obtained uniquely from the new dacards $(d+k, G_1-x')$ and $(d+1, G_1-y)$. Now the vertex z in G_1 is identifiable as the unique vertex of degree at least $d + k + 1$. Consequently, G can be obtained uniquely from G_1 by deleting the vertex z. In other words, G can be obtained uniquely from the two dacards $(d + k, G - x')$ and $(d, G - y)$ and hence $drn(G) = 2$.

Theorem 5. *If G is a biregular bipartite graph such that all but one vertex in a partite set have degree d, then $drn(G) = 2$.*

Proof. Let Y be such a partite set. We proceed by two cases.

Case 1. X contains at least two $(d + k)$-vertices.
Here we consider the dacards $(d, G{-}x)$ (or $(d{+}k, G{-}x')$) and $(d{+}k, G{-}y')$, where x (or x') $\notin N(y')$. In $G - y'$, exactly one partite set is regular. In $H(d, G - x)$, if v were joined to all the $(d - 1)$-vertices, H would be isomorphic to G; otherwise, any $(d + k)$-vertex deleted dacard of H contains irregular partite sets and so it is not isomorphic to $G - y'$.

Case 2. X contains exactly one $(d + k)$-vertex.
Case 2.1. Both "$|X| = |Y| = d + k + 1$" and $y' \notin N(x')$ hold.
In $(d + k, G - x')$, exactly one partite set is regular and the unique $(d + k)$-vertex is adjacent to all the d-vertices. In $H(d, G - x)$, if v were joined to all the $(d - 1)$-vertices, then H would be isomorphic to G; otherwise, any $(d + k)$-vertex deleted dacard of H contains at least one vertex of degree $d - 2$, no vertex of degree $d + k$, or the dacard has irregular partite sets. Therefore no dacard of H is isomorphic to $G - x'$.

Case 2.2. Either $|X| = |Y| = d + k + 1$ or $y' \notin N(x')$ does not hold.
If G is disconnected, then we consider the dacards $(d, G - x)$ and $(d + k, G - y')$, where x and y' belong to different components of G. In $G - y'$, there is a unique component such that one of the partite sets of the component must contain $(d - 1)$-vertices. In $H(d, G - x)$, if v were joined to all the $(d - 1)$-vertices, then H would be isomorphic to G; otherwise, any $(d + k)$-vertex deleted dacard of H contains at least two components containing $(d - 1)$-vertices or at least two partite sets containing $(d - 1)$-vertices and so it is not isomorphic to $G - y'$.

Now we assume that G is connected. Consider the dacards $(d{+}k, G{-}y')$ and $(d, G - x)$, where $x \notin N(y')$. Exactly one partite set of $G - y'$, say Y_1 is regular. Similarly, the dacard $G - x$ contains exactly one partite set, say X_1 such that every vertex in X_1 has degree at least d. Now we construct two new dacards of a supergraph say G_1 obtained from G by adding a new vertex z to the partite set Y and joining it to all the vertices in the partite set X and therefore the new vertex attains the degree at least $d{+}k{+}1$. By adding a new vertex w_1 and joining it to the neighbours of every vertex in Y_1 of $G - y'$ gives the new dacard $(d + k, G_1 - y')$. Note that degree of w_1 is at least $d + k + 1$ in the supergraph G_1. Similarly, by adding a new vertex w_2 and joining it to all the vertices in X_1 of $G - x$ gives the new dacard $(d + 1, G_1 - x)$.

In the extension $H_1(d{+}1, G_1{-}x)$, if v were joined to all the $(d{-}1)$-vertices and to w_2, then the resulting extension H_1 would be isomorphic to G_1. Otherwise, any $(d{+}k)$-vertex deleted dacard of the extension H_1 is connected and containing no vertex that is adjacent to all the vertices of the other partite set, or the dacard has a unique partite set having a $(d-1)$-vertex and a vertex (say w) such that w is adjacent to all the vertices of the other partite set. Hence no $(d+k)$-vertex deleted dacard of H_1 is isomorphic to $G_1 - y'$ and $drn(G_1) = 2$, which means that G_1 can be obtained uniquely from the new dacards $(d+1, G_1 - x)$ and $(d+k, G_1 - y')$. Now the vertex z in G_1 is identifiable as the unique vertex of degree at least $d+k+1$. Consequently, G can be obtained uniquely from G_1 by deleting the vertex z. In other words, G can be obtained uniquely from the two dacards $(d+1, G_1 - x)$ and $(d+k, G_1 - y')$ and so $drn(G) = 2$.

Theorem 6. *If G is a biregular bipartite graph such that every partite set contains at least two vertices of degree d as well as $d+k$, where $k \geq 2$, then $drn(G) = 2$.*

Proof. Assume first that G is connected. Let X have m_1 vertices of degree d and n_1 vertices of degree $d+k$. Let that in Y be m_2 and n_2 respectively.

Case 1. $m_1 \neq m_2$.
Without loss of generality, let us take that $m_1 > m_2$. Here we use the two dacards $(d, G-y)$ and $(d+k, G-x')$. In $G-x'$, exactly one partite set contains m_1 vertices of degree d and the rest of them in the partite set have degree $d+k$. In $H(d, G-y)$, if v were joined to all the vertices of degree $d-1$ or $d+k-1$ then H would be isomorphic to G; otherwise, any $(d+k)$-vertex deleted dacard of H contains at most $m_1 - 1$ vertices of degree d in each partite set, or exactly one partite set has m_1 vertices of degree d and has at least one vertex of degree $d-1$ or $d+k-1$. Thus no dacard of H would be isomorphic to $G-x'$.

Case 2. $n_1 \neq n_2$.
Without loss of generality, let us take that $n_1 > n_2$. Here we use the two dacards $(d+k, G-y')$ and $(d, G-x)$. In $G-x$, exactly one partite set contains n_1 vertices of degree $d+k$ and the rest of them in the partite set have of degree d. In $H(d+k, G-y')$, if v were joined to all the vertices of degree $d-1$ or $d+k-1$ then H would be isomorphic to G; otherwise, any d-vertex deleted dacard of H contains at most $n_1 - 1$ vertices of degree $d+k$ in each partite set, or exactly one partite set has n_1 vertices of degree $d+k$ and has at least one vertex of degree $d-1$ or $d+k-1$. Thus no dacard of H would be isomorphic to $G-x$.

Case 3. $m_1 = m_2$ and $n_1 = n_2$.
Consider the dacards $(d, G-y)$ and $(d+k, G-x')$. The dacard $G-y$ (respectively $G-x'$) contains a unique partite set, say Z such that every vertex in it has degree $d-1$ or $d+k-1$. Now we construct two new dacards of a supergraph say G_1 obtained from G by adding a new vertex z to the partite set Z and joining it to all the vertices in the partite set Y and therefore the new vertex attains the degree at least $d+k+1$ in G_1. By adding a new vertex w_1 to $G-y$ and joining it to all the neighbours of every vertex in Z of $G-y$ gives the new dacard $(d+1, G_1 - y)$. Note that the degree of w_1 is at least $d+k+1$ in the graph G_1. Similarly, by adding a new vertex w_2 to $G-x'$ and joining it to all the vertices in Z of $G-x'$ gives the other new dacard $(d+k, G_1 - x')$.

In the extension $H_1(d+1, G_1 - y)$, if v were joined to w_2 and to all the vertices of degree $d-1$ or $d+k-1$, then the resulting extension H_1 would be isomorphic to G_1. Otherwise, any $(d+k)$-vertex deleted dacard of the extension H_1 is connected and it contains no vertex adjacent to all the vertices of the other partite set, or exactly one partite set has a $(d-1)$-vertex and a vertex (say w) such that w is adjacent to all the vertices of the other partite set. Hence no dacard of H_1 is isomorphic to $(d+k, G_1 - x')$ and so $drn(G_1) = 2$, which means that G_1 can be obtained uniquely from the new dacards $(d+1, G_1 - y)$ and $(d+k, G_1 - x')$. Now the vertex z in G_1 is identifiable as the unique vertex of degree at least $d+k+1$. Consequently, G can be obtained uniquely from G_1 by deleting the vertex z. In other words, G can be obtained uniquely from the two dacards $(d+1, G_1 - y)$ and $(d+k, G_1 - x')$ and hence $drn(G) = 2$, which completes the case that G is connected.

Now G is assumed to be disconnected; let $G_1, G_2, ..., G_n$ be the components of G. Let (X_i, Y_i) be the bipartition of G_i, $i = 1, 2, \ldots, n$. Suppose that X_i contains m'_i (respectively n'_i) vertices of degree d (respectively $d+k$) and Y_i contains m''_i (respectively n''_i) vertices of degree d (respectively $d{+}k$). If $m'_i > m''_i$ (or $n'_i > n''_i$) for some i, then consider any one component of G and proceeding as in Case 1 (or Case 2), we get $drn(G) = 2$. So, assume that $m'_i = m''_i$ and $n'_i = n''_i$ for all i.

Suppose that $m'_i \geq m'_j$ or $n'_i \geq n'_j$ for some $i \neq j$. If $m'_i > m'_j$ and $n'_i = n'_j$ for some $i \neq j$, then we consider the two dacards $(d{+}k, G{-}x'_{n'_i})$ and $(d, G{-}x_{n'_j})$. In $G - x_{n'_j}$, there is a component containing exactly one partite set with n'_i vertices of degree $d+k$ and all the vertices in the other partite set have degree d or $d+k$. In $H(d+k, G - x'_{n'_i})$, if v were joined to all the vertices of degree $d-1$ or $d+1$, then H would be isomorphic to G; otherwise any d-vertex deleted dacard of H contains a component such that one of the partite sets of the component contains m'_i vertices of degree d and all the vertices in the other partite set have degree $d-1$ or $d+k-1$. Therefore no dacard of H would be isomorphic to $G - x_{n'_j}$.

If $m'_i = m'_j$ and $n'_i > n'_j$ for some $i \neq j$, then consider the two dacards $(d, G - x_{n'_i})$ and $(d+k, G - x'_{n'_j})$. In $G - x'_{n'_j}$, there is a component containing exactly one partite set having n'_i vertices of degree $d+k$ and all the vertices in the other partite set have degree d or $d+k$. In $H(d, G - x_{n'_i})$, if v were joined to the pairs of vertices of degrees $d-1$ and $d+k-1$, then H would be isomorphic to G; otherwise, any $(d+k)$-vertex deleted dacard of H contains a component such that one of the partite sets of it contains m'_i vertices of degree d and all the vertices in other partite set have degree $d-1$ or $d+k-1$. Thus no dacard of H would be isomorphic to $G - x'_{n'_j}$. Finally, if $m'_i = m'_j$ and $n'_i = n'_j$ for all $i \neq j$; then consider any one component of G and by proceeding as in Case 3, we get $drn(G) = 2$.

From Theorems 3 to 6, we have the following main result.

Theorem 7. *If G is a biregular bipartite graph with degrees d and $d+k$, where $k \geq 2$, then $drn(G) = 2$.*

Acknowledgments. The work reported here is supported by the Research Project EMR/2016/000157 awarded to the second author by SERB, Government of India, New Delhi.

References

1. Anu, A., Monikandan, S.: Degree associated reconstruction number of certain connected graphs with unique end vertex and a vertex of degree $n-2$. Discrete Math. Algorithms Appl. **8**(4), 1650068-1–1650068-13 (2016)
2. Anusha Devi, P., Monikandan, S.: Degree associated reconstruction number of graphs with regular pruned graph. Ars Combin. (to appear)
3. Anusha Devi, P., Monikandan, S.: Degree associated reconstruction number of connected digraphs with unique end vertex. Australas. J. Combin. **66**(3), 365–377 (2016)
4. Asciak, K.J., Francalanza, M.A., Lauri, J., Myrvold, W.: A survey of some open questions in reconstruction numbers. Ars Combin. **97**, 443–456 (2010)
5. Barrus, M.D., West, D.B.: Degree-associated reconstruction number of graphs. Discrete Math. **310**, 2600–2612 (2010)
6. Bondy, J.A.: A graph reconstructors manual, in Surveys in Combinatorics. In: Proceedings of 13th British Combinatorial Conference London Mathematical Society, Lecture Note Series, vol. 166, pp. 221–252 (1991)
7. Harary, F.: Graph Theory. Addison Wesley, Reading (1969)
8. Harary, F.: On the reconstruction of a graph from a collection of subgraphs. In: Fieldler, M. (ed.) Theory of Graphs and its Applications, pp. 47–52. Academic Press, New York (1964)
9. Harary, F., Plantholt, M.: The graph reconstruction number. J. Graph Theory **9**, 451–454 (1985)
10. Kelly, P.J.: On isometric transformations. Ph.D. Thesis, University of Wisconsin Madison (1942)
11. Ma, M., Shi, H., Spinoza, H., West, D.B.: Degree-associated reconstruction parameters of complete multipartite graphs and their complements. Taiwanese J. Math. **19**(4), 1271–1284 (2015)
12. Ramachandran, S.: On a new digraph reconstruction conjecture. J. Combin. Theory Ser. B. **31**, 143–149 (1981)
13. Ramachandran, S.: Degree associated reconstruction number of graphs and digraphs. Mano. Int. J. Math. Scis. **1**, 41–53 (2000)
14. Spinoza, H.: Degree-associated reconstruction parameters of some families of trees (2014). (submitted)
15. Stockmeyer, P.K.: The falsity of the reconstruction conjecture for tournaments. J. Graph Theory. **1**, 19–25 (1977)
16. Ulam, S.M.: A Collection of Mathematical Problems. Interscience Tracts in Pure and Applied Mathematics 8. Interscience Publishers, New York (1960)

1-Normal DRA for Insertion Languages

Abhisek Midya[1(✉)], Lakshmanan Kuppusamy[2], V.S. Sumitha[3], and Alok Kumar Pani[3]

[1] Computer Science & Engineering, ICFAI Tech School, Hyderabad 501203, India
abhisekmidyacse@gmail.com

[2] Theoretical Computer Science Lab, School of Computer Science and Engineering, VIT University, Vellore 632014, India
klakshma@vit.ac.in

[3] Computer Science and Engineering, Christ University Faculty of Engineering, Bangalore 560074, India
{sumitha.vs,alok.kumar}@christuniversity.in

Abstract. Restarting automaton is a type of regulated rewriting system, introduced as a model for analysis by reduction. It is a linguistically motivated method for checking the correctness of a sentence. In this paper, we introduce a new definition of normal restarting automaton in which only one substring is removed using the DEL operation in a cycle. This DEL operation is applied to reverse the insertion operation in an insertion grammar. We use this 1-normal restarting automaton to solve the membership problem of insertion languages. Further, we introduce some interesting closure properties of 1-normal restarting automata.

Keywords: Insertion grammars · Membership problem · Restarting automaton

1 Introduction

The restarting automaton was introduced by Petr Jancar et al. in 1995 in order to model the 'analysis by reduction', which is a technique being used in linguistics to analyze sentences of *natural languages*. Analysis by reduction consists of step wise simplifications (reductions) of a given (lexically disambiguate) extended sentence until a correct simple sentence is obtained. It is accepted, until an error is found and the input is rejected. Each simplification replaces a short part of the sentence by an even shorter one.

A restarting automaton contains a finite control unit, a head with a look-ahead window attached to a tape. At several points it does cut-off substrings from the look-ahead window using DEL operation followed by *restart* (RST) operation. The head moves right along the tape until it takes any RST operation. RST implies that the restarting automaton places the look-ahead window over the left border of the tape and it completes one *cycle*. After performing a DEL/RST operation, the restarting automaton is unable to remember any step of computation that was performed already. We can say that it is a modification

S. Arumugam et al. (Eds.): ICTCSDM 2016, LNCS 10398, pp. 10–19, 2017.
DOI: 10.1007/978-3-319-64419-6_2

of the *list automaton* [7] and *forgetting automaton* [8]. Further, when each time the DEL operation is performed, the tape becomes smaller and smaller. A word u can be reduced to a word v if there is a cycle starting with u and ending with v. The computation ends by halting in an accepting or a rejecting state.

Insertion operations are introduced in [3] and based on these operations, insertion grammars are introduced in [4,5] and further studied in [6]. The motivation for insertion grammar comes from linguistic and as well from DNA processing and RNA editing. Informally, the insertion operation is defined as follows: If a string x is inserted between two parts w_1 and w_2 of a string w_1w_2 to get w_1xw_2, we call the operation *insertion*. The working nature of insertion grammar is counterpart to the functionality of contextual grammar [9], where based on the *selector* present in a string as a substring, the *contexts* are adjoined left and right of the substring.

In [1], it has been shown that restarting automaton with delete (simply, DRA) can represent the analyzer for characterizing the class of *contextual grammars with regular selector* (CGR). Also [2] showed that restarting automata recognize a family of languages which can be generated by certain type of contextual grammars, called *regular prefix contextual grammars with bounded infix* (RPCGBI). In this paper, we make a relationship between *restarting automaton* and *insertion languages*.

The *membership problem* for a language is defined as follows: Given a grammar G and a string w, whether w belongs to the language generated by G or not? In this paper, we introduce *1-normal DRA*. With the existing automaton - DRA, we introduce a variant of normal DRA where the DEL operation can be taken only once followed by restart in a cycle. We can say that 1-normal DRA is similar to *clearing restarting automata* [12].

The paper is organized as follows. Section 2 is Preliminaries that recall mainly the restarting automaton with delete operation (DRA) and insertion grammars. Section 3 introduces 1-normal DRA and discusses some properties of it. Section 4 discusses the relationship between the 1-normal DRA and insertion grammars. Section 5 discusses about some interesting properties of 1 Normal - DRA. Section 6 concludes the paper with some future work.

2 Preliminaries

Throughout the paper we will use the following notations. If Σ is an alphabet, then Σ^* denotes the set of all strings over Σ. For a string w, $|w|$ is the length of the string, sometimes called size of the string and $\emptyset$ denotes empty set. Any consecutive symbols of a string is called a substring. If a string x is a substring of y, then it is denoted by $x \in sub(y)$. A string $x \in \Sigma^*$ is called a scattered substring of a string $y \in \Sigma^*$ where $|x| \geq |y|$, then x can be obtained by omitting some symbols from y but maintaining the relative order of the remaining ones. For an automaton, the language accepted by M is denoted by $L(M)$ and for a given grammar γ, the language generated by γ is denoted by $L(\gamma)$.

2.1 Restarting Automaton with Delete Operation (DRA)

A restarting automaton with delete (denoted by DR-automaton or by *DRA*) is $M = (Q, \Sigma, \triangleleft, \triangleright, q_0, k, \delta)$ where Q is a finite set of states, Σ is the input alphabet, $\triangleleft, \triangleright$ are left and right borders respectively and $\triangleleft, \triangleright \notin \Sigma$, k is the size of the read-write window ($k \geq 1$).

The transition relation δ describes different types of transition steps which are given below. u' is assumed to be the content of the look-ahead window (and not necessarily the content of the tape).

- *MVR* - $(q', MVR) \in \delta(q, u')$, if M is in state q and sees a string u' where $u' \neq \triangleright$ in its look-ahead window, then this *MVR* step shifts the look-ahead window one position to the right and M enters into the state q'.
- *DEL* - $(q', v') \in \delta(q, u')$, if M is in state q and sees a string u' in its look-ahead window, deleting an item from the look-ahead window. u' is replaced by its scattered substring v' such that $|v'| < |u'|$. The border markers $\triangleleft, \triangleright$ must not disappear from the tape. After using the DEL operation the automaton can still read the remaining part of the tape also the automaton can place its head to the right of the just rewritten (deleted) string [1].
- *RST* - Restart. It causes M to move its look-ahead window to the left border marker $\triangleleft$ and re-enters into the initial state q_0.
- *ACCEPT* - *Accept* $\in \delta(q, u')$ where $q \in Q$. It gets into an accepting state.
- *REJECT* - If $\delta(q, u') = \emptyset$ (i.e., when δ is undefined), then M will reject.

A configuration of the automaton M is (u', q, v'), where $u' \in \{\triangleleft\Sigma^* \cup \lambda\}$ is the content from the *left border* till the position of the head, $q \in Q$ is the current state and $v' \in \{\triangleleft\Sigma^* \triangleright \cup \Sigma^* \triangleright\}$ is the content of the working list from the position of the head and to the right till the right end of the tape. In the initial configuration on an input word w, the control unit is in the fixed *initial state* $q_0 \in Q$, and the head is attached to the left border $\triangleleft$, i.e. $(\lambda, q_0, \triangleleft w \triangleright)$-scanning $\triangleleft$ and looking at the next $k - 1$ symbols. We suppose that the states Q of the finite control are divided into two classes: the non-halting states (at least one instruction must be there which is applicable when the unit is in such a state) and the halting states (any computation ends by entering such a state), the halting states are further divided into the *accepting state* and the *rejecting state*.

In general, the restarting automaton is *non-deterministic*, i.e. there can be two or more instructions for a $\delta(q, u')$, it suggests that there can be more than one computation for an input string. Otherwise the automaton is said to be deterministic. Any finite computation of a DRA consists of certain phases. A phase, called a *cycle*, starts in a restarting configuration, the head moves along the tape and performing MVR, DEL operations until a RST operation is performed and thus a new restarting configuration is reached. If no further RST operation is performed, any finite computation necessarily finishes in a halting configuration -such phase is called *tail*.

---

[1] in our paper, we assume that after every DEL operation is immediately followed by RST, its forming DEL-RST.

The notation $u' \Rightarrow_M v'$ indicates that there exists a cycle in M starting in the initial configuration with the word u' and ending in the configuration having the word v', the relation $\Rightarrow_M^*$ is the reflexive and transitive closure of $\Rightarrow_M$. We say that u' becomes v' by M (or u' is reduced to v' by M) if $u' \Rightarrow_M v'$, we are certain that the word v' is strictly shorter than u' (v' is the scattered subword of u'). An input word w is accepted by M if there is a computation which starts in the initial configuration with w (bounded by borders $\triangleleft, \triangleright$) on the list and finishes in an accepting configuration where the control unit is in one of the accepting states. $L(M)$ denotes the language consisting of all words accepted by M and we say that M recognizes the language $L(M)$.

A DEL step of an DRA may remove an arbitrary number of factors from the actual content of the look-ahead window. Therefore the following restriction has been included in DRA [1,2].

Definition 1 (Normal DRA). *A DRA is called normal if all the DEL operations are in the form $(q', v') \in \delta(q, u')$ where v' is a scattered substring of u', there exist words $x_1, x_2, x_3, x_4, x_5 \in \Sigma^*$ such that $u' = x_1x_2x_3x_4x_5$ and $v' = x_1x_3x_5$, that is two substrings of u' can be deleted.*

Proposition 1 (Error preserving property of DRA). *If $u' \Rightarrow_M^* v'$ and $u' \notin L(M)$ then $v' \notin L(M)$.*

2.2 Insertion Grammars

An Insertion grammar $\gamma = (T, A, I)$, where T is an *alphabet* set, A is a finite set of *strings* over T called *axioms*, I is the set of *insertion rules* of the form $(u, \lambda/x, v)$ where $u, v \in T^*$ and $x \in T^+$ which corresponds to the *rewriting rule* $uv \rightarrow uxv$,
Here u, v are called contexts and x is called inserted string for an insertion rule. As usual, $\Rightarrow^*$ denotes the *reflexive transitive closure* of $\Rightarrow$. A language $L(\gamma)$ generated by γ is defined by $L(\gamma) = \{w \in T^* \mid y \in A : y \Rightarrow^* w\}$.

3 1-Normal DRA

We first define 1-normal DRA. The functionality and the accepting configurations defined for DRA are the same for 1-normal DRA except the following changes. Normal DRA can delete at most two substrings from the current string but in this version at most one substring is deleted using DEL operation then it takes RST (restart) immediately without reading the remaining part of the tape, thus forming a new operation DEL-RST.

Definition 2 (1 − Normal DRA). *A restarting automata is called 1-normal DRA if all the DEL operations are in the form $(q', v') \in \delta(q, u')$ where v' is a scattered substring of u', there exist words $x_1, x_3 \in \Sigma^*, x_2 \in \Sigma^+$ such that $u' = x_1x_2x_3$ and $v' = x_1x_3$. In a cycle one substring can be deleted using DEL operation and RST is followed immediately.*

As insertion grammars do not contain non terminals, 1-normal DRA do not need to use any non terminal, so the error preserving property is satisfied for 1-normal DRA and correctness preserving property is satisfied for deterministic 1-normal DRA.

Before we go to analyze the relationship between 1-normal DRA and insertion grammars which is the objective of the paper, we first need to understand the relationship of DRA with contextual grammars [1]. External contextual grammars are introduced by S. Marcus in 1969 [9]. Internal contextual grammars [10] produce strings starting from an *axiom* and in each step *left context* and *right context* are adjoined to the string based on certain string called selector present as a substring in the derived string. u, v are called *left context* and *right context* respectively. For more details on contextual grammars, we refer to [11]. We recall that in insertion grammar, looking at the context (u, v), the string x is inserted. The selector in a contextual grammar can be of arbitrary type in nature, like regular, context free etc., but the strings u, v are finite. In insertion grammars all the strings u, v, x are finite. Normal DRA works in the opposite way of contextual grammars in accepting strings [1]. In a normal DRA M, w is given as an input. It checks the items of the look-ahead window with the contextual grammar G that any given rule P in G has been used or not. If it finds that any rule has been used then the automaton deletes the left and right context u, v and takes the RST operation, otherwise takes MVR and checks whether any rule in G can be applied. In this way, the automaton simulates the derivation of contextual grammar in reverse order and if the input string can be reduced back to the axiom z, it implies that the string w can be generated using the given grammar G, thus $w \in L(G)$. Here the *size* of the tape of the automaton M is same as the size of the string w. Step by step, the automaton M only deletes substrings of w, so the size of the tape becomes smaller and smaller. Tape size of 1-normal DRA will be $|w| + 2$ where the 2 is added for the left border $\triangleleft$ and the right border $\triangleright$.

4 1-Normal DRA and Insertion Grammar

In this section we shall establish the relationship between 1-normal DRA and insertion grammars. We show that the membership problem for insertion languages can be solvable by the introduced 1-normal DRA. The paradigm of this version of 1-normal DRA is closely related to insertion grammars. Insertion grammar works just in the *opposite direction* of 1-normal DRA. The connection is established based on the following observation.

- For an insertion rule $(u, \lambda/x, v)$ where $u, v \in T^*$ and $x \in T^+$, the 1-normal DRA has to delete the substring x between u and v (this means that uxv is occurred as a substring in the given input string and the machine deletes this substring x). In that case, we informally say that an insertion rule is found/used in the look-ahead window as a substring.

Let M be 1-normal DRA. A reduction system induced by M is $RS(M) = (\Sigma^*, \Rightarrow_M)$. For each insertion grammar γ, we define a reduction system induced by γ as $RS(\gamma) = (T^*, \Rightarrow_\gamma^{-1})$ where $(u \Rightarrow_\gamma^{-1} v)$ iff $(v \Rightarrow_\gamma u), u, v \in T^*$.

With the above detail we will construct a 1-normal DRA M in such a way that if $z \Rightarrow_\gamma^* w$ then $w \Rightarrow_M^* z$ for $w, z \in T^*$, z-axiom, thus $RS(\gamma) = RS(M)$.

Let w be the input string given to 1-normal DRA. The automaton M checks the string of the look-ahead window of (size k) with the given grammar γ that any insertion rule from I has been found or not as a substring. If any insertion rule from I is found in the look-ahead window as a substring (uxv) then the automaton M deletes the inserted string $x \in T^+$ using the DEL operation.

Theorem 1. *For an insertion grammar γ, a 1-normal DRA M can be constructed in such a way that $RS(\gamma) = RS(M)$ and $L(\gamma) = L(M)$.*

Proof. Given an insertion-grammar $\gamma = (T, A, R)$, we have to construct a 1-normal DRA $M = (Q, \Sigma, \triangleleft, \triangleright, q_0, k, \delta)$, that accepts $L(\gamma)$ where

- $Q = \{q_0 = q', q, Accept, Reject\}$
- $\Sigma = T$ is the input alphabet
- $\triangleleft, \triangleright$ are left and right borders respectively and $\triangleleft, \triangleright \notin \Sigma$
- k is the size of the look-ahead window ($k \geq 1$).
- δ is defined as follows:

For an insertion rule of the form: $(x_1, \lambda/x_2, x_3)$ where $x_1, x_3 \in T^*, x_2 \in T^+$ (which offers a rewriting rule $x_1x_3 \rightarrow_\gamma x_1x_2x_3$), the instruction of the 1-normal DRA M will be $(q', v') \in \delta(q, u')$ where $u' = x_1x_2x_3, v' = x_1x_3$. Here u' is replaced by $v' : |v'| < |u'|$ where v' is a scattered substring of u', immediately followed by a RST instruction: RST $\in \delta(q, u')$ for any possible contents u' of the look-ahead window. If no insertion rule does belong to look-ahead window as a subword($uxv \notin u'$) and $\triangleright$ does not belong to look-ahead ($\triangleright \notin u'$) then the automaton takes MVR operation.

- ACCEPT- *Accept* $\in \delta(q, u')$ where $u' = \triangleleft z \triangleright, z \in A$.
- REJECT - $\delta(q, u') = \emptyset$. That is when δ is undefined. In other words, when 1-normal DRA is unable to take any of the DEL, MVR operations then the transition becomes undefined.

Size of the Look-ahead Window:
Size of the look-ahead window of M will be $k = max(k_c, k_b + 2)$ where k_c is the maximum length of the inserted string with its contexts - $k_c = max\{(|u| + |x| + |v|)$ where $(u, \lambda/x, v) \in I, u, v =$ contexts$\}$. k_b is the maximum axiom size- $k_b = max\{|z| : z \in A\}$. 2 is added there for the left border $\triangleleft$ and the right border $\triangleright$. The reason for 2 is added with k_b is to satisfy the accepting condition - *Accept* $\in \delta(q, u')$ where $u' = \triangleleft z \triangleright$ where $\triangleleft z \triangleright \leq k$.

1-normal DRA simulates the derivation of insertion-deletion grammar in reverse order, in case of insertion rule it deletes the inserted string using DEL instruction which is defined above. For insertion grammar the derivation starts from the axiom to the generated string, the automaton starts the reduction from the generated string to the axiom. If $z \Rightarrow_{\gamma}^{*} w$ then $w \Rightarrow_{M}^{*} z$ where $w, z \in T^{*}$, z-axiom, thus $RS(\gamma) = RS(M)$.

We have the following important result and the proof is obvious from Theorem 1, and from the discussions of above paragraphs of Theorem 1.

Theorem 2. *The membership problem for insertion languages can be solved by 1-normal DRA.*

5 The Power of 1-Normal DRA

In this section, we discuss the power of 1-normal DRA and discuss some interesting properties.

Theorem 3. *All regular languages can be recognized by 1-normal DRA, i.e. for each regular language L there exists a M such that $L(M) = L \cup \lambda$.*

Proof. Given a regular grammar $\gamma = (N, T, P, S)$ where N is the finite set of non-terminals,T is the finite set of terminals, P is the finite set of production rules of the following form:$A \rightarrow aB$ where $A \in N$, $A \rightarrow b$ where $b \in T$, S is the starting symbol. Now we have to construct a 1-normal DRA $M = (Q, \Sigma, \triangleleft, \triangleright, q_0, k, \delta)$, that accepts $L(\gamma)$ where $Q = \{q_0 = q', q, Accept, Reject\}$, $\Sigma = N \cup T$ is the input alphabet, $\triangleleft, \triangleright$ are left and right borders respectively and $\triangleleft, \triangleright \notin \Sigma$, k is the size of the look-ahead window ($k \geq 1$), δ is defined as follows: For a regular grammar rule is of the form: $A \rightarrow aB$ where $A \in N$, the instruction of the 1-normal DRA M will be $(q', v') \in \delta(q, u')$ where $u' = aB, v' = A$. Here u' is replaced by $v' : |v'| < |u'|$ where v' is a scattered substring of u', immediately followed by a RST instruction: RST $\in \delta(q, u')$ for any possible contents u' of the look-ahead window. Also, if a regular grammar rule is $A \rightarrow b$ where $b \in T$ then the instruction of 1-normal DRA will be $(q', v') \in \delta(q, u')$ where $u' = b, v' = A$. In the same way, u' is replaced by $v' : |v'| < |u'|$ where v' is a scattered substring of u', immediately followed by a RST instruction: RST $\in \delta(q, u')$.
If no regular rule does belong to look-ahead window as a subword ($aB \notin u'$ or $b \notin u'$) and $\triangleright$ does not belong to look-ahead ($\triangleright \notin u'$) then the automaton takes MVR operation.

- ACCEPT- $Accept \in \delta(q, u')$ where $u' = \triangleleft S \triangleright$, S is the starting symbol.
- REJECT - $\delta(q, u') = \emptyset$. That is when δ is undefined. In other words, when 1-normal DRA is unable to take any of the DEL, MVR operations then the transition becomes undefined.

Size of the Look-ahead Window:
Size of the look-ahead window of M will be $k = max(k_c, k_b + 2)$ where k_c is the length of the right-hand side of the production, $k_c = 2$. k_b is the size of the axiom - $k_b = 1$. 2 is added there in order to satisfy the accepting condition - $Accept \in \delta(q, u')$ where $u' = \triangleleft S \triangleright$.

Theorem 4. *There are context free languages which cannot be recognized by 1-normal DRA.*

Proof. We conclude Theorem 4 by focusing on Lemmas 1 and 2.

Lemma 1. *The language* $L_1 = \{p^n r q^n \mid n \geq 0\} \cup \{\lambda\}$ *cannot be recognized by* 1*-normal DRA.*

Proof. 1-normal DRA can delete at most one string in a cycle, so it will delete all p in first n cycles and from $(n+1)th$ cycle it will start deleting q. Actually in this case 1-normal DRA cannot delete two substrings in a same cycle, so it is unable to keep track of the equality of $p's$ and $q's$.

As we have seen in Lemma 1 that not all context-free languages are recognized by a 1-normal DRA. We still could characterize CFL using 1-normal DRA using inverse homomorphism and Greibach's hardest context-free language [2]. Greibach constructed a context-free language H [12], such that:

- Any context-free language can be parsed in whatever time or space it takes to recognize H.
- Any context-free language L can be obtained from H by an inverse homomorphism. That is, for each context-free language $L \subseteq \Sigma^*$, there exist a homomorphism ρ so that $L = \rho^{-1}(H)$. The definition of the Greibach's language follows. Let $\Sigma = \{x_1, x_2, \bar{x_1}, \bar{x_2}, \#, c\}$.
  Define $H = \{\lambda \cup \{a_1 c b_1 c z_1 d ... a_n c b_n c z_n d \mid n \geq 1, b_1 ... b_n \in \#D, a_i, z_i \in \Sigma^*\}$, for all i, $1 \leq i \leq n, b_1 \in \{x_1, x_2, \bar{x_1}, \bar{x_2}\}^*, b_i \in \{x_1, x_2, \bar{x_1}, \bar{x_2}\}^*$, for all $i \geq 2\}$.
- (Note that a_i and z_i can contain c and $\#$). D is a semi-Dyck language over the alphabet $\{x_1, x_2, \bar{x_1}, \bar{x_2}$, generated by the grammar with one non terminal S and the set of rules: $S \Rightarrow \lambda \mid SS \mid a_1 S \bar{a_1} \mid a_2 s \bar{a_2}$. Clearly it is a context free language.

Lemma 2. H *is not accepted by* 1*-normal DRA.*

Proof. Consider the language H as given above. Then, H cannot be accepted by any 1-normal DRA. The main problem of recognition H by 1-normal DRA is selection of $b_1, b_2, ..b_n$(see the formal definition of H). Unfortunately no 1-normal DRA can recognize H. We need to construct $M = (Q, \Sigma, \triangleleft, \triangleright, q_0, k, \delta)$, such that $L(M) = H$. Apparently, $w = c\#x_1^m cd c\bar{x}_1^m cd \in H$. Let the accepting computation will $w_1 \Rightarrow_M w' \Rightarrow_M ... \Rightarrow_M$ Axiom. Firstly, there must be a deleted substring of the form $x_1^r cdc\bar{x}_2^s$ for some $0 \leq r, s \leq m$. Here it is easy to see that $r = s$. The first applied instruction of 1-normal DRA in order to recognize will be $(q', v') \in \delta(q, u')$ where $u' = x_1^\alpha x_1^r cdc\bar{x}_1^r \bar{x}_1^\beta, v' = x_1^\alpha \bar{x}_1^\beta$. Now consider the word $w = c\#x_1^{m+1} x_1^r cdc\bar{x}_1^r \bar{x}_1^m c\#x_1\bar{x}_1 cd$. From here we can easily conclude $w \notin H$, but $w' = c\#x_1^{m+1} \bar{x}_1^m c\#x_1\bar{x}_1 cd$ is in H. So, contradiction of error preserving property.

Corollary 1. *(a)* $\mathcal{L}$*(1-normal DRA)*$\subset \mathcal{L}$*(Normal* $-$ *DRA)*
(b) $CFL - \mathcal{L}$*(1-normal DRA)*$\neq \emptyset$

Proof. In Lemma 1, the language $L_1 = \{p^n r q^n \mid n \geq 0\} \cup \{\lambda\}$ can be recognized by normal DRA, so from this fact easily we can conclude corollary(a).

Let $L_2 = \{p^n q^n \mid n > 0\}$ and $L_3 = \{p^n q^{2n} \mid n > 0\}$ be two sample languages. It is easy to see from the Theorem 2 that both L_2, L_3 are recognized by 1-Normal DRA, in this way we can conclude corollary(b).

Theorem 5. *$\mathcal{L}$(1-normal DRA) is not closed under union and concatenation.*

Proof. Languages $L_2 \cup L_3$ and $L_2 \cdot L_3$ cannot be recognized by 1-Normal DRA. For a contradiction let us suppose that, there exist a 1-Normal DRA M such that $L(M) = L_2 \cup L_3$ ($L(M) = L_2 \cdot L_3$, respectively). Let K be the size of the look-ahead window. Let $p^n q^n \Rightarrow_M p^{n-s} b^{n-t}$ be the first step of an accepting condition, where $s, t > 0, s + t > 0$. Suppose when $d = p$ comes in the look-ahead window then 1-Normal DRA takes DEL-RST operation and delete q where $|d| \leq |k|$. Since we set n arbitrarily large where s, t are constants we get necessarily $s = t$. Now if $d = p^\alpha p^\eta q^\eta q^\beta$ then 1-Normal DRA takes DEL-RST operation and delete $p^\eta q^\eta$ where $\alpha + \beta + 2\eta \leq |k|$. Then $p^{m+\eta} q^{2m+\eta} \Rightarrow_M p^m q^{2m}$ which is a contradiction with the error preserving property as $p^{m+\eta} q^{2m+\eta}$ is not in $L_2 \cup L_3$ (and not in $L_2 \cdot L_3$ respectively).

Theorem 6. *There are non context free languages which can be recognized by 1-normal DRA.*

Proof. 1-normal DRA. M recognizing a language that is not context-free ($\{(ab)^{2^m} \mid m \geq 0\}$). The instructions of 1 - normal DRA are given below.

- : DEL-RST :$(q', v') \in \delta(q, u')$ where $u' = abb, v' = ab$.
- : DEL-RST :$(q', v') \in \delta(q, u')$ where $u' = bab, v' = bb$.
- : *Accept* $\in \delta(q, u')$ where $u' = \lhd z \rhd$ where $z = ab$

The computation as follows, $\lhd ab\underline{a}babababababab\rhd \Rightarrow_M \lhd a\underline{b}bababababababab\rhd \Rightarrow_M \lhd ab\underline{a}babababababab\lhd \Rightarrow_M \lhd a\underline{b}babababababab\lhd \Rightarrow_M \lhd ab\underline{a}babababab\rhd \Rightarrow_M \lhd a\underline{b}babababab\rhd \Rightarrow_M \lhd ab\underline{a}bababab\rhd \Rightarrow_M \lhd a\underline{b}bababab\rhd \Rightarrow_M \lhd ab\underline{a}babab\rhd \Rightarrow_M \lhd a\underline{b}babab\rhd \Rightarrow_M \lhd ab\underline{a}bab\rhd \Rightarrow_M \lhd a\underline{b}bab\rhd \Rightarrow_M \lhd ab\underline{a}b\rhd \Rightarrow_M \lhd a\underline{b}b\rhd \Rightarrow_M \lhd ab\rhd$, accepted.

Theorem 7. *$\mathcal{L}$(1-normal DRA) is not closed under homomorphism.*

Proof. Consider $L = \{p^n q^n r^m s^{2m} \mid n, m \geq 0\}$ recognized by $\mathcal{L}(1\!-\!normalDRA)$ and the homomorphism $h : p \mapsto p, q \mapsto q, r \mapsto r, s \mapsto q$.

It is easy to see that each of the following languages:
$L_4 = \{p^n r q^n \mid n \geq 0\} \cup \{p^n q^n \mid n \geq 0\}, L_5 = \{p^n r q^m \mid n, m \geq 0,\} \cup \{\lambda\}$, $L_6 = \{p^m q^m \mid m \geq 0\}$ can be recognized by a 1-normal DRA. Using these languages we can show several non-closure properties of 1-normal DRA.

Theorem 8. *$\mathcal{L}$(1-normal DRA) is not closed under intersection, intersection with a regular language, and set difference.*

Proof. Intersection part follows the equality $L_1 = L_4 \cap L_5$ and Lemma 1. In order to proof the next part, just notice that L_5 is a regular language. Set differences the part follows the equality $L_3 = (L_4 - L_6) \cup \{\lambda\}$.

6 Conclusion

In this paper, we have introduced 1-normal DRA. We have solved the membership problem of insertion languages. We saw that s insertion grammars do not contain non terminals, 1-normal DRA do not need to use any non terminal, so the error preserving property is satisfied for 1-normal DRA and correctness preserving property is satisfied for deterministic 1-normal DRA.

There is scope of future work. H in Lemma 2, can be accepted by introducing auxiliary symbol to 1-normal DRA. Also, in case of running time, as here we are using non deterministic 1-normal DRA, we are unable to comment about the polynomial time complexity solution. We can solve the membership problem of insertion languages in polynomial time by extending our work.

Acknowledgments. Tool Development - Poorna Chandra Tejasvi, Btech third year student of Computer Science and Engineering, Christ University Faculty of Engineering, Bangalore.

References

1. Jancar, P., Mraz, F., Platek, M., Prochazka, M., Vogel, J.: Deleting automata with a restart operation. In: Bozapalidis, S. (ed.) Proceedings of Developments in Language language theory III(97), Thessaloniki, pp. 191–202 (1998)
2. Jancar, P., Mraz, F., Platek, M., Prochazka, M., Vogel, J.: Restarting automata, marcus grammars and context-free languages. In: Dassow, J., Rozenberg, G., Salomaa, A. (eds.) Developments in Language Theory, pp. 102–111. World Scientific Publishing (1996)
3. Haussler, D.: Insertion and Iterated Insertion as Operations on Formal Languages. Ph.D. Thesis, University of Colorado, Boulder (1982)
4. Haussler, D.: Insertion languages. Inf. Sci. **131**(1), 77–89 (1983)
5. Galiukschov, B.S.: Semicontextual grammars (in Russian). MAT. Logica i Mat. Ling. 38–50 (1981)
6. Kari, L., Thierrin, G.: Contextual insertions/deletions and computability. Inf. Comput. **131**(1), 47–61 (1996)
7. Chytil, M.P., Platek, M., Vogel, J.: A note on the Chomsky hierarchy. Bull. EATCS **27**, 23–30 (1985)
8. Jancar, P., Mráz, F., Plátek, M.: A taxonomy of forgetting automata. In: Borzyszkowski, A.M., Sokołowski, S. (eds.) MFCS 1993. LNCS, vol. 711, pp. 527–536. Springer, Heidelberg (1993). doi:10.1007/3-540-57182-5_44
9. Marcus, S.: Contextual Grammars. Revue Roumane de Mathematiques Pures et Appliques 14(10), 1525–1534 (1969)
10. Paun, G., Nguyen, X.M.: On the inner contextual grammars. Rev. Roum. Pures. Appl. **25**, 641–651 (1980)
11. Paun, G.: Marcus Contextual Grammars. Kluwer Academic Publishers, Dordrecht (1997)
12. Cerno, P., Mráz, F.: Clearing restarting automata. Fundam. Inform. **104**, 17–54 (2010)

Formal Language Representation and Modelling Structures Underlying RNA Folding Process

Anand Mahendran(✉) and Lakshmanan Kuppusamy

School of Computer Science and Engineering, VIT University,
Vellore 632014, India
{manand,klakshma}@vit.ac.in

Abstract. The biological sequences that occur in DNA, RNA and proteins can be considered as strings formed over the well defined chemical alphabets. Such gene sequences form structure based on the complementary pair and the structures can be interpreted as languages. Matrix insertion-deletion system has been introduced a few years back that modelled several bio-molecular structures occur at intramolecular and intermolecular level. In this paper, we identify some structures that are frequently noticed during RNA folding process such as *double bulge loop*, *extended internal loop*, *triple stem and loop* and we give the corresponding formal language representation to such structures. Further, we model the structures using Matrix insertion-deletion systems. This work is pioneering to give the language representation and modelling the structures of RNA folding process using formal grammar.

Keywords: Gene sequences · Bio-molecular structures · Matrix grammars · Insertion-deletion systems · Folding process

1 Introduction

Insertion-deletion systems are inspired from the insertion and deletion operations that take place in gene sequences. These operations frequently occur in DNA processing and RNA editing. In [4], the insertion operation was first studied. The deletion operation was first studied in [7] from a formal language perspective. Insertion and deletion operations together were introduced in [6]. The corresponding grammatical mechanism is called *insertion-deletion system* (abbreviated as ins-del system). Informally, insertion operation means inserting a string η in between the strings w_1 and w_2 to obtain $w_1\eta w_2$ whereas deletion operation means deleting a substring δ from the string $w_1\delta w_2$ and obtain w_1w_2.

The DNA molecule consists of sequences that are built of nucleotides, which are of in four forms $a(adenine)$, $t(thymine)$, $g(guanine)$, $c(cytosine)$. The RNA molecule consists of sequences that are built of nucleotides, which are of in four forms a, $u(uracil)$, g, c. The complementary pair for RNA (DNA) is given as $\bar{a} = u(t)$, $\bar{u}(\bar{t}) = a$, $\bar{g} = c$ and $\bar{c} = g$. The patterns are formed in the bio-molecules based on the complementary pairs and other biological constraints. Such patterns

S. Arumugam et al. (Eds.): ICTCSDM 2016, LNCS 10398, pp. 20–29, 2017.
DOI: 10.1007/978-3-319-64419-6_3

can be considered as structures. These structures play an important role in governing the functionality and behavior of the bio-molecules [1,16].

We now discuss briefly about the bio-molecular structures that are commonly noticed in bio-molecules such as DNA, RNA and protein. First, we will look the *stem and loop*, *hairpin*, *pseudoknot* and *attenuator* structures which are shown in Figs. 1 and 2. The strings are obtained by reading the symbols as per directed dotted lines. The string *cuucaucagaaaaugac* resembles the stem and loop language (refer Fig. 1(a)) and the string *atcgcgat* resembles the hairpin language (refer Fig. 1(b)). The string *gcucgcga* (refer Fig. 2(a)) represents pseudoknot structure which resembles the crossed dependency pattern and the string *gucgacgucgac* (refer Fig. 2(b)) represents attenuator structure which resembles the *copy language* (which is a non-context-free language) pattern respectively. The above examples clearly depict the correlation between gene sequence and natural language constructions. The bio-molecular structures stem and loop and hairpin (as shown in Fig. 1) can be modelled by context-free (shortly, cf) grammars [15]. The bio-molecular structures pseudoknot and attenuator (as shown in Fig. 2) which are beyond the power of cf grammars. The connection between non-context-free constructs, such as *multiple agreement, crossed dependencies*, and gene sequences were carried out in [16,18]. For more details on linguistic behaviour of gene sequences and genomic structures, we refer to [1,15,18].

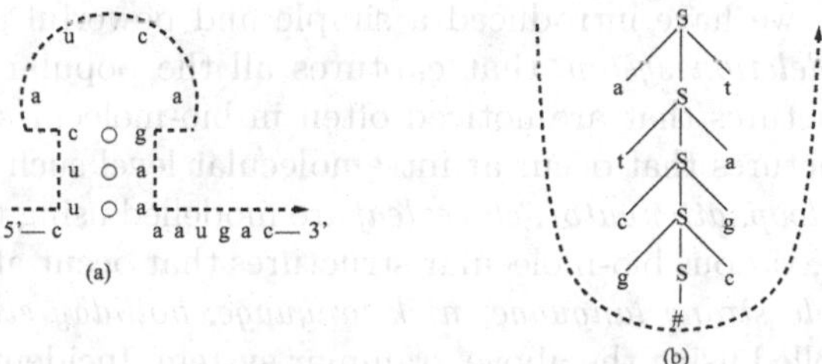

Fig. 1. Bio-molecular structures: (a) stem and loop (∘ stands for complementary pair) (b) hairpin (S is a non-terminal of the cf grammar and # denotes the empty string)

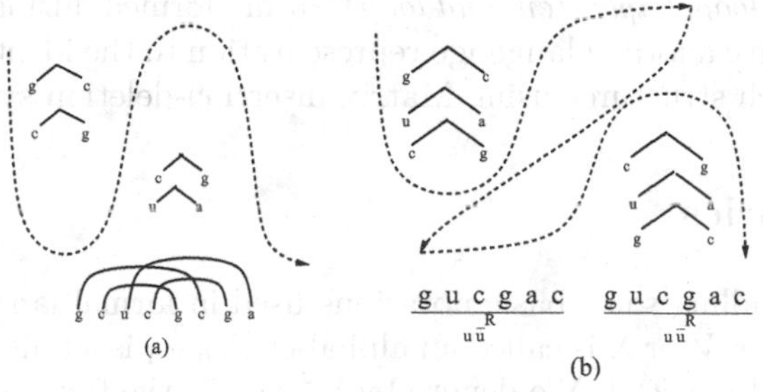

Fig. 2. (a) Pseudoknot (b) attenuator

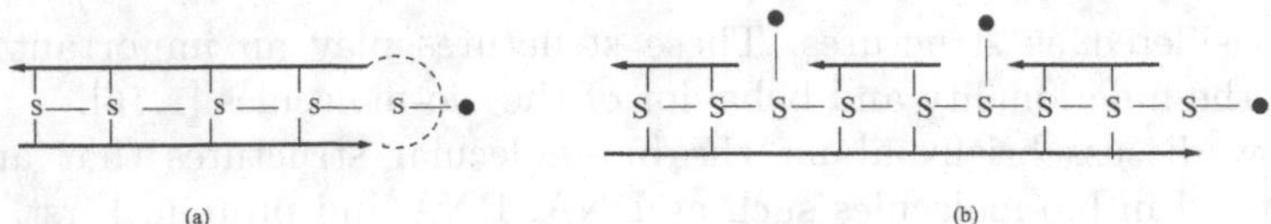

Fig. 3. Intermolecular structures: (a) double stranded molecule (S is a non-terminal for the cf grammar) (b) nick language (S is a non-terminal for the cf grammar)

Now, we shall look into some intermolecular structures found in RNA [17] and is shown in Fig. 3: (a) *double strand language* and (b) *nick language* where the cut takes place at arbitrary positions represented by a $\bullet$. In [1,5], an initial attempt was carried out on how formal grammars can be used for analyzing the linguistic behavior of biological sequences. Later, the study was extended by David Searls in [15,16]. In the last three decades, so many attempts have been made to establish the linguistic behavior of biological sequences starting by looking into regular, context-free and by defining new grammar formalisms like *crossed-interaction grammar* [13], *cut grammars, ligation grammars* [15,16], *simple* and *extended simple linear tree adjoining grammars* [19].

However, there was no unique grammar model that encapsulates all the above discussed bio-molecular structures. For example, double copy language cannot be modelled by a simple linear tree adjoining grammar [19]. To overcome this failure, in [8], we have introduced a simple and powerful grammar model *Matrix insertion-deletion system* that captures all the popular and important bio-molecular structures that are noticed often in bio-molecules. In [8], various bio-molecular structures that occur at intramolecular level such as *pseudo knot, hairpin, stem and loop, attenuator, cloverleaf* are modelled using the above grammar system. In [9], various bio-molecular structures that occur at intermolecular level such as *double strand language, nick language, holliday structure, replication fork* are modelled using the above grammar system. Incidentally in [12], the same system has been introduced from formal language theory perspective by Ion Petre and Sergey Verlan and a few computational completeness results of the system were discussed in the paper.

In this paper, we first identify some structures such as *double bulge loop, extended internal loop, triple stem and loop* that are formed during the RNA folding process. We give a formal language representation to the identified structures and we model such structures using Matrix insertion-deletion systems.

2 Preliminaries

We start with recalling some basic notations used in formal language theory. A finite non-empty set V or Σ is called an alphabet. Σ_{RNA} is a finite non-empty set over the symbols $\{a, u, g, c\}$. We denote by V^* or Σ^*, the free monoid generated by V or Σ, by λ its identity or the empty string, and by V^+ or Σ^+ the set $V^* - \{\lambda\}$ or $\Sigma^* - \{\lambda\}$. The elements of V^* or Σ^* are called *words* or *strings*.

A language L is defined as $L \subseteq \Sigma^*$. For any word $w \in V^*$ or Σ^*, we denote the length of w by $|w|$. For more details on formal language theory, we refer to [14].

Next, we recall the basic definition of insertion-deletion systems. Given an insertion-deletion system $\gamma = (V, T, A, R)$, where V is an alphabet (set of non-terminal and terminal symbols), $T \subseteq V$ (set of terminal symbols), A is a finite language over V, R is a set of finite triples of the form $(u, \alpha/\beta, v)$, where $(u, v) \in V^* \times V^*, (\alpha, \beta) \in (V^+ \times \{\lambda\}) \cup (\{\lambda\} \times V^+)$. The pair (u, v) are called contexts which will be used in deletion/insertion rules. Insertion rule is of the form $(u, \beta, v)_{ins}$ which means that β is inserted between u and v. Deletion rule is of the form $(u, \alpha, v)_{del}$, which means that α is deleted between u and v. In other words, (u, β, v) corresponds to the rewriting rule $uv \rightarrow u\beta v$, and $(u, \alpha/\lambda, v)$ corresponds to the rewriting rule $u\alpha v \rightarrow uv$.

Consequently, for $x, y \in V^*$ we can write $x \Longrightarrow^* y$, if y can be obtained from x by using either an insertion rule or a deletion rule which is given as follows: (the down arrow $\downarrow$ indicates the position where the string is inserted, the down arrow $\Downarrow$ indicates the position where the string is deleted and the underlined string indicates the string inserted)

1. $x = x_1 u^{\downarrow} v x_2,\ y = x_1 u \underline{\beta} v x_2$, for some $x_1, x_2 \in V^*$ and $(u, \beta, v)_{ins} \in R$.
2. $x = x_1 u \alpha v x_2,\ y = x_1 u^{\Downarrow} v x_2$, for some $x_1, x_2 \in V^*$ and $(u, \alpha, v)_{del} \in R$.

The language generated by γ is defined by $L(\gamma) = \{w \in T^* \mid x \Longrightarrow^* w, \text{ for some } x \in A\}$, where $\Longrightarrow^*$ is the reflexive and transitive closure of the relation $\Longrightarrow$.

2.1 Matrix Insertion-Deletion Systems

In this subsection, we describe the *matrix insertion-deletion systems* introduced in [8,12].

Definition 1. *A matrix insertion-deletion system is a construct* $\Gamma = (V, T, A, R)$ *where* V *is an alphabet,* $T \subseteq V$*,* A *is a finite language over* V*,* R *is a finite set of matrices* $\{R_1, R_2, \ldots R_l\}$*, where each* r_i*,* $1 \le i \le l$*, is a matrix of the form* $R_i = [(u_1, \alpha_1, v_1)_{t_1}, (u_2, \alpha_2, v_2)_{t_2}, \ldots, (u_k, \alpha_k, v_k)_{t_k}]$ *with* $t_j \in \{ins, del\}$*,* $1 \le j \le k$.

For $1 \le j \le k$, the triple $(u_j, \alpha_j, v_j)_{t_j}$ is an ins-del rule. Consequently, for $x, y \in V^*$ we write $x \Rightarrow x' \Rightarrow x'' \Rightarrow \ldots \Rightarrow y$, if y can be obtained from x by applying all the rules of a matrix R_i, $1 \le i \le l$, in order; in this case, we write $x \Longrightarrow_{r_i} y$. Note that the string w is collected after applying all the rules in a matrix and also $w \in T^*$. At this point, we make a note that in a derivation, the rules of a matrix are applied sequentially one after another in order and no rule is in *appearance checking*. By $w \Longrightarrow_* z$, we denote the relation $w \Longrightarrow_{R_{i_1}} w_1 \Longrightarrow_{R_{i_2}} \cdots \Longrightarrow_{R_{i_k}} z$, where for all j, $1 \le j \le k$, we have $1 \le i_j \le l$. The language generated by Γ is defined as $L(\Gamma) = \{w \in T^* \mid x \Longrightarrow_* w, \text{ for some } x \in A\}$.

3 Modelling Bio-Molecular Structures

In this section, we first discuss the structures that are formed during the RNA folding process. We next give the interpretation for each structure with a formal language representation (as shown in Table 1) and model them using Matrix insertion-deletion systems. The bio-molecular structures that are commonly noticed in RNA structures can be categorized as primary, secondary and tertiary structures. In addition to the RNA folding process, protein folding process also plays a major role in the biological functions of a living cell.

The primary structure of a nucleic acid molecule represents the exact sequence of nucleotides that forms the complete molecule. The secondary structure (two dimensional) is a two dimensional representation formed by folding back onto itself with base pairing of complementary nucleotides (Watson-Crick pairs) which may form loops. The tertiary structures are three dimensional structure and study of such structures is very difficult. The evolution of RNA sequence needs to satisfy three requirements: folding, structure, and function [11]. As folding is to be considered as one of the important requirements in RNA sequence, identifying the structure during RNA folding process deserves a special attention. The commonly noticed looping structures during this process are stem and loop, internal loop, bulge loop and multi branch loop.

Table 1. Bio-molecular structure and formal language representation

Fig. no.	Bio-molecular structure formal language representation
Figure 4(a) (Left)	Double bulge loop $L_{dbl} = \{u_1 u_2 v_1 \bar{u_2}^R v_2 \bar{u_1}^R\}$
Figure 4(a) (Middle)	Extended internal loop $L_{eil} = \{u_1 v_1 u_2 v_2 \bar{u_2}^R u_3 v_3 \bar{u_3}^R \bar{u_1}^R\}$
Figure 4(a) (Right)	Double stem and loop $L_{dsl} = \{v_1 u_1 v_2 \bar{u_1}^R v_3 u_2 v_4 \bar{u_2}^R v_5\}$
Figure 4(b) (Left)	Hybrid loop $L_{hl} = \{u_1 v_1 u_2 v_2 \bar{u_2}^R v_3 u_3 v_4 u_4 v_5 \bar{u_4}^R v_6 \bar{u_3}^R v_7 \bar{u_1}^R\}$
Figure 4(b) (Right)	Quadruple stem and loop $L_{qsl} = \{u_1 u_2 v_1 \bar{u_2}^R u_3 v_2 \bar{u_3}^R u_4 v_3 \bar{u_4}^R \bar{u_1}^R \#\}$
Figure 4(c) (Left)	Triple stem and loop $L_{tsl} = \{u_1 v_1 \bar{u_1}^R u_2 v_2 \bar{u_2}^R u_3 v_3 \bar{u_3}^R\}$
Figure 4(c) (Right)	Extended double bulge loop $L_{edbl} = \{u_1 A_1 u_2 v_1 \bar{u_2}^R B v_2 \bar{B} A_2 \bar{u_1}^R\}$

For more details on RNA secondary structures and its prediction, we refer to [2,3,13]. The bio-molecular structures found in RNA folding process represented in Figs. 4(a) through (c) can be given in terms of languages as shown in Table 1 if the strings are collected as per the dotted directed lines. We now model each of the above language by Matrix insertion-deletion system. In most of the following derivations, at each derivation step, we directly write the resultant string

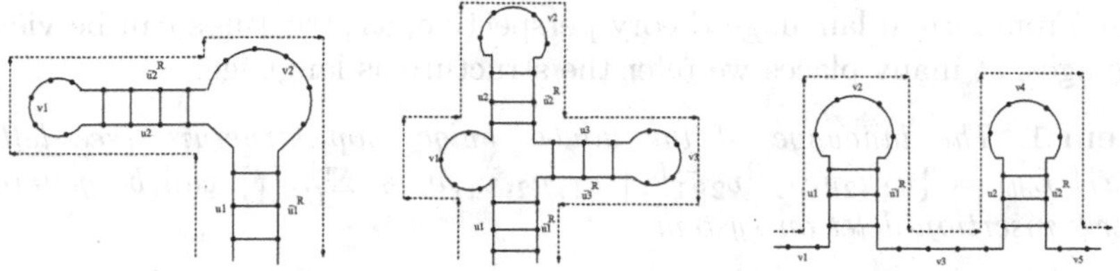

(a) Left: Double bulge loop, Middle: Extended internal loop, Right: Double stem and loop

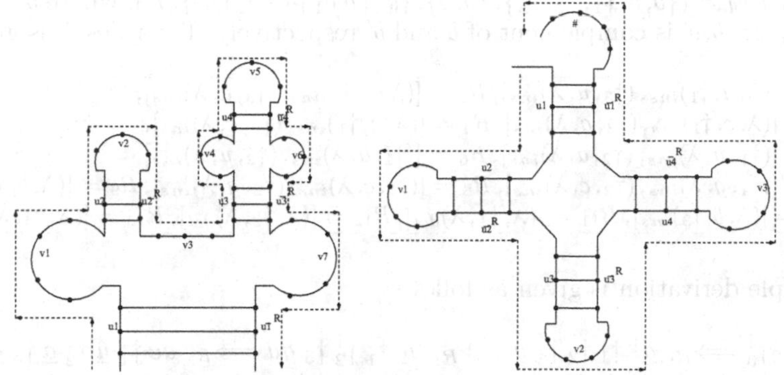

(b) Left: Hybrid loop, Right: Quadruple stem and loop

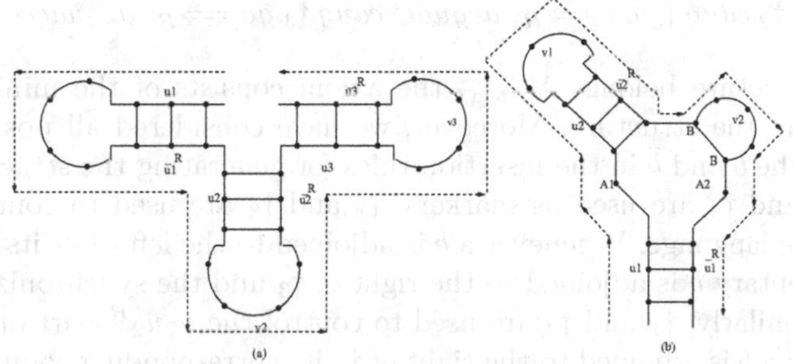

(c) Left: Triple stem and loop, Right: Extended double bulge loop

Fig. 4. Some bio-molecular structures found during RNA folding process

obtained by applying all the rules in a matrix. In all the propositions, we have adopted the method of proof by construction in modelling the bio-molecular structures using Matrix insertion-deletion systems. In the derivation step, the rule at the suffix of $\Longrightarrow$ denotes the corresponding matrix rule applied. In all the structures, the loop (v_i) and the stem (u_i and its complementary pair $\bar{u_i}^R$) need not be empty string. If needed, all possible combinations of each v_i, u_i and its complementary pair $\bar{u_i}^R$ of string length one can be included in axiom itself. An important objective is to have a minimum (length) axiom, wherever possible. From formal language theory perspective, as structures can be viewed as languages, at many places we refer the structure as language.

Theorem 1. *The language of the double bulge loop structure (see left of Fig. 4(a))* $L_{dbl} = \{u_1 u_2 v_1 \bar{u_2}^R v_2 \bar{u_1}^R \mid u_1, u_2, v_1, v_2 \in \Sigma_{RNA}^+\}$ *can be generated by Matrix insertion-deletion system.*

Proof. The language L_{dbl} can be generated by the Matrix insertion-deletion system $\Upsilon_{dbl} = (\{b, \bar{b}, \dagger_1, \dagger_2, \dagger_3\}, \{b, \bar{b}\}, \{b \dagger_1 b' b_1 \dagger_2 \bar{b'} b_2 \dagger_3 \bar{b}\}, R)$, where $b, b', b_1, b_2 \in \{a, u, g, c\}$, $\bar{b}, \bar{b'}$ is complement of b and b' respectively. The rules R is given by:

$R_1 = [(\lambda, a, \dagger_1)_{ins}, (\dagger_3, u, \lambda)_{ins}], R_2 = [(\lambda, u, \dagger_1)_{ins}, (\dagger_3, a, \lambda)_{ins}],$
$R_3 = [(\lambda, c, \dagger_1)_{ins}, (\dagger_3, g, \lambda)_{ins}], R_4 = [(\lambda, g, \dagger_1)_{ins}, (\dagger_3, c, \lambda)_{ins}],$
$R_5 = [(\dagger_1, u, \lambda)_{ins}, (\dagger_2, a, \lambda)_{ins}], R_6 = [(\dagger_1, a, \lambda)_{ins}, (\dagger_2, u, \lambda)_{ins}],$
$R_7 = [(\dagger_1, g, \lambda)_{ins}, (\dagger_2, c, \lambda)_{ins}], R_8 = [(\dagger_1, c, \lambda)_{ins}, (\dagger_2, g, \lambda)_{ins}], R_9 = [(\lambda, b, \dagger_2)_{ins}],$
$R_{10} = [(\lambda, b, \dagger_3)_{ins}], R_{11} = [(\lambda, \dagger_1, \lambda)_{del}], R_{12} = [(\lambda, \dagger_2, \lambda)_{del}], R_{13} = [(\lambda, \dagger_3, \lambda)_{del}].$

A sample derivation is given as follows:

$$^{\downarrow} \dagger_1 \dagger_2 \dagger_3^{\downarrow} \Longrightarrow_{R_1} \underline{a}^{\downarrow} \dagger_1 \dagger_2 \dagger_3^{\downarrow} \underline{u} \Longrightarrow_{R_1} a\underline{c} \dagger_1^{\downarrow} \dagger_2^{\downarrow} \dagger_3 \underline{g}u \Longrightarrow_{R_2} ac \dagger_1^{\downarrow} \underline{u} \dagger_2^{\downarrow} \underline{a} \dagger_3 gu$$
$$\Longrightarrow_{R_2} ac \dagger_1 \underline{g}u^{\downarrow} \dagger_2 \underline{c}a \dagger_3 gu \Longrightarrow_{R_3} ac \dagger_1 gu\underline{a}^{\downarrow} \dagger_2 ca \dagger_3 gu \Longrightarrow_{R_3} ac \dagger_1 gua\underline{c} \dagger_2$$
$$ca^{\downarrow} \dagger_3 gu \Longrightarrow_{R_4} ac \dagger_1 guac \dagger_2 ca\underline{u}^{\downarrow} \dagger_3 gu \Longrightarrow_{R_4} ac \dagger_1 guac \dagger_2 cau\underline{g} \dagger_3 gu \Longrightarrow_{R_5}$$
$$ac^{\Downarrow} guac \dagger_2 caug \dagger_3 gu \Longrightarrow_{R_6} acguac^{\Downarrow} caug \dagger_3 gu \Longrightarrow_{R_7} acguaccaug^{\Downarrow} gu.$$

As the structure belongs Σ_{RNA}^+, the axiom consists of the minimum possible string from the structure. Moreover, we have considered all possible combinations for the b and $\bar{b}$ in the insertion rules for generating the structure. The idea is $\dagger_1$, $\dagger_2$ and $\dagger_3$ are used as markers. $\dagger_1$ and $\dagger_3$ are used to control the $u_1 \bar{u_1}^R$ part of the language. Whenever a b is adjoined to the left of $\dagger_1$ its corresponding complementary $\bar{b}$ is adjoined to the right of $\dagger_3$ and the synchronization is maintained. Similarly, $\dagger_1$ and $\dagger_2$ are used to control the $u_2 \bar{u_2}^R$ part of the language. Whenever a b is adjoined to the right of $\dagger_1$ its corresponding complementary $\bar{b}$ is adjoined to the right of $\dagger_2$ such that the synchronization is maintained. $\dagger_2$ and $\dagger_3$ are used to control the v_1 and v_2 part of the language respectively.

From the construction of the system, it is easy to see that the usage of markers guarantee that the system Υ_{dbl} generates only the language L_{dbl}.

Remark 1. We can note from the previous system that (especially, the matrix rules), a gene and its complimentary gene are specified with the possible combinations in the insertion rules, the system looks cumbersome and congested. In order to avoid this, for the rest of the propositions, we generalize a gene and its compliment gene (or pair) in insertion rules using b and $\bar{b}$ where $b \in \{a, t, g, c\}$ and if $b = a$ then, $\bar{b}$ refers to its complimentary pair t. For example, if there is an insertion rule $R_I = [(\lambda, b, \dagger_1)_{ins}, (\dagger_5, \bar{b}, \lambda)_{ins}]$, then it refers to the following four insertion rules: (*i*) $R_1 = [(\lambda, a, \dagger_1)_{ins}, (\dagger_3, u, \lambda)_{ins}]$, (*ii*) $R_2 = [(\lambda, u, \dagger_1)_{ins}, (\dagger_3, a, \lambda)_{ins}]$, (*iii*) $R_3 = [(\lambda, c, \dagger_1)_{ins}, (\dagger_3, g, \lambda)_{ins}]$, (*iv*) $R_4 = [(\lambda, g, \dagger_1)_{ins}, (\dagger_3, c, \lambda)_{ins}]$.

Using this generalization, we have the following on similar lines.

Theorem 2. *The language of the extended internal loop structure and double stem and loop structure (see middle and right of Fig. 4(a)) can be generated by Matrix insertion-deletion system.*

Proof. The proof is similar to that of Theorem 1 and the details are omitted.

Theorem 3. *The language of the triple stem and loop structure and extended double bulge loop structure (see left and right of Fig. 4(c)) can be generated by Matrix insertion deletion system.*

Proof. The language $L_{tsl} = \{u_1v_1\bar{u_1}^R u_2v_2\bar{u_2}^R u_3v_3\bar{u_3}^R \mid u_1, u_2, u_3, v_1, v_2, v_3 \in \Sigma^+_{RNA}\}$ representing of the triple stem loop structure can be generated by the Matrix insertion-deletion system $\Upsilon_{tsl} = (\{b, \bar{b}, \dagger_1, \dagger_2, \dagger_3, \dagger_4, \dagger_5\}, \{b, \bar{b}\}, \{b \dagger_1 b_1\bar{b} \dagger_2 b'b_2 \dagger_3 \bar{b'}b'' \dagger_4 b_3\bar{b''}\dagger_5\}, R)$, where $b, b', b'', b_1, b_2, b_3 \in \{a, u, g, c\}$, $\bar{b}, \bar{b'}, \bar{b''}$ is complement of b, b' and b'' respectively. The rules R is given as follows:

$$R_1 = [(\lambda, b, \dagger_1)_{ins}, (\lambda, \bar{b}, \dagger_2)_{ins}],\ R_2 = [(\dagger_2, b, \lambda)_{ins}, (\dagger_3, \bar{b}, \lambda)_{ins}],$$
$$R_3 = [(\lambda, b, \dagger_4)_{ins}, (\lambda, \bar{b}, \dagger_5)_{ins}],\ R_4 = [(\dagger_1, b, \lambda)_{ins}],\ R_5 = [(\lambda, b, \dagger_3)_{ins}],$$
$$R_6 = [(\dagger_4, b, \lambda)_{ins}],\ R'_i = [(\lambda, \dagger_i, \lambda)_{del}] \mid 1 \le i \le 5.$$

Similarly, the language $L_{edbl} = \{u_1A_1u_2v_1\bar{u_2}^R Bv_2\bar{B}A_2\bar{u_1}^R \mid u_1, u_2, v_1, v_2 \in \Sigma^+_{RNA}, A_1, A_2, B, \bar{B} \in \Sigma_{RNA}\}$ representing the extended double bulge loop structure can be generated by the Matrix insertion-deletion system given by $\Upsilon_{edbl} = (\{b, \bar{b}, \dagger_1, \dagger_2, \dagger_3, \dagger_4, \dagger_5\}, \{b, \bar{b}\}, \{b\dagger_1 A_1b' \dagger_2 b_1\bar{b'} \dagger_3 Bb_2 \dagger_4 \bar{B}A_2\bar{b}\dagger_5\}, R)$, where the symbols $b, b', b_1, b_2, A_1, A_2, B, \bar{B} \in \{a, u, g, c\}$, $\bar{b}, \bar{b'}, B$ is complement of b, b' and B respectively. The rules R is given as follows:

$$R_1 = [(\lambda, b, \dagger_1)_{ins}, (\lambda, \bar{b}, \dagger_5)_{ins}],\ R_2 = [(\lambda, b, \dagger_2)_{ins}, (\lambda, \bar{b}, \dagger_3)_{ins}],$$
$$R_3 = [(\dagger_2, b, \lambda)_{ins}],\ R_4 = [(\lambda, b, \dagger_4)_{ins}],\ R'_i = [(\lambda, \dagger_i, \lambda)_{del}] \mid 1 \le i \le 5.$$

From the construction of the system, one can check that the usage of markers guarantee that the systems Υ_{tsl}, Υ_{edbl} generate only the languages L_{tsl}, L_{edbl} respectively.

Theorem 4. *The language of hybrid loop as well as quadruple stem and loop structure (see left and right of Fig. 4(a)) can be generated by Matrix insertion-deletion system.*

Proof. For $u_1, u_2, u_3, u_4, v_1, v_2, v_3, v_4, v_5, v_6, v_7 \in \Sigma^+_{RNA}$, the language of the hybrid loop structure given by $L_{hl} = \{u_1v_1u_2v_2\bar{u_2}^R v_3u_3v_4u_4v_5\bar{u_4}^R v_6\bar{u_3}^R v_7\bar{u_1}^R\}$ can be generated by Matrix insertion-deletion system $\Upsilon_{hl} = (\{b, \bar{b}, \dagger_1, \dagger_2, \cdots, \dagger_8\}, \{b, \bar{b}\}, \{b\dagger_1 b_1b' \dagger_2 b_2\bar{b'} \dagger_3 b_3b'' \dagger_4 b''' \dagger_5 b_4\bar{b'''} \dagger_6 b_5\bar{b''} \dagger_7 b_6 \dagger_8 \bar{b}\}, R)$, where the symbols $b, b', b'', b''', b_1, b_2, b_3, b_4, b_5, b_6 \in \{a, u, g, c\}$, $\bar{b}, \bar{b'}, \bar{b''}, \bar{b'''}$ is complement of b, b', b'', b''' respectively. The rules R is given as follows:

$$R_1 = [(\lambda, b, \dagger_1)_{ins}, (\dagger_8, \bar{b}, \lambda)_{ins}],\ R_2 = [(\lambda, b, \dagger_2)_{ins}, (\lambda, \bar{b}, \dagger_3)_{ins}],$$
$$R_3 = [(\lambda, b, \dagger_4)_{ins}, (\lambda, \bar{b}, \dagger_7)_{ins}],\ R_4 = [(\lambda, b, \dagger_5)_{ins}, (\lambda, \bar{b}, \dagger_6)_{ins}],$$
$$R_5 = [(\dagger_1, b, \lambda)_{ins}],\ R_6 = [(\dagger_2, b, \lambda)_{ins}],\ R_7 = [(\dagger_3, b, \lambda)_{ins}],\ R_8 = [(\dagger_5, b, \lambda)_{ins}],$$
$$R_9 = [(\dagger_6, b, \lambda)_{ins}],\ R_{10} = [(\dagger_7, b, \lambda)_{ins}],\ R'_i = [(\lambda, \dagger_i, \lambda)_{del}] \mid 1 \le i \le 8.$$

Similarly, the language L_{qsl} can be generated by the Matrix insertion-deletion system $\Upsilon_{qsl} = (\{b, \bar{b}, \dagger_1, \dagger_2, \dagger_3, \dagger_4, \dagger_5, \dagger_6\}, \{b, \bar{b}\}, \{b \dagger_1 b'b_1 \dagger_2 \bar{b'}b'' \dagger_3 b_2\bar{b''} \dagger_4 b'''b_3 \dagger_5 \bar{b'''}\bar{b} \dagger_6 \#\}, R)$, where $b, b', b'', b''', b_1, b_2, b_3, \# \in \{a, u, g, c\}$, $\bar{b}, \bar{b'}, \bar{b''}, \bar{b'''}$ is complement of b, b', b'', b''' respectively. The rules R is given as follows:

$$\boxed{\begin{array}{l} R_1 = [(\lambda, b, \dagger_1)_{ins}, (\lambda, \bar{b}, \dagger_6)_{ins}],\ R_2 = [(\dagger_1, b, \lambda)_{ins}, (\dagger_2, \bar{b}, \lambda)_{ins}], \\ R_3 = [(\lambda, b, \dagger_3)_{ins}, (\lambda, \bar{b}, \dagger_4)_{ins}],\ R_4 = [(\dagger_4, b, \lambda)_{ins}, (\dagger_5, \bar{b}, \lambda)_{ins}], \\ R_5 = [(\lambda, b, \dagger_2)_{ins}],\ R_6 = [(\dagger_3, b, \lambda)_{ins}], \\ R_7 = [(\lambda, b, \dagger_5)_{ins}],\ R'_i = [(\lambda, \dagger_i, \lambda)_{del}] \mid 1 \leq i \leq 6. \end{array}}$$

From the construction of the system, one can check that the usage of markers guarantee that the systems Υ_{hl}, Υ_{qsl} generate the languages L_{hl}, L_{qsl} respectively.

4 Conclusion

In this paper, using the Matrix insertion-deletion system we have modelled several bio-molecular structures that occur at RNA folding process. We remark that in this paper, to model all the bio-molecular structures, we used matrix of insertion rules and matrix of deletion rules separately (i.e., the system has no insertion rule and deletion rule together in a matrix), thus forming a new subclass. In the systems we have considered here, the insertion rules does not use any context but only the deletion rules use contexts. This can be viewed as the insertion operation works in a context-sensitive manner and the deletion operation works in a context-free manner. Thus, the system uses both the nature of context-sensitiveness and context-freeness and it seems to be a promising model for application to various domains including natural language processing. More specifically, In the vein of the paper, modelling tertiary structures, protein secondary structures [10] using this system are left as future research work.

References

1. Brendel, V., Busse, H.G.: Genome structure described by formal languages. Nucleic Acids Res. **12**(5), 2561–2568 (1984)
2. Brown, M., Wilson, C.: RNA Pseudoknot modelling using intersections of stochastic CFG with applications to database search. In: Proceedings of the Pacific Symposium on Biocomputing, Hawaii, USA, pp. 109–125 (1995)
3. Cai, L., Russell, L., Wu, Y.: Stochastic modelling of RNA pseudoknotted structures: a grammatical approach. Bioinformatics **19**(1), 66–73 (2003)
4. Galiukschov, B.S.: Semicontextual grammars (in Russian). Matem. Logica i Matem. Lingvistika, pp. 38–50 (1981)
5. Head, T.: Formal language theory and DNA: an analysis of the generative capacity of specific recombinant behaviors. Bull. Math. Biol. **49**(6), 737–750 (1987)
6. Kari, L., Thierrin, G.: Contextual insertions/deletions and computability. Inf. Comput. **131**(1), 47–61 (1996)
7. Kari, L.: On insertion and deletion in formal languages. Ph.D. Thesis, University of Turku (1991)

8. Kuppusamy, L., Mahendran, A., Krishna, S.N.: Matrix insertion-deletion systems for bio-molecular structures. In: Natarajan, R., Ojo, A. (eds.) ICDCIT 2011. LNCS, vol. 6536, pp. 301–312. Springer, Heidelberg (2011). doi:10.1007/978-3-642-19056-8_23
9. Lakshmanan, K., Anand, M., Clergerie, E.V.: Modelling intermolecular structures and defining ambiguity in gene sequences using matrix insertion-deletion systems. In: Enguix, G.B., Dahl, V., Dolores Jimenez Lopez, M. (eds.) Biology, Computation and Linguistics, New Interdisciplinary Paradigms, pp. 71–85. IOS Press, Amsterdam (2011)
10. Mamitsuka, H., Abe, N.: Prediction of beta-sheet structures using stochastic tree grammars. In: Proceedings of Fifth Workshop on Genome Informatics, pp. 19–28. Universal Academy Press, Yokohama (1994)
11. Pan, T., Sosnick, T.: RNA folding during transcription. Annu. Rev. Biophys. Biomol. Struct. **35**, 161–175 (2006)
12. Petre, I., Verlan, S.: Matrix insertion-deletion systems. Theoret. Comput. Sci. **456**, 80–88 (2012)
13. Rivas, E., Eddy, S.R.: The language of RNA: a formal grammar that includes pseudoknots. Bioinformatics **16**(4), 334–340 (2000)
14. Rozenberg, G., Salomaa, A.: Handbook of Formal Languages. Springer, New York (1996). doi:10.1007/978-3-642-59126-6
15. Searls, D.B.: Representing genetic information with formal grammars. In: Proceedings of the National Conference on Artificial Intelligence, Saint Paul, Minnesota, pp. 386–391 (1988)
16. Searls, D.B.: The computational linguistics of biological sequences. In: Hunter, L. (ed.) Artificial Intelligence and Molecular Biology, pp. 47–120. AAAI Press, Menlo Park (1993)
17. Searls, D.B.: Formal grammars for intermolecular structures. In: First International IEEE Symposium on Intelligence and Biological Systems, Washington, USA, pp. 30–37 (1995)
18. Searls, D.B.: The language of genes. Nature **420**(6912), 211–217 (2002)
19. Uemura, Y., Hasegawa, A., Kobayashi, S., Yokomori, T.: TAG for RNA structure prediction. Theoret. Comput. Sci. **210**(2), 277–303 (1999)

Homometric Number of a Graph and Some Related Concepts

Anu V.[1(✉)] and Aparna Lakshmanan S.[2]

[1] Department of Mathematics, St. Peter's College, Kolenchery 682 311, Kerala, India
anusaji1980@gmail.com

[2] Department of Mathematics, St. Xavier's College for Women, Aluva 683 101, Kerala, India
aparnaren@gmail.com

Abstract. Given a graph $G = (V, E)$, two subsets S_1 and S_2 of the vertex set V are homometric, if their distance multisets are equal. The homometric number $h(G)$ of a graph G is the largest integer k such that there exist two disjoint homometric subsets of cardinality k. We prove that the homometric number of the Cartesian product of two graphs is at least twice the product of the homometric numbers of the individual graphs. We also prove that the homometric number of the kth-power graph of a graph G is always greater than or equal to that of G. The homometric number of some classes of graphs are also obtained. A lower bound for the homometric number of triangle-free regular graphs is obtained and two graph parameters; weak homometric number and twin number, which are related to homometric number are also discussed.

Keywords: Homometric number · Weak Homometric number · Twin Number

1 Introduction

Let $G = (V(G), E(G))$ be a graph with vertex set $V(G)$ and edge set $E(G)$. If there is no ambiguity in the choice of G, then we write $V(G)$ and $E(G)$ as V and E respectively. For any set $S \subseteq V$, the cardinality of S is denoted by $|S|$. The distance multiset of S, denoted by $DM_G(S)$ or simply by $DM(S)$, is the multiset of all pair-wise distances between any two vertices of S. Two subsets S_1 and S_2 of the vertex set V are said to be homometric, if their distance multisets are equal. The homometric number $h(G)$ of a graph G is the largest integer k such that there exist two disjoint homometric subsets, S_1 and S_2 of the vertex set V, each of cardinality k. Clearly, $h(G) \leqslant \lfloor \frac{n}{2} \rfloor$, where $\lfloor x \rfloor$ denotes the greatest integer less than or equal to x. Even though there is a concept of infinite distance in the case of disconnected graphs, to avoid ambiguity we consider only connected graphs. For a family $\mathcal{G}$ of graphs, $h(\mathcal{G}) = inf\{h(G) : G \in \mathcal{G}\}$. If $\mathcal{G}_n$ denotes the class of all graphs on n vertices, then $h(\mathcal{G}_n)$ is denoted by $h(n)$.

S. Arumugam et al. (Eds.): ICTCSDM 2016, LNCS 10398, pp. 30–37, 2017.
DOI: 10.1007/978-3-319-64419-6_4

In 2010, Albertson et al. [1] initiated the study of homometric sets in graphs. They proved that $\frac{c\ logn}{log\ logn} \leqslant h(n) \leqslant \frac{n}{4}$, for $n > 3$. Axenovich and Özkahya [4] gave a better lower bound on the maximal size of homometric sets in trees. They showed that every tree on n vertices contain homometric sets of size at least $\sqrt[3]{n}$. A haircomb tree on n vertices contains homometric sets of size at least $\frac{\sqrt{n}}{2}$. They also proved that, for any graph G of diameter d, $h(G) \geqslant cn^{\frac{1}{2d-2}}$. Recently, Fulek and Mitrović [8] improved the result on haircomb trees by proving that there exist disjoint homometric sets of size at least $cn^{\frac{2}{3}}$, for a constant c. Lemke et al. [10] showed that if G is a cycle of length $2n$ then every subset of $V(G)$ with n vertices and its complement are homometric sets.

1.1 Basic Definitions and Preliminaries

For any graph G the number of vertices in G is denoted by $n(G)$ or simply by n. The distance between any two vertices u and v in V is the length of a shortest path joining u and v in G and is denoted by $d_G(u, v)$ or simply by $d(u, v)$. The maximum distance between any pair of vertices in G is the diameter of the graph G and is denoted by $diam(G)$. Any induced path $P = u_1, u_2, \ldots, u_l$ in G where $l = diam(G) + 1$ is called a diametral path with end vertices u_1 and u_l. Since $\{u_1, u_2, \ldots, u_{\lfloor \frac{l}{2} \rfloor}\}$ and $\{u_{\lfloor \frac{l}{2} \rfloor + 1}, \ldots, u_{2\lfloor \frac{l}{2} \rfloor}\}$ are disjoint homometric subsets, $\lceil \frac{diam(G)}{2} \rceil \leqslant h(G)$, where $\lceil x \rceil$ denotes the least integer greater than or equal to x.

For any set $S \subseteq V(G)$ the distance set of S, denoted by $D_G(S)$ or simply by $D(S)$, is the set of all pair-wise distances between any two vertices of S. Two subsets S_1 and S_2 of the vertex set V are said to be weakly homometric if their distance sets are equal [11]. The weak homometric number of a graph G is the largest integer k such that there exist two disjoint weakly homometric subsets S_1 and S_2 of the vertex set V each of cardinality k and it is denoted by $h_w(G)$. Clearly $h(G) \leqslant h_w(G) \leqslant \lfloor \frac{n}{2} \rfloor$ [2].

For a graph G, two disjoint subsets of vertices are called twins if they have the same cardinality and induced subgraphs with the same number of edges [5]. The twin number $t(G)$ is the largest k such that there are twins A and B in G with $|A| = |B| = k$. *i.e.*, $DM(A)$ and $DM(B)$ contains equal number of one's. Hence, $t(G) \geqslant h(G)$.

The Cartesian product of two graphs G and H, denoted by $G \Box H$ is the graph with vertex set $V(G) \times V(H)$ and any two vertices (u_1, v_1) and (u_2, v_2) are adjacent in $G \Box H$ if $u_1 = u_2$ and $v_1 v_2 \in E(H)$ or $u_1 u_2 \in E(G)$ and $v_1 = v_2$. It is known that [9], if (u_1, v_1) and (u_2, v_2) are two vertices in $G \Box H$, then $d_{G \Box H}((u_1, v_1), (u_2, v_2)) = d_G(u_1, u_2) + d_H(v_1, v_2)$.

The kth-power graph of a graph G, denoted by G^k, is the graph obtained from G by adding edges between any two vertices of G of distance less than or equal to k. The join of two graphs G and H, denoted by $G \vee H$ is defined as the graph with $V(G \vee H) = V(G) \cup V(H)$ and $E(G \vee H) = E(G) \cup E(H) \cup \{uv : u \in V(G) \text{ and } v \in V(H)\}$. Let K_n and C_n denote the complete graph and cycle on n vertices respectively. The join of K_1 and C_{n-1} is called the wheel, denoted by W_n.

A distance-hereditary graph is a graph in which the distances in any connected induced subgraph are the same as they are in the original graph. A universal vertex is a vertex adjacent to all the other vertices of the graph. A vertex with degree one is called a pendant vertex. The (n, m)-*kite* is the graph constructed by taking a copy of K_n and a path on m vertices and adding an edge between a vertex in K_n to a pendant vertex in the path [1]. The friendship graph F_k is a graph which consists of k triangles with a common vertex.

A graph G is almost irregular if it has exactly one pair of vertices of the same degree. The construction of almost irregular graph is explained in [3]. Two vertices u and v are said to be false twins if $N(u) = N(v)$, where $N(u)$ denotes the set of adjacent vertices of u and true twins if $N[u] = N[v]$ where $N[u] = N(u) \cup \{u\}$.

Throughout this paper, we consider only simple graphs. For any graph theoretic terminology and notations the readers may refer to [6].

2 Homometric Number

Theorem 1. *If G is a distance-hereditary graph and H is a connected induced subgraph of G, then $h(H) \leqslant h(G)$.*

Proof. Let S_1 and S_2 be two disjoint homometric sets in H with $|S_1| = |S_2| = h(H)$. For any two vertices u, v in S_1, $d_H(u, v) = d_G(u, v)$. This is true for the corresponding vertices u' and v' in S_2 with $d_H(u, v) = d_H(u', v')$. Thus S_1 and S_2 will be disjoint homometric sets in G also. Hence $h(G) \geqslant h(H)$.

Theorem 2. *For any two connected graphs G and H, $h(G \Box H) \geqslant 2\, h(G)h(H)$.*

Proof. Let $S_1 = \{u_1, u_2, \ldots, u_k\}$ and $S_2 = \{u_{1'}, u_{2'}, \ldots, u_{k'}\}$ be two disjoint homometric subsets of $V(G)$ and $T_1 = \{v_1, v_2, \ldots, v_s\}$ and $T_2 = \{v_{1'}, v_{2'}, \ldots, v_{s'}\}$ be two disjoint homometric subsets of $V(H)$. Hence corresponding to any two vertices u_i and u_j in S_1 there exist two vertices u_i' and u_j' in S_2 such that $d_G(u_i, u_j) = d_G(u_i', u_j')$. Similarly corresponding to any two vertices v_a and v_b in T_1 there exist two vertices v_a' and v_b' in T_2 such that $d_H(v_a, v_b) = d_H(v_a', v_b')$. Consider $(S_1 \times T_1) \cup (S_1 \times T_2)$ and $(S_2 \times T_1) \cup (S_2 \times T_2)$. Clearly, they are two disjoint subsets of $V(G \Box H)$ of the same cardinality. Let (u_i, v_a) and (u_j, v_b) be any two vertices in $(S_1 \times T_1) \cup (S_1 \times T_2)$. Then $d_{G \Box H}((u_i, v_a), (u_j, v_b)) = d_G(u_i, u_j) + d_H(v_a, v_b)$.

Case 1. Both (u_i, v_a) and (u_j, v_b) are in $S_1 \times T_1$.

If $u_i \neq u_j$ and $v_a \neq v_b$, then there exist $(u_i', v_a'), (u_j', v_b') \in S_2 \times T_2$ such that $d_{G \Box H}((u_i, v_a), (u_j, v_b)) = d_{G \Box H}((u_i', v_a'), (u_j', v_b'))$. If $u_i = u_j$, then take $u_{i'} = u_{j'} \in S_2$ so that $(u_{i'}, v_a'), (u_{j'}, v_b') \in S_2 \times T_2$ and $d_{G \Box H}((u_i, v_a), (u_j, v_b)) = d_H(v_a', v_b') = d_{G \Box H}((u_{i'}, v_a'), (u_{j'}, v_b'))$. If $v_a = v_b$, then take $v_{a'} = v_{b'} \in T_2$ so that $(u_i', v_{a'}), (u_j', v_{b'}) \in S_2 \times T_2$ and $d_{G \Box H}((u_i, v_a), (u_j, v_a)) = d_{G \Box H}((u_i', v_{a'}), (u_j', v_{b'}))$.

Case 2. Both (u_i, v_a) and (u_j, v_b) are in $S_1 \times T_2$.

A similar argument as in Case 1 shows that corresponding to any two vertices in $S_1 \times T_2$, there exists a pair of vertices in $S_2 \times T_1$ such that the distance is preserved.

Case 3. (u_i, v_a) is in $S_1 \times T_1$ and (u_j, v_b) is in $S_1 \times T_2$.

If $u_i \neq u_j$, then there must exist $u'_i, u'_j \in S_2$ such that $d_G(u_i, u_j) = d_G(u'_i, u'_j)$. Thus there exist $(u'_i, v_a) \in S_2 \times T_1$ and $(u'_j, v_b) \in S_2 \times T_2$ such that $d_{G\Box H}((u_i, v_a), (u_j, v_b)) = d_{G\Box H}((u'_i, v_a), (u'_j, v_b))$. If $u_i = u_j$, then take $u_{i'} = u_{j'} \in S_2$ so that $(u_{i'}, v_a) \in S_2 \times T_1$ and $(u_{j'}, v_b) \in S_2 \times T_2$ and $d_{G\Box H}((u_i, v_a), (u_j, v_b)) = d_{G\Box H}((u_{i'}, v_a), (u_{j'}, v_b))$. Here v_a cannot be equal to v_b.

Hence, we have proved that corresponding to any two vertices in $(S_1 \times T_1) \cup (S_1 \times T_2)$, there exists a pair of vertices in $(S_2 \times T_1) \cup (S_2 \times T_2)$ such that the distance is preserved. Thus $(S_1 \times T_1) \cup (S_1 \times T_2)$ and $(S_2 \times T_1) \cup (S_2 \times T_2)$ are two disjoint homometric subsets of $V(G\Box H)$. Therefore, $h(G\Box H) \geqslant 2\, h(G)h(H)$.

Remark 1. In $G\Box H$, there are $n(H)$ copies of G and $n(G)$ copies of H. So the above theorem can be modified as follows: *For any two connected graphs G and H, $h(G\Box H) \geqslant max\{2h(G)h(H), n(G), n(H)\}$.* But the bound cannot be improved since $h(P_n \Box P_n) = \frac{n^2}{2} = 2\, h(P_n)h(P_n)$, if n is even.

Theorem 3. *The homometric number of kth-power graph of a graph G is greater than or equal to that of G. i.e., $h(G^k) \geqslant h(G)$.*

Proof. Let S_1 and S_2 be two disjoint homometric subsets of G with $|S_1| = |S_2| = h(G)$. Each distance d in $DM_G(S_1) = DM_G(S_2)$ becomes $\lceil \frac{d}{k} \rceil$ while considering in G^k. Therefore S_1 and S_2 will be two disjoint homometric subsets of G^k and hence $h(G^k) \geqslant h(G)$.

Remark 2. There are graphs G such that $h(G^2) = 2\, h(G)$. For example, let G be the $(m, m-2)$-*kite*, where m is odd, with vertices $v_1, v_2, \ldots, v_{2m-2}$; where $v_1, v_2, \ldots, v_m$ are the vertices of K_m and $v_{m+1}, v_{m+2}, \ldots, v_{2m-2}$ are the vertices of the path and the edge is added between v_m and v_{m+1}. In [1], it is proved that $h(G) = \frac{m-1}{2}$. In G^2, $S_1 = \{v_1, v_3, \ldots, v_{2m-3}\}$ and $S_2 = \{v_2, v_4, \ldots, v_{2m-2}\}$ are two disjoint homometric sets so that $h(G^2) = m - 1 = 2\, h(G)$. But the bound cannot be improved since $h(C_n^2) = h(C_n) = \lfloor \frac{n}{2} \rfloor$.

Theorem 4. *If G is a wheel graph W_{n+1}, the complement of a path or a cycle, a friendship graph or an almost irregular graph on n vertices, then $h(G) = \lfloor \frac{n}{2} \rfloor$.*

Proof. **Case 1.** G is the wheel graph $W_{n+1} = K_1 \vee C_n$.

Let $v_1, v_2, \ldots, v_n$ be the vertices of C_n and v be the vertex of K_1. Put $v_1, v_2, \ldots, v_{\lfloor \frac{n}{2} \rfloor}$ in S_1 and $v_{\lfloor \frac{n}{2} \rfloor + 1}, \ldots, v_{2\lfloor \frac{n}{2} \rfloor}$ in S_2. Then $DM(S_1) = DM(S_2)$ consists of $\lfloor \frac{n}{2} \rfloor - 1$ one's and ${}^{\lfloor \frac{n}{2} \rfloor}C_2 - (\lfloor \frac{n}{2} \rfloor - 1)$ two's. Thus $h(W_{n+1}) \geqslant \lfloor \frac{n}{2} \rfloor$. If $n+1$ is odd, this is the maximum possible value. If $n+1$ is even, to increase the homometric number, we have to put v in any of S_1 and S_2. Let it be in S_1.

Then the number of one's in $DM(S_1)$ will be greater than that of $DM(S_2)$. (The other case also follows similarly.) So $h(W_{n+1}) = \lfloor \frac{n}{2} \rfloor$.

Case 2. G is the complement of the path P_n.

Let $P_n : v_1, v_2, \ldots, v_n$ be a path on n vertices. If n is even, $S_1 = \{v_1, v_3, v_5, \ldots, v_{n-1}\}$ and $S_2 = \{v_2, v_4, \ldots, v_n\}$ are independent sets in P_n. So they are cliques in $\overline{P_n}$. Hence $DM(S_1) = DM(S_2)$ contains ${}^{\lfloor \frac{n}{2} \rfloor}C_2$ one's in $\overline{P_n}$. So $h(\overline{P_n}) = \lfloor \frac{n}{2} \rfloor$. If n is odd, take $S_1 = \{v_1, v_3, v_5, \ldots, v_{n-2}\}$ and $S_2 = \{v_2, v_4, \ldots, v_{n-1}\}$. As in the above case $DM(S_1) = DM(S_2)$ in $\overline{P_n}$. Hence $h(\overline{P_n}) = \lfloor \frac{n}{2} \rfloor$.

If G is the complement of a cycle, the proof is similar to that of a path.

Case 3. G is the friendship graph F_k.

Let $u_1, v_1, u_2, v_2, \ldots, u_k, v_k, u$ be the vertices of F_k, where u is an universal vertex and u_iv_i is an edge for each $i = 1, 2, \ldots, k$. Then $S_1 = \{u_1, u_2, \ldots, u_k\}$ and $S_2 = \{v_1, v_2, \ldots, v_k\}$ will be two disjoint homometric sets and hence homometric number is k.

Case 4. G is an almost irregular graph.

We can directly verify the result for $n \leqslant 3$. Suppose $n \geqslant 4$. Let G be a graph with vertices $v_1, v_2, \ldots, v_n$. If n is even, let

$$d(v_i) = \begin{cases} i, & for \;\; i = 1, 2, \ldots, \frac{n}{2}. \\ i-1, & for \;\; i = \frac{n}{2}+1, \ldots, n. \end{cases}$$

Then $N(v_i) = \{v_{n-j}, j = 0, 1, \ldots, i-1;\ i \neq n-j\}$. If $n = 4k+2$, take $S_1 = \{v_i/\text{odd } i \text{ with } i \leqslant \frac{n}{2} - 2 \text{ and even } i \text{ with } i \geqslant \frac{n}{2} + 1\}$ and $S_2 = \{v_i/\text{even } i \text{ with } i \leqslant \frac{n}{2} - 1 \text{ and odd } i \text{ with } i \geqslant \frac{n}{2}$. If $n = 4k$, take $S_1 = \{v_i/\text{odd } i \text{ with } i \leqslant \frac{n}{2} - 1 \text{ and even } i \text{ with } i \geqslant \frac{n}{2} + 2\}$ and $S_2 = \{v_i/\text{even } i \text{ with } i \leqslant \frac{n}{2} \text{ and odd } i \text{ with } i \geqslant \frac{n}{2} + 1$.

If n is odd, let $n = 2k+1$, and

$$d(v_i) = \begin{cases} i, & for \;\; i = 1, 2, \ldots, \frac{n-1}{2}. \\ i-1, & for \;\; i = \frac{n+1}{2}, \ldots, n. \end{cases}$$

Take $S_1 = \{v_i/i \text{ is odd and } i \neq k+1 \; or \; k+2\}$ and $S_2 = \{v_i/i \text{ is even}\}$.

In all cases, arrange the vertices in S_1 (similarly in S_2) in increasing order of their suffix. Then the distance between any two vertices in S_1 will be same as that of corresponding vertices in the same position in S_2. Hence S_1 and S_2 will be two disjoint homometric sets and $h(G) = \lfloor \frac{n}{2} \rfloor$.

3 Regular Graphs

In this section we discuss the homometric number of regular graphs. The only 1-regular connected graph is K_2 and the only 2-regular connected graphs are the cycles. Therefore in this section we consider graphs with regularity greater than or equal to 3.

Lemma 1. *If G is a k-regular graph, then $h(G) \geqslant 2$.*

Proof. Let G be a k-regular graph, $k \geqslant 3$. Then $n \geqslant 4$. Let u and v be two adjacent vertices in G. Let w be a vertex distinct from u and v. Since $d(w) = k \geqslant 3$, there is a vertex z adjacent to w distinct from u and v. Take $S_1 = \{u, v\}$ and $S_2 = \{w, z\}$. Then $DM(S_1) = DM(S_2) = \{1\}$ and hence $h(G) \geqslant 2$.

Proposition 1. *Let G be a k-regular triangle free graph. Then $h(G) \geqslant \lfloor \frac{k}{2} \rfloor$.*

Proof. If G is complete, then the proof is trivial. Now suppose u and v are two non adjacent vertices. If k is even, partition $N(u) = S_1 \cup S_2$ such that and $|S_1| = |S_2| = \frac{k}{2}$. Then S_1 and S_2 will form two disjoint homometric sets whose distance multisets containing only two's and hence $h(G) \geqslant \frac{k}{2}$.

If k is odd, put $\frac{k-1}{2}$ neighbours of u together with u in S_1 and $\frac{k-1}{2}$ neighbours of v together with v in S_2 so that $S_1 \cap S_2 = \phi$. Then S_1 and S_2 will form two disjoint homometric sets with cardinality $\frac{k+1}{2}$ and hence $h(G) \geqslant \frac{k+1}{2}$.

Theorem 5. *If $b \geqslant \lceil \frac{k}{2} \rceil$, then there exists a k-regular graph with homometric number b.*

Proof. Consider the cycle C_{2b} with vertex set $\{v_1, v_2, \ldots, v_{2b}\}$. If k is even, make v_i adjacent to v_j for every $d(v_i, v_j) \leqslant \frac{k}{2}$. If k is odd make v_i adjacent to v_j for every $d(v_i, v_j) \leqslant \frac{k}{2}$ and v_i adjacent to v_{i+b} for every $i = 1, 2, \ldots, b$. Then $S_1 = \{v_1, v_2, \ldots, v_b\}$ and $S_2 = \{v_{b+1}, v_{b+2}, \ldots, v_{2b}\}$ will be two disjoint homometric subsets and hence $h(G) = b$.

4 Some Related Graph Parameters

In this section we find the weak homometric number of the complete bipartite graph and the wheel graph. We also find that the homometric number and the twin number are equal for any graph G with diameter 2.

Theorem 6. *For the complete bipartite graph $K_{m,n}$, with $m \leqslant n$,*

$$h_w(K_{m,n}) = \begin{cases} \lfloor \frac{n}{2} \rfloor, & if \ \ m = 1, n \geqslant 2, \\ \lfloor \frac{m+n}{2} \rfloor, & otherwise. \end{cases}$$

Proof. Let $V = (X, Y)$ be a bipartition of $K_{m,n}$ with $|X| = m$ and $|Y| = n$.

Case 1. $m = 1, n \geqslant 2$.

Put the vertices of Y in S_1 and S_2 so that $S_1 \cap S_2 = \phi$ and $|S_1| = |S_2| = \lfloor \frac{n}{2} \rfloor$. Therefore, in this case $h_w(K_{m,n}) \geqslant \lfloor \frac{n}{2} \rfloor$. If n is even, this is the maximum possible value. If n is odd, in order to increase the weak homometric number, we have to put the universal vertex in any of S_1 or S_2. Then the distance set of that set will contain one's but not that of other. Hence $h_w(K_{m,n}) = \lfloor \frac{n}{2} \rfloor$.

Case 2. $m = n = 1$ or $m, n \geqslant 2$.

Let $X = \{u_1, u_2, \ldots, u_m\}$ and $Y = \{v_1, v_2, \ldots, v_n\}$. If $m = n$, $S_1 = X$ and $S_2 = Y$ will form a weak homometric partition in G. If $m = 2$ and $n = 3$, then $S_1 = \{u_1, u_2\}$ and $S_2 = \{v_1, v_2\}$ will be two disjoint weak homometric sets in G. In all other cases, put u_1, v_1 and v_2 in S_1 and u_2, v_3 and v_4 in S_2. Thus $D(S_1) = D(S_2) = \{1, 2\}$. Put the remaining vertices of V in S_1 and S_2 so that $S_1 \cap S_2 = \phi$ and $|S_1| = |S_2| = \lfloor \frac{m+n}{2} \rfloor$. Hence, in this case $h_w(K_{m,n}) = \lfloor \frac{m+n}{2} \rfloor$.

Theorem 7. *For the wheel graph* W_n*,*

$$h_w(W_n) = \begin{cases} \lfloor \frac{n}{2} \rfloor, & if \;\; n \neq 6, \\ 2, & if \;\; n = 6. \end{cases}$$

Proof. $W_n = K_1 \vee C_{n-1}$. Let $v_1, v_2, \ldots, v_{n-1}$ be the vertices of C_{n-1} and v be the vertex of K_1. We can directly verify the result up to W_6. Suppose $n \geqslant 7$. Put v_1, v_2 and v_3 in S_1 and v_4, v_5 and v_6 in S_2 so that $D(S_1)$ and $D(S_2)$ contains both 1 and 2. Put the remaining vertices in S_1 and S_2 in such a way that $|S_1| = |S_2|$ and $S_1 \cap S_2 = \phi$. Hence $h_w(W_n) = \lfloor \frac{n}{2} \rfloor$, if $n \neq 6$.

Theorem 8. *For any graph G with diameter* 2*, the homometric number and the twin number are equal.*

Proof. Since diameter is 2, distance multiset of any subset of vertices contains 1 and 2 only. Let S_1 and S_2 be twins with $|S_1| = |S_2| = t(G)$. Then S_1 and S_2 induce subgraphs with the same number of edges and hence number of pair of non adjacent vertices in S_1 and S_2 are also equal. So $DM(S_1)$ and $DM(S_2)$ contains equal number of one's and two's. Hence S_1 and S_2 are two disjoint homometric sets. Therefore $h(G) = |S_1| = |S_2| = t(G)$.

Note 1. A graph G is a cograph if and only if G does not contain P_4 as an induced subgraph [7]. Hence two vertices of a connected cograph are a distance at most two apart and by above theorem homometric number and twin number are equal for any connected cograph.

Acknowledgments. The first author thanks University Grants Commission for granting fellowship under Faculty Development Programme (FDP).

References

1. Albertson, M.O., Pach, J., Young, M.E.: Disjoint homometric sets in graphs. Ars Mathematica Contemporanea **4**(1), 1–4 (2011)
2. Aparna, L.S., Menon, M.K., Anu, V.: Homometric number of graphs (Communicated)
3. Benjamin, A., Chartrand, G., Zhang, P.: The Fascinating World of Graph Theory. Princeton University Press, Princeton (2015)
4. Axenovich, M., Özkahya, L.: On homometric sets in graphs. Australas. J. Combin. **55**, 175–187 (2013)

5. Axenovich, M., Martin, R., Ueckerdt, T.: Twins in graphs. Eur. J. Comb. **39**, 188–197 (2014)
6. Balakrishnan, R., Ranganathan, K.: A Textbook of Graph Theory. Springer, New York (2000)
7. Corneil, D.G., Lerchs, H., Stewart Burlingham, L.: Complement reducible graphs. Discret. Appl. Math. **3**(3), 163–174 (1981)
8. Fulek, R., Mitrović, S.: Homometric sets in trees. Eur. J. Comb. **35**, 256–263 (2014)
9. Hammack, R., Imrich, W., Klavžar, S.: Handbook of Product Graphs. CRC Press, Boca Raton (2011)
10. Lemke, P., Skiena, S.S., Smith, W.D.: Reconstructing sets from interpoint distances. In: Aronov, B., Basu, S., Pach, J., Sharir, M. (eds.) Discrete and Computational Geometry. Algorithms and Combinatorics, vol. 25, pp. 507–631. Springer, Heidelberg (2003). doi:10.1007/978-3-642-55566-4_27
11. Senechal, M.: A point set puzzle revisited. Eur. J. Comb. **29**, 1933–1944 (2008)

Forbidden Subgraphs of Bigraphs of Ferrers Dimension 2

Ashok Kumar Das(✉) and Ritapa Chakraborty

Department of Pure Mathematics, University of Calcutta, Kolkata, India
ashokdas.cu@gmail.com, ritapa.chakroborty@gmail.com

Abstract. A bipartite graph B with bipartion X, Y is called a Ferrers bigraph if the neighbor sets of the vertices of X (or equivalently Y) are linearly ordered by set inclusion. The Ferrers dimension of B is the minimum number of Ferrers bigraphs whose intersection is B. In this paper we present a new approach of finding the forbidden subgraphs of bigraphs of Ferrers dimension 2 when it contains a strong bisimplicial edge.

Keywords: ATE · Ferrers dimension · Strong bisimplicial edge · Forbidden subgraphs

1 Introduction

Ferrers bigraphs (the bipartite analogue of Ferrers digraphs) were introduced independently by Guttman [4] and Riguet [7].

A bipartite graph (in short, bigraph) $B = (X, Y, E)$ is a *Ferrers bigraph* if it satisfies any of the following equivalent conditions:

(i) The neighbors of the vertices of X (or equivalently of Y) are linearly ordered by inclusion.
(ii) The rows and columns of the biadjacency matrix can be permuted (independently) so that the 1's cluster in the upper right (or lower left) as a Ferrers diagram.
(iii) The biadjacency matrix has no 2-by-2 permutation matrix $\begin{pmatrix} 1 & 0 \\ 0 & 1 \end{pmatrix}$ or $\begin{pmatrix} 0 & 1 \\ 1 & 0 \end{pmatrix}$ as a submatrix.

The biadjacency matrix is the submatrix of the adjacency matrix whose rows are indexed by one partite set and columns by the other. The biadjacency matrix of B is called a *Ferrers matrix*.

The inclusion condition on the verices of X partite set and the vertices of Y partite set induced two natural partitions of the vertices of $X(Y)$ partite sets associated with a Ferrers bigraph. These different disjoint subsets into which the vertices of $X(Y)$ partite set of a Ferrers bigraph is being partitioned are called *partition classes.*

S. Arumugam et al. (Eds.): ICTCSDM 2016, LNCS 10398, pp. 38–49, 2017.
DOI: 10.1007/978-3-319-64419-6_5

The *Ferrers dimension* of a bigraph B, written $f(B)$, is defined to be the minimum number of Ferrers bigraphs whose intersection is B. The bigraphs with Ferrers dimension 2 have been characterized, by Cogis [1] and others. Cogis [1] introduced the *associated graph* $H(B)$ for B whose vertices are the 0s of the biadjacency matrix $A(B)$ of B, with two vertices are adjacent in $H(B)$ if and only if they are the 0s of a 2-by-2 permutation submatrix of $A(B)$ and proved that $f(B) = 2$ if and only if $H(B)$ is bipartite. In [8] Sen et al. translate the Cogis's condition to an adjacency matrix condition for bigraphs of Ferrers dimension 2 in the following theorem.

Theorem 1. [1,8] *The following conditions are equivalent:*

(i) B has Ferrers dimension at most 2;
(ii) The rows and the columns of $A(B)$ can be permuted independently, so that no 0 has a 1 both below it and to its right;
(iii) The associated graph $H(B)$ of B is bipartite.

A *two clique circular arc graph* is a circular arc graph whose vertices can be covered by two disjoint cliques. Trotter and Moore [9] characterized two clique circular arc graph in terms of forbidden subgraphs. They presented the forbidden families as a set system and proved that B is a circular arc graph if and only if its complement $\overline{B}$ contains no induced subgraphs of the form G_1, G_2, G_3 and several infinite families C_i, T_i, W_i, M_i, N_i $(i \geq 1)$.

Huang [5] proved that a bipartite graph B is of Ferrers dimension two if and only if its complement is a two clique circular arc graph. Therefore a bigraph B is of Ferrers dimension two if and only if B contains none of the graphs of the families C_i, T_i, W_i, D_i, M_i, N_i $(i \geq 1)$ and the graphs G_1, G_2 and G_3 as induced subgraphs.

Now it can be observed that the graphs IV, VI, V of Fig. 1 are respectively the graphs G_1, G_2 and G_3. The class C_i is the class of even cycles of length ≥ 6. The classes T_i, W_i, D_i are respectively the classes of graphs III_n, I_n and II_n of Fig. 1. The class M_i has exactly one strong bisimplicial edge. The class N_i has no strong bisimplicial edges. In this paper using condition *(ii)* of the Theorem 1 we alternatively determine the class M_i of forbidden subgraphs of bigraphs of Ferrers dimension 2.

A pair of edges of a bigraph B is *separable* if they induce the subgraph $2K_2$ in B. A bigraph B is *separable* if it contains $2K_2$ as an induced subgraph, otherwise it is called *non-separable.* Obviously the biadjacency matrix of a separable bigraph contains 2×2 permutation submatrix and hence a non-separable bigraph is a Ferrers bigraph. A bigraph is *chordal bipartite* or *bichordal* if it does not contain any chordless cycle of length ≥ 6. As every chordless cycle of length ≥ 6 is of Ferrers dimension ≥ 3, so a bigraph B having Ferrers dimension at most 2 is necessarily bichordal. Three mutually separable edges e_1, e_2, e_3 of a graph G are said to form an *asteroidal triple of edges(ATE)* [3,6], if for any two of them, there is a path from the vertex set of one to the vertex set of another that avoids the neighbors of the third edge. According to Das and Sen [3] if a bichordal graph B contains an ATE then its Ferrers dimension is greater than 2.

Das and Sen [2] have determined a minimal set of bichordal graphs (see Fig. 1) with the property that any bichordal graph has an ATE if and only if it contains a graph of this set as an induced subgraph.

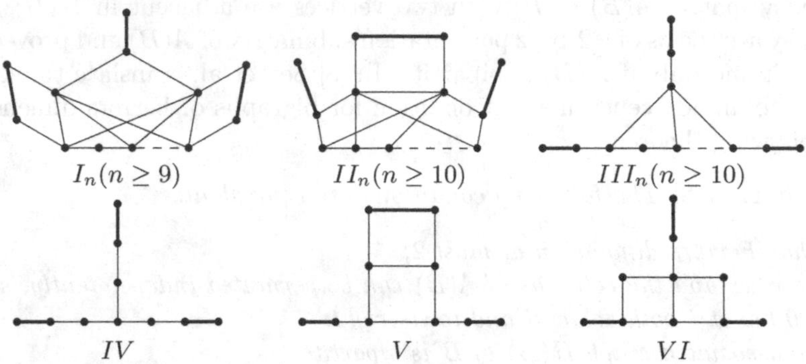

Fig. 1. List of ATE's.

2 Forbidden Induced Subgraphs of ATE - Free Bigraphs of Ferrers Dimension 2

In [3] Das and Sen showed that the graph B^1 of Fig. 2 is bichordal and ATE-free but of Ferrers dimension 3 as the associated graph $H(B^1)$ of the graph has an odd cycle. Actually it is the first graph of M_i class of Trotter and Moore.

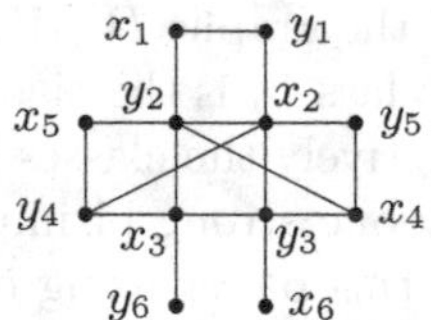

Fig. 2. The bigraph B^1.

Definition 1. [3] *Let $e = xy$ be an edge of a bipartite graph $B = (X, Y, E)$. Also let $B(e) = B(xy)$ denote the subgraph induced by $adj(x) + adj(y)$. An edge $e = xy$ of the bipartite graph B is bisimplicial if $B(e)$ or $B(xy)$ is complete. A bisimplicial edge $e = xy$ of B is said to be strong if $B \backslash B(e)$ is connected; otherwise it is weak.*

It can be observed that bigraph of Fig. 2 contains the strong bisimplicial edge x_1y_1. The biadjacency matrix of this bigraph can be arranged as follows (Fig. 3).

Motivated by the structure of the above matrix we state below the following theorem which is the central result of this paper.

	y_1	y_2	y_3	y_4	y_5	y_6
x_1	1	1	0	0	0	0
x_2	1	1	1	1	1	0
x_3	0	1	1	1	0	1
x_4	0	1	1	0	1	0
x_5	0	1	0	1	0	0
x_6	0	0	1	0	0	0

Fig. 3. Biadjacency matrix of B^1

Theorem 2. *Let a bipartite graph $B = (X, Y, E)$ be bichordal and ATE - free and contains a strong bisimplicial edge. Then either $f(B) = 2$ or the biadjacency matrix $A(B)$ of B contains one of the matrix of the infinite class $\mathcal{M} = \{M_1, M_2, M_3, ...\}$ of matrices as a submatrix, where M_1, M_2, M_3 etc. are given in the* Fig. 4.

1	1	0	0	0	0
1	1	1	1	1	0
0	1	1	1	0	1
0	1	1	0	1	0
0	1	0	1	0	0
0	0	1	0	0	0

1	1	0	0	0	0	0	0
1	1	1	1	1	1	1	0
0	1	1	1	1	1	0	1
0	1	1	1	1	0	1	0
0	1	1	1	0	1	0	0
0	1	1	0	1	0	0	0
0	1	0	1	0	0	0	0
0	0	1	0	0	0	0	0

1	1	0	0	0	0	0	0	0	0
1	1	1	1	1	1	1	1	1	0
0	1	1	1	1	1	1	1	0	1
0	1	1	1	1	1	1	0	1	0
0	1	1	1	1	1	0	1	0	0
0	1	1	1	1	0	1	0	0	0
0	1	1	1	0	1	0	0	0	0
0	1	1	0	1	0	0	0	0	0
0	1	0	1	0	0	0	0	0	0
0	0	1	0	0	0	0	0	0	0

Fig. 4. The matrices M_1, M_2, M_3 respectively

Now we state the notations used in the proof.

Let $e = xy$ be a bisimplicial edge of a bipartite graph $B = (X, Y, E)$. We write $B' = (X', Y', E') = B \backslash \{x, y\}$, $B_1 = (X_1, Y_1, E_1) = B(xy) \backslash \{x, y\}$, $B_2 = (X_2, Y_2, E_2) = B \backslash B(xy) = B' \backslash B_1$. $N(e) =$ Vertex set of $B(e)$ i.e., the set of neighbors of x and y. Let $X_2{}'$ be the set of those members of X_2 which are adjacent to some members of Y_1 and $X_2{}'' = X_2 - X_2{}'$. So, $X_2 = X_2{}' \bigcup X_2{}''$. Similarly we can define $Y_2{}'$ and $Y_2{}''$ so that $Y_2 = Y_2{}' \bigcup Y_2{}''$.

Next we denote the subgraphs induced by the vertices $X_1 \bigcup Y_2{}'$ and $X_2{}' \bigcup Y_1$ by P and Q respectively.

The proof of the theorem is very long and requires a careful reading. Here we consider the bigraphs as copy-free i.e. no two vertices have the same neighbor. To prove Theorem 2 we first prove the following lemma.

Lemma 1. *Let B be a bichordal and ATE-free bigraph and let $e = xy$ be a strong bisimplicial edge of B. Then the induced subgraphs P and Q as defined above are non-separable.*

Proof of the Lemma. If possible, let the induced subgraph P have separable edges $x_i y_i$ and $x_j y_j$ where $x_i, x_j \in X_1$ and $y_i, y_j \in Y_2{}'$. Since B_2 is connected, it must contain a path between y_i and y_j. Now if this path is of length 2, say $y_i x' y_j \in X_1$, then we have a six cycle $x_i y x_j y_j x' y_i x_i$ in B. On the other hand if the path is of length > 2, say $y_i x' ... x'' y_j$, then three edges $e = xy$, $e' = x' y_i$ and $e'' = x'' y_j$ are mutually separable and constitute an ATE of B; for we have paths $y x_i y_i$ between e and e' and $y x_j y_j$ between e and e'' which avoid $N(e'')$ and $N(e')$ respectively and $N(e)$ does not contain any vertex of the bigraph B_2. Similarly it can be shown that the subgraph Q is also non-separable.

Proof of Theorem 2. We recall that no vertex of B is a copy of another. By the above lemma, the subgraphs P and Q are non - separable (i.e., their biadjacency matrices are Ferrers bigraph) so we can order the vertices of $Y_2{}'$ and $X_2{}'$ such that, $Adj(B')$, the biadjacency matrix of B' has the following configuration (Fig. 5).

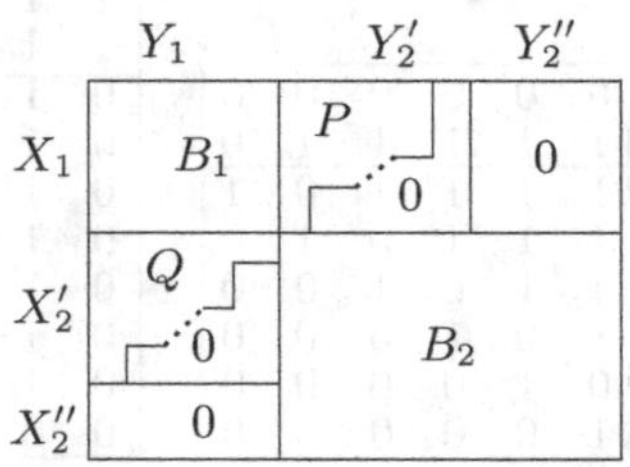

Fig. 5. Biadjacency matrix of B'

We recall that the vertices $X_2 = X_2{}' \bigcup X_2{}''$ and $Y_2 = Y_2{}' \bigcup Y_2{}''$ induced the subgraph B_2 of B'. Consequently $Adj(B_2)$, the biadjacency matrix of B_2 is a submatrix of Fig. 5, where the row and column arrangements are the same as in $Adj(B')$ (i.e., of Fig. 5). Also it is to be noted that in the $Adj(B_2)$ we can permute the rows (columns) of $X_2{}''$ ($Y_2{}''$), and the rows (columns) of $X_2{}'$ ($Y_2{}'$) which belongs to the same partitioned class of the Ferrers bigraphs $Q(P)$ without changing the structure of Fig. 5. These permutations will be referred to as the *permissible permutations.*

We will show that when a bigraph B satisfies the given conditions of the theorem and its biadjacency matrix is free from the matrices of the infinite class $\mathcal{M}$ of binary matrices, $Adj(B')$ with its rows and columns arranged in Fig. 5 will exhibit the characteristics of a bigraph of Ferrers dimension 2 (Theorem 1) and once this is established, we place the x-row and y-column of the bisimplicial edge $e = xy$ to the top and extreme left of $Adj(B')$ and this will prove the theorem. This will be established if we can show that $Adj(B_2)$ has the property that no 0 has a 1 both to its right and below it. For this we will show below that, if $Adj(B_2)$ for any permissible permutation of its rows and columns, contains the configuration (i) of the form,

	y_i	y_j
x_i	0	1
x_j	1	–

......(i)

Where '–' position is either 0 or 1, then the bigraph B contains either an ATE or a 6-cycle or its biadjacency matrix contains any of the matrices of the class $\mathcal{M}$.

To complete the proof of the theorem, we need to consider following cases:

Case 1. Neither P nor Q is complete bipartite graph;
Case 2. P is complete but Q is not;
Case 3. Q is complete but P is not;
Case 4. Both P and Q are complete.

We will observe that while a 6-cycle or an ATE will be present in all the four cases the class of forbidden matrices $\mathcal{M}$ will occur only in case 4.

Case 1. Suppose for any permissible permutation of its rows and columns, *Adj*(B_2) contains the configuration (i).

This case is to be divided again into four subcases subject to whether the vertices x_i, x_j and the vertices y_i, y_j belong to the same partitioned class or to distinct partitioned classes.

Subcase 1a. x_i and x_j belong to two distinct partitioned classes of X_2 and so do y_i and y_j belong to two distinct classes of Y_2, where $x_1, y_1 \in V(B_1) = X_1 \bigcup Y_1$.

In this case clearly *Adj*(B') contains a configuration.

	y_1	y_i	y_j
x_1	1	1	0
x_i	1	0	1
x_j	0	1	–

Now if the '–' position is 1, then the above configuration is 6-cycle. So we suppose that '–' position is 0. Then the three edges $e = xy, e_1 = x_iy_j, e_2 = x_jy_i$ of B are mutually separable. Also we have path xy_1x_i between e and e_1 and path yx_1y_i between e and e_2 which avoid respectively $N(e_2)$ and $N(e_1)$. And there exists path between e_1 and e_2 which avoid $N(e)$ (since e_1 and e_2 are two edges of the connected component B_2 and no vertex of it is adjacent to e). So $\{e, e_1, e_2\}$ constitute an ATE of B.

Subcase 1b. y_i and y_j belong to the same partitioned class, whereas x_i and x_j belong to two distinct classes.

In this case *Adj*(B') contains a configuration.

	y_1	y_i	y_j
x_1	1	–	–
x_i	1	0	1
x_j	0	1	–

Let $x_1 \in X_1$, $y_1 \in Y_1$, and x_1y_i, x_1y_j positions are both 1 or both 0.

It is possible that by permuting the vertices y_i and y_j, we can get a F_2-matrix, except of course when we are confronted with a vertex of x_k belonging to still another partitioned class of X_2 and having the following configuration of $Adj(B')$.

	y_1	y_2	y_i	y_j
x_1	1	1	–	–
x_i	1	1	0	1
x_j	1	0	1	0
x_k	0	0	0	1

Now, let $x_1 \in X_1$ and $y_1, y_2 \in Y_1$.

Here also we find three mutually separable edges $e = xy, e_1 = x_jy_i$ and $e_2 = x_ky_j$. And these three edges constitute an ATE of B, for the paths $xy_1x_j, xy_2x_iy_j$ between e, e_1 and e, e_2 avoid the neighbour of e_2 and e_1 respectively and because B_2 is connected the path between e_1 and e_2 avoids neighbours of e.

Subcase 1c. x_i and x_j belong to the same partitioned class of X_2 whereas y_i and y_j belong to two distinct classes of Y_2.

This is similar to case 1*b* and so is omitted.

Subcase 1d. x_i and x_j belong to the same partitioned class and y_i and y_j belong to the same partitioned class.

First suppose that $x_i, x_j \in X_2'$ and $y_i, y_j \in Y_2'$. Now one possibility is that we can permute x_i, x_j and/or y_i, y_j and get $Adj(B')$ as F_2-matrix straightway without facing any obstruction elsewhere. Otherwise, the four configurations are the instances in $Adj(B')$, where there are vertices x_k, belonging to a class other than that of x_i, x_j and y_k, belonging to a class other than that of y_i and y_j, when we fail to derive the F_2-matrix directly (Fig. 6).

	y_1	y_i	y_j	y_k
x_1	1	1	1	0
x_i	1	0	1	0
x_j	1	1	0	1
x_k	0	0	1	–

	y_1	y_k	y_i	y_j
x_1	1	1	0	0
x_k	1	–	1	0
x_i	0	1	0	1
x_j	0	0	1	–

	y_1	y_i	y_j	y_k
x_1	1	1	1	0
x_k	1	1	0	–
x_i	0	0	1	0
x_j	0	1	0	1

	y_1	y_k	y_i	y_j
x_1	1	1	0	0
x_i	1	1	0	1
x_j	1	0	1	0
x_k	0	–	0	1

Fig. 6.

In all these cases it can be seen through a careful and exhaustive scrutiny that B contains either an ATE or 6-cycle.

Case 2 and 3. Proof in these cases are similar to the case 1 (i.e., if, for any permissible permutation of its rows and columns, $Adj(B_2)$ contains the configuration (i) as before, then the bigraph B contains either an ATE or a 6-cycle) and so are omitted.

Case 4. Here both P and Q are complete bipartite graphs. Since no vertex of B is a copy of another, it is clear that X_1 and Y_1 are singleton sets, let $X_1 = \{x_1\}$ and $Y_1 = \{y_1\}$. So $Adj(B)$ has the following structure (Fig. 7):

If possible let $Adj(B_2)$ contains a configuration (i). Here we will show that $\mathcal{M}$ is the only class of matrices of Ferrers dimension > 2 but the corresponding bigraph is free from a 6-cycle or an ATE. We observe that the structure,

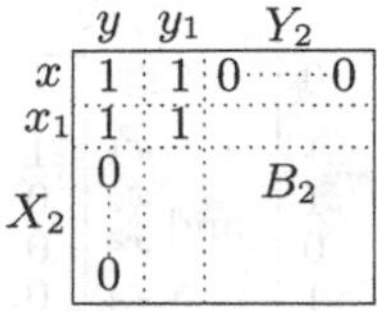

Fig. 7. Biadjacency matrix of B.

	y	y_1	y_i	y_j
x	1	1	0	0
x_1	1	1	1	0
x_i	0	1	0	1
x_j	0	0	1	–

of $Adj(B)$ with xy as strong bisimplicial edge implies that B must contain an ATE or C_6 according as '–' position is a 0 or 1. So if $Adj(B)$ contains the configuration (i), (and symmetry of $Adj(B)$ shows that) $Adj(B)$ must have one of the following structures:

	y	y_1	y_i	y_j
x	1	1	0	0
x_1	1	1	–	–
x_i	0	1	0	1
x_j	0	1	1	–

or

	y	y_1	y_i	y_j
x	1	1	0	0
x_1	1	1	–	–
x_i	0	1	0	1
x_j	0	0	1	–

The positions x_1y_i and x_1y_j are both 0 or both 1.

Here we suppose that both the positions x_1y_i and x_1y_j are 1. (We are not considering the possibility that x_1y_i and x_1y_j are 0, since later we add all possible row and/or column to the above matrices).

Note that the x_jy_j position in either of the matrices is 0 or 1. To facilitate the matter we replace the x_j row by two rows x_{j_1} and x_{j_2} to the matrices, one row taking the value '0' and the other taking the value '1' in the corresponding y_j column. So we get the matrices.

	y	y_1	y_i	y_j
x	1	1	0	0
x_1	1	1	1	1
x_i	0	1	0	1
x_{j1}	0	1	1	0
x_{j2}	0	1	1	1

and

	y	y_1	y_i	y_j
x	1	1	0	0
x_1	1	1	1	1
x_i	0	1	0	1
x_{j1}	0	0	1	0
x_{j2}	0	0	1	1

Fig. 8.

We label the vertices x_i, x_{j_1}, x_{j_2} by x_3, x_4 and x_2 and the vertices y_i, y_j by y_3, y_2 respectively in both the figures for the sake of convenience.

Clearly, by permuting the x_3, x_4 and x_2 rows and y_3, y_2 columns of Fig. 8(i), $Adj(B)$ gets the following F_2-matrix structure (Fig. 9(i)). Also permuting the rows and columns of the matrix in Fig. 8(ii) we get F_2-matrix of Fig. 9(ii).

	y	y_1	y_2	y_3
x	1	1	0	0
x_1	1	1	1	1
x_2	0	1	1	1
x_3	0	1	1	0
x_4	0	1	0	1

and

	y	y_1	y_2	y_3
x	1	1	0	0
x_1	1	1	1	1
x_2	0	1	1	0
x_3	0	0	1	1
x_4	0	0	0	1

Fig. 9.

Naturally, the questions arises; is it possible that by adding a row/column to the rearranged matrices of the above figures we will get a matrix which forbids its F_2 representation ? And we have to address this important question every time, whenever we come across a matrix having F_2-characteristics.

We answer this question through a very long and exhaustive searching process, where we will show that:

In any of the case of Fig. 9(i) or Fig. 9(ii), this attempts lead in addition to a 6-cycle or an ATE, the only $\mathcal{M}$ class of forbidden matrices.

Now we will prove our point for the matrix of Fig. 9(i) through a detailed study. The proof for the matrix of Fig. 9(ii) is of similar nature and so will be omitted.

We first take into account the particular means of adding suitable rows and columns to Fig. 9(i) (to forbid F_2-matrix) that yields the bigraph B^1.

To the matrix of Fig. 9(i), there are several alternatives for adding rows among these, we consider the particular row (name it x_5).

	y	y_1	y_2	y_3
x_5	0	0	0	1

Then the new matrix gets the following F_2-representation (Fig. 10).

	y	y_1	y_2	y_3
x	1	1	0	0
x_1	1	1	1	1
x_2	0	1	1	1
x_3	0	1	1	0
x_4	0	1	0	1
x_5	0	0	0	1

Fig. 10.

	y	y_1	y_2	y_3	y_4	y_5
x	1	1	0	0	0	0
x_1	1	1	1	1	1	0
x_2	0	1	1	1	–	–
x_3	0	1	1	0	–	–
x_4	0	1	0	1	–	–
x_5	0	0	0	1	–	–

Fig. 11.

To Fig. 10, we add two new columns, say y_4 and y_5 to get the matrix (Fig. 11).

First we suppose that x_3y_4 position is a 1. Then we have the matrix of the left side and permuting the rows and columns of that matrix we have the matrix of the right side.

	y	y_1	y_2	y_3	y_4	y_5
x	1	1	0	0	0	0
x_1	1	1	1	1	1	0
x_2	0	1	1	1	–	–
x_3	0	1	1	0	1	–
x_4	0	1	0	1	–	–
x_5	0	0	0	1	–	–

	y	y_1	y_2	y_3	y_4	y_5
x	1	1	0	0	0	0
x_1	1	1	1	1	1	0
x_3	0	1	1	1	0	–
x_2	0	1	1	–	1	–
x_4	0	1	0	–	1	–
x_5	0	0	0	–	1	–

If this matrix has a configuration (i), then the corresponding bigraph B has an ATE or C_6 according as '–' entry is a 0 or 1. Otherwise it is a F_2 matrix. So we consider x_3y_4 position is a 0. Then if x_3y_5 is a 1 then we have an ATE or C_6 according as x_5y_5 position is a 0 or 1. Thus both the entries x_3y_4 and x_3y_5 are 0. So we have the matrix of Fig. 12.

	y	y_1	y_2	y_3	y_4	y_5
x	1	1	0	0	0	0
x_1	1	1	1	1	1	0
x_2	0	1	1	1	–	–
x_3	0	1	1	0	0	0
x_4	0	1	0	1	–	–
x_5	0	0	0	1	–	–

Fig. 12.

	y	y_1	y_2	y_3	y_4	y_5
x	1	1	0	0	0	0
x_1	1	1	1	1	1	0
x_2	0	1	1	1	0	1
x_3	0	1	1	0	0	0
x_4	0	1	0	1	1	0
x_5	0	0	0	1	0	0

Fig. 13. The biadjacency matrix of B^1

Now it is a matter of verification that when y_4 and y_5 column of the above matrix of Fig. 12 have the structure as in Fig. 13 then we get the matrix which is the biadjacency matrix of the bigraph B^1 (after suitable labeling the vertices). For the other structures of y_4 and y_5 columns, when Fig. 12 contains the submatrix.

	y_4	y_5
x_i	0	1
x_j	1	–

where x_i, x_j are any two among the x_2, x_4, x_5 rows, then it can be checked that its corresponding bigraph B must contains either an ATE or a 6-cycle.

Next rearranging the rows and columns of the matrix of Fig. 13 and renaming y_3 and y_2 as y_2 and y_3 respectively also renaming x_4 and x_3 as x_3 and x_4 respectively we have the matrix (Fig. 14), which is also the biadjacency matrix of B^1.

Motivated by the above structure of the matrix of Fig. 14 we consider the matrix of Fig. 15.

Now rearranging the rows and columns of the above matrix in Fig. 15 we have the matrix of Fig. 16, which is actually a F_2-matrix.

Now if the position x_2y_2 is 0 then we have an ATE. If the position $x_2y_4 = 0$ then deleting the x_3 row and y_5 column we have the matrix of Fig. 14 i.e., the biadjacency matrix of B^1. If $x_2y_3 = 0$ then we have a F_2-matrix. Next if the

	y	y_1	y_2	y_3	y_4	y_5
x	1	1	0	0	0	0
x_1	1	1	1	1	1	0
x_2	0	1	1	1	0	1
x_3	0	1	1	0	1	0
x_4	0	1	0	1	0	0
x_5	0	0	1	0	0	0

Fig. 14.

	y	y_1	y_2	y_3	y_4	y_5	y_6
x	1	1	0	0	0	0	0
x_1	1	1	1	1	1	1	0
x_2	0	1	1	1	1	0	1
x_3	0	1	1	1	0	1	0
x_4	0	1	1	0	1	0	0
x_5	0	1	0	1	0	0	0
x_6	0	0	1	0	0	0	0

Fig. 15.

	y	y_1	y_3	y_2	y_5	y_4	y_6
x	1	1	0	0	0	0	0
x_1	1	1	1	1	1	1	0
x_3	0	1	1	1	1	0	0
x_2	0	1	1	1	0	1	1
x_5	0	1	1	0	0	0	0
x_4	0	1	0	1	0	1	0
x_6	0	0	0	1	0	0	0

Fig. 16.

position x_3y_2 or x_3y_5 is 0 then it retains its F_2-matrix structure. And if x_3y_3 position is 0 then deleting x_4 row and y_4 column we have again a matrix which has same structure of Fig. 14. Again it can be observed that if any of the positions x_6y_3, x_5y_4, x_4y_5, x_3y_6 is a 1 then we have an induced C_6 since in each case we have the matrix $\begin{pmatrix} 1\,1\,0 \\ 1\,0\,1 \\ 0\,1\,1 \end{pmatrix}$ as a submatrix. And if the positions x_6y_5 or x_5y_6 is 1 the we have an ATE. Finally, if any of the positions x_6y_4,x_6y_6,x_5y_5 or x_4y_6 is 1 the matrix of Fig. 15 retains its F_2-matrix structure. Thus we do not have any new forbidden bigraph from Fig. 15 which is free from ATE or C_6 but of Ferrers dimension > 2.

Next we consider the matrix of Fig. 17. The bigraph corresponding to M_2 is bichordal, ATE free and have xy as a strong bisimplicial edge but its associated graph is not a bipartite graph and hence the Ferrers dimension of $M_2 > 2$. Now we can observe that no position among x_3y_7, x_4y_6, x_5y_5, x_6y_4 and x_7y_3 should be 1. Since in each case we have a submatrix which is the biadjacency matrix of C_6.

	y	y_1	y_2	y_3	y_4	y_5	y_6	y_7
x	1	1	0	0	0	0	0	0
x_1	1	1	1	1	1	1	1	0
x_2	0	1	1	1	1	1	0	1
x_3	0	1	1	1	1	0	1	0
x_4	0	1	1	1	0	1	0	0
x_5	0	1	1	0	1	0	0	0
x_6	0	1	0	1	0	0	0	0
x_7	0	0	1	0	0	0	0	0

Fig. 17. The matrix M_2

Next if $x_6y_7 = 1$, then xy, x_6y_7, x_7y_2 forms an ATE. If $x_5y_7 = 1$ then x_5y_7,x_3y_6 and x_4y_5 form an ATE. If $x_7y_7 = 1$ then we have the following matrix (which is M_1) as a submatrix of M_2.

	y_7	y_2	y_3	y_4	y_5	y_6
x_7	1	1	0	0	0	0
x_2	1	1	1	1	1	0
x_3	0	1	1	1	0	1
x_4	0	1	1	0	1	0
x_5	0	1	0	1	0	0
x_6	0	0	1	0	0	0

Similarly we can verify that if $x_4y_7 = 1$ or $x_5y_6 = 1$ then we have the matrix M_1 as a submatrix of M_2 and if $x_6y_6 = 1$ then we have an ATE.

Also it can be verified that if we replace any 1 by a 0 in the matrix M_2 then it either contains an ATE, a C_6, the matrix M_1 or becomes a F_2-matrix. Thus the bigraph corresponding to M_2 is minimal bichordal, ATE free graph but of Ferrers dimension > 2.

Similarly we can verify that bigraphs corresponding to M_3 and other matrices of the class $\mathcal{M}$ are the only minimal bichordal, ATE free bigraphs but of Ferrers dimension > 2. This completes the proof.

References

1. Cogis, O.: A characterization of digraphs with Ferrers dimension 2. Rapport de Recherche, no. 19, G.R. CNRS no. 22, Paris (1979)
2. Das, A.K., Sen, M.K.: Asteroidal triple of edges in bichordal graphs: a complete list. Electr. Notes Discrete Math. **15**, 68–70 (2003)
3. Das, A.K., Sen, M.K.: Bigraphs/digraphs of Ferrers dimension 2 and asteroidal triple of edges. Discrete Math. **295**, 191–195 (2005)
4. Guttman, L.: A basis for scaling quantitative data. Ann. Social Rev. **9**, 139–150 (1944)
5. Huang, J.: Representation characterization of chordal bipartite graphs. J. Combin Theory Series B **96**, 673–683 (2006)
6. Muller, H.: Recognition Interval digraphs and interval bigraphs in polynomial time. Disc. Appl. Math. **78**, 189–205 (1997)
7. Riguet, J.: Les relation de Ferrers. C. R. Acad. Sci. Paris **232**, 1729–1730 (1951)
8. Sen, M., Das, S., Roy, A.B., West, D.B.: Interval digraphs: an analogue of interval graphs. J. Graph Theory. **13**, 189–203 (1989)
9. Trotter, W.T., Moore, J.I.: Characterization problem for graphs, partially ordered sets, lattice and family of sets. Disc. Math. **16**, 361–380 (1976)

Global Secure Domination in Graphs

S.V. Divya Rashmi[1(✉)], S. Arumugam[2], and A. Somasundaram[3]

[1] Department of Mathematics, Vidyavardhaka College of Engineering, Mysuru 570002, Karnataka, India
rashmi.divya@gmail.com

[2] Kalasalingam University, Anand Nagar, Krishnankoil 626126, Tamil Nadu, India
s.arumugam.klu@gmail.com

[3] Department of Mathematics, BITS, Pilani-Dubai, Dubai International Academic City, Dubai, UAE
somu_raji@yahoo.co.in

Abstract. Let $G = (V, E)$ be a graph. A subset S of V is called a dominating set of G if every vertex in $V \backslash S$ is adjacent to a vertex in S. A dominating set S is called a secure dominating set if for every vertex $v \in V - S$, there exists $u \in S$ such that $uv \in E$ and $(S - \{u\}) \cup \{v\}$ is a dominating set of G. If S is a secure dominating set of both G and its complement $\overline{G}$, then S is called a global secure dominating set (gsd-set) of G. The minimum cardinality of a gsd-set of G is called the global secure domination number of G and is denoted by $\gamma_{gs}(G)$. In this paper we present several basic results on $\gamma_{gs}(G)$ and interesting problems for further investigation.

Keywords: Domination · Global domination · Secure domination · Global secure domination

1 Introduction

By a graph $G = (V, E)$ we mean a finite, undirected graph with neither loops nor multiple edges. The order $|V|$ and the size $|E|$ are denoted by n and m respectively. For graph theoretic terminology we refer to Chartrand and Lesniak [1].

Let $G = (V, E)$ be a graph. A subset S of V is called a dominating set of G if every vertex $v \in V - S$ is adjacent to a vertex is S. The domination number γ of G in the minimum cardinality of a dominating set of G. For an excellent treatment of the fundamentals of domination we refer to the book by Haynes et al. [3]. A survey of several advanced topics in domination is given in the book edited by Haynes et al. [4]. Sampathkumar [5] introduced the concept of global domination in graphs.

Definition 1. *A subset S of V is called a global dominating set of G if S is a dominating set of both G and its complement $\overline{G}$. The global domination number γ_g of G is the minimum cardinality of a global dominating set of G.*

S. Arumugam et al. (Eds.): ICTCSDM 2016, LNCS 10398, pp. 50–54, 2017.
DOI: 10.1007/978-3-319-64419-6_6

Strategies for protection of a graph $G = (V, E)$ by placing one or more guards at every vertex of a subset S of V, where a guard at a vertex can protect all vertices in its closed neighborhood have resulted in the study of several concepts such as Roman domination, weak Roman domination and secure domination. The concept of secure domination is motivated by the following situation and was introduced by Cockayne et al. [2]. Given a graph $G = (V, E)$, we wish to place one guard at each vertex of a subset S of V in such a way that S is a dominating set of G and if a guard at v moves along an edge to protect an unguarded vertex u, then the resulting configuration of guards also forms a dominating set. This leads to the concept of secure domination.

Definition 2 [2]. *A dominating set S of G is called a secure dominating set of G if for each $u \in V - S$, there exists $v \in S$ such that u is adjacent to v and $(S - \{v\}) \cup \{u\}$ is a dominating set of G. In this case we say that u is S-defended by v or v S-defends u. The secure domination number $\gamma_s(G)$ is the minimum cardinality of a secure dominating set of G.*

In this paper we combine the concepts of global domination and secure domination which arises naturally and present several results on global secure domination number of a graph.

We need the following definitions and theorems.

Definition 3. *The corona of two graphs G_1 and G_2, denoted by $G_1 \circ G_2$, is the graph obtained by taking $|V(G_1)|$ copies of G_2 and joining the ith vertex of G_1 to every vertex in the ith copy of G_2.*

Definition 4. *Let $G = (V, E)$ be graph, $S \subseteq V$ and $v \in S$. A vertex $u \in V$ is an S-private neighbor of v if $N(u) \cap S = \{v\}$. The set of all S-private neighbors of v is denoted by $PN(v, S)$. If further $u \in V \setminus S$, then u is called an S-external private neighbor or S-epn of v.*

Theorem 1 [2]. *For the path P_n we have $\gamma_s(P_n) = \lceil \frac{3n}{7} \rceil$ for all $n \geq 4$.*

Theorem 2 [2]. *For the cycle C_n we have $\gamma_s(C_n) = \lceil \frac{3n}{7} \rceil$ for all $n \geq 4$.*

Theorem 3 [5]. *If G is a graph with $\delta = 1$, then $\gamma_g \leq \gamma + 1$.*

2 Main Results

Definition 5. *Let $G = (V, E)$ be a graph. A subset S of V is a global secure dominating set (g.s.d. set) of G if S is a secure dominating set of G and its complement $\overline{G}$. The minimum cardinality of a global secure dominating set of G is called the global secure domination number of G and is denoted by $\gamma_{gs}(G)$.*

Observation 1. *If S is a γ_s-set of G, then S is a secure dominating set of $\overline{G}$ if for every vertex $v \in V - S$, there exists a vertex u in S such that $uv \notin E(G)$ and $(S - \{u\}) \cup \{v\}$ is a dominating set of $\overline{G}$.*

Since any vertex v which is isolated either in G or in $\overline{G}$ lies in every g.s.d set of G, we confine ourselves to graphs G for which G and $\overline{G}$ have no isolated vertices.

Theorem 4. *Let S be a global dominating set of G, which is a secure dominating set of G. Let $u \in V - S$. Then u is not S-defended in $\overline{G}$ if and only if for every vertex v in S adjacent to u in $\overline{G}$, there exists a vertex x in $V - S$, not adjacent to u such that x is an S-epn of v in $\overline{G}$.*

Proof. Suppose u is not S-defended in $\overline{G}$. Hence for every $v \in S$ such that v is adjacent to u in $\overline{G}$, the set $S_1 = (S - \{v\}) \cup \{u\}$ is not a dominating set of $\overline{G}$. Let x be a vertex in $V - S_1$ which is not dominated by S_1 in $\overline{G}$. Therefore x is an S-epn of v in $\overline{G}$. Conversely if x is an S-epn of v in $\overline{G}$, then x is not dominated by any vertex of S_1 in $\overline{G}$. Hence u is not S-defended in $\overline{G}$.

We proceed to determine $\gamma_{gs}(G)$ for some standard graphs. If $G = K_{n_1,n_2,\ldots,n_r}$ where each $n_i \geq 2$ and $r \geq 3$, then $\overline{G} = K_{n_1} \cup K_{n_2} \cup \cdots \cup K_{n_r}$. Hence it follows that $\gamma_{gs}(G) = r$.

We observe that $\gamma_{gs}(G) \geq \max\{\gamma_s(G), \gamma_s(\overline{G})\}$. The following theorem shows that equality holds for paths P_n and cycles C_n, for all $n \geq 6$.

Theorem 5

1. $\gamma_{gs}(C_n) = \gamma_s(C_n) = \begin{cases} 2 & if\ n = 4 \\ 3 & if\ n = 5 \\ \lceil \frac{3n}{7} \rceil & if\ n \geq 6. \end{cases}$
2. $\gamma_{gs}(P_n) = \gamma_s(P_n) = \lceil \frac{3n}{7} \rceil$ *for* $n \geq 4$.

Proof. Let $C_n = (v_1, v_2, v_3, \ldots, v_n, v_1)$. Obviously $\gamma_{gs}(C_4) = 2$ and $\gamma_{gs}(C_5) = 3$. Now let $n \geq 6$. For any i, with $1 \leq i \leq n$, $\{v_i, v_j\}$ where $j \neq i+2, i+(n-2)$ (where addition is taken modulo n) is a secure dominating set of $\overline{C_n}$. Hence any subset S of $V(C_n)$ with $|S| \geq 3$ is a secure dominating set of $\overline{C_n}$. So $\gamma_{gs}(C_n) = \max\{\gamma_s(C_n), \gamma_s(\overline{C_n})\} = \gamma_s(C_n) = \lceil \frac{3n}{7} \rceil$.

The proof is similar for P_n.

Theorem 6. *Let p and q be two integers with $2 \leq p \leq q$ and let $G = K_{p,q}$. Then*

$$\gamma_{gs}(G) = \begin{cases} 2\ if\ p = 2\ and\ q = 2 \\ 3\ if\ p = 2\ and\ q > 2 \\ 3\ if\ p = 3 \\ 4\ if\ p \geq\ 4. \end{cases}$$

Proof. Let V_1, V_2 be the bipartition of G with $|V_1| = p$ and $|V_2| = q$. If $p = q = 2$, then $S = \{v_1, v_2\}$ where $v_1 \in V_1$ and $v_2 \in V_2$ is a global secure dominating set. Therefore $\gamma_{gs}(G) = 2$. If $p = 2, q > 2$, then $S = V_1 \cup \{x\}$ where $x \in V_2$, is a global secure dominating set of G and hence $\gamma_s(G) \leq 3$. Further any minimum secure dominating set of $\overline{G}$ contains exactly one vertex from V_1 and one vertex from V_2 and this is not a secure dominating set of G. Hence $\gamma_{gs}(G) = 3$. When $p = 3, \gamma_s(G) = 3$ and if S is any minimum secure dominating set of G, then $S \cap V_1 \neq \emptyset$ and $S \cap V_2 \neq \emptyset$. Therefore $\gamma_{gs}(G) = 3$. The proof is similar for $p \geq 4$.

Observation 2. *Let $G = K_{2,q}$ or $K_{2,q} - e$ where $q \geq 3$. Then $\gamma_s(G) = \gamma_s(\overline{G}) = 2$. However $\gamma_{gs}(G) = 3 > \max\{\gamma_s(G), \gamma_s(\overline{G})\}$. Also if $G = G_1 \cup G_2$ where $G_1 = K_2$ or K_3 and $G_2 = K_n - e$ with $n \geq 5$, then $\gamma_s(G) = 3, \gamma_s(\overline{G}) = 2$ if $G_1 = K_2$ and $\gamma_s(\overline{G}) = 3$ if $G_1 = K_3$, but $\gamma_{gs}(G) = 4 > \max\{\gamma_s(G), \gamma_s(\overline{G})\}$.*

Hence the following problem arises.

Property 1. Characterize the class of graphs G for which

$$\gamma_{gs}(G) = \max\{\gamma_s(G), \gamma_s(\overline{G})\}.$$

Theorem 5 shows that path P_n and cycle C_n where $n \geq 6$ satisfy the above equation. In the following two theorems we give two infinite families of graphs satisfying the above equation.

Theorem 7. *Let T be any tree with $\Delta < n - 1$. Then $\gamma_{gs}(T) = \max\{\gamma_s(T), \gamma_s(\overline{T})\}$.*

Proof. Let V_1, V_2 be the bipartition of T. Then $\langle V_1 \rangle$ and $\langle V_2 \rangle$ are cliques in $\overline{T}$ and hence $\gamma_s(\overline{T}) = 2$. Also $\gamma_s(T) \geq 2$ and hence $\gamma_{gs}(T) = \max\{\gamma_s(T), \gamma_s(\overline{T})\} = \gamma_s(T)$.

Theorem 8. *If either G or $\overline{G}$ is disconnected having at least three components, then $\gamma_{gs}(G) = \max\{\gamma_s(G), \gamma_s(\overline{G})\}$.*

Proof. Assume that G is disconnected. Then any secure dominating set S of G contains at least one vertex from each component of G and hence S is also a secure dominating set of $\overline{G}$. Hence $\gamma_{gs}(G) = \gamma_s(G) = \max\{\gamma_s(G), \gamma_s(\overline{G})\}$.

In the following theorem we determine the global secure domination number of corona of two graphs.

Theorem 9. *Let G_1 and G_2 be two connected graphs of order n_1 and n_2 respectively with $n_1 \geq 3$. Then $\gamma_{gs}(G_1 \circ G_2) = n_1 \gamma_s(G_2 + K_1)$.*

Proof. Let $V(G_1) = \{v_1, v_2, \ldots, v_{n_1}\}$. Let S_i be a γ_s-set of $H_i = G_2 + \{v_i\}$. Then $S = \bigcup_{i=1}^{n_1} S_i$ is a secure dominating set of $G_1 \circ G_2$ and hence $\gamma_s(G_1 \circ G_2) \leq n_1 \gamma_s(G_2 + K_1)$. Now let D be any secure dominating set of $G_1 \circ G_2$. Then $D \cap V(H_i)$ is a secure dominating set of H_i. Thus $|D \cap V(H_i)| \geq \gamma_s(G_2 + K_1)$ and hence $|D| = \sum_{i=1}^{n_1} |D \cap V(H_i)| \geq n_1 \gamma_s(G_2 + K_1)$. Therefore $\gamma_s(G_1 \circ G_2) = n_1 \gamma_s(G_2 + K_1)$. Further any γ_s-set of $G_1 \circ G_2$ is a secure dominating set of its complement and hence $\gamma_{gs}(G_1 \circ G_2) = n_1 \gamma_s(G_2 + K_1)$.

Observation 3. *Let $C_1, C_2, C_3, \ldots, C_{2r}$ be the cyclic Hamiltonian decomposition of K_{4r+1}. Let G be the subgraph induced by the cycles $C_1, C_2, \ldots, C_r$. Then $\gamma_s(G) = \gamma_{gs}(G)$.*

Observation 4. *It follows from Theorem 3 that if* $\delta = 1$, *then* $\gamma_g \leq \gamma + 1$. *Such a result is not true for secure domination. Given a positive integer* k, *there exists a graph* G *such that* $\delta = 1$ *and* $\gamma_{gs} = \gamma_s + k$. *For example, consider the graph* G *obtained by adding a vertex* u *and joining* u *to the vertex* v *of* $K_{k+3} - e$ *where* $d(v) = k + 1$. *Then* $\gamma_s(G) = 2$ *and* $\gamma_{gs}(G) = 2 + k$.

Property 2. Investigate graphs for which $\gamma_{gs} = \gamma_s + 1$.

Theorem 10. *Let* G *be a connected bipartite graph with bipartition* X, Y *such that* $|X| \leq |Y|$. *Then* $\gamma_{gs}(G) = \gamma_s(G)$ *or* $\gamma_s(G) + 1$. *Further* $\gamma_{gs}(G) = \gamma_s(G) + 1$ *if and only if* $\gamma_s(G) = |X|, X$ *is the only* γ_s*-set of* G *and there exists a vertex* y *in* Y *which is adjacent to all vertices of* X.

Proof. If there exists a γ_s-set S of G such that $S \cap X \neq \emptyset$ and $S \cap Y \neq \emptyset$, then S is secure dominating set of $\overline{G}$ and hence $\gamma_{gs}(G) = \gamma_s(G)$. Otherwise X is the only γ_s-set of G. Now if every vertex in Y is non-adjacent to a vertex in X, then X is a secure dominating set of $\overline{G}$ and hence $\gamma_{gs}(G) = \gamma_s(G) = |X|$. If there exists a $y \in Y$ such that y is adjacent to all vertices in X, then X is not a secure dominating set of $\overline{G}$ and $X \cup \{y\}$ is a secure dominating set of $\overline{G}$. Thus $\gamma_{gs}(G) = \gamma_s(G) + 1$.

Conjecture 1. For the n-dimensional hypercube Q_n, we have $\gamma_{gs}(Q_n) = \gamma_s(Q_n)$.

3 Conclusion and Scope

In this paper we have initiated a study of global secure domination in graphs. For any graph $G, \max\{\gamma_s, \overline{\gamma}_s\} \leq \gamma_{gs}(G) \leq \gamma_s + \overline{\gamma}_s$. Hence the following problem arises naturally.

Property 3. Given three integers a, b, c with $\max\{a, b\} \leq c \leq a + b$, does there exist a graph G with $\gamma_s = a, \overline{\gamma}_s = b$ and $\gamma_{gs} = c$?

For any γ_s-set S of G, let $X(S) = \{v \in V - S$ such that either $(V - N(v)) \cap S = \emptyset$ or for any $u \in (V - N(v)) \cap S, (S - \{u\}) \cup \{v\}$ is not a dominating set of $\overline{G}\}$. Choose a γ_s-set S of G for which $|X(S)|$ is minimum.

Conjecture 2. $\gamma_{gs}(G) \leq |S \cup X(S)|$.

References

1. Chartrand, G., Lesniak, L.: Graphs & Digraphs. Chapman & Hall/CRC, Boca Raton (2005)
2. Cockayne, E.J., Grobler, P.J.P., Gründlingh, W.R., Munganga, J., van Vuuren, J.H.: Protection of a graph. Util. Math. **67**, 19–32 (2005)
3. Haynes, T.W., Hedetniemi, S.T., Slater, P.J.: Fundamentals of Domination in Graphs. Marcel Dekker, Inc., New York (1998)
4. Haynes, T.W., Hedetniemi, S.T., Slater, P.J.: Domination in Graphs-Advanced Topics. Marcel Dekker Inc., New York (1998)
5. Sampathkumar, E.: The global domination number of a graph. Jour. Math. Phy. Sci. **23**(5), 377–385 (1980)

A Novel Reversible Data Hiding Method in Teleradiology to Maximize Data Capacity in Medical Images

S. Fepslin Athish Mon[1], K. Suthendran[2(✉)], K. Arjun[1], and S. Arumugam[2]

[1] Department of Computer Science and Engineering, Royal College of Engineering and Technology, Thirusur, Kerala, India
fepslin@gmail.com, arjunkppc@gmail.com

[2] Kalasalingam University, Anand Nagar, Krishnankoil 626126, Tamil Nadu, India
{k.suthendran,s.arumugam}@klu.ac.in

Abstract. Teleradiology is one of the emerging technologies used for sharing medical images through networks such as LAN, WAN, Cloud, etc. for better treatment, expert suggestions and research purpose. The images and medical related records of a patient are accessed by any physician at any time from any location. Medical field is one of the main areas for privacy violation and security threats. This work aims to protect the privacy of patients, hospitals and centralized server which store the patient's information by combining both patients image and medical records into a single image file using data hiding. In such a sensitive area data loss is not acceptable. In this paper, a novel method is proposed for solving the above problem which enable us to retrieve original medical image without any data loss. We hide the patient's details in the medical image and encrypt the cover medical image. Once image is decrypted the concerned person can only extract the patient's information from the cover medical image. The proposed system uses two keys; one for encryption and another for data hiding. If the third party knows both the keys, the patient's information can be retrieved. Our proposed method provides double protection and achieves the security in sharing the medical image with patient's record. This proposed method provides the advantage over other existing method in terms of improved data capacity and zero error rates, and maintains PSNR above 51 dB.

Keywords: Medical image · Data hiding · Reversibility

1 Introduction

Data hiding is an invisible communication but gives more importance to the information rather than medical image. Patient's information is secure and completely recoverable, but medical image is not the same as the original image after the data is extracted. Hence a normal data hiding technique is not suited for this work. A small change in medical image may violate image properties. Since

S. Arumugam et al. (Eds.): ICTCSDM 2016, LNCS 10398, pp. 55–62, 2017.
DOI: 10.1007/978-3-319-64419-6_7

medical image is diagnosed visually or by analyzing small values in the cover medical image, loss in sensitive medical image is not acceptable. So we use the technology called reversible data hiding. Reversible data hiding is a data hiding technique, which can recover the original medical image without any distortion from the embedded medical image after the data is extracted. Reversible data hiding is guaranteed for reversible, lossless, distortion free recovery of both cover medical image and patient information.

Arjun [1] proposed a method for improving the data capacity of the cover images called duplicating peak pair values. Yun [2] proposed a new method for cover image recovery, namely Reversible Data Hiding (RDH). In this paper, the author describes two methods: LSB method and Quantization Index Modulation (QIM) method. In LSB method, we replace LSB bits of a pixel with data. This replaced LSB values are not memorized further. In Quantization Index Modulation (QIM) method, normally the quantization error occurs. Due to these reasons complete cover image recovery gets impossible in effect. In [1,2], authors achieved good data capacity, but is not useful for an application with large amount of data and data capacity is depending on the cover image.

Wen [3] introduced a method for Reversible Data Hiding for complete cover image recovery. This method keeps a separate record for change of the selected minimum points. This method improves the data capacity and reached the goal of data and cover image recovery. In this method also the authors could not achieve large amount of data capacity. Yongjian [4] has introduced a new method namely even and odd number based embedding method. Ho [5] proposed two methods, difference image histogram and the transform coefficient histogram. The method introduced by Masoumeh [6] utilizes the difference of the pixel values of the host image and the zero or the minimum points of a histogram of the different image. It then modifies the pixel gray scale value slightly to embed secret data into image. In this histogram work [3–6], the data capacity depends only on the cover image and could not increase the data capacity to a maximum.

Qiminget [7] proposed a method that uses JPEG images as cover image. Data are embedded on the compressed data of the image. It does not require decompression of the JPEG cover images. In this work, the data capacity depends on the compression method and image. This work may lead to error in the data extraction.

A new method proposed by Hsiang [8] uses hierarchical relationships of original cover images. The result shows that better performance can be obtained in enhanced image quality and embedding capacity. This hierarchical relationship breaks at some point of data extraction. This may lead to increase in the error rate.

In this paper we propose a novel reversible data hiding method which achieves better quality after data embedding, high PSNR value, maximum data embedding capacity and complete recoverability of both cover medical image and patient's information. There is no error after extracting the data. The file size remains intact and data embedding capacity does not depend on the cover image property. We get uniform embedding capacity for all the images (1bpp). This novel technique seems to be the best method that works extremely well and improves all the parameters of data hiding.

2 Proposed Method

The new Reversible Data Hiding Method meets all the measurements like improved security, data embedding capacity, PSNR value and also keeps the file size same as original. The block diagram of the method is given in Fig. 1.

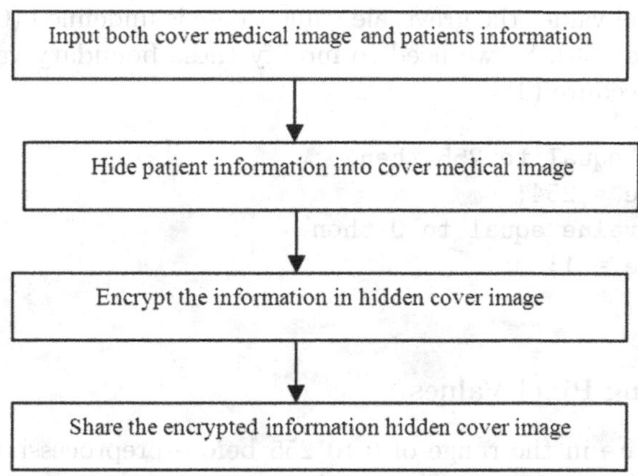

Fig. 1. Block diagram of hiding and encryption of cover medical image

2.1 Image Encryption

The input medical cover image is in uncompressed format and each pixel with gray value ranging from 0 to 255 is represented by 8 bits. Denote the bits of a pixel as $b_(i, j, 0)$, $b_(i, j, 1) \ldots b_(i, j, 7)$ where (i, j) indicates the pixel position of the medical image, and the gray value is $p_(i, j)$.

$$\text{Let } b_(i, j, r) = \frac{p_(i, j)}{2^r} (mod\ 2), \text{ where } r = 0, 1, \ldots, 7 \tag{1}$$

and

$$P_(i, j, r) = \sum_{u=0}^{7} b_(i, j, r).2^r \tag{2}$$

For providing security on the cover medical image we perform encryption using simple XOR (exclusive-or) operation.

$$\text{Thus } B_(i, j, r) = b_(i, j, r) \oplus k_(i, j, r) \tag{3}$$

where $k_(i, j, r)$ are the encryption key using a standard stream cipher. Then $B_(i, j, k)$ are the encrypted medical image with data. The medical image with data is encrypted by using XOR operation. Without knowing the key, the third party cannot understand medical image as well as data.

2.2 Pre-processing and Location Map

Our primary aim is to keep the boundary values of cover medical image same as original. For that we perform preprocessing. Preprocessing is necessary for the medical images because any one pixel value changed after the data extraction is waste and not used for diagnosing.

In grayscale medical image boundary values are 0 and 255. If we add any data on these pixel values the grayscale value range is undefined (0 and 255 may change to −1 and 256). So we need to modify these boundary values by using the following procedure (4).

```
If Pixel value equal to 255 then
New Pixel value = 254;
else if Pixel value equal to 0 then
New Pixel value = 1;
end
```

2.3 Duplicating Pixel Values

All pixel values are in the range of 0 to 255 before preprocessing and contain original values of cover image. The aim is to increase the image data capacity (1 bit per pixel). To get maximum data capacity, we set all the pixel values in the cover image into two peak gray scale values (I_R, I_S). Limiting gray scale to two peak value (I_R, I_S) in this method gets maximum frequency of the peak pair. This is the core idea of the proposed extended XOR method. Limiting peak gray scale values to two is a step by step procedure and is carried out by six rounds of XOR operation.

2.4 Data Embedding, Extracting and Recovery

In embedding process, generate a histogram from cover image and find highest two peak values IS and IR, where $(IS < IR)$. The data for embedding is *bk*. Then perform the following data embedding algorithm.

```
If Pixel value < IS, set Pixel IS-1;
If Pixel value = IS, set IS  bk;
If IS< Pixel value < IR, no change in pixel value;
If Pixel value = IR, set IR +bk;
If Pixel value > IR, set Pixel IR +1;
```

Finally last 16 pixel LSB values of the image are replaced with IS and IR value for extraction. If the data to be embedded is not complete, then the above procedure is repeated iteratively. After the data embedding phase the output will be embedded on cover image.

The data extraction phase is actually the reverse process of data embedding. First extract two peak values from the last 16 bit LSBs and perform the following procedure for data extraction.

```
If Pixel value= IS  1, bk is 1;
If Pixel value= IS, bk is 0;
If Pixel value= IR, bk is 0;
If Pixel value= IR + 1, bk is 1;
```

Finally perform the cover image recovery phase, because in RDH both cover image and data are equally important. Cover image recovery, completely recovers the cover image that is same as original cover input image. The recovery of cover image operation is performed by using the following procedure.

```
If Pixel value < IS 1, set pixel value IS+ 1.
If Pixel value = IS - 1 or IS, set pixel value IS;
If Pixel value = IR or IR + 1, set pixel value IR.
If Pixel value > IR + 1, set pixel value IR? 1.
```

Finally, compare both original image and recovered image. If the result is same, it means that reversible data hiding (RDH) is successful. We tested all the images in the test set and we got a completely recovered image, which is the same as the original image before embedding.

3 Experimental Results

The aim for this extended XOR method is to get fixed range of data capacity for all image data set, to get the maximum data embedding capacity (1 bit per pixel) and improving PSNR value. This method also keeps the file size same as original.

This method is extended from duplicating peak pairs by XOR method and is focused on the data embedding capacity and PSNR value. This method works well and the result shows that the data embedding capacity is fixed for all image sets. The embedding capacity get 1 bpp (bit per pixel) and PSNR value is above 51 dB.

The Proposed method has been tested four data sets which are normal, aerial, medical and sequence images. Each test set shows the fixed data capacity, improvement in PSNR value and improved security. The file size is preserved and is same as that of original image.

An example of a medical image, informations embedded image, encrypted image and recovered image is given in Fig. 2.

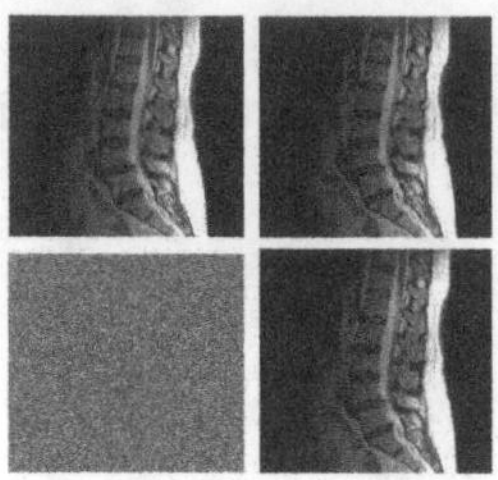

Fig. 2. The original and embedded images of Medical image 3. (a) Original cover image of "Medical image 1". (b) Information embedded image. (c) Encrypted image. (d) Recovered medical image.

3.1 6-Rounds Duplicating for Improving Data Capacity

In the proposed method, 6-rounds duplication of pixel values into peak pair values are performed using XOR operation. The result shows that this proposed method works well in maximizing the data embedding capacity, PSNR value, security and also keeps file cover image size as original. This method is focused on the data embedding capacity and PSNR value. Table 1 shows the analysis of the data hiding parameters is kept. The parameters are PSNR (Peak Signal to Noise Ratio), bit per pixel, and pure payload length. The result shows that the proposed method works well for all the parameters.

Table 1. Analysis of data hiding in medical images.

Test image name	PSNR	Bit per pixel	Pure payload
Medical image 1	51.1375	1	262144
Medical image 2	51.0593	1	262144
Medical image 3	50.9464	1	262144
Medical image 4	51.1266	1	262144
Medical image 5	51.0012	1	262144
Medical image 6	51.0254	1	262144
Medical image 7	51.6812	1	262144
Medical image 8	51.0050	1	262144

Figure 3 gives the comparison of medical images PSNR and data embedding capacity.

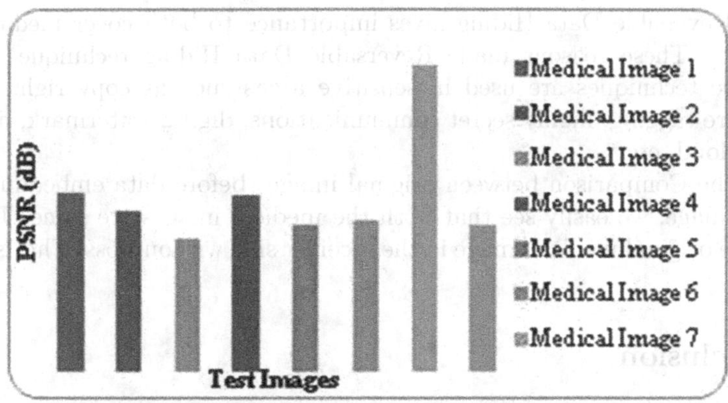

Fig. 3. PSNR comparison of medical images

3.2 File Size

One of the main properties of cover image is its size. The image size is calculated by M * N* L, where M and N are the number of rows and columns of the image respectively and L is the number of gray levels in the image. In the proposed method the gray levels are not changed. For example, if the dimension of image is 512 * 512 and the number of gray levels is 8, then the image size is $M * N * L = 512 * 512 * 8 = 262.2\,\text{KB}$.

3.3 Reversibility

In most cases of data hiding, the cover image will experience some distortion due to data hiding and cannot revert back to the original cover object. Some parameter distortion to the cover image even after the hidden data has been extracted

Table 2. Analysis of reversibility

Test image name	Before Data embedding (Original Medical image) PSNR (dB)	PSNR (dB) After Data embedding PSNR (dB)	After data Extracted PSNR (dB)
Medical image 1	99	51.1375	99
Medical image 2	99	51.0593	99
Medical image 3	99	50.9464	99
Medical image 4	99	51.1266	99
Medical image 5	99	51.0012	99
Medical image 6	99	51.0254	99
Medical image 7	99	51.6812	99
Medical image 8	99	51.0050	99

out. The Reversible Data Hiding gives importance to both cover medium and information. These reasons made Reversable Data Hiding technique popular now. These techniques are used in sensitive areas such as copy right protection, feature tagging, highly secret communications, digital watermark, medical, military, cloud, etc.

From the Comparison between original image (before data embedding) and retrieved image, we easily see that both the medical images are same. Thus we can get the original medical image in the receiver side without loss. This is shown in Table 2.

4 Conclusion

The method proposed in this paper gives better results in all parameters like data capacity, PSNR value and file size. The proposed data hiding method keeps the original image without loss after the data extraction and achieves maximum data capacity (1 bit per pixel). This method is also tested with many medical images and we achieved maximum data embedding capacity and PSNR value above 51 dB.

References

1. Arjun, K.P., Fepslin Athis Mon, S., Arunvinodh, C.: Reversible data hiding in images with imporved data capacity by duplicating peak pair values. Int. J. Adv. Inf. Sci. Technol. **5**(7), 62–67 (2016)
2. Shi, Y.Q.: Reversible data hiding. In: Cox, I.J., Kalker, T., Lee, H.-K. (eds.) IWDW 2004. LNCS, vol. 3304, pp. 1–12. Springer, Heidelberg (2005). doi:10.1007/978-3-540-31805-7_1
3. Kuo, W.-C., Jiang, D.-J., Huang, Y.-C.: Reversible data hiding based on histogram. In: Huang, D.-S., Heutte, L., Loog, M. (eds.) ICIC 2007. LNCS (LNAI), vol. 4682, pp. 1152–1161. Springer, Heidelberg (2007). doi:10.1007/978-3-540-74205-0_119
4. Hu, Y., Lee, H.-K., Li, J., Chen, K.: Reversible data hiding using prediction error values embedding. In: Shi, Y.Q., Kim, H.-J., Katzenbeisser, S. (eds.) IWDW 2007. LNCS, vol. 5041, pp. 244–253. Springer, Heidelberg (2008). doi:10.1007/978-3-540-92238-4_19
5. Thom, H.T.H., Canh, H., Tien, T.N.: Steganalysis for reversible data hiding. In: Ślęzak, D., Kim, T., Zhang, Y., Ma, J., Chung, K. (eds.) DTA 2009. CCIS, vol. 64, pp. 1–8. Springer, Heidelberg (2009). doi:10.1007/978-3-642-10583-8_1
6. Khodaei, M., Faez, K.: Histogram-based reversible data hiding. In: Qiu, G., Lam, K.M., Kiya, H., Xue, X.-Y., Kuo, C.-C.J., Lew, M.S. (eds.) PCM 2010. LNCS, vol. 6297, pp. 677–684. Springer, Heidelberg (2010). doi:10.1007/978-3-642-15702-8_62
7. Li, Q., Wu, Y., Bao, F.: A reversible data hiding scheme for JPEG images. In: Qiu, G., Lam, K.M., Kiya, H., Xue, X.-Y., Kuo, C.-C.J., Lew, M.S. (eds.) PCM 2010. LNCS, vol. 6297, pp. 653–664. Springer, Heidelberg (2010). doi:10.1007/978-3-642-15702-8_60
8. Huang, H.-C., Huang, K.-Y., Chang, F.-C.: Reversible data hiding with hierarchical relationships. In: Pan, J.-S., Chen, S.-M., Nguyen, N.T. (eds.) ACIIDS 2012. LNCS (LNAI), vol. 7198, pp. 446–455. Springer, Heidelberg (2012). doi:10.1007/978-3-642-28493-9_47

Indicated Coloring of Cartesian Product of Graphs

P. Francis(✉) and S. Francis Raj

Department of Mathematics, Pondicherry University, Puducherry 605014, India
selvafrancis@gmail.com, francisraj_s@yahoo.com

Abstract. Indicated coloring of a graph G is a coloring in which there are two players Ann and Ben, Ann picks a vertex and Ben chooses a color for this vertex. The aim of Ann is to achieve a proper coloring of the whole graph G, while Ben tries to block the same. The smallest number of colors required for Ann to win the game on a graph G is called the indicated chromatic number of G and is denoted by $\chi_i(G)$. In this paper, we prove that $T\square C_n, T\square K_{n_1,n_2,\ldots,n_m}$ and $K_{n_1,n_2,\ldots,n_m}\square C_m$ are k-indicated colorable for all k greater than or equal to the indicated chromatic number of their corresponding Cartesian product, where T is any tree. Also we prove that $\chi_i(K_{k_1,k_2,\ldots,k_m}\square K_{l_1,l_2,\ldots,l_n}) = \chi(K_{k_1,k_2,\ldots,k_m}\square K_{l_1,l_2,\ldots,l_n})$. Finally we have given non-trivial examples of graphs G and H for which $\chi_i(G\square H) > \chi(G\square H)$.

Keywords: Game chromatic number · Indicated chromatic number · Cartesian product

1 Introduction

All graphs considered in this paper are simple, finite and undirected. A game coloring of a graph is a coloring in which two players Ann and Ben are jointly coloring the graph G by using a fixed set of colors C. The motive of Ann is to get a proper coloring of the whole graph, where as Ben is trying to prevent the realization of this project. The minimum number of colors required for Ann to win the game on a graph G irrespective of Ben's strategy is called the game chromatic number of the graph G and it is denoted by $\chi_g(G)$. The idea of indicated coloring was introduced by A. Grzesik in [3] as a slight variant of the game coloring in the following way: in each round Ann is only picking a vertex while Ben is choosing a color for this vertex. The aim of Ann as in indicated coloring is to achieve a proper coloring of the whole graph G, while Ben tries to "block" some vertex. A *block* vertex means an uncolored vertex which has all colors from C on its neighbors. The smallest number of colors required for Ann to win the game on a graph G is called the indicated chromatic number of G and is denoted by $\chi_i(G)$. Clearly from the definition we see that $\omega(G) \leq \chi(G) \leq \chi_i(G) \leq \Delta(G) + 1$. If Ann has a winning strategy using k colors for a graph G then we say that G is k-indicated colorable. Let $st_k(G)$ denote a winning strategy of Ann while using k colors. The coloring number of a graph G, denoted by col(G) is defined by $\text{col}(G) = 1 + \max_{H\subseteq G} \delta(H)$. By Szekeres-Wilf inequality [6], $\chi(G) \leq \text{col}(G)$.

Zhu in [9] has asked the following question for game coloring. Whether increasing the number of colors will favor Ann? That is, if Ann has a winning strategy using

S. Arumugam et al. (Eds.): ICTCSDM 2016, LNCS 10398, pp. 63–68, 2017.
DOI: 10.1007/978-3-319-64419-6_8

k colors, will Ann have a winning strategy using $k+1$ colors? The same question was asked by Grzesik for indicated coloring. Also he showed by an example that the increase in number of colors does make life simple for Ann rather it makes it much harder. There has been already some partial answers to this question. For instance, Pandiya Raj et al. [2,5] showed that chordal graphs, cographs, complement of bipartite graphs, $\{P_5, K_3\}$-free graphs, $\{P_5, \text{paw}\}$-free graphs, $\{P_5, C_5, K_4-e\}$-free graphs and $\{P_5, K_4-e\}$-free graphs having induced C_5 are k-indicated colorable for all $k \geq \chi(G)$. In addition Lason in [4] has obtained the indicated chromatic number of matroids. In this paper, we obtain $T\square C_n$, $T\square K_{n_1,n_2,\ldots,n_m}$ and $K_{n_1,n_2,\ldots,n_m}\square C_m$ are k-indicated colorable for all k greater than or equal to the indicated chromatic number of their corresponding Cartesian product, where T is any tree. In addition, we have prove that $\chi_i(K_{k_1,k_2,\ldots,k_m}\square K_{l_1,l_2,\ldots,l_n}) = \chi(K_{k_1,k_2,\ldots,k_m}\square K_{l_1,l_2,\ldots,l_n})$. Finally we have given non-trivial examples of graphs G and H for which $\chi_i(G\square H) > \chi(G\square H)$.

Notations and terminologies not mentioned here are as in [8].

2 Indicated Coloring on Cartesian Product of Graphs

The Cartesian product of two graphs G and H, denoted by $G\square H$, is a graph whose vertex set $V(G) \times V(H) = \{(x, y) : x \in V(G) \text{ and } y \in V(H)\}$ and two vertices (x_1, y_1) and (x_2, y_2) of $G\square H$ are adjacent if and only if either $x_1 = x_2$ and $y_1y_2 \in E(H)$, or $y_1 = y_2$ and $x_1x_2 \in E(G)$. Vizing [7] proved that $\chi(G\square H) = \max\{\chi(G), \chi(H)\}$. Note that while considering the cartesian product $G\square H$, for each $v \in V(G)$, $\langle v \times V(H)\rangle$ (for $S \subseteq V(G)$, $\langle S\rangle$ denotes the induced subgraph of S in G) is a copy of H and for each $u \in V(H)$, $\langle V(G) \times u\rangle$ is a copy of G. Also if S is an independent set in G and T is an independent set in H, then $S\square T$ is an independent set in $G\square H$.

Our main focus in Sect. 2 is to see whether the following is true. If G is k-indicated colorable for all $k \geq \chi_i(G)$ and H is k-indicated colorable for all $k \geq \chi_i(H)$, will $G\square H$ be k-indicated colorable for all $k \geq \chi_i(G\square H)$? As a first step, we have considered a few families for which this works out. In fact this also gives some partial answer to the question raised by Grzesik in [3]. Let us recall a few results done in [3,5].

Theorem 1. [5] *Any graph G is k-indicated colorable for all* $k \geq col(G)$.

Theorem 2. [3] *Every bipartite graphs is k-indicated colorable for every* $k \geq 2$.

An immediate consequence of Theorem 2 is the following.

Corollary 1. *Let G and H be two non-trivial graphs. Then G and H are bipartite if and only if* $G\square H$ *is k-indicated colorable for all* $k \geq 2$.

Proof. We know that $G\square H$ is bipartite if and only if G and H are bipartite. Suppose G and H are bipartite, by using Theorem 2, $G\square H$ is k-indicated colorable for all $k \geq 2$. Suppose $G\square H$ is k-indicated colorable for all $k \geq 2$, then $2 \leq \chi(G\square H)) \leq \chi_i(G\square H) = 2$. Hence $G\square H$ is bipartite.

By using Corollary 1, we see that if both m and n are even, then $C_m \square C_n$ is k-indicated colorable for all $k \geq 2$. While considering the case when either m or n (or both) is odd, the $\text{col}(C_m \square C_n) = 5$. Thus for showing that $C_m \square C_n$ is k-indicated colorable for all $k \geq 3$, it is enough to prove that $C_m \square C_n$ is 3 and 4-indicated colorable. This still remains an open problem.

Let us next recall a strategy used in [1].

Definition 1. *While coloring a graph G by using k colors, let $\mathcal{N}_{un}(v)$ denote the number of uncolored neighbors of v in G and $C(v)$ denote the number of available colors for v in G. A vertex v is said to be of type1 if $C(v) > \mathcal{N}_{un}(v)$ and of type2 if $C(v) = \mathcal{N}_{un}(v)$.*

Lemma 1. *Let Ann and Ben plays an indicated coloring game on graph G with $k \geq \chi(G)$ colors. In certain stage, if all the uncolored vertices in G can be partitioned into disjoint paths such that one end of each path is of type1 and all the other vertices are of type2, then Ann has a winning strategy.*

Proof. Let the color set be $\{1, 2, \ldots, k \geq \chi(G)\}$. By our assumption, let $P_1, P_2, \ldots, P_l$ be a partition of the uncolored vertices with the property that one end of each P_i, $1 \leq i \leq l$, is of type1 and all the other vertices in P_i are of type2. Let $P_1 = v_{11}, v_{12}, \ldots, v_{1j}$ for some $j \geq 1$ and let v_{1j} be of type1 and v_{1i}, $1 \leq i \leq j-1$ be of type2. Clearly there is always an available color for v_{1j}. Now let Ann present the vertices of P_1 in the order $v_{11}, v_{12}, \ldots, v_{1j}$ (same order of the path P_1). Since $C(v_{1i}) = \mathcal{N}_{un}(v_{1i})$, for every i, $1 \leq i \leq j-1$ and one of the neighbor of v_{1i}, namely $v_{1(i+1)}$ is presented after v_{1i}. Thus Ben has an available color for each v_{1i}, $1 \leq i \leq j-1$. Since all the uncolored vertices where only of type1 or type2, Ben cannot create a block vertex in any of the paths. Thus a similarly technique can be applied by Ann for all the other paths to yield an indicated coloring using k colors.

Theorem 3. *Let T be any tree. Then*

(i) $T \square C_m$ is k-indicated colorable for all $k \geq \chi_i(T \square C_m) = \chi(T \square C_m)$
(ii) $T \square K_{n_1,n_2,\ldots,n_m}$ is k-indicated colorable for all $k \geq \chi_i(T \square K_{n_1,n_2,\ldots,n_m}) = m$.

Proof. Let v_0 be a center of T. Let V_i be the set of all vertices of T which are at a distance i from v_0, $1 \leq i \leq r$ where r is the radius of the tree. Let us label the vertices of T as $v_0, v_1, v_2, \ldots, v_{n-1}$ such that the vertices of V_i are to the left of the vertices of V_j for every i, j such that $1 \leq i < j \leq r$. Let $v_{ij} = (v_i, u_j)$ be the vertex of $T \square G$ where $v_i \in T$, $u_j \in G$, $0 \leq i \leq n-1$ and $0 \leq j \leq |V(G)| - 1$. Let $c(v)$ denote the color given by Ben to the vertex v and if v is uncolored then assume that $c(v) = \emptyset$. In $T \square G$, let $H_0, H_1, \ldots, H_{n-1}$ be the copies of G corresponding to the vertices $v_0, v_1, \ldots, v_{n-1}$ of T respectively. If v_i and v_j are non-adjacent vertices in T then $\langle V(H_i), V(H_j) \rangle = \emptyset$ in $T \square G$. If v_i and v_j are adjacent vertices in T then $\langle V(H_i), V(H_j) \rangle = \{v_{il}v_{jl} : 0 \leq l \leq |V(G)| - 1\}$ in $T \square G$.

(i) Let us consider the graph $G = C_m$. Suppose m is even, $T \square C_m$ is bipartite and by using Theorem 2, $T \square C_m$ is k-indicated colorable for all $k \geq 2 = \chi_i(T \square C_m)$. Now let us consider m to be odd. It is easy to observe that $\chi(T \square C_m) = 3$ and $\text{col}(T \square C_m) = 4$. Hence by using Theorem 1, it is enough to show that $T \square C_m$ is 3-indicated colorable. Let the color set be $\{1, 2, 3\}$. Let Ann present the vertices of H_0 in any order. Since the col(H_0) $= 3$, Ben always has an available color for each vertex of H_0. Irrespective of Ben's strategy, there exist a vertex v_{0j} of H_0 having two different colors in its neighbor, namely $v_{0(j-1)}$ and $v_{0(j+1)}$ where $0 \leq j \leq m-1$ and j is taken mod n. Now consider the subgraph $G_1 = \langle \{v_0 \cup V_1\} \rangle \Box C_m$ in $T \Box C_m$. Let H_i be the C_m copy of the vertex $v_i \in V_1$, $1 \leq i \leq |V_1|$. Since $\langle V(H_0), V(H_i) \rangle = \{v_{0j}v_{ij} : 0 \leq j \leq m-1\}$ for all $1 \leq i \leq |V_1|$ and there are 3 colors, the uncolored vertices of G_1 are the vertices of H_i which are the vertices of type2. Now Ann will present the vertex v_{ij} of H_i for all $1 \leq i \leq |V_1|$. Suppose Ben color v_{ij} with the color of $v_{0(j-1)}$ then the vertex $v_{i(j-1)}$ is a vertex of type1, otherwise $v_{i(j+1)}$ is a vertex of type1 where $1 \leq i \leq |V_1|$, $0 \leq j \leq m-1$ and j is taken mod n. By using Lemma 1, Ann have an winning strategy on G_1.

Let us consider the subgraph $G_i = \langle \{V_{i-1} \cup V_i\} \rangle \Box C_m$ where $2 \leq i \leq r$. Similarly Ann follow the same procedure to presents the uncolored vertices of G_i, and thus Ann has a winning strategy for G_i, $2 \leq i \leq r$. This yields a winning strategy for Ann on the graph $T \Box C_m$ with 3 colors.

(ii) It is easy to observe that $\chi(T \Box K_{n_1,n_2,\ldots,n_m}) = m$ and $\text{col}(T \Box K_{n_1,n_2,\ldots,n_m}) = m + 1$. Hence by using Theorem 1, it is enough to show that $T \Box K_{n_1,n_2,\ldots,n_m}$ is m-indicated colorable. Let the color set be $\{1, 2, \ldots, m\}$. Let U_i, $1 \leq i \leq m$ be the m-partites of the graph $K_{n_1,n_2,\ldots,n_m}$. Let us consider the subgraph $G_0 = \langle \{u_1, u_2, \ldots, u_m\} \rangle$ in $K_{n_1,n_2,\ldots,n_m}$, where $u_i \in U_i$, $1 \leq i \leq m$. Clearly $G_0 \cong K_m$ and $\omega(K_{n_1,n_2,\ldots,n_m}) = m$. Now consider the graph $T \Box G_0$. Let $J_0, J_1, \ldots, J_{n-1}$ be the copies of G_0 corresponding to the vertices $v_0, v_1, \ldots, v_{n-1}$ of T respectively.

Ann starts presenting the vertices of J_0 in any order. Since col(J_0) $= m$, Ben have an available color for each vertex of J_0. Now consider the subgraph $G_1 = \langle \{v_0 \cup V_1\} \rangle \Box G_0$ in $T \Box G_0$. Since $\langle V(J_0), V(J_i) \rangle = \{v_{0j}v_{ij} : 1 \leq j \leq m\}$ for all $1 \leq i \leq |V_1|$ and there are m colors, the uncolored vertices of G_1 are the vertices of J_i which are the vertices of type2. Now Ann will present the vertex v_{i1} of J_i for all $1 \leq i \leq |V_1|$. Ben should color v_{i1} with one of the color from $\{1, 2, \ldots, m\} \backslash \{c(v_{01})\}$ and let it be c_i, $1 \leq i \leq |V_1|$. For each J_i, $1 \leq i \leq |V_1|$ there is a vertex $v_{ii'}$ which is adjacent to the color c_i of J_0 where $1 \leq i' \leq m$ and thus the vertex $v_{ii'}$ is a vertex of type1. By using Lemma 1, Ann has a winning strategy on G_1.

Let us next consider the subgraph $G_i = \langle \{V_{i-1} \cup V_i\} \rangle \Box G_0$ where $2 \leq i \leq r$ in $T \Box G_0$. Let Ann follow a similar procedure as done in G_1, for presenting the uncolored vertices of G_i. This will give Ann a winning strategy for G_i, $2 \leq i \leq r$, and hence a winning strategy for $T \Box G_0$ with m colors. Let it be $st_m(T \Box G_0)$. Now consider the graph $T \Box K_{n_1,n_2,\ldots,n_m}$. The strategy $st_m(T \Box G_0)$ makes Ben to color exactly one vertex of U_i, $1 \leq i \leq m$ in each of the copies of $K_{n_1,n_2,\ldots,n_m}$ corresponding to the vertices of T. Hence for the remaining vertices in each of the U_i in the copies of $K_{n_1,n_2,\ldots,n_m}$ corresponding to the vertices of T, Ben will be forced to give the color given to vertex in U_i that is already colored. Thus Ann can presents the remaining vertices in any order and this will yield an m-indicated coloring for $T \Box K_{n_1,n_2,\ldots,n_m}$.

An immediate consequence of Theorem 3 is the following.

Corollary 2. *For all $m \geq 2$, the graph $T \Box K_m$ is k-indicated colorable for all $k \geq m$.*

In a similar fashion but with a little more involved arguments, we have showed Theorems 4 and 5.

Theorem 4. *For all $m \geq 3$ and $n \geq 3$, the graph $K_{n_1,n_2,\ldots,n_m} \square C_n$ is k-indicated colorable for all $k \geq m$.*

Theorem 5. *For all $m \geq 2$ and $n \geq 2$, $\chi_i(K_{k_1,k_2,\ldots,k_m} \square K_{l_1,l_2,\ldots,l_n}) = \chi(K_{k_1,k_2,\ldots,k_m} \square K_{l_1,l_2,\ldots,l_n})$.*

An immediate consequence of Theorems 4 and 5 is the following.

Corollary 3. *For all $m \geq 3$ and $n \geq 3$, the graph $K_m \square C_n$ is k-indicated colorable for all $k \geq m$ and $\chi_i(K_m \square K_n) = \chi(K_m \square K_n)$.*

By the definition of indicated coloring, $\chi_i(G) \geq \chi(G)$ and thus $\chi_i(G \square H) \geq \chi(G \square H)$. The families of graphs considered for our discussion till now are examples of graphs for which $\chi_i(G \square H) = \chi(G \square H)$. But we do have examples of non-trivial graphs G and H for which $\chi_i(G \square H) > \chi(G \square H)$. This is done in Proposition 1 and Theorem 6.

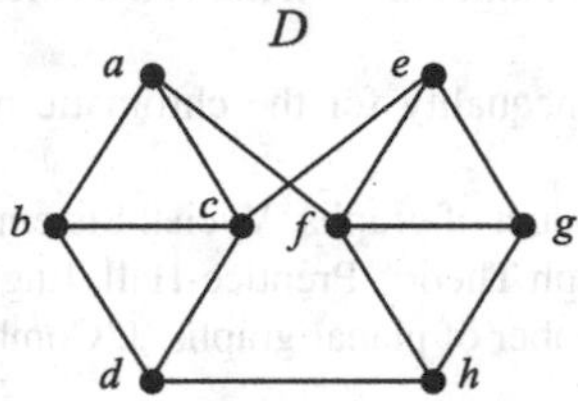

Fig. 1. Graph with $\chi_i(D) = \chi(D) + 1 = 4$.

Proposition 1. *Let D be the graph given in Fig. 1. Then $\chi_i(D \square K_2) > \chi(D \square K_2)$.*

Proof. Let us consider the graph D given in Fig. 1. Clearly D is a uniquely colorable graph such that $\chi(D) = 3$ and $\chi_i(D) = 4$ (see, [3]). Let us consider the graph $D \square K_2$. Clearly $D \square K_2$ contains two copies of D. Let us denote these copies by D_1 and D_2. Let the vertices of D_1 be $a, b, \ldots, h$ as shown in Fig. 1, and its corresponding vertices of D_2 be denoted by $a', b', \ldots, h'$ respectively. By the definition of Cartesian product, $aa', bb', \ldots, hh' \in E(D \square K_2)$. It is clear that $\chi(D \square K_2) = 3$. Let the colors set be $\{1, 2, 3\}$. We have to show that there is no winning strategy for Ann using 3 colors. In any 3-coloring of $D \square K_2$ the vertices $\{a, d, g\}$, $\{b, e, h\}$ and $\{c, f\}$ should receive the same color c_1, c_2 and c_3 respectively and the vertices $\{a', d', g'\}$, $\{b', e', h'\}$ and $\{c', f'\}$ should receive the same color c_2, c_3 and c_1 respectively or c_3, c_1 and c_2 respectively such that $\{c_1, c_2, c_3\} = \{1, 2, 3\}$. Let Ann start by presenting the vertex a in D_1 and let the color given by Ben be 1. Let the following be the strategy followed by Ben.

(i) color the vertex d with 2 or 3, (or) color the vertex d' with 1.
(ii) color any one of the vertex of $\{f, g, e', h'\}$ with 2.

If Ben is able to accomplish one of the above, then clearly Ann does not have a winning strategy. In order to avoid (i), she has to present the vertices in the order b, c, d, d'. But even in this case Ann cannot prevent Ben from applying (ii). Thus Ben wins the game on $D \square K_2$ with 3 colors and hence $\chi_i(D \square K_2) > \chi(D \square K_2)$.

This idea for $D \square K_2$ can be generalised to $D \square T$ where T is any tree.

Theorem 6. *Let D be the graph given in Fig. 1 and T be any tree. Then* $\chi_i(D\square T) > \chi(G\square T)$.

Acknowledgement. For the first author, this research was supported by the Council of Scientific and Industrial Research, Government of India, File no: 09/559(0096)/2012-EMR-I.

References

1. Francis, P., Francis Raj, S.: Indicated Coloring of Graphs (Preprint)
2. Francis Raj, S., Pandiya Raj, R., Patil, H.P.: On indicated chromatic number of graphs. Graphs Combin. **33**, 203–219 (2017)
3. Grzesik, A.: Indicated coloring of graphs. Discrete Math. **312**, 3467–3472 (2012)
4. Lason, M.: Indicated coloring of matroids. Discrete Appl. Math. **179**, 241–243 (2014)
5. Pandiya Raj, R., Francis Raj, S., Patil, H.P.: On indicated coloring of graphs. Graphs Combin. **31**, 2357–2367 (2015)
6. Szekeres, G., Wilf, H.S.: An inequality for the chromatic number of a graph. J. Combin. Theory **4**, 1–3 (1968)
7. Vizing, V.G.: The Cartesian product of graphs. Vycisl. Sistemy. **9**, 30–43 (1963)
8. West, D.B.: Introduction to Graph Theory. Prentice-Hall, Englewood Cliffs (2000)
9. Zhu, X.: The game coloring number of planar graphs. J. Combin. Theory Ser. B. **75**, 245–258 (1999)

A Bi-level Security Mechanism for Efficient Protection on Graphs in Online Social Network

D. Franklin Vinod(✉) and V. Vasudevan

Department of Information Technology, Kalasalingam University, Krishnankoil 626126, Tamil Nadu, India
datafranklin@gmail.com, vasudevan_klu@yahoo.co.in

Abstract. The Big Data plays a valuable role in large scale information management that overshoots the potential of traditional data processing technologies. The importance of volume, velocity, variety, veracity and value of big data made researchers to put efforts to handle them efficiently. On considering the 5V's of big data, if the veracity characteristic is not well focused, then the idea of big data will not be widely recognized. Advances in technology allow users to extract and utilize the big data which make data privacy violations in maximum cases. Also the data used for big data analytics may include restricted information. So it is necessary to protect and notice whether this kind of data is used with certain principles. In this paper we formalize a Bi-level security mechanism called Cosine Similarity with P-Stability for data privacy and graph protection in one of the big data environment called Online Social Network.

1 Introduction

The data is flooded unconditionally due to the development of multichannel business environment which led to Big Data. The big data is in account because of 5V's named Volume, Velocity, Variety, Veracity and Value [1]. The Volume parameter of big data indicates the massive quantity of data originating every second. It is not in terms of terabytes but in terms of zettabytes and Yottabytes. The Velocity parameter of big data indicates the speed at which data originated, handled and distributed. This data will move viral in seconds but today's technology allows us to analyze the data while it's being originated before putting it into the database. The Variety parameter of big data indicates the dissimilar types of data that all can now use. In the earlier days all concentrated only on the structured data which will be attached into tables or relational databases but in today's world eighty percentage of data available are unstructured. The big data technology allows us to analyze all dissimilar data together. The Veracity parameter of big data indicates the disorder or messiness or trustworthiness of data. Another major issue is reliability and accuracy of data. The Value parameter of big data indicates that there should be a business value from the data extracted.

From the above mentioned parameters of big data, our work is to concentrate on veracity parameter to identify privacy issues when data is used for analysis

S. Arumugam et al. (Eds.): ICTCSDM 2016, LNCS 10398, pp. 69–75, 2017.
DOI: 10.1007/978-3-319-64419-6_9

and when data network is in public. If the privacy of data is not well focused, then the idea of big data will not be widely recognized [2]. For our work online social network is taken as a model and the privacy issues and protection methods are analyzed. Also since an online social network is expressed by means of graphs, it is necessary for graph protection to maintain data privacy. The social networking sites will collect and store all the information about our personal life and our relationship in society. Then they recycle our personal information for business credits [3]. Advances in technology allow users to extract and utilize the big data which may cause data privacy violations. So it is necessary to protect and notice whether private data are used with certain principles. Already there are many privacy and security mechanisms for traditional data but they are insufficient for big data era. Therefore different efforts are needed to identify proper mechanism for securing big data.

In this paper, we map out the basic privacy requirements of big data analytics in Sect. 2. The existing mechanism for data privacy and graph protection is discussed in Sect. 3. In Sect. 4, we propose a bi-level security mechanism called Cosine Similarity with P-Stability for data privacy and graph protection. The Performance Evaluation of our proposed mechanism is shown in Sect. 5.

2 Privacy Requirements of Big Data

The research on privacy in big data is in its early stage. In general, if data privacy is not applicable, then the big data is not authentic. While big data generates extensive values for profitable extension and technical revolution, we are already conscious that the severe flood of data also leads to new privacy concerns. In this paper we concentrate on big data privacy and locate the privacy requirements of big data analytics.

Big Data Collection privacy requirements: The collection of big data takes place universally, so there is possibility of intrusion and the data may be dripped unexpectedly. Since the data collected is personal and sensitive, we should head for physical protection mechanisms and information security mechanisms for data privacy before it is securely stored.

Big Data Storage privacy requirements: In comparison to Intrusion on individual's data during the big data collection, compromising of big data storage system is more injurious [4]. Hence it is necessary to make sure the confidentiality of data stored in storage system or data center.

Big Data Processing privacy requirements: The big data processing is one of the significant parts of big data analytics, as it extracts valuable knowledge for profitable growth and technical revolution. Since the efficiency of big data processing is an essential measure for the success of big data, the big data processing privacy requirements will be very demanding. The big data processing efficiency can't be compromised for big data privacy and should ensure efficiency while protecting individual privacy. Maintaining big data privacy is the most demanding issue in big data processing.

In the current era, numbers of data privacy mechanisms are proposed already [5]. But they are used to handle only traditional analytics data privacy. The traditional data privacy mechanisms are not adequate to overcome the privacy issues related to the big data analytics. Traditional mechanisms can provide only single level of data privacy. In the next section we will discuss about existing mechanisms for data privacy and graph protection.

3 Existing Mechanisms for Data Privacy and Graph Protection

The existing mechanism for data privacy and graph protection are huge in number. Some of them are discussed in this section.

Privacy-preserving aggregation: This mechanism is favored for data collection which is designed on some homomorphic encryption [6]. Here distinct sources use the same public key for encrypting their individual data when they want to share group of individual data together. Then the ciphertext from every individual is aggregated and shared. The authorized user can decrypt the data using respective private key. This privacy preserving aggregation mechanism can give security for individual data privacy during big data collection and storing. But the issue here is the mechanism is purpose specific. The ciphertext aggregated for one purpose can't be used for other purpose. Hence the Privacy preserving aggregation mechanism is inadequate for big data analytics.

Operations over encrypted data: For securing individual sensitive data [7], the data and their related keywords are encrypted and maintained in the storage. When the user wants to view the data, query can be executed and data can be retrieved from the storage. Now the data privacy is ensured. While analyzing operation over encrypted data, there is possibility of using this individual privacy in big data analytics but the problem here is it takes long computing duration and is complex [13]. Since big data analytics deals with huge volume and need to process the analysis in a timely manner, this mechanism is inadequate for big data analytics.

De-identification: This mechanism is also traditional one used for protecting privacy. Here to provide individual privacy, first data should be disinfected and second some values should not be delivered to all. While comparing with privacy-preserving aggregation and operation over encrypted data, de-identification will be more successful and flexible for data analytics [8]. But the hacker will be able to get exterior information support for de-identification in the big data world. Thus this mechanism is also inadequate for big data privacy protection.

The discussion above indicates that the data privacy protection will be still demanding one because of the risks involved. But in contrast with privacy-preserving aggregation and operation over encrypted data, the de-identification mechanism will be more suitable for providing data privacy in big data analytics. This is possible if we eliminate risks on de-identification.

Protection of Graphs: There are various mechanisms for the protection of graphs. Such as modifying the edges of the original graph, perturbation of the original graph, and creating supernodes which crumple set of vertices in the original graph. There are some other mechanisms that focus on the content attached to the vertices of the graph and the aim is to protect identity disclosure and attribute disclosure [9].

In the next section, we propose bi-level security mechanism for protecting data privacy for both individuals and for graphs using Cosine Similarity with P-Stability.

4 Cosine Similarity with P-Stability

The online social network is represented as a graph where individuals are the vertices and relationships between individuals are edges. From every user the data are collected by data analyst for big data analysis which may lead to data privacy issues. Thus security on data privacy is required. Also once every individual makes a relationship with 'n' different individuals, and based on the relationship maintained there will be several sub-graphs in a single main graph called online social network. So there is a possibility of hacking data by individuals within one subgraph or between different subgraphs. Hence to overcome both individuals and graph data privacy issues we propose bi-level security mechanism called Cosine Similarity with P-Stability.

The first level of security is provided by Cosine Similarity computing algorithm [10]. The Cosine Similarity algorithm has two types. One type reveals information between each other and the other type won't reveal information between each other. Our algorithm will identify the similarity between two individuals without revealing information between them. For every pair of individuals the Cosine Similarity (1) is measured and finally all values are aggregated.

The steps of Cosine Similarity Computing algorithm are as follows

$$
\begin{aligned}
I_A &= Individual\ A \\
\overrightarrow{a} &= (a_1, a_2, \ldots, a_n) \\
I_B &= Individual\ B \\
\overrightarrow{b} &= (b_1, b_2, \ldots, b_n)
\end{aligned}
$$

Step 1: I_A Calculation:
Consider the security parameters k_1, k_2, k_3, k_4
Select two large prime numbers α, p
Assign $| p |= k_1, | \alpha |= k_2$
Set $a_{n+1} = a_{n+2} = 0$
Identify a large random number $s \in Z_p$
and $n + 2$ random numbers $c_x, x = 1, 2, \ldots n + 2$, with $| c_x |= k_3$.
For each $a_x, x = 1, 2, \ldots n + 2$.

$C_x = \begin{cases} s(a_x.a + c_x) \; mod \; p, a_x \neq 0 \\ s.c_x \; mod \; p, a_x = 0 \end{cases}$
End for
Calculate $A = \sum_{x=1}^{n} a_x^2$
Keep s^{-1} mod p as secret
Send $(\alpha, p, C_1, \ldots C_{n+2})$ to I_B

Step 2: I_B Calculation:
Set $b_{n+1} = b_{n+2} = 0$
For each $b_x, x = 1, 2, \ldots n + 2$
$D_x = \begin{cases} b_x.\alpha.C_x \; mod \; p, b_x \neq 0 \\ r_x.C_x \; mod \; p, b_x = 0 \end{cases}$ where r_x is a random number with $| r_x |= k_4$
End for
$B = \sum_{x=1}^{n} b_x^2$ and
$D = \sum_{x=1}^{n+2} D_x \; mod \; p$
Send (B, D) to I_A

Step 3: I_A Calculation:
Calculate $E = s^{-1}.D \; mod \; p$
Calculate $\overrightarrow{a}.\overrightarrow{b} = \sum_{x=1}^{n} a_x.b_x = \frac{E-(E \; mod \; \alpha^2)}{\alpha^2}$

$$Cos(\overrightarrow{a}, \overrightarrow{b}) = \frac{\overrightarrow{a}.\overrightarrow{b}}{\sqrt{A}\sqrt{B}} \tag{1}$$

The second level of security is provided by P-Stability for graphs [9]. When we publish original graph (OSN) in the public, there will be a problem in data protection. So the original graph should be replaced with anonymous graph. For this, first we have to identify the degree sequence of original graph. Then with this identified degree sequence multiple possible graphs are generated. Now the original graph is replaced with generated anonymous graph and the modified graph will look like original to the public. This transformation will confuse the hackers or analyst since the original graph is replaced by another. A graph with same degree sequence as original graph is referred as P-Stable graph and if a graph is P-Stable then it is protected. Also the graphs can be P-Stable if it satisfies the Theorems 1, 2 and 3.

For a graph with n vertices, m edges and degree sequence d, we define m, Δ and δ as follows:

$$m = \text{number of edges in } G$$
$$\Delta = max\{d_i : 1 \leq i \leq n\}$$
$$\delta = min\{d_i : 1 \leq i \leq n\}$$

Theorem 1 [11]. *The Class of all graphs with degree sequences d which satisfy* $m > \Delta(\Delta - 1)$ *is P-stable.*

Theorem 2 [12]. *The class of all graphs with degree sequences* $d = (d_1, \ldots d_n)$ *which satisfy* $(\Delta - \delta + 1)^2 \leq 4\delta(n - \Delta - 1)$ *where* δ *and* Δ *are defined as above, is P-stable.*

Theorem 3 [12]. *The class of all graphs with degree sequences* $d = (d_1, \ldots d_n)$ *which satisfies* $\{(2m-n\delta)-(\Delta-\delta)\delta\}\{(n\Delta-2m)-(\Delta-\delta)(n-\Delta-1)\} \leq (\Delta-\delta)^2\delta(n-\Delta-1)$ *where* δ, Δ *and* m *are defined as above, is P-stable.*

5 Performance Evaluation

To measure the security level of our proposed Cosine Similarity with P-Stability mechanism, we collected hundred individuals information. With this information the similarity between individuals are identified using Cosine Similarity mechanism. With this a graph representing Online Social Network model is framed. The graph protection is implemented using P-Stability for graphs. The experimental result proves that, of the number of existing mechanisms available today some only are applicable for big data. In these the applicable existing traditional mechanisms also provide only individuals data privacy and it fails to provide graph protection in online social network. But our proposed mechanism Cosine Similarity with P-Stability will provide both individuals and graph data privacy for online social network. Figure 1 shows the comparison of existing and proposed mechanisms.

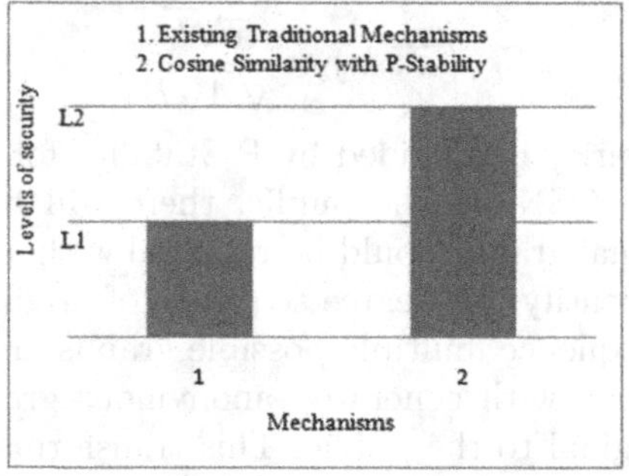

Fig. 1. Comparison of existing traditional mechanisms with our proposed Cosine Similarity with P-Stability mechanism. In this figure L1 indicates Level 1 and the L2 indicates Level 2.

6 Conclusion

The volume of data is growing daily and it is difficult to visualize the next age applications without providing data privacy algorithms. In this paper, we explore the privacy demands in the big data analytics environment. We explored privacy demands in every phase of big data life cycle and examined some advantages and disadvantages of existing data privacy mechanisms in the context of big data. We proposed a bi-level security mechanism called Cosine similarity with P-Stability for data privacy and graph protection in one of the big data environment Online

Social Network. The experimental result proves that our proposed mechanism provide dual security in data privacy for both individuals and graphs which the existing traditional mechanisms couldn't do.

Acknowledgments. The first author would like to thank the management of Kalasalingam University for providing financial assistance under the University Research Fellowship.

References

1. Xue-Wen, C., Xiaotong, L.: Big data deep learning: challenges and perspectives. IEEE Access **2**, 514–525 (2014)
2. Abid, M., Iynkaran, N., Yong, X., Guang, H., Song, G.: Protection of big data privacy. IEEE Access **4**, 1821–1834 (2016)
3. Avita, K., Mohammad, W., Goudar, R.H.: Big data: issues, challenges, tools and good practices. In: The Proceedings of the Sixth International Conference on Contemporary Computing (IC3), Noida, pp. 404–409 (2013)
4. Han, H., Yonggang, W., Tat Seng, C., Xuelong, L.: Toward scalable systems for big data analytics: a technology tutorial. IEEE Access **2**, 652–687 (2014)
5. Lei, X., Chunxiao, J., Jian, W., Jian, Y., Yong, R.: Information security in big data: privacy and data mining. IEEE Access **2**, 1149–1176 (2014)
6. Paillier, P.: Public-key cryptosystems based on composite degree residuosity classes. In: Stern, J. (ed.) EUROCRYPT 1999. LNCS, vol. 1592, pp. 223–238. Springer, Heidelberg (1999). doi:10.1007/3-540-48910-X_16
7. Ming, L., Shucheng, Y., Kui, R., Wenjing, L., Thomas, H.Y.: Toward privacy-assured and searchable cloud data storage services. IEEE Netw. **27**, 56–62 (2013)
8. Ann, C., Jeff, J.: Privacy by Design in the Age of Big Data. Information and Privacy Commissioner, Canada (2012)
9. Vicenc, T., Termeh, S., Julian, S.: Data protection for online social networks and P-Stability for graphs. IEEE Trans. Emerg. Top. Comput. **4**, 374–381 (2016)
10. Rongxing, L., Hui, Z., Ximeng, L., Joseph, K.L., Jun, S.: Toward efficient and privacy-preserving computing in big data era. IEEE Netw. **28**, 46–50 (2014)
11. Mark, J., Alistair, S.: Fast uniform generation of regular graphs. Theoret. Comput. Sci. **73**, 91–100 (1990)
12. Jerrum, M., Sinclair, A., McKay, B.: When is a graphical sequence stable? In: Frieze, A., Luczak, T. (eds.) Random Graphs, vol. 2, pp. 101–115. Wiley (1992)
13. Prabu Kanna, G., Vasudevan, V.: Enhancing the security of user data using the keyword encryption and hybrid cryptographic algorithm in cloud. In: Proceedings of the International Conference on Electrical, Electronics and Optimization Techniques (ICEEOT), pp. 3688–3693 (2016)

On Nearly Distance Magic Graphs

Aloysius Godinho[1(✉)], T. Singh[1], and S. Arumugam[2]

[1] Department of Mathematics, Birla Institute of Technology and Science Pilani, K K Birla Goa Campus, NH-17B, Zuarinagar, Sancoale 403726, Goa, India
{p2014001,tksingh}@goa.bits-pilani.ac.in

[2] National Centre for Advanced Research in Discrete Mathematics, Kalasalingam University, Anandnagar, Krishnankoil 626126, Tamil Nadu, India
s.arumugam.klu@gmail.com

Abstract. Let $G = (V, E)$ be a graph on n vertices. A bijection $f : V \to \{1, 2, \ldots, n\}$ is called a *nearly distance magic labeling* of G if there exists a positive integer k such that $\sum_{x \in N(v)} f(x) = k$ *or* $k+1$ for every $v \in V$. The constant k is called a magic constant of the graph and any graph which admits such a labeling is called a *nearly distance magic graph*. In this paper we present several basic results on nearly distance magic graphs and compute the magic constant k in terms of the fractional total domination number of the graph.

Keywords: Distance magic graphs · Nearly distance magic graphs

1 Introduction

All graphs in this paper are simple graphs without isolated vertices. For graph theoretic terminology and notation we refer to Chartrand and Lesniak [3].

By a graph labeling we mean an assignment of numbers to graph elements such as vertices or edges or both. Different types of labelings have been defined by various researchers by imposing different conditions on such an assignment. For an overview on graph labelings and its recent developments we refer to the dynamic survey by Gallian [4].

Let $G = (V, E)$ be a graph of order n. Let $f : V \to \{1, 2, \ldots, n\}$ be a bijection. The weight $w_f(v)$ of a vertex v is defined by $w_f(v) = \sum_{x \in N(v)} f(x)$, where $N(v)$ is the open neighbourhood of v. If $w_f(v) = k$ (a constant) for every $v \in V$, then f is said to be a *distance magic labeling* of the graph G. A graph which admits a distance magic labeling is called a *distance magic graph*. The constant k is called the *distance magic constant*. The concept of distance magic labeling was originally introduced by Vilfred [9]. For an overview of known results on distance magic labeling we refer to Arumugam *et al.* [1] and Rupnow [8]. Two distance magic graphs of order 7 with magic constants 21 and 7 respectively are shown in Fig. 1.

S. Arumugam et al. (Eds.): ICTCSDM 2016, LNCS 10398, pp. 76–82, 2017.
DOI: 10.1007/978-3-319-64419-6_10

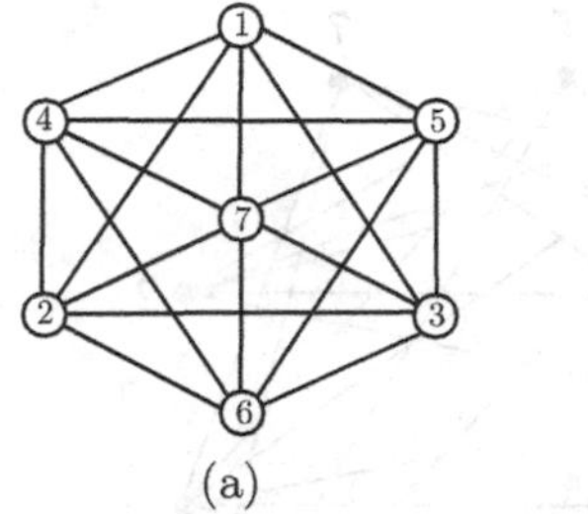

(a)

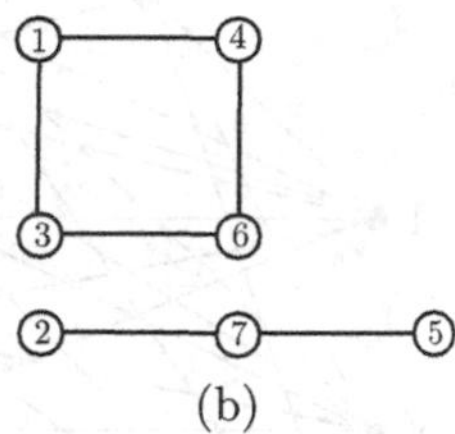

(b)

Fig. 1. Distance magic graphs

Definition 1. *Let $G = (V, E)$ be a graph without isolated vertices. A function $f : V \to [0, 1]$ is said to be a fractional total dominating function of G if for every vertex $v \in V$, $\sum_{x \in N(v)} f(x) \geq 1$. The fractional total domination number $\gamma_{ft}(G)$ is defined as $\gamma_{ft}(G) = \min\{|f| : f$ is a fractional total dominating function of $G\}$, where $|f| = \sum_{v \in V} f(v)$.*

At the International workshop on graph labeling (IWOGL-2010) Arumugam posed the following problem: *For a distance magic graph G of order n, is it possible to obtain two distance magic labelings f_1, f_2 with distinct magic constants k_1, k_2?* Arumugam *et al.* [2] later solved this problem by obtaining a formula for the magic constant k in terms of the fractional total domination number of the graph, thereby showing that the magic constant is independent of the labeling f. This result was also independently proved by Slater *et al.* [7].

Theorem 1. *If G is a distance magic graph of order n, then the distance magic constant k of G is given by*

$$k = \frac{n(n+1)}{2\gamma_{ft}(G)} \tag{1}$$

Kamatchi [5] showed that the integers 4, 6, 8 and 12 do not appear as magic constants for any distance magic graph. He posed the following problem: *Determine the set $\mathcal{S}$ of positive integers which appear as magic constants of some distance magic graph.* Froncek *et al.* [6] proved that for every $t \geq 6$ there exists a 4−regular distance magic graph with magic constant 2^t.

2 Main Results

Consider the complete tripartite graph $G = K_{4,3,3}$. One can easily prove that $\gamma_{ft}(G) = \frac{3}{2}$. If f is a distance magic labeling of G, then by (1) the magic constant $k = \frac{110}{3} \notin \mathbb{N}$. Hence the graph G is not distance magic. However, we can find a bijection $f : V \to \{1, \ldots, 10\}$ for which the weights $w_f(v) = 36$ *or* 37 (see Fig. 2). It is interesting to note that the vertex weights satisfy the inequality $36 < \frac{110}{3} < 37$. This motivates the following concept of nearly distance magic labeling.

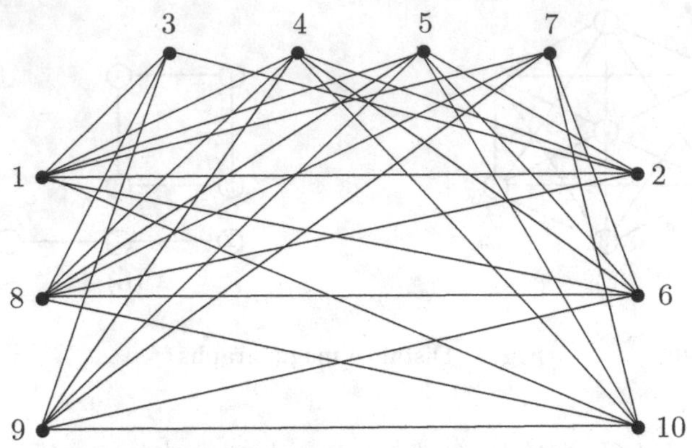

Fig. 2. Nearly distance magic labeling of $K_{4,3,3}$

Definition 2. *Let $G = (V, E)$ be a graph on n vertices. A bijection $f : V \to \{1, 2, \ldots, n\}$ is called a nearly distance magic labeling of G if there exists a positive integer k such that $\sum_{x \in N(v)} f(x) = k$ or $k+1$ for every $v \in V$. The constant k is called the magic constant of the graph and the graph which admits such a labeling is called a nearly distance magic graph.*

The following theorem is a generalisation of Theorem 1.

Theorem 2. *If G is a nearly distance magic graph with magic constant k, then*

$$k = \left\lfloor \frac{n(n+1)}{2\gamma_{ft}(G)} \right\rfloor. \tag{2}$$

Proof. Let f be a nearly distance magic labeling of G with magic constant k. Let $V = \{v_1, \ldots, v_r, \ldots v_n\}$ be the vertex set of G such that $w(v_i) = k$ for $1 \le i \le r$ and $w(v_i) = k+1$ for $r+1 \le i \le n$. Let $A = (a_{i,j})_{n \times n}$ be the adjacency matrix of G. Let $X = (f(v_1), f(v_2), \ldots, f(v_n))^T$. Then,

$$AX = (\underbrace{k, k, \ldots, k}_{\text{r terms}}, k+1, k+1, \ldots, k+1)^T$$

Let g be a fractional total dominating function of G such that $|g| = \sum_{i=1}^{n} g(v_i) = \gamma_{ft}(G)$.

Let $Y = (g(v_1), g(v_2), \ldots g(v_n))^T$ and $AY = (l_1, l_2, \ldots l_n)^T$ with each $l_i \ge 1$

Since $X^T AY$ is a 1×1 matrix, $X^T AY = (X^T AY)^T = Y^T AX$.

$$\begin{aligned}
\text{Now } Y^T AX &= Y^T (AX) \\
&= (g(v_1), g(v_2), \ldots g(v_n))(k, k, \ldots, k, k+1, k+1, \ldots, k+1)^T \\
&= k(g(v_1) + g(v_2) + \ldots + g(v_r)) + (k+1)(g(v_{r+1}) + g(v_{r+2}) + \ldots + g(v_n)) \\
&= k\gamma_{ft}(G) + \sum_{i=r+1}^{n} g(v_i).
\end{aligned}$$

Similarly,

$$\begin{aligned} X^T AY &= X^T(AY) \\ &= (f(v_1), f(v_2), \dots f(v_n))(l_1, l_2, \dots l_n)^T \\ &= \sum_{i=1}^{n} f(v_i)l_i \geq \sum_{i=1}^{n} f(v_i) = \frac{n(n+1)}{2}. \end{aligned}$$

Since $X^T AY = Y^T AX$ we get $k\gamma_{ft}(G) + \sum\limits_{i=r+1}^{n} g(v_i) \geq \frac{n(n+1)}{2}$. It follows that

$$k + \frac{\sum\limits_{i=r+1}^{n} g(v_i)}{\gamma_{ft}(G)} \geq \frac{n(n+1)}{2\gamma_{ft}(G)} \tag{3}$$

Now define $\theta : V(G) \longrightarrow [0,1]$ by, $\theta(v) = min\ \{1, \frac{f(v)}{k}\}$. Since f is a nearly distance magic labeling with magic constant k it follows that $\sum_{x\in N(v)} \theta(x) \geq 1$. Hence θ is a fractional total dominating function. Therefore we have,

$$\gamma_{ft}(G) \leq |\theta| = \sum_{v\in V(G)} \theta(v) \leq \frac{1}{k} \sum_{v\in V(G)} f(v) = \frac{n(n+1)}{2k}.$$

Therefore

$$k \leq \frac{n(n+1)}{2\gamma_{ft}(G)}. \tag{4}$$

From (3) and (4) we have $k \leq \frac{n(n+1)}{2\gamma_{ft}(G)} \leq k + \frac{\sum\limits_{i=r+1}^{n} g(v_i)}{\gamma_{ft}(G)}$ and $0 \leq \frac{\sum\limits_{i=r+1}^{n} g(v_i)}{\gamma_{ft}(G)} < 1$. Hence $k = \left\lfloor \frac{n(n+1)}{2\gamma_{ft}(G)} \right\rfloor$

Observation 1. *If G is a nearly distance magic graph of order n with magic constants k_1 and k_2, $k_1 \leq k_2$ then $k_1 \geq n-1$ and $k_2 \geq n$.*

Theorem 3. *The graph $G = nC_4 \cup P_2$, where $n \geq 1$ is nearly distance magic with magic constant $4n+1$.*

Proof. Let $(u_{i1}, u_{i2}, u_{i3}, u_{i4}, u_{i1})$, $1 \leq i \leq n$, be the n copies of C_4 in G and let $\{v_1, v_2\}$ be the vertices of P_2. Define $f : V \to \{1, 2, \dots, 4n+2\}$ by

$$f(v_1) = 4n+2, \quad f(v_2) = 4n+1$$

$$\text{and } f(u_{ij}) = \begin{cases} i + n(j-1) & \text{if } 1 \leq i \leq n \text{ and } j = 1, 2 \\ (7-j)n + 1 - i & \text{if } 1 \leq i \leq n \text{ and } j = 3, 4. \end{cases} \tag{5}$$

Clearly f is a bijection. $w(v_2) = 4n+2$ and the weight of the remaining vertices is $4n+1$. Therefore f is a nearly distance magic labeling with magic constant $4n+1$.

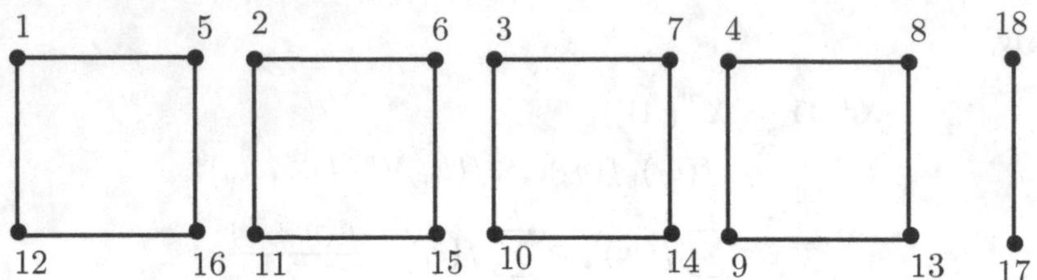

Fig. 3. Nearly distance magic labeling of $4C_4 \cup P_2$

A nearly distance magic labeling of $4C_4 \cup P_2$ is given in Fig. 3.

Theorem 4. *The graph $G = nC_4 \cup K_{2,3}$ is nearly distance magic with magic constant $4n + 7$.*

Proof. Let $(u_{i1}, u_{i2}, u_{i3}, u_{i4}, u_{i1})$, $1 \leq i \leq n$, be the n copies of C_4 in G and $\{u_{(n+1)1}, u_{(n+1)3}\}$, $\{u_{(n+1)2}, u_{(n+1)4}, u_{(n+1)5}\}$ be the bipartition of $K_{2,3}$. Define $g : V \to \{1, 2, \ldots, 4n + 5\}$ by,

$$g(u_{ij}) = \begin{cases} i + 1 + (n + 1)(j - 1) & \text{if } 1 \leq i \leq n + 1 \text{ and } j = 1, 2 \\ (7 - j)(n + 1) + 2 - i & \text{if } 1 \leq i \leq n + 1 \text{ and } j = 3, 4 \\ 1 & \text{if } i = n + 1 \text{ and } j = 5. \end{cases} \tag{6}$$

Then $w(x) = 4(n + 2)$ for $x \in \{u_{(n+1)1}, u_{(n+1)3}\}$. The weight of the remaining vertices is $4n+7$. Hence g is a nearly distance magic labeling with magic constant $4n + 7$

Lemma 1. *There is no nearly distance magic graph with magic constant 4.*

Proof. Suppose there exists a nearly distance magic graph G with magic constant 4 then $|V(G)| = 4$ *or* 5. Let $|V(G)| = 4$ and v be the vertex in G with label 4. Then all vertices adjacent to v are pendent vertices and hence $w(v)$ cannot be 4 or 5, which is a contradiction. The proof is similar if $|V(G)| = 5$.

Observation 2. *It follows from Theorems 3 and 4 that for every odd integer $k \geq 5$ there exists a nearly distance magic graph with magic constant k.*

Observation 2, Theorems 3, 4 and Lemma 1 lead to the following problem:

Problem 1. For any even integer $k \neq 4$, does there exist a nearly distance magic graph G with magic constant k?

Theorem 5. *Let G be a r-regular graph which is not distance magic. Suppose G admits a nearly distance magic labeling f with magic constant k. Then r is odd and there are exactly $\frac{n}{2}$ vertices with weight k.*

Proof. Let f be a nearly distance magic labeling of G. Let t be the number of vertices with weight k. Hence $\sum_{v\in V} w(v) = tk + (n-t)(k+1) = n(k+1) - t$. Since G is r-regular it follows that $\sum_{v\in V} w(v) = \sum_{v\in V}\sum_{x\in N(v)} f(v) = r\frac{n(n+1)}{2}$. Hence

$$r\frac{n(n+1)}{2} = n(k+1) - t.$$

Therefore $k = \frac{r(n+1)}{2} + \frac{t}{n} - 1$. Since k is an integer and $t < n$ it follows that r is odd, n is even and $t = \frac{n}{2}$.

Theorem 6. *A tree T is nearly distance magic if and only if $T \cong P_2$ or P_3.*

Proof. Let f be a nearly distance magic labeling of a tree T of order n. Suppose u_1 and u_2 are pendent vertices such that $N(u_1) = \{v_1\}$, $N(u_2) = \{v_2\}$ and $v_1 \neq v_2$. Then it follows that $f(v_1) = n$ and $f(v_2) = n - 1$. Now any vertex adjacent to v_1 is a pendent vertex and hence the component which contains v_1 is a star. Therefore T is not connected which is a contradiction. Hence $T \cong K_{1,n-1}$ and it follows that $T \cong P_2$ *or* P_3.

3 Conclusion and Scope

Since any distance magic graph is nearly distance magic, an interesting problem for further investigation is the construction of graphs which are nearly distance magic but not distance magic. Another possible direction for further investigation is characterisation of specific families of graphs which are nearly distance magic.

Acknowledgement. The first two authors are thankful to the Department of Science and Technology, New Delhi for financial support through the project No. SR/S4/MS-734/11.

References

1. Arumugam, S., Frončec, D., Kamatchi, N.: Distance magic graphs-A survey. J. Indones. Math. Soc., 11–26 (2011). Special Edition
2. Arumugam, S., Kamatchi, N., Vijayakumar, G.R.: On the uniqueness of d-vertex magic constant. Discussiones Math. Graph Theory **34**, 279–286 (2014)
3. Chartrand, G., Lesniak, L.: Graphs & Digraphs. Chapman and Hall, CRC, Boca Raton (2005)
4. Gallian J.A.: A dynamic survey of graph labeling. Electron. J. Comb., 1–408 (2016). DS #6
5. Kamatchi, N.: Distance magic and antimagic labelings of graphs. Ph.D. Thesis, Kalaslingam University, Tamil Nadu, India (2013)
6. Kovar, P., Frončec, D., Kovarova, T.: A note on 4-regular distance magic graphs. Australas. J. Combin. **54**, 127–132 (2012)

7. O'Neal, A., Slater, P.J.: Uniqueness of vertex magic constants. SIAM J. Discrete Math. **27**, 708–716 (2013)
8. Rupnow, R.: A survey on distance magic graphs. Master's report, Michigan Technological University (2014)
9. Vilfred, V.: Σ-labelled graph and Circulant Graphs. Ph.D. Thesis, University of Kerala, Trivandrum, India (1994)

Evenly Partite Directed Bigraph Factorization of Wreath Product of Graphs

P. Hemalatha(✉)

Department of Mathematics, Vellalar College for Women(Autonomous), Erode 638012, Tamilnadu, India
dr.hemalatha@gmail.com

Abstract. In this paper, some necessary and sufficient conditions for the existence of an evenly partite directed bigraph ($\overrightarrow{K}_{p,q\oplus q}$) factorization in the product graphs $(C_m \circ \overline{K}_n)^*$ and $(K_m \circ \overline{K}_n)^*$, for $m \geq 3$, $n, p, q \geq 2$ are obtained.

Keywords: $\overrightarrow{K}_{p,q\oplus q}$-factorization · m-partite graph · Wreath product of graphs · Symmetric digraph

1 Introduction

Let G be a graph. Then, for any positive integer s, sG denotes s disjoint copies of G, $G(s)$ denotes the graph obtained from G by replacing each edge by s edges and G^* is the *symmetric digraph* of G obtained by replacing every edge of G by a symmetric pair of arcs. An *m-partite graph* G has the partition of the vertex set V into m subsets such that uv is an edge of G if and only if u and v belong to different partite sets. The graph with vertex set V having partite sets $V_1, V_2, \ldots, V_m$ such that $|V_i| = n_i$ and edge set $E = \{(u, v) \in V_i \times V_j,$ $i, j \in \{1, 2, \ldots, m\}$ and $i \neq j\}$ is called a *complete m-partite graph* and is denoted by $K_{n_1,n_2,\ldots,n_m}$. $\overrightarrow{K}_{p,q\oplus q}$ denotes an *evenly partite directed bigraph* having partite sets V_1, V_2 and V_3 with $|V_1| = p$, $|V_2| = |V_3| = q$ such that all arcs are oriented from the p vertices (tails) at V_1 towards q vertices (heads) at V_2 and V_3 and there is no arc between V_2 and V_3. *Decomposition* of G is a partition of G into edge - disjoint subgraphs $G_1, G_2, ..., G_l$ such that $E(G) = E(G_1) \cup E(G_2) \cup ... \cup E(G_l)$; in this case we express $G = \bigoplus_{i=1}^{l} G_i$. In particular, if F is any graph and $G_i \cong F$, then it is called an *F-decomposition* of G and is denoted by $F|G$. A spanning subgraph of G is called an *F - factor* of G, if each component of G is isomorphic to F. Decomposition of G into F - factors is called an *F - factorization* of G and we denote it by $F\|G$. The *wreath product* of the graphs G and H denoted by $G \circ H$, has vertex set $V(G) \times V(H)$ in which two vertices (u_1, v_1) and (u_2, v_2) are adjacent whenever $u_1u_2 \in E(G)$ or $u_1 = u_2$ and $v_1v_2 \in E(H)$. For other definitions which are not mentioned here, see [1].

S. Arumugam et al. (Eds.): ICTCSDM 2016, LNCS 10398, pp. 83–89, 2017.
DOI: 10.1007/978-3-319-64419-6_11

In 2000, Ushio [4] obtained a necessary and sufficient condition for the existence of an $\overline{S}_k(=\overrightarrow{K}_{1,\frac{k-1}{2}\oplus\frac{k-1}{2}})$-factorization of $K^*_{n,n,n}$. In 2002, Ushio [5] proved that the necessary condition for the existence of an $\overline{S}_k$-factorization in $K^*_{n,n,n}(\lambda)$ is also sufficient. Hemalatha and Muthusamy [3] proved that the necessary condition is also sufficient for the existence of an $\overline{S}_k$-factorization in $(C_m \circ \overline{K}_n)^*$. Further, some necessary and sufficient conditions for the existence of an $\overline{S}_k$-factorization of $K^*_{n_1,n_2,...,n_m}$ have been established. Hemalatha and Muthusamy [2] obtained some necessary and sufficient conditions for the existence of an $\widetilde{S}_k \cong \overrightarrow{K}_{1,\frac{k-1}{2}\oplus\frac{k}{2}}$-factorization in $(C_m \circ \overline{K}_n)^*$ and $(K_m \circ \overline{K}_n)^*$.

In this paper, some necessary and sufficient conditions for the existence of $\overrightarrow{K}_{p,q\oplus q}$-factorizations in $(C_m \circ \overline{K}_n)^*$ and $(K_m \circ \overline{K}_n)^*$, for $m \geq 3$, $n, p, q \geq 2$ are obtained.

2 Evenly Partite Directed Bigraph Factorization of $(C_m \circ \overline{K}_n)^*$

Necessary Condition

Theorem 1. *$m \geq 3$ and $n, p, q \geq 2$ be given integers. If $(C_m \circ \overline{K}_n)^*$ has an $\overrightarrow{K}_{p,q\oplus q}$-factorization, then*

(a) $mn \equiv 0(mod\, 3p)$ where $p = q$
(b) $n \equiv 0(mod\, \frac{p+2q}{3d}\frac{pq}{d})$, where $m = 3, p + 2q \equiv 0(mod\, 3)$ and $d = (p, q)$
(c) $n \equiv 0(mod\, \frac{(p+2q)pq}{d})$ where $m = 3$, $p + 2q \not\equiv 0(mod\, 3)$ and $d = (p, q)$
(d) $n \equiv 0(mod\, \frac{(p+2q)pq}{d})$ where $m > 3$, $p \neq q$ and $d = (p, q)$.

Proof. Let $V_1, V_2, ..., V_m$ be the partite sets of the m-partite digraph $(C_m \circ \overline{K}_n)^*$. Assume that $(C_m \circ \overline{K}_n)^*$ has an $\overrightarrow{K}_{p,q\oplus q}$-factorization. Let r be the number of $\overrightarrow{K}_{p,q\oplus q}$ - factors, s be the number of components in each $\overrightarrow{K}_{p,q\oplus q}$ - factor and b be the total number of components in the $\overrightarrow{K}_{p,q\oplus q}$-factorization.

$$s = \frac{mn}{p+2q} \tag{1}$$

$$b = \frac{mn^2}{pq} \tag{2}$$

$$r = \frac{b}{s} = \frac{n(p+2q)}{pq}. \tag{3}$$

When $p = q$, $mn = 3ps$ and hence (a) follows.

Now due to the structure of $\overrightarrow{K}_{p,q\oplus q}$, the p tails at V, requires q heads at V_{i+1} and V_{i-1}. Thus $p|n$ and $q|n$ and hence if $d = gcd(p, q)$, V_i then

$$n \equiv 0(mod\, \frac{pq}{d}) \tag{4}$$

From Eq. (1), $n = \frac{s(p+2q)}{m}$. Now we will deal the theorem into 2 cases as $m = 3$ and $m > 3$.

Case(i). $m = 3$.

It divides either s or $p + 2q$ since it is a prime. If $3|s$, then

$$n \equiv 0(mod\, p + 2q) \tag{5}$$

If $3|p + 2q$, then

$$n \equiv 0(mod\,(p + 2q)/3)). \tag{6}$$

Thus, (b) and (c) follows from Eqs. (4), (5) and (6).

Case(ii). $m > 3$. By the definition of $\overrightarrow{K}_{p,q\oplus q}$, the number of tail vertices of a $\overrightarrow{K}_{p,q\oplus q}$-factor from any V_i must be a multiple of p. Also, due to the structure of $(C_m \circ \overline{K}_n)^*$ it is clear that, all the m - parts of $V((C_m \circ \overline{K}_n)^*)$ should have equal number of tail vertices since otherwise if s_i is the number of components in an $\overrightarrow{K}_{p,q\oplus q}$,-factor having tail at V_i, then $s_i = pt_i$ and hence it requires pqt_i head vertices from V_{i-1} and V_{i+1}. If t_i is different for each V_i then $|V_i| \neq n$ for all $i = 1, 2, ..., m$. Thus,

$$n \equiv 0(mod\; p + 2q). \tag{7}$$

(d) follows from Eqs. (4) and (7). Hence the proof.

Sufficient Conditions

Notation: Let v_{ij} denotes jth vertex at V_i, where $i, j \in 1, 2..., m$. A $\overrightarrow{K}_{p,q\oplus q}$ with tail at $v_{1i}, i = 1, 2, ..., p$ and q heads at $v_{2j}, j = 1, 2, ..., q$ and $v_{3j}, j = 1, 2, ..., q$ is denoted by $[v_{11}, v_{12}, ..., v_{1p}; v_{21}, v_{22}, ..., v_{2q}, v_{31}, v_{32}, ..., v_{3q}]$.

The following lemmas are required to prove the main result.

Lemma 1. *For given integers $m \geq 3$, p, $q \geq 2$ and an m-partite digraph G, if G has an $\overrightarrow{K}_{p,q\oplus q}$-factorization, then sG also has an $\overrightarrow{K}_{p,q\oplus q}$-factorization for every positive integer s.*

Proof. sG is nothing but s disjoint copies of G and hence the proof follows.

Lemma 2. *For any integer $s > 0$, $\overrightarrow{K}_{p,q\oplus q}\|\overrightarrow{K}_{sp,sq\oplus sq}$.*

Lemma 3. *If $(C_m \circ \overline{K}_n)^*$ and $(K_m \circ \overline{K}_n)^*$ have $\overrightarrow{K}_{p,q\oplus q}$-factorizations, then so do $(C_m \circ \overline{K}_{sn})^*$ and $(K_m \circ \overline{K}_{sn})^*$, for every positive integer s.*

Proof. If $(C_m \circ \overline{K}_n)^*$ and $(K_m \circ \overline{K}_n)^*$ have $\overrightarrow{K}_{p,q\oplus q}$-factorizations, then $(C_m \circ \overline{K}_{sn})^*$ and $(K_m \circ \overline{K}_{sn})^*$, have $\overrightarrow{K}_{sp,sq\oplus sq}$-factorizations. But, by Lemma 2, $\overrightarrow{K}_{sp,sq\oplus sq}$ has a$\overrightarrow{K}_{p,q\oplus q}$ - factorization. Hence the proof.

Theorem 2. *If $n \equiv 0(mod\, p)$ and $m = 3t$, $t \geq 1$ then $\overrightarrow{K}_{p,p\oplus p}\|(C_m \circ \overline{K}_n)^*$.*

Proof. Let $n = ps$ and $s = 1$. Let the $m = 3t$ vertex partite sets of $(C_m \circ \overline{K}_p)^*$ be $V_1 = \{1^1, 2^1, ..., p^1\}$, $V_2 = \{1^2, 2^2, ..., p^2\}$, $\ldots$ $V_{3t} = \{1^{3t}, 2^{3t}, ..., p^{3t}\}$. Then the 3 $\overrightarrow{K}_{p,p\oplus p}$ -factors F_0, F_1 and F_2 of $(C_m \circ \overline{K}_p)^*$ are as follows:

$$F_j = \bigoplus_{i=0}^{t-1} \{[1^{3i+j+1}, 2^{3i+j+1}, ..., p^{3i+j+1}; \quad 1^{3i+j+2}, 2^{3i+j+2}, ..., p^{3i+j+2}, \quad 1^{3i+j}, 2^{3i+j}, ...p^{3i+j}]\}, j = 0, 1, 2,$$

where the superscripts are taken addition modulo $3t$ with residues $1, 2, ..., 3t$. Thus, F_0, F_1 and F_2 together comprise an $\overrightarrow{K}_{p,p\oplus p}$ - factorization of $(C_m \circ \overline{K}_p)^*$ and hence by Lemma 3, $\overrightarrow{K}_{p,p\oplus p} \| (C_m \circ \overline{K}_n)^*$.

Theorem 3. *Let odd $m \geq 3$ and $p, q \geq 2$ be given positive integers such that $gcd(p, q) = 1$. If $n \equiv 0(mod\ pq(p + 2q))$ then $\overrightarrow{K}_{p,q\oplus q} \| (C_m \circ \overline{K}_n)^*$.*

Proof. Let $n = pq(p+2q)s$. If $s = 1$, then $n = pq(p+2q)$. Let the m-partite sets of $(C_m \circ \overline{K}_{pq(p+2q)})^*$ be $V_1 = \{1^1, 2^1, ..., (pq(p + 2q))^1\}$, $V_2 = \{1^2, 2^2, ..., (pq(p + 2q))^2\}$, ..., $V_m = \{1^m, 2^m, ..., (pq(p + 2q))^m\}$. Now for $j = 0, 1, 2, ..., (p + 2q) - 1$ and $r = 0, 1, 2, ..., (p + 2q) - 1$, we can construct $(p + 2q)^2$ $\overrightarrow{K}_{p,q\oplus q}$-factors F_{jr} of $(C_m \circ \overline{K}_{pq(p+2q)})^*$ as follows:

$$
\begin{aligned}
F_{jr} = \bigoplus_{i=0}^{pq-1} \{&[(pqj + ip + 1)^1, (pqj + ip + 2)^1, ..., (pqj + ip + p)^1;\\
&(q(r + i) + 1)^2, (q(r + i) + 2)^2, ..., (q(r + i) + q)^2,\\
&(q(pq + r + j) + 1)^m, (q(pq + r + j) + 2)^m, ..., (q(pq + r + j) + q)^m],\\
&[q(pq + r) + ip + 1)^2, (q(pq + r) + ip + 2)^2, ..., (q(pq + r) + ip + p)^2;\\
&(pq(j + p + q) + iq + 1)^3,\\
&(pq(j + p + q) + iq + 2)^3, ..., (pq(j + p + q) + iq + q)^3,\\
&(pq(j + p) + iq + 1)^1, (pq(j + p) + iq + 2)^1 ..., (pq(j + p) + iq + q)^1],\\
&[(pqj + ip + 1)^3, (pqj + ip + 2)^3, ..., (pqj + ip + p)^3;\\
&(q(r + i) + 1)^4, (q(r + i) + 2)^4, ..., (q(r + i) + q)^4,\\
&(pq(p + q) + (r + i)q + 1)^2,\\
&(pq(p + q) + (r + i)q + 2)^2, ..., (pq(p + q) + (r + i)q + q)^2],\\
&[q(pq + r) + ip + 1)^4, (q(pq + r) + ip + 2)^4, ..., (q(pq + r) + ip + p)^4;\\
&(pq(j + p + q) + iq + 1)^5, (pq(j + p + q) + iq + 2)^5, ..., (pq(j + p + q) + iq + q)5,\\
&(pq(j + p) + iq + 1)^3, (pq(j + p) + iq + 2)^3 ..., (pq(j + p) + iq + q)^3],\\
&[(pqj + ip + 1)^5, (pqj + ip + 2)^5, ..., (pqj + ip + p)^5;\\
&(q(r + i) + 1)^6, (q(r + i) + 2)^6, ..., (q(r + i) + q)^6, (pq(p + q) + (r + i)q + 1)^4,\\
&(pq(p + q) + (r + i)q + 2)^4, ..., (pq(p + q) + (r + i)q + q)^4],\\
&\ldots\\
&[q(pq + r) + ip + 1)^{(m-1)}, (q(pq + r) + ip + 2)^{(m-1)}, ...,\\
&(q(pq + r) + ip + p)^{(m-1)};\\
&(pq^2 + q(p + r - i - 1) + 1)^m, (pq^2 + q(p + r - i - 1) + 2)^m, ...,
\end{aligned}
$$

$(pq^2+q(p+r-i-1)+q)^m,$
$(pq(j+p)+iq+1)^{(m-2)}, (pq(j+p)+iq+2)^{(m-2)}...,$
$(pq(j+p)+iq+q)^{(m-2)}],$
$[(pq(p+2q+r)+ip+1)^m, (pq(p+2q+r)+ip+2)^m, ...,$
$(pq(p+2q+r)+ip+p)^m;$
$(pq(p+q)+(r+i)q+1)^{(m-1)}, (pq(p+q)+(r+i)q+2)^{(m-1)}, ...,$
$(pq(p+q)+(r+i)q+q)^{(m-1)},$
$(pq(p+q)+(j+i)q+1)^1, (pq(p+q)+(j+i)q+2)^1...,$
$(pq(p+q)+(j+i)q+q)^1]\},$

where the additions in the bases are taken modulo $pq(p+2q)$ with residues $1, 2, ..., pq(p+2q)$. i.e., by dividing each partite set into $p+2q$ parts consisting of pq vertices and keeping the first p^2q vertices at V_1 as fixed tail we can form $p+2q$ $\overrightarrow{K}_{p,q\oplus q}$-factors. When j, r varies over $1, 2, ..., p+2q$, we have $(p+2q)^2$ $\overrightarrow{K}_{p,q\oplus q}$-factors of $(C_m \circ \overline{K}_{pq(p+2q)})^*$. Due to symmetry of arcs, we have another$(p+2q)^2$ $\overrightarrow{K}_{p,q\oplus q}$-factors as above. All these $2(p+2q)^2$ $\overrightarrow{K}_{p,q\oplus q}$-factors together comprise an $\overrightarrow{K}_{p,q\oplus q}$-factorization of $(C_m \circ \overline{K}_{pq(p+2q)})^*$. By Lemma 3, $\overrightarrow{K}_{p,q\oplus q} \| (C_m \circ \bar{K}_{pq(p+2q)s})^*$ and hence $\overrightarrow{K}_{p,q\oplus q} \| (C_m \circ \overline{K}_n)^*$.

3 Evenly Partite Directed Bigraph Factorization of $(K_m \circ \overline{K}_n)^*$

In this section, some necessary and sufficient conditions for the existence of an $\overrightarrow{K}_{p,q\oplus q}$-factorization of $(K_m \circ \overline{K}_n)^*$ for $m \geq 3$, $n, p, q \geq 2$ have been obtained.

Necessary Condition

Theorem 4. *Let $m \geq 3$ and $n, p, q \geq 2$ be given integers. If $(K_m \circ \overline{K}_n)^*$ has an $\overrightarrow{K}_{p,q\oplus q}$-factorization, then*
(a) $mn \equiv 0(mod\, 3p)$ where $p = q$
(b) $n \equiv 0(mod\, \frac{p+2q}{3}\frac{pq}{d})$, where $m = 3, p+2q \equiv 0(mod\, 3)$ and $d = (p, q)$
(c) $n \equiv 0(mod\, \frac{pq(p+2q)}{d})$, where $m = 3$, $p+2q \not\equiv 0(mod\, 3)$ and $d = (p, q)$
(d) $n \equiv 0(mod\, \frac{pq}{d}\frac{(p+2q)}{d})$, where $m > 3$, $p \neq q$ and $d = (p, q)$.

Proof. Let $V_1, V_2, ..., V_m$ be the partite sets of the m-partite digraph $(K_m \circ \overline{K}_n)^*$. Assume that $(K_m \circ \overline{K}_n)^*$ has an $\overrightarrow{K}_{p,q\oplus q}$-factorization. Let r be the number of $\overrightarrow{K}_{p,q\oplus q}$ - factors, s be the number of components in each $\overrightarrow{K}_{p,q\oplus q}$ - factor and b be the total number of components in the $\overrightarrow{K}_{p,q\oplus q}$-factorization.

$$s = \frac{mn}{p+2q} \tag{8}$$

$$b = \frac{m(m-1)n^2}{2pq} \tag{9}$$

$$r = \frac{b}{s} = \frac{n(m-1)(p+2q)}{2pq}. \tag{10}$$

When $p = q$, $mn = 3ps$ and hence (a) follows.
From Eq. (8) for $m > 3$ and $d = gcd(p, q)$, we have,

$$n \equiv 0(mod\,\frac{p+2q}{d}). \tag{11}$$

Now due to the structure of $\overrightarrow{K}_{p,q\oplus q}$, the p tails at V_i, requires q heads at V_{i+1} and V_{i-1}. Thus $p|n$ and $q|n$ and hence if $d = gcd(p, q)$, then

$$n \equiv 0(mod\,\frac{pq}{d}) \tag{12}$$

(d) follows from Eqs. (11) and (12).
When $m = 3$, (b) and (c) follows from Eqs. (8) and (12).

Sufficient Conditions

Theorem 5. *For given integers $p, q \geq 2$, if $q = ps$ then $\overrightarrow{K}_{p,q\oplus q} \| (K_{2s+1} \circ \overline{K}_q)^*$, for some positive integer s.*

Proof. Let $V_1 = \{1^{(1)}, 2^{(1)}, ..., (q)^{(1)}, \}$, $V_2 = \{1^{(2)}, 2^{(2)}, ..., (q)^{(2)}\}$, ..., $V_{2s+1} = \{1^{(2s+1)}, 2^{(2s+1)}, ..., (q)^{(2s+1)}\}$ be the $2s+1$-partite sets of $(K_{2s+1} \circ \overline{K}_q)^*$. We now construct $s(2s+1)$ $\overrightarrow{K}_{p,q\oplus q}$ - factors as follows: Let

$$\begin{aligned} F_{jr} = \bigoplus_{i=0}^{s-1} \{[(p(i+r)+1)^{(j)}, (p(i+r)+2)^{(j)}, ..., \\ (p(i+r+1))^{(j)}; 1^{(2i+j+1)}, 2^{(2i+j+1)}, ..., (q)^{(2i+j+1)}, \\ 1^{(2i+j+2)}, 2^{(2i+j+2)}, ..., (q)^{(2i+j+2)}]\}, \end{aligned}$$

$j = 1, 2, ..., 2s+1$, $r = 0, 1, 2, ..., s-1$, where the superscripts are taken modulo $2s+1$ with residues $1, 2, ..., 2s+1$ and the additoins in the bases are taken modulo q with residues $1, 2, ..., q$. i.e., by fixing the $q = ps$ vertices in one partite set as centers and the vertices of the remaining partite sets as end vertices, we can form s $\overrightarrow{K}_{p,q\oplus q}$-factors. By shifting the center through the $2s+1$ -partite sets we have all $s(2s+1)$ $\overrightarrow{K}_{p,q\oplus q}$-factors F_{jr}, $j = 1, 2, ..., 2s+1$, $r = 0, 1, 2, ..., s-1$ which together give an $\overrightarrow{K}_{p,q\oplus q}$ - factorization of $(K_{2s+1} \circ \overline{K}_q)^*$. Thus, $\overrightarrow{K}_{p,q\oplus q} \| (K_{2s+1} \circ \overline{K}_q)^*$.

Theorem 6. *For given positive integers odd $m \geq 3$ and $p, q \geq 2$ such that $gcd(p, q) = 1$, if $n \equiv 0(mod\ pq(p+2q))$, then $\overrightarrow{K}_{p,q\oplus q} \| (K_m \circ \overline{K}_n)^*$.*

Proof. By Walecki's construction, when $m \geq 3$ is odd, $(K_m \circ \overline{K}_n)^* \cong \bigoplus_{i=1}^{\frac{m-1}{2}} (H_i \circ \overline{K}_n)^*$, where H_i is a Hamiltonian cycle of $(K_m \circ \overline{K}_n)^*$. But by Theorem 3, $\overrightarrow{K}_{p,q\oplus q} \| (H_i \circ \overline{K}_n)^*$ when $n \equiv 0(mod\ pq(p+2q))$ and hence by Lemma 1, $\overrightarrow{K}_{p,q\oplus q} \| (K_m \circ \overline{K}_n)^*$. Hence the theorem.

4 Conclusion

In this paper, some necessary conditions for the existence of an $\overrightarrow{K}_{p,q\oplus q}$-factorization of $(C_m \circ \overline{K}_n)^*$ are obtained. Further, it is proved that the necessary conditions are also sufficient for the existence of an $\overrightarrow{K}_{p,p\oplus p}$-factorization in $(C_m \circ \overline{K}_n)^*$ if (i). $m = 3t, t \geq 1$ and $n \equiv 0(mod\, p)$ and (ii). $m \geq 3$ is odd and $n \equiv 0(mod\ pq(p+2q))$. In Sect. 3, some necessary conditions for the existence of an $\overrightarrow{K}_{p,q\oplus q}$-factorization of $(K_m \circ \overline{K}_n)^*$ are obtained and the sufficiency is proved when (i). $q = ps$ and $m = 2s+1$, for some positive integer s and (ii). $m \geq 3$ is odd and $n \equiv 0(mod\ pq(p+2q))$.

Acknowledgement. The investigator thanks the University Grants Commission for its financial support for the Minor Research Project entitled " Decomposition/ Factorizations of Product Graphs into Complete Bipartite Graphs" vide Link No. UGC/SERO/MRP/4950, dated March 2014.

References

1. Bondy, J.A., Murty, U.S.R.: Graph Theory with Applications. MacMillan, New York (1976)
2. Hemalatha, P., Muthusamy, A.: $\widetilde{S}_k$ - factorization of the symmetric digraph of wreath product of graphs. Discrete Math. Algorithms Appl. **6**(4), 1450048-1–1450048-12 (2014)
3. Hemalatha, P., Muthusamy, A.: Evenly partite star factorization of symmetric digraph of wreath product of graphs. AKCE Int. J. Graphs Combin. **14**(1), 54–62 (2017)
4. Ushio, K.: $\overline{S}_k$ - factorization of symmetric complete tripartite digraphs. Discrete Math. **211**, 281–286 (2000)
5. Ushio, K.: Evenly partite star-factorization of symmetric complete tripartite multi-digraphs. Electr. Notes Discrete Math. **11**, 608–611 (2002)

Minimum Layout of Circulant Graphs into Certain Height Balanced Trees

Jessie Abraham(✉) and Micheal Arockiaraj

Department of Mathematics, Loyola College, Chennai 600034, India
jessie.abrt@gmail.com, marockiaraj@gmail.com

Abstract. A graph embedding comprises of an ordered pair of injective maps $\prec f, p \succ$ from a guest graph $G = (V(G), E(G))$ to a host graph $H = (V(H), E(H))$ which is formulated as follows: f is a mapping from $V(G)$ to $V(H)$ and p assigns to each edge (a, b) of G, a shortest path $p(a, b)$ in H. The minimum layout problem is to find an embedding $\prec f, p \succ$ from a graph G into a graph H such that $\sum_{e \in E(H)} EC_{\prec f,p \succ}(e) = \sum |(a, b) \in E(G) : e \in E(p(a, b))|$ is minimized. In this paper we develop an algorithm to find the minimum layout of embedding the circulant graph into certain height balanced trees like Fibonacci tree and wounded lobster.

Keywords: Height balanced tree · Layout · Circulant graph · Fibonacci tree

1 Introduction

Graph embedding has been an integral tool in efficient implementation of parallel algorithms on parallel computers with minimal communication overhead. A graph embedding comprises of an ordered pair of injective maps $\prec f, p \succ$ from a guest graph $G = (V(G), E(G))$ to a host graph $H = (V(H), E(H))$ which is formulated as follows: f is a mapping from $V(G)$ to $V(H)$ and p assigns to each edge (a, b) of G, a shortest path $p(a, b)$ in H [1,7]. Figure 1 illustrates a graph embedding. The edge congestion of an embedding is defined by $EC_{\prec f,p \succ}(e) = |(a, b) \in E(G) : e \in E(p(a, b))|$ [6].

The layout $L_{\prec f,p \succ}(G, H)$ of an embedding is defined as the sum of edge congestion of all the edges of H [3,5]. The minimum layout of G into H is given by $L(G, H) = \min L_{\prec f,p \succ}(G, H)$. The minimum layout problem is to find the embedding that induces $L(G, H)$. When the host graph is a tree, the layout problem finds application in graph drawing, data structures and representations and networks for parallel systems [5,10].

Maximum Induced Subgraph Problem [3]: Let $G = (V(G), E(G))$ and $S \subseteq V(G)$. Let $I_G(S) = \{(u, v) \in E(G) : u \in S \text{ and } v \in S\}$ and for $1 \le k \le |V(G)|$, let $I_G(k) = \max_{S \subseteq V, |S| = k} |I_G(S)|$. Then the problem is to find $S \subseteq V(G)$ with

S. Arumugam et al. (Eds.): ICTCSDM 2016, LNCS 10398, pp. 90–97, 2017.
DOI: 10.1007/978-3-319-64419-6_12

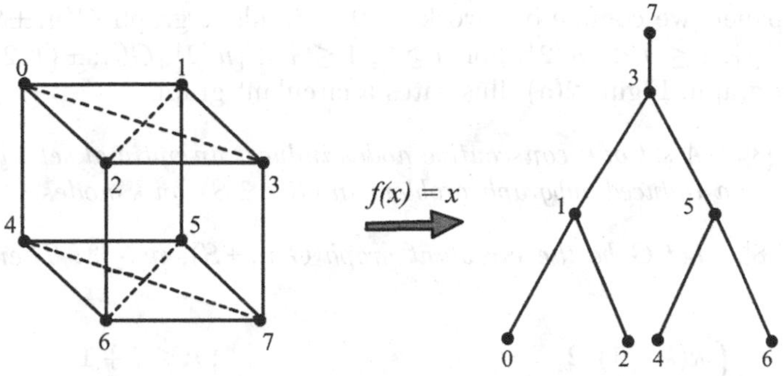

Fig. 1. Embedding of an enhanced hypercube G into rooted complete binary tree H

$|S| = k$ such that $I_G(k) = |I_G(S)|$. Such a set S is called an optimal set with respect to the maximum induced subgraph problem.

Min-cut Problem [3]: Let $G = (V(G), E(G))$ and $S \subseteq V(G)$. Let $\Theta_G(S) = \{(u,v) \in E : u \in S \text{ and } v \notin S\}$ and for $1 \leq k \leq |V(G)|$, let $\Theta_G(k) = \min_{S \subseteq V,\, |S|=k} |\Theta_G(S)|$. Then the problem is to find $S \subseteq V(G)$ with $|S| = k$ such that $\Theta_G(k) = |\Theta_G(S)|$. Such a set S is said to be optimal with respect to the min-cut problem. For any graph G, $\Theta_G(V - S) = \Theta_G(S)$ for all $S \subseteq V(G)$. If G is an r-regular graph, then $\Theta_G(k) = rk - 2I_G(k)$ for every $k \in \{1, 2, \ldots, |V(G)|\}$.

The following results provide a method for partitioning the edges of the host graph which in turn can be effectively used to solve the minimum layout problem.

Lemma 1 (Congestion Lemma) [6]. *Let G be an r-regular graph and $\prec f, p \succ$ be an embedding of G into H. Let S be an edge cut of H such that the removal of edges of S splits H into 2 components H_1 and H_2 and $EC_{\prec f,p\succ}(S)$ denote the sum of edge congestion over all the edges in S. Let $G_1 = G[f^{-1}(H_1)]$ and $G_2 = G[f^{-1}(H_2)]$. Suppose the following conditions hold.*

1. *For every edge $(a, b) \in G_i$, $i = 1, 2$, $p(a, b)$ has no edges in S.*
2. *For every edge (a, b) in G with $a \in G_1$ and $b \in G_2$, $p(a, b)$ has exactly one edge in S.*
3. *G_1 is optimal with respect to the maximum induced subgraph problem.*

Then $EC_{\prec f,p\succ}(S)$ is minimum and $EC_{\prec f,p\succ}(S) = \Theta_G(|V(G_1)|) = \Theta_G(|V(G_2)|)$.

Lemma 2 [6]. *Let $\prec f, p \succ$ be an embedding from G into H. Let $\{S_1, S_2, \ldots, S_p\}$ be a partition of $E(H)$ such that $EC_{\prec f,p\succ}(S_i)$ is minimum for all i. Then $L_{\prec f,p\succ}(G, H)$ is minimum and $L_{\prec f,p\succ}(G, H) = \sum_{i=1}^{p} EC_{\prec f,p\succ}(S_i)$.*

Definition 1 [10]. *A circulant undirected graph $G(n; \pm S)$, $S \subseteq \{1, \ldots, \lfloor n/2 \rfloor\}$, $n \geq 3$ is defined as a graph consisting of the node set $V = \{0, 1, \ldots, n-1\}$ and the edge set $E = \{(i, j) : |j - i| \equiv s(mod\ n),\ s \in \pm S\}$.*

In this paper, we confine our work to the circulant graph $G(n; \pm S)$, where $S = \{1, 2, ..., j\}$, $1 \leq j < \lfloor n/2 \rfloor$. For $n \geq 3$, $1 \leq j < \lfloor n/2 \rfloor$, $G(n; \pm\{1, 2, ..., j\})$ is a $2j$-regular graph. Figure 2(a) illustrates a circulant graph.

Lemma 3 [8]. *A set of k consecutive nodes induces an optimal set with respect to the maximum induced subgraph problem in $G(n; \pm S)$ on k nodes.*

Lemma 4 [8]. *Let G be the circulant graph $G(n; \pm S)$, $n \geq 3$. Then for $1 \leq k \leq n$,*

$$I_G(k) = \begin{cases} k(k-1)/2 & ; k \leq j+1 \\ kj - j(j+1)/2 & ; j+1 < k \leq n-j \\ \frac{1}{2}\{(n-k)^2 + (4j+1)k - (2j+1)n\} & ; n-j < k \leq n\ . \end{cases}$$

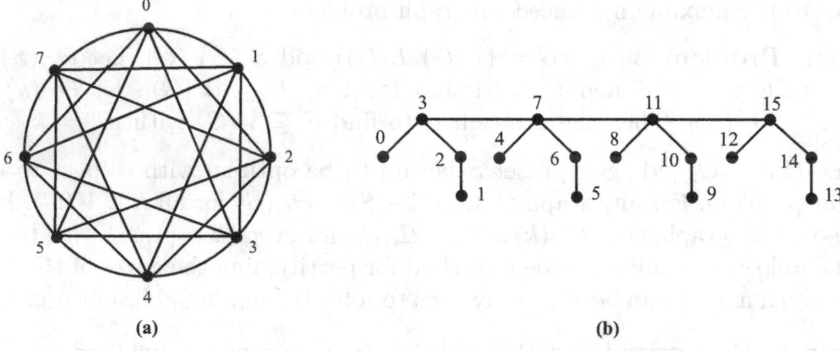

Fig. 2. (a) Circulant graph $G(8; \pm\{1, 2, 3\})$ (b) Wounded lobster L_4

A height balanced tree T is a rooted binary tree in which for every node v, the difference between the heights of the left and right child denoted as v_1 and v_2 respectively is at most one [2].

Fibonacci trees are a type of height balanced trees which are built recursively in one of the following two ways.

Fibonacci Tree f_h [4]: The trees f_1 and f_2 consists of only the root node. For $h \geq 3$, f_h is constructed by taking a new root node and attaching f_{h-1} on the left side and f_{h-2} on the right side of the root node by an edge as shown in Fig. 3(a).

Fibonacci Tree $f_h^{'}$ [2]: The tree f_1^1 consists of only the root node and $f_2^{'}$ is formed by attaching a pendant node to the root node. For $h \geq 3$, the left subtree of $f_h^{'}$ is $f_{h-1}^{'}$ and its right subtree is $f_{h-2}^{'}$. Figure 3(b) illustrates $f_h^{'}$ for $h = 1, 2, \ldots 5$.

Let $|V(f_h)| = m_h$ and $|V(f_h^{'})| = m_h^{'}$. Then, $m_h = 2F_h - 1$ and $m_h^{'} = F_{h+2} - 1$, where F_h denotes the Fibonacci number.

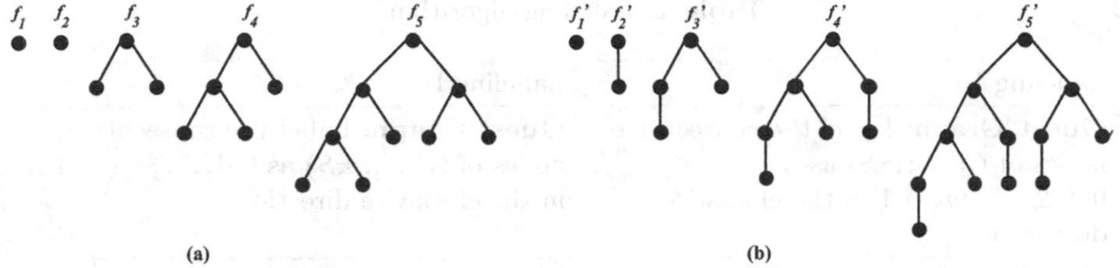

Fig. 3. (a) f_h type Fibonacci trees (b) $f_h^{'}$ type Fibonacci trees

Definition 2 [9]. *A lobster is a tree with the property that the removal of pendant nodes leaves a caterpillar. A wounded lobster L_n is a lobster satisfying the following conditions:*

(i) There are 2^{n-2} spine nodes and every spine node is adjacent to exactly one node of degree 2 and one node of degree 1.
(ii) Removal of pendant nodes incident at nodes of degree 2 leaves a caterpillar.

Figure 2(b) illustrates a wounded lobster.

There are several techniques for traversing the nodes of a tree according to the order in which the nodes are visited. In this paper we confine our study to postorder.

Algorithm 1. Postorder Tree Traversal Algorithm

Do the following recursively until all nodes are traversed:
Step 1 - Traverse left subtree.
Step 2 - Traverse right subtree.
Step 3 - Visit root node.

2 Main Results

In this section we embed the circulant graph into Fibonacci trees and wounded lobster to minimize their layouts.

Theorem 1. *The minimum layout of circulant graphs $G = G(m_h; \pm S)$ and $G^{'} = G(m_h^{'}; \pm S)$ into the Fibonacci trees is given by (a) $L(G, f_h) = F_{h-2}.\Theta_G(m_3) + F_{h-3}.\Theta_G(m_4) + \ldots + F_2.\Theta_G(m_{h-1}) + 2|S|$ and (b)$L(G', f_h^{'}) = F_{h-1}.\Theta_{G'}(m_2^{'}) + F_{h-2}.\Theta_{G'}(m_3^{'}) + \ldots + 2\Theta_{G'}(m_{h-2}^{'}) + \Theta_{G'}(m_{h-1}^{'}) + 2|S|$.*

Proof. We split the proof into three parts comprising of labeling the guest and host graphs, followed by the proposal of embedding and layout computation.

Guest and Host Labeling: Label the circulant graph and the two types of Fibonacci trees as in the pattern given in Table 1.

Table 1. Labeling algorithm

Labeling I	Labeling II
Guest Graph: Label the consecutive nodes of $G(m_h; \pm S)$ as $0, 1, 2, \ldots, m_h - 1$ in the clockwise direction	**Guest Graph:** Label the consecutive nodes of $G(m_h^{'}; \pm S)$ as $0, 1, \ldots, m_h^{'} - 1$ in the clockwise direction
Host Graph: Label the nodes of f_h by postorder tree traversal from 0 to $m_h - 1$	**Host Graph:** Label the nodes of $f_h^{'}$ by postorder tree traversal from 0 to $m_h^{'} - 1$

Proposed Embedding: Define an embedding $\prec f, p \succ$ from $G(m_h; \pm S)$ into f_h and $G(m_h^{'}; \pm S)$ into $f_h^{'}$ such that $f(x) = x$.

Layout Computation: We split the proof into two cases.

Proof for (a): For $1 \leq i \leq m_h - 1$, let S_i be an edge cut of f_h such that its removal disengages f_h into two components X_i and $\overline{X}_i$ as shown in Fig. 4(a), with the node set V_i of X_i being as follows.

For $1 \leq i \leq m_{h-1}$,

$$V_i = \begin{cases} \{0, 1, \ldots, i-1\}, & if\ i = m_g, 1 \leq g \leq h-1 \\ \{m_a, m_a + 1, \ldots, i-1\}, & if\ i = m_a + m_b, 1 \leq b < a \leq h-1 \\ \{i-1\}, & otherwise. \end{cases}$$

For $m_{h-1} + 1 \leq i \leq m_h - 1$,

$$V_i = \begin{cases} \{m_{h-1}, m_{h-1} + 1, \ldots, i-1\}, if\ i = m_{h-1} + m_g, 1 \leq g \leq h-1 \\ \{m_{h-1} + m_a, m_{h-1} + m_a + 1, \ldots, i-1\}, if\ i = m_{h-1} + m_a + m_b, \\ \qquad\qquad 1 \leq b < a \leq h-1 \\ \{i-1\}, \quad otherwise. \end{cases}$$

Let G_i be the graph induced by $\{f^{-1}(u) : u \in V_i\}$. It can be noted that X_i is consecutively labeled for all i and hence by Lemma 3, V_i is an optimal set with respect to the maximum induced subgraph problem. S_i also satisfies conditions (i) and (ii) of Lemma 1. In addition, $\{S_i\}_{i=1}^{m_h-1}$ forms a partition of $E(f_h)$. Hence by Lemma 2, $L_{\prec f,p \succ}(G, f_h)$ is minimum.

Let $m_h - 1 = F_h + F_{h-2} + F_{h-3} + F_{h-4} + \ldots F_2$, where $m_h - 1$ represents the number of edge cuts of f_h and $F_h, F_{h-2}, F_{h-3}, \ldots, F_3, F_2$ denote the number of node sets V_i of cardinality $m_2, m_3, \ldots, m_{h-2}$ and m_{h-1} respectively.

Layout: $L(G, f_h) = \sum_{i=1}^{m_h-1} EC_{\prec f,p \succ}(S_i) = \sum_{i=1}^{m_h-1} \Theta_G(|V_i|) = \sum_{i=1}^{F_h} \Theta_G(m_2) + \sum_{i=1}^{F_{h-2}} \Theta_G\ (m_3) + \sum_{i=1}^{F_{h-3}} \Theta_G(m_4) + \ldots + \sum_{i=1}^{F_3} \Theta_G(m_{h-2}) + \sum_{i=1}^{F_2} \Theta_G(m_{h-1}) = F_{h-2}.\Theta_G(m_3) + F_{h-3}.\Theta_G(m_4) + \ldots + F_2.\Theta_G(m_{h-1}) + 2|S|$.

Proof for (b): Let $S_i^{'}$, $1 \leq i \leq m_h - 1$ be an edge cut of $f_h^{'}$ such that removal of $S_i^{'}$ disconnects $f_h^{'}$ into two components Y_i and $\overline{Y}_i$ as depicted in Fig. 4(b) where the node set $V_i^{'}$ of Y_i is defined by replacing m_g, m_a, m_b and m_{h-1} in V_i of case (a) by $m_g^{'}, m_a^{'}, m_b^{'}$ and $m_{h-1}^{'}$ respectively.

Let $G_i^{'}$ be the graph induced by $\{f^{-1}(a) : a \in V_i^{'}\}$. Clearly X_i is labeled consecutively for all i and hence by Lemma 3, $V_i^{'}$ is an optimal set with respect to the maximum induced subgraph problem. $S_i^{'}$ also satisfies the remaining two conditions of Lemma 1. In addition, $\{S_i^{'}\}_{i=1}^{m_h^{'}-1}$ forms a partition of $E(f_h^{'})$. Hence by Lemma 2, $L_{\prec f,p\succ}(G, f_h^{'}) = L(G, f_h^{'})$.

Let $m_h^{'} - 1 = F_h + F_{h-1} + F_{h-2} + \ldots F_2$, where $F_h, F_{h-1}, F_{h-2}, \ldots, F_3, F_2$ denote the number of nodes sets $V_i^{'}$ of cardinality $m_1^{'}, m_2^{'}, \ldots, m_{h-2}^{'}$ and $m_{h-1}^{'}$ respectively.

Layout: $L(G^{'}, f_h^{'}) = \sum_{i=1}^{m_h^{'}-1} EC_{\prec f,p\succ}(S_i^{'}) = \sum_{i=1}^{m_h^{'}-1} \Theta_{G^{'}}(|V_i^{'}|) = \sum_{i=1}^{F_h} \Theta_{G^{'}}(m_1^{'}) + \sum_{i=1}^{F_{h-1}} \Theta_{G^{'}}(m_2^{'}) + \ldots + \sum_{i=1}^{F_3} \Theta_{G^{'}}(m_{h-2}^{'}) + \sum_{i=1}^{F_2} \Theta_{G^{'}}(m_{h-1}^{'}) = F_{h-1}.\Theta_{G^{'}}(m_2^{'}) + F_{h-2}.\Theta_{G^{'}}(m_3^{'}) + \ldots + 2\Theta_{G^{'}}(m_{h-2}^{'}) + \Theta_{G^{'}}(m_{h-1}^{'}) + 2|S|$.

Theorem 2. *The minimum layout of* $G = G(2^n; \pm S)$ *into the wounded lobster* L_n *is given by* $L(G, L_n) = \frac{1}{3}\{2^{n-1}(12j(2^{n-4}+1) + 2^{n-3}(3-2^n) - 7)\}$.

Proof. **Guest and Host Labeling:** Label $G(2^n; \pm S)$ in the clockwise direction as described in Table 1. Label L_n using postorder tree traversal order from 0 to $2^n - 1$.

Proposed Embedding: Define an embedding $\prec f, p \succ$ from $G(2^n; \pm S)$ into L_n such that $f(x) = x$.

Layout Computation: Table 2 gives three sets of edge cuts covering $E(L_n)$ and the node set of the components obtained by the removal of these edge cuts as depicted in Fig. 4(c).

Let G_r, $G_r^{'}$ and $G_r^{''}$ be the inverse image of Y_r, $Y_r^{'}$ and $Y_r^{''}$ respectively under $\prec f, p \succ$. By Lemma 3, the node set of all the three inverse images are optimal in G with respect to the maximum induced subgraph problem. All three edge cuts S_r, $S_r^{'}$

Table 2. Edge cuts of L_n

Edge Cuts	Components	V(Component)
S_r $r = 1, 2, \ldots, 2^{n-1}$	$Y_r, \overline{Y}_r$	$V(Y_r) = \begin{cases} \{4(r-1)\} & if\ r\ is\ odd \\ \{2(r-2)+1\} & if\ r\ is\ even \end{cases}$
$S_r^{'}$ $r = 1, 2, \ldots, 2^{n-2}$	$Y_r^{'}, \overline{Y}_r^{'}$	$V(Y_r^{'}) = \{4(r-1)+1, 4(r-1)+2\}$
$S_r^{''}$ $r = 1, 2, \ldots, 2^{n-2} - 1$	$Y_r^{''}, \overline{Y}_r^{''}$	$V(Y_r^{''}) = \{4(r-1)+0, \ldots, 4(r-1)+3\}$

and $S_r^{''}$ also satisfy the remaining two conditions of Lemma 1. In addition, $\{S_r, r = 1, 2, \ldots, 2^{n-1}\} \cup \{S_r^{'}, r = 1, 2, \ldots, 2^{n-2}\} \cup \{S_r^{''}, r = 1, 2, \ldots, 2^{n-2} - 1\}$ forms a partition of $E(L_n)$. Hence by Lemma 2, the layout induced by the embedding $\prec f, p \succ$ is minimum.

From Lemmas 2 and 4, $L(G, L_n) = \sum_{r=1}^{2^{n-1}} EC_{\prec f,p\succ}(S_r) + \sum_{r=1}^{2^{n-2}} EC_{\prec f,p\succ}(S_r^{'}) + \sum_{r=1}^{2^{n-2}-1} EC_{\prec f,p\succ}(S_r^{''}) = \sum_{r=1}^{2^{n-1}} \Theta_G(|V(Y_r)|) + \sum_{r=1}^{2^{n-2}} \Theta_G(|V(Y_r^{'})|) + \left\{ \sum_{r=1}^{2^{n-3}} \Theta_G (|V(Y_r^{''})|) + \sum_{r=2^{n-3}+1}^{2^{n-2}-1} \Theta_G(|V(Y_r^{''})|) \right\} = \frac{2^{n-1}}{3} \{ 12j(2^{n-4}+1) + 2^{n-3}(3 - 2^n) - 7 \}$.

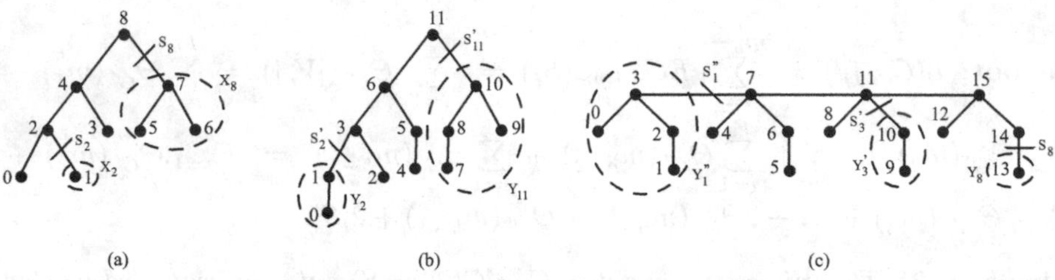

Fig. 4. Edge cuts of (a) f_5 (b) $f_5^{'}$ (c) L_4

3 Conclusion

In this paper we have embedded and found the minimum layout of the circulant graph into certain classes of height balanced trees like Fibonacci trees and wounded lobster by using edge partitioning techniques and isoperimetric methods.

Acknowledgement. This work was supported by Project No. 5LCTOI14MAT002, Loyola College - Times of India, Chennai, India.

References

1. Bezrukov, S., Monien, B., Unger, W., Wechsung, G.: Embedding ladders and caterpillars into the hypercube. Discrete Appl. Math. **83**(1–3), 21–29 (1998)
2. Choudham, S.A., Raman, I.: Embedding height balanced trees and Fibonacci trees in hypercubes. J. Appl. Math. Comput. **30**, 39–52 (2009)
3. Harper, L.H.: Global Methods for Combinatorial Isoperimetric Problems. Cambridge University Press, London (2004)
4. Horibe, Y.: Notes on Fibonacci trees and their optimality. Fibonacci Quart. **21**(2), 118–128 (1983)

5. Lai, Y.L., Williams, K.: A survey of solved problems and applications on bandwidth, edgesum and profile of graphs. J. Graph Theor. **31**, 75–94 (1999)
6. Manuel, P., Rajasingh, I., Rajan, B., Mercy, H.: Exact wirelength of hypercubes on a grid. Discrete Appl. Math. **157**(7), 1486–1495 (2009)
7. Opatrny, J., Sotteau, D.: Embeddings of complete binary trees into grids and extended grids with total vertex-congestion 1. Discrete Appl. Math. **98**, 237–254 (2000)
8. Rajasingh, I., Manuel, P., Arockiaraj, M., Rajan, B.: Embeddings of circulant networks. J. Comb. Optim. **26**(1), 135–151 (2013)
9. Rajasingh, I., Rajan, B., Mercy, H., Manuel, P.: Exact wirelength of hypercube and enhanced hypercube layout on wounded lobsters. In: 4th International Multiconference on Computer Science and Information Technology, CSIT2006, pp. 449–454, Amman, Jordan (2006)
10. Xu, J.-M.: Topological Structure and Analysis of Interconnection Networks. Kluwer Academic Publishers, Amsterdam (2001)

Dissecting Power of Certain Matrix Languages

J. Julie[1(✉)], J. Baskar Babujee[2], and V. Masilamani[3]

[1] Department of Mathematics, Easwari Engineering College, Chennai 600 089, India
juliemathematics@gmail.com

[2] Department of Mathematics, Anna University, MIT Campus, Chennai 600 044, India
baskarbabujee@yahoo.com

[3] Department of Computer Science and Engineering, Indian Institute of Information Technology, Design and Manufacturing, Kancheepuram, Chennai 600 127, India
masila@iiitdm.ac.in

Abstract. In formal language theory, the Siromoney matrix grammars generate matrix languages. They are two dimensional languages which are $m \times n$ arrays of terminals. In string languages, the ability of a regular language to dissect an infinite language into two partitions of infinite size has already been studied under the dissecting power of regular languages. In this paper we extend this special dissecting capacity of certain classes of string languages to matrix languages. The results demonstrate the matrix dissectibility of certain classes of matrix languages like infinite recursive matrix languages, constantly growing matrix languages (CGML), languages that are not CGML immune and CF:CF Siromoney matrix languages. In this paper the objectives of the study, extension methodology and results are discussed in detail.

Keywords: Theory of computing · Formal languages · Regular languages · Matrix languages · Regular matrix languages · Semi-linear · Constantly growing · Dissectible

1 Introduction

In formal language theory, the study of the structural properties of different families of languages has been a vital tool in understanding and solving many problems. In this line of study, another structural property which deals with the ability of an infinite language to partition an infinite set into two infinite sets called dissectibility was studied by Tomoyuki Yamakami and Yuichi Kato [9]. This study motivated us to extend the concept of dissectibility to the language of arrays. If the dissecting set was regular it was called REG-dissectibility and in general if it was an infinite language C it was called C-dissectibility.

Till now, the infinite language which dissects and the one which is being dissected have been string languages.

S. Arumugam et al. (Eds.): ICTCSDM 2016, LNCS 10398, pp. 98–105, 2017.
DOI: 10.1007/978-3-319-64419-6_13

Now, we consider matrix languages that have been introduced by Siromoney et al. [6]. These matrix languages are languages whose sentences are matrices and are two dimensional languages ($m \times n$ rectangular arrays of terminals). More about these languages can be seen in [2,4,7]. This version of matrix language is also called as Siromoney matrix language [8]. We extend the concept of dissectibility to these kind of matrix languages.

2 Preliminaries

Let Σ be a finite non-empty set of symbols. A string over Σ is a finite sequence of symbols in Σ. The set of all strings over Σ is Σ^*. The empty string is λ. The set $\Sigma^+ = \Sigma^* - \{\lambda\}$. For basic definitions and notations we follow [1,5].

If A and B are two sets, their difference $A - B = \{x \mid x \in A, x \notin B\}$. $|A| = \infty$ means A is infinite and $|A| < \infty$ means A is finite. If A and B are two countable sets $A \subseteq_{ae} B$ means $|A - B| < \infty$. ae means almost everywhere. $A =_{ae} B$ whenever both $A \subseteq_{ae} B$ and $B \subseteq_{ae} A$ hold. $\overline{A}$ denotes the complement of A.

Definition 1. [9] *An infinite language L is regular dissectible (REG-dissectible) if there exists a regular language C such that $|L \cap C| = \infty$ and $|L \cap \overline{C}| = \infty$.*

Definition 2. [9] *A non-empty language family $\mathcal{F}$ is REG-dissectible if and only if every infinite language in $\mathcal{F}$ is REG-dissectible.*

In notations, REG-DISSECT denotes the collection of all infinite REG-dissectible languages.

Instead of C being regular if it is an arbitrary non-empty language family then it yields the general case which is termed as C-dissectibilty.

We require the concepts of linear and semilinear as defined by Parikh [3].

Definition 3. *Let $\mathbb{N}$ denote the non-negative integers and let $\mathbb{N}^n$ be the cartesian product of $\mathbb{N}$ with itself n times. For elements $x = (x_1, x_2, \ldots, x_n)$ and $y = (y_1, y_2, \ldots, y_n)$ in $\mathbb{N}^n$, let $x + y = (x_1 + y_1, x_2 + y_2, \ldots, x_n + y_n)$ and $c(x_1, x_2, \ldots, x_n) = (cx_1, cx_2, \ldots, cx_n)$, c in $\mathbb{N}$. A subset A of $\mathbb{N}^n$ is said to be linear if there exist elements $v, v_1, v_2, \ldots, v_m$ in $\mathbb{N}^n$ such that $A = \{x \mid x = v + k_1v_1 + \cdots + k_mv_m$, each k_i in $\mathbb{N}\}$. A subset of $\mathbb{N}^n$ is said to be semi-linear if it is a finite union of linear sets.*

Definition 4. *The Parikh mapping is a monoid morphism $\psi : \Sigma^* \to \mathbb{N}^k$ where N denotes non-negative integers. For any string $x \in \Sigma^*$, $\Sigma = \{a_1, a_2, \ldots, a_k\}$ the parikh image of x denoted by $\psi(x) = (\#_{a_1}(x), \#_{a_2}(x), \ldots, \#_{a_k}(x))$ where $\#_{a_k}(x)$ denotes the number of occurrences of a_k in x.*

$\psi(L)$ of a language L over Σ is $\{\psi(x) \mid x \in L\}$. A language L is semilinear whenever $\psi(L)$ is semilinear.

A classic property of context free languages is the semilinearity of their length sets.

Definition 5. *A phrase-structure matrix grammar (abbreviated PSMG), (context-sensitive matrix grammar (CSMG), context-free matrix grammar (CFMG), right-linear matrix grammar (RLMG)) is a pair* $G = (G_1, G_2)$ *where* $G_1 = (V_1, I_1, P_1, S)$ *is a phrase-structure grammar (PSG), (context sensitive grammar (CSG), context free grammar (CFG), right linear grammar (RLG)) with*

V_1 = *a finite set of horizontal non-terminals*
I_1 = *a finite set of intermediates* $= \{S_1, \ldots, S_k\}$
P_1 = *a finite set of PSG (CSG, CFG, RLG) production rules called horizontal production rules and*
S = *the start symbol,* $S \in V_1$, $V_1 \cap I_1 = \phi$. $G_2 = \bigcup_{i=1}^{k} G_{2i}$ *where* $G_{2i} = (V_{2i}, I_2, P_{2i}, S_i)$, $i = 1, 2, \ldots, k$ *are k-right linear grammars with*
I_2 = *a finite set of terminals*
V_{2i} = *finite set of vertical non-terminals*
S_i = *the start symbol and* P_{2i} *finite set of right linear production rules,* $V_{2i} \cap V_{2j} = \phi$ *if* $i \neq j$.

The derivations are obtained by first applying the horizontal productions and then the vertical productions.

If both G_1 *and* G_2 *are context free grammars then the matrix languages they generate are called CF:CF Siromoney matrix languages* [8].

Definition 6. *The set of all matrices generated by G is defined to be* $M(G) = \{m \times n \text{ arrays } [a_{ij}] \mid i = 1, 2, \ldots, m, j = 1, 2, \ldots, n,\ m, n \geq 1,\ S \overset{*}{\underset{G_1}{\Rightarrow}} S_1 \ldots S_n \Downarrow^{*}_{G_2} [a_{ij}]\}$ *and* $M(G)$ *is called a phrase-structure matrix language (PSML), (context-sensitive matrix language (CSML), context free matrix language (CFML), regular matrix language (RML)) if G is a PSMG (CSMG, CFMG, RLMG).*

Definition 7. *A set* $S \subseteq \mathbb{N} \times \mathbb{N}$ *is called double-semilinear* [8] *if and only if it is a finite union of Cartesian products of the form* $S_1 \times S_2$ *where* $S_1, S_2 \subseteq \mathbb{N}$ *are semilinear.*

The length set of a CF:CF Siromoney matrix language is double semilinear [8].

In notations, if L is the language generated by G_1 and $R_1, R_2, \ldots, R_k$ (the subsets of) the regular sets corresponding to G_{2i}, $i = 1, 2, \ldots, k$ we write $M(G) = (L) :: (R_1, \ldots, R_k)$.

Definition 8. $\mathcal{P}$ *is the family of all polynomial time decidable languages. A language L is said to be in* $\mathcal{P}$ *iff there exists a deterministic Turing machine that runs in polynomial time on all inputs and outputs 1 if* $w \in L$ *and outputs 0 if* $w \notin L$. $\mathcal{P}$ *can also be viewed as a uniform family of boolean circuits. A language L is in* $\mathcal{P}$ *iff there exists an uniform family of Boolean circuits* $\{c_n \mid n \in \mathbb{N}\}$ *such that (i) for all* $n \in \mathbb{N}$, c_n *takes n bits as input and output only one bit (ii) for all* $x \in L$, $c(x) = 1$ *(iii) for all* $x \notin L$, $c(x) = 0$.

Definition 9. *A non-empty language L is said to be constantly growing if there exists a constant $p > 0$ and a finite subset $k \subseteq N^+$ that satisfies the following condition: for every string x in L with $|x| \geq p$, there exists a string $y \in L$ and a constant $c \in K$ for which $|x| = |y| + c$ holds. This property is called constant growth property and the languages which satisfy it are called constantly growing languages and are denoted by CGL.*

Definition 10. [9] *Given any family $\mathcal{F}$ of languages, a language S is said to be $\mathcal{F}$ immune if S is infinite and S has no infinite subset belonging to $\mathcal{F}$.*

Definition 11. *A language L is CGL immune if L is infinite and L has no infinite subset belonging to CGL.*

3 Dissecting Matrix Languages

We extend the concept of dissectibility to matrix languages as follows.

Two dimensional rectangular pictures or images are generalizations of words (A word is a row in a picture).

For an alphabet Σ, let $\Sigma^{m \times n}$ denote the set of pictures (matrices) of size (m, n) i.e., pictures with m rows and n columns over Σ. Let Σ^{**} denote the set of all rectangular pictures over Σ.

Each member of Σ^{**} is called as image or picture. Each member of $\Sigma^{m \times n}$ is called as picture of image of size $m \times n$.

$L \subseteq \Sigma^{**}$ is called a matrix language.

Definition 12. *Let Σ be a finite set of symbols. Let Σ^{**} be the set of all images over Σ. Let L be a matrix language over Σ. $L \subseteq \Sigma^{**}$. L is said to be REG MAT-dissectible if there exists a regular matrix language C, $C \subseteq \Sigma^{**}$ such that $|L \cap C| = \infty$ and $|L \cap \overline{C}| = \infty$ (Fig. 1).*

Definition 13. *In general, an infinite matrix language S is C-MAT-dissectible if there exists an infinite matrix language C such that $|S \cap C| = \infty$ and $|S \cap \overline{C}| = \infty$.*

Definition 14. *A non-empty matrix language family $\mathcal{F}$ is REG-MAT dissectible if and only if every infinite matrix language in $\mathcal{F}$ is REG-MAT dissectible.*

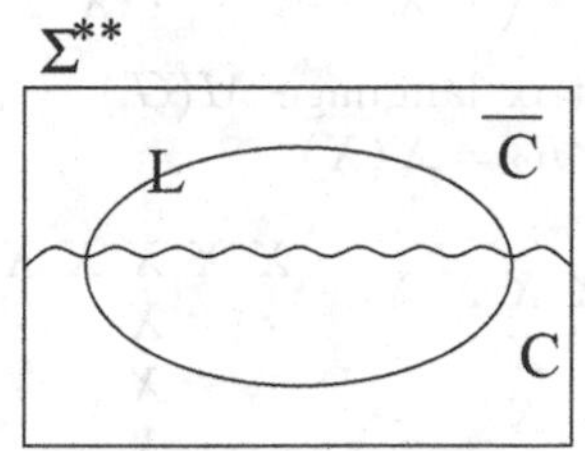

Fig. 1. C MAT-DISSECTS *L*

Theorem 1. *The family of regular matrix dissectible regular matrix languages is non-empty.*

Proof. Consider the regular matrix language $M(G) = (L) :: (S_1, S_2)$ where $M(G) = L = S_1 S_2^+$, $S_1 = (X)^+$, $S_2 = (\cdot)^+ X$ which generates the token L (every element of the language looks like L).

$$M(G) = L = \left(\begin{array}{cccccc} X & \cdot & \cdot & \cdot & \cdot & \cdot \\ X & \cdot & \cdot & \cdot & \cdot & \cdot \\ X & \cdot & \cdot & \cdot & \cdot & \cdot \\ X & \cdot & \cdot & \cdot & \cdot & \cdot \\ X & \cdot & \cdot & \cdot & \cdot & \cdot \\ X & \cdot & \cdot & \cdot & \cdot & \cdot \\ X & X & X & X & X & X \end{array}, \begin{array}{cccc} X & \cdot & \cdot & \cdot \\ X & \cdot & \cdot & \cdot \\ X & \cdot & \cdot & \cdot \\ X & X & X & X \end{array}, \ldots \right)$$

Consider the regular matrix language $M(G') = (C) :: (R_1, R_2)$ where $C = S_1(S_2^2)^+$, $R_1 = (X)^+$, $R_2 = (\cdot)^+ X$

$$C = \left(\begin{array}{ccc} X & \cdot & \cdot \\ X & \cdot & \cdot \\ X & \cdot & \cdot \\ X & X & X \end{array}, \begin{array}{ccccc} X & \cdot & \cdot & \cdot & \cdot \\ X & \cdot & \cdot & \cdot & \cdot \\ X & \cdot & \cdot & \cdot & \cdot \\ X & \cdot & \cdot & \cdot & \cdot \\ X & X & X & X & X \end{array}, \ldots \right).$$

Hence $|L \cap C| = \infty$ and $|L \cap \overline{C}| = \infty$ and L is a regular matrix language that is REG-MAT dissectible.

Theorem 2. *The class of regular dissectible context free matrix languages is non-empty.*

Proof. Consider the context free matrix language $M(G) = (L) :: (S_1, S_2)$ where $M(G) = L = \{S_1^n S_2 S_1^n \mid n \geq 1\}$, $S_1 = X(\cdot)^+$, $S_2 = X(X)^+$ which generates the token T (every element of the language looks like T).

$$M(G) = L = \left(\begin{array}{ccc} X & X & X \\ \cdot & X & \cdot \\ \cdot & X & \cdot \\ \cdot & X & \cdot \\ \cdot & X & \cdot \end{array}, \begin{array}{ccccc} X & X & X & X & X \\ \cdot & \cdot & X & \cdot & \cdot \\ \cdot & \cdot & X & \cdot & \cdot \\ \cdot & \cdot & X & \cdot & \cdot \\ \cdot & \cdot & X & \cdot & \cdot \end{array}, \ldots \right)$$

Consider the regular matrix language $M(G') = (C) :: (R_1, R_2)$ where $C = \{S_1 S_1 S_2 S_1 S_1\}$, $R_1 = X(\cdot)^+$, $R_2 = X(X)^+$,

$$C = \left(\begin{array}{ccccc} X & X & X & X & X \\ \cdot & \cdot & X & \cdot & \cdot \\ \cdot & \cdot & X & \cdot & \cdot \end{array}, \begin{array}{ccccc} X & X & X & X & X \\ \cdot & \cdot & X & \cdot & \cdot \\ \cdot & \cdot & X & \cdot & \cdot \\ \cdot & \cdot & X & \cdot & \cdot \end{array}, \ldots \right).$$

Here L is REG-MAT dissectible.

Definition 15. *A matrix language L is said to be a recursive matrix language if there is a Turing machine matrix automata TMMA that halts on all inputs. If $X \in L$, it accepts X and halts. If $X \notin L$, it rejects X and halts.*

Definition 16. *The family of all polynomial time decidable matrix languages is denoted by $\mathcal{P}$-MAT.*

Theorem 3. *Every infinite recursive matrix language is $\mathcal{P}$-MAT dissectible.*

Proof. Consider $\Sigma = \{0, 1\}$. Let L be an infinite recursive matrix language. By definition there exists a turing machine matrix automaton M that halts on all inputs. If $X \in L$, then M accepts X and halts. If $X \notin L$ then M rejects X and halts.

Case (i). $L =_{ae} \Sigma^{**}$

L is infinite. Consider

$$C = \{X \mid X = (x_{ij}) \text{ and } x_{ij} = \begin{cases} 0; & \text{if } i = j = 1 \\ 0 \text{ or } 1; & \text{otherwise} \end{cases}$$

$$= \left\{ (0), \begin{pmatrix} 0\,0 \\ 0\,0 \end{pmatrix}, \begin{pmatrix} 0\,0 \\ 0\,1 \end{pmatrix}, \begin{pmatrix} 0\,0 \\ 1\,0 \end{pmatrix}, \begin{pmatrix} 0\,0 \\ 1\,1 \end{pmatrix}, \right.$$

$$\left. \begin{pmatrix} 0\,1 \\ 0\,0 \end{pmatrix}, \begin{pmatrix} 0\,1 \\ 0\,1 \end{pmatrix}, \begin{pmatrix} 0\,1 \\ 1\,0 \end{pmatrix}, \begin{pmatrix} 0\,1 \\ 1\,1 \end{pmatrix}, \ldots \right\}$$

Clearly $|L \cap C| = \infty$ and $|L \cap \overline{C}| = \infty$.

This C easily dissects L.

Case (ii). $L \neq_{ae} \Sigma^{**}$

By definition, $|\Sigma^{**} - L|$ is infinite. Let $z_0, z_1, z_2, \ldots$ be the lexicographic ordering of all matrices over Σ^{**}. To define lexicographic ordering on matrices we consider matrix as concatenation of rows to form a sequence and then we use lexicographic ordering of words. For each $X_{m \times n}$, to determine the value of $C(X)$ we use the following procedure P from round 0 to mn. Initially we set $A = R = \phi$. At round i, we compute $C(z_i)$ by calling P recursively round by round. We then stimulate the turing machine matrix automaton M on the input z_i within mn steps. When M accepts z_1 we update A to $A \cup \{i\}$ if $C(z_i) = 1$ and R to $R \cup \{i\}$ if $C(z_i) = 0$. On the contrary, when either $M(z_i) = 0$ or $M(z_i)$ is not obtained within mn steps we do nothing. After round mn, if $|A| > |R|$ then define $C(X) = 0$; otherwise define $C(X) = 1$.

Now $C = \{X \in \Sigma^{**} | C(X) = 1\}$. Clearly C is in $\mathcal{P}$. $|L \cap C| = \infty$ and $|L \cap \overline{C}| = \infty$. Therefore every infinite recursive matrix language can be dissected by an appropriate matrix language in $\mathcal{P}$.

Definition 17. *A matrix language L is said to be constantly growing matrix language (CGML) if there exists constant $p > 0$ and $K \subseteq \mathbb{N}^+$ such that for every matrix $X_{m_1 \times n_1} \in L$ with $m_1 n_1 \geq p$ there exists a matrix $Y_{m_2 \times n_2} \in L$ and a constant $C \in K$ for which $n_1 = n_2 + C$ holds.*

Let CGML denote the family of all constantly growing matrix languages.

Theorem 4. *Every constantly growing matrix language is REG-MAT-dissectible.*

Proof. Let L be a constantly growing matrix language over Σ. $L \subseteq \Sigma^{**}$. By definition, there exists a constant $p > 0$ and $K \subseteq N^+$ such that for every matrix $X_{m_1 \times n_1} \in L$ there exists a matrix $Y_{m_2 \times n_2} \in L$ and $C_0 \in K$ for which $n_1 = n_2 + C_0$. By notation, let $[c] = \{1, 2, 3, \ldots, c\}$. Let c be the maximal element in K. Let $c' = c + 1$. Let $L_i = \{X_{m_1 \times n_1} \in L \mid n_1 \equiv i \pmod{c'}\}$. $L = \bigcup_{i \in [C]} L_i$.

We prove that there exists at least 2 indices i_1 and i_2 that make L_{i_1} and L_{i_2} infinite. The proof is by the method of contradiction. Assuming the contradiction let i be the only index that makes L_i infinite.

$$L_i = \{X_{m_1 \times n_1} \in L \mid n_1 \equiv i \pmod{c'}\}.$$

Let $S_{ij} = \{Y_{m_2 \times n_2} \in L \mid \text{there exists } X_{m_1 \times n_1} \in L \text{ such that } n_1 = n_2 + C_0\}$. Since L is a constantly growing language, there exists j such that S_{ij} is infinite for some j. $S_{ij} \subseteq L_\ell$ where $\ell = i - j \pmod{c'}$. This implies that L_ℓ if infinite. This is a contradiction to our assumption that i is the only index that makes L_ℓ infinite. Therefore there exists 2 indices i_1 and i_2 such that $|L_{i_1}| = \infty$ and $|L_{i_2}| = \infty$. We define C as follows

$C = \{X_{m_1 \times n_1} \in \Sigma^{**} \mid n_1 \equiv i_1 \pmod{c'}\}$ which is a regular matrix language.

Since $L_{i_1} \subseteq C$ and $L_{i_2} \subseteq \overline{C}$, $|L \cap C| = \infty$ and $|L \cap \overline{C}| = \infty$. Therefore C dissects L.

Definition 18. *A matrix language is said to be constantly growing matrix language immune (CGML immune) if L is infinite and L has no infinite subset belonging to CGML.*

Theorem 5. *Every language that is not CGML-immune is REG-MAT dissectible.*

Proof. Let L be a matrix language that is not CGML immune. By definition, L has an infinite subset S belonging to CGML. By the transitive closure property of REG-MAT dissectibility, for any two infinite matrix languages A and B if A is REG-MAT dissectible and $A \subseteq B$, then B is also *REG-MAT* dissectible. Here $S \subseteq L$ and S is in CGML. Therefore S is also REG-MAT dissectibile. This implies that L is REG-MAT dissectible.

Theorem 6. *Every CF:CF Siromoney matrix language is REG-MAT dissectible.*

Proof. Every CF:CF Siromoney matrix language is double semilinear. Hence every such matrix language can be defined by a set of linear equations which conveys the existence of constant growth property. By Theorem 4 every constantly growing matrix language is REG-MAT dissectible. Therefore every such CF:CF Siromoney matrix language is REG-MAT dissectible.

4 Conclusion

The two dimensional picture languages called Siromoney matrix languages are generalization of string languages in two dimensions. The structural property of dissecting an infinite language into two infinite sets has played a vital part in learning more about string languages and it has been extended to matrix languages in this paper.

In the paper, the $\mathcal{P}$-MAT dissectibility of infinite recursive matrix languages and REG-MAT dissectibility of constantly growing matrix languages (CGML), languages that are not CGML immune and CF:CF Siromoney matrix languages have been demonstrated. The work reported in this paper is highly significant because it introduces MAT-dissectibility and dissects certain classes of infinite matrix languages using their classical fundamental properties.

References

1. Hopcroft, J., Motwani, R., Ullman, J.: Introduction to Automata Theory. Languages and Computation, 3rd edn. Addison-Wesley, Boston (2006)
2. Krithivasan, K., Siromoney, R.: Characterizations of regular and context free matrices. Int. J. Comput. Math. **4**(1–4), 229–445 (1974)
3. Parikh, R.: On context free languages. J. ACM. B. **13**(4), 570–581 (1961)
4. Radhakrishnan, V., Chakravarthy, V., Krithivasan, K.: Pattern matching in matrix grammars. J. Autom. Lang. Comb. **3**(1), 59–76 (1998)
5. Salomaa, A.: Formal Languages. Academic Press, New York (1973)
6. Siromoney, G., Siromoney, R., Krithivasan, K.: Abstract family of matrices and picture language. Comput. Graph. Image Process. **1**, 284–307 (1972)
7. Siromoney, G., Siromoney, R., Krithivasan, K.: Picture languages with array rewriting rules. Inf. Control **22**, 447–470 (1973)
8. Steibe, R.: Slender siromoney matrix languages. Inf. Comput. **206**, 1248–1258 (2008)
9. Yamakami, T., Kato, Y.: The dissecting power of regular languages. Inf. Process. Lett. **113**, 116–122 (2013)

Degree Associated Reconstruction Number of Split Graphs with Regular Independent Set

N. Kalai Mathi and S. Monikandan(✉)

Department of Mathematics, Manonmaniam Sundaranar University, Tirunelveli 627 012, India
kalaimathijan20@gmail.com, monikandans@gmail.com

Abstract. A vertex-deleted subgraph of a graph G with which the degree of the deleted vertex is given is called a *degree associated card* of G. The *degree associated reconstruction number* (or *drn*) of a graph G is the size of the smallest collection of the degree associated cards of G that uniquely determines G. A *split graph* G is a graph in which the vertices can be partitioned into an independent set and a clique. We prove that the *drn* is 1 or 2 for all split graphs such that all the vertices in the independent set have equal degree, except four graphs on six vertices and for these exceptional graphs, the *drn* is 3.

Keywords: Isomorphism · Reconstruction number · Split graph

1 Introduction

All graphs considered are simple and finite. We shall mostly follow the graph theoretic terminology of [8]. A *vertex-deleted subgraph* or *card* $G - v$ of a graph (digraph) G is the unlabeled graph obtained from G by deleting the vertex v and all edges incident with v. The *deck* of a graph (digraph) G is the collection of all its cards. Following the formulation in [7], a graph (digraph) G is *reconstructible* if it can be uniquely determined from its deck. The well-known Reconstruction Conjecture (RC) of Kelly [11] and Ulam [20] has been open for more than 50 years. It asserts that every graph G with at least three vertices is reconstructible. The conjecture has been proved for many special classes, and many properties of G may be deduced from its deck. Nevertheless, the full conjecture remains open. Surveys of results on RC and related problems include [7,16]. For a reconstructible graph G, Harary and Plantholt [10] defined the reconstruction number of a graph G, denoted by $rn(G)$, to be the minimum number of cards which can only belong to the deck of G and not to the deck of any other graph H, $H \ncong G$, these cards thus uniquely identifying G. Reconstruction number is known for only few classes of graphs [5].

An extension of RC to digraphs is the *Digraph Reconstruction Conjecture* (DRC), proposed by Harary [9]. The DRC was disproved by Stockmeyer [19]

S. Monikandan—Research is supported by the SERB, Govt. of India, Grant no. EMR/2016/000157.

S. Arumugam et al. (Eds.): ICTCSDM 2016, LNCS 10398, pp. 106–112, 2017.
DOI: 10.1007/978-3-319-64419-6_14

by exhibiting several infinite families of counter-examples. Ramachandran then proposed a variation in the DRC and introduced the degree associated reconstruction [14] and the degree associated reconstruction number [15] of graphs (digraphs).

The ordered triple (a, b, c) where a, b and c are respectively the number of unpaired outarcs, unpaired inarcs and symmetric pair of arcs incident with v in a digraph D is called the *degree triple of* v. The *degree associated card* or *dacard* of a digraph (graph) is a pair (d, C) consisting of a card C and the degree triple (degree) d of the deleted vertex. The *degree associated deck* (or *dadeck*) of a graph (digraph) is the collection of all its dacards. A digraph is said to be *N-reconstructible* if it can be uniquely determined from its dadeck. The *new digraph reconstruction conjecture* (NDRC) asserts that all digraphs are N-reconstructible. The *degree (degree triple) associated reconstruction number* of a graph (digraph) G is the size of the smallest subcollection of the dadeck of G which is not contained in the dadeck of any other graph H, $H \ncong G$, this subcollection of dacards thus uniquely identifying G. Articles [1–4,6,12,13,18] are recent papers on this parameter.

A *split graph* G is a graph in which the vertices can be partitioned into an independent set (say X) and a clique (say Y). Throughout this paper, we use the notation G, X and Y in the sense of this definition. The independent set X is said to regular if all the vertices in it have equal degree in G. Ramachandran and Monikandan proved [17] that the validity of the RC for all graphs is equivalent to the validity of the RC for all 2-connected graphs G with $diam(G) = 2$ or $diam(G) = diam(\overline{G}) = 3$. As many split graphs belong to this class of 2-connected graphs, to determine any reconstruction parameter for split graphs assumes important. In this paper, we prove that $drn(G) = 1$ or 2 for all split graphs G with regular independent set except four graphs on six vertices (Fig. 1) and for these exceptional four graphs, the drn is 3.

2 Drn of Split Graphs

The next theorem, due to Barrus and West [6], characterizes all graphs G with $drn(G) = 1$.

Theorem 1. *The dacard (C, d) belongs to the dadeck of only one graph (up to isomorphism) if and only if one of the following holds:*

(1) *$d = 0$ or $d = |V(C)|$;*
(2) *$d = 1$ or $d = |V(C) - 1|$, and C is vertex-transitive;*
(3) *C is complete or edgeless.*

In a graph G of order ν, a vertex with degree d is called a *d-vertex* and a $(\nu - 1)$-vertex is called a *complete vertex*. By $m(d(v), G - v)$, we mean m dacards each isomorphic to $(d(v), G - v)$. The *bistar* $B_{m,n}$ is the tree with $m + n + 2$ vertices whose central vertices have m and n leaf neighbours respectively. An *s-blocking set* of a graph G is a family $\mathscr{F}$ of graphs not isomorphic to G such

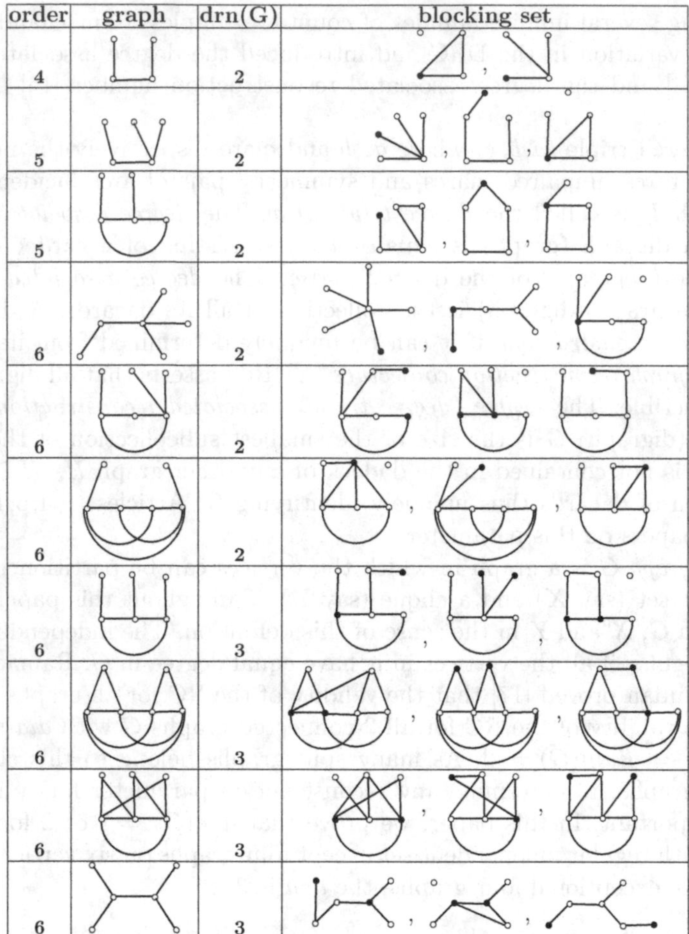

order	graph	drn(G)	blocking set
4		2	
5		2	
5		2	
6		2	
6		2	
6		2	
6		3	
6		3	
6		3	
6		3	

Fig. 1. Split graphs of order at most 6 with regular independent set and having *drn* 2 or 3.

that every collection of s dacards of G will appear in the dadeck of some graph of $\mathscr{F}$ and every graph in $\mathscr{F}$ will have s dacards in common with G.

Let $|X| = m > 0$, $|Y| = n > 0$ and let X be r-regular. Then clearly $0 \leq r \leq n$ and if r were 0 or n, then G would contain an isolated vertex or a complete vertex, which implies $drn(G) = 1$ by Theorem 1. Thus $1 \leq r \leq n - 1$. All split graphs G on at most six vertices with regular independent set, except the ten graphs given in the table in Fig. 1, must contain a complete vertex or an isolated vertex and so $drn(G) = 1$. The *drn* of these ten graphs is two or three (dark vertex of graphs given in Fig. 1 denotes the vertex whose removal results in a dacard common with G). So, we assume that *all split graphs G consider hereafter have order at least seven and, by Theorem* 1, *no isolated as well as complete vertices.*

Let Y_i denote the set of vertices in Y that are adjacent to exactly i vertices in X for $i = 0, 1, ..., m$. Then, in G, the degree of a vertex $v \in Y_i$ is $n - 1 + i$ for $i = 0, 1, ..., m$. Let $k_1, k_2, ..., k_t$ be integers, where $0 \leq k_1 < k_2 < ... < k_t \leq m$, such that $Y_{k_i} \neq \phi$ for all $i = 1, 2, ..., t$. Thus Y can be written as $\cup_{i=1}^{t} Y_{k_i}$.

An *extension* of a dacard $(d(v), G - v)$ of G is a graph obtained from the dacard by adding a new vertex w and joining it to $d(v)$ vertices of the dacard and it is denoted by $H(d(v), G - v)$ (or simply by H). Throughout this paper, H and w are used in the sense of this definition.

Theorem 2. *If G is a split graph with $r = n - 1$, then $drn(G) = 2$.*

Proof. We proceed on the value of k_1, which is the smallest integer such that Y_{k_1} is non empty.

If k_1 were equal to 0, then $|Y_0|$ would be equal to 1 (because Y_0 can contain at most only one vertex as $r = n-1$) and since $n \geq 2$ and $r = n - 1$, it follows that Y_m would be nonempty, so G would contain a complete vertex, which is excluded.

Case 1. $k_1 = 1$.
If $n > 2$, then the vertex in Y_1 is adjacent to exactly one vertex, say s in X. Also, since $r = n - 1$, every other vertex in X is adjacent to all the vertices in $Y \setminus Y_1$. Moreover, the vertex s is non-adjacent to exactly one vertex in Y. Thus Y can be written as $Y = Y_1 \cup Y_{m-1} \cup Y_m$, where $|Y_1| = |Y_{m-1}| = 1$, which implies that $Y_m \neq \phi$ as $n > 2$. Hence G has a complete vertex, which is excluded.

If $n = 2$, then assume $m > 4$ (as otherwise $\nu \leq 6$, which is excluded). Clearly the partite set Y can be written as $Y = Y_1 \cup Y_{m-1}$, where $|Y_1| = |Y_{m-1}| = 1$. The dadeck of G consists of only the dacards $(m, K_2 \cup (m-1)K_1)$, $(2, K_{1,m-1} \cup K_1)$, $(1, K_{1,m})$ and $(m-1)(1, B_{m-2,1})$. Now consider the two dacards $(m, K_2 \cup (m-1)K_1)$ and $(1, K_{1,m})$. To get an extension $H(m, K_2 \cup (m-1)K_1)$ non-isomorphic to G, add a new vertex and join it to the two vertices of positive degree. But then every one-vertex deleted dacard of H must contain a cycle and so it is non-isomorphic to $(1, K_{1,m})$. Thus no graph ($\not\cong G$) has both the dacards $(m, K_2 \cup (m-1)K_1)$ and $(1, K_{1,m})$ in its dadeck and hence $drn(G) \leq 2$.

Case 2. $k_1 = 2$.
Clearly a vertex in Y_2 is adjacent to exactly two vertices, say s, t in X. Also, since $r = n - 1$, every vertex in X, other than s and t, is adjacent to all the vertices in $Y \setminus Y_2$ and so every vertex in $Y \setminus Y_2$ gets at least $m - 2$ neighbours in X. Since $r = n-1$, the vertex s (respectively t) is nonadjacent to exactly one vertex, say s' (respectively t') in $Y \setminus Y_2$.

If $s' \neq t'$ (this happens when $n \geq 3$), then every vertex in $Y \setminus Y_2$ gets at least $m-1$ neighbours in X and hence $Y = Y_2 \cup Y_{m-1} \cup Y_m$, where $|Y_2| = 1$ and $|Y_{m-1}| = 2$. We can take that $n = 3$ (as otherwise n would be at least four and G would contain a complete vertex). Since G has order at least seven, we have $m \geq 4$. Now consider the dacards $(m + 1, G - v)$ and $(2, G - u)$, where $v \in Y_{m-1}, u \in X$, and $uv \in E(G)$. The dacard $G - u$ contains exactly $m - 1$ vertices of degree two. To get an extension $H(m+1, G-v)$, join the newly added vertex w to all but one vertex, say z in $G-v$. If z were the unique 2-vertex, then H would be isomorphic to G. If z is not the unique 2-vertex, then every 2-vertex deleted dacard of the resulting extension H must contain at most $m-2$ vertices of degree two and so it is not isomorphic to $G-u$. Hence $drn(G) \leq 2$.

Now assume $s^{'} = t^{'}$ (this happens when $n \geq 2$). Then $Y = Y_2 \cup Y_{m-2} \cup Y_m$ where $|Y_2| = 1$ and $|Y_{m-2}| = 1$. We can take that $n = 2$ (as otherwise n would be at least three and G would contain a complete vertex). Since G has order at least seven, we have $m \geq 5$. Hence, in this case, the graph G is isomorphic to the bistar $B_{2,m-2}$ whose drn is proved (Barrus and West [6]) to be 2.

Case 3. $k_1 > 2$.
Consider the dacards $(n-1+k_t, G-v)$ and $(n-1, G-u)$, where $v \in Y_{k_t}$, $u \in X$ and $uv \notin E(G)$. The dacard $G-u$ contains no n-vertices. To get an extension $H(n-1+k_t, G-v)$, add a new vertex w to $G-v$ and join it to some set of vertices (say $Y^{'}$) in $Y \setminus \{v\}$ and some set of vertices (say $X^{'}$) in X.

Suppose $|X^{'}| = k_t$ and $Y^{'} = Y \setminus \{v\}$. If $X^{'}$ consists of only $(n-2)$-vertices, then $H \cong G$. If every vertex in $X^{'}$ has degree $n-1$, then $m - k_t = k_t$ and the resulting extension has no $(n-1)$-vertex and so it has no dacard isomorphic to $(n-1, G-u)$. Therefore we assume that $X^{'}$ contains vertices of degree $n-1$, $n-2$ and that it contains no vertices of other degree. But then any $(n-1)$-vertex deleted dacard of the resulting extension must contain an n-vertex and so it is not isomorphic to $G-u$.

Suppose $X^{'} = X$ and $|Y^{'}| = n-1+k_t-m$. Then any $(n-1)$-vertex deleted dacard of the extension H must contain an n-vertex and so it is not isomorphic to $G-u$.

We now assume the only remaining case that $\phi \neq X^{'} \subsetneq X$ and $\phi \neq Y^{'} \subsetneq Y \setminus \{v\}$. Then $|Y^{'}| < n-2$, which implies $|X^{'}| > k_t$ because the associated degree of $G-v$ is $n-1+k_t$. Since $G-v$ has exactly k_t vertices of degree $n-2$ (in X of it), it follows that $X^{'}$ must contain at least one $(n-1)$-vertex. Now this vertex must occur as an n-vertex in any $(n-1)$-vertex deleted dacard of the resulting extension H and so such a dacard is not isomorphic to $G-u$. Hence $drn(G) \leq 2$ and by Theorem 1, $drn(G) = 2$.

Theorem 3. *If G is a split graph with $r \leq n-2$, then $drn(G) = 2$.*

Proof. We proceed by two cases depending upon the value of r as below.

Case 1. $r \leq n-3$.
Now $n \geq 4$ and $k_t \leq m-1$. Consider the dacards $(n-1+k_t, G-v)$ and $(r, G-u)$, where $v \in Y_{k_t}$ and $u \in X$. Clearly the dacard $G-u$ contains no vertices of degree $r-1$ or $r+1$. To get an extension $H(G-v)$, add a new vertex w to $G-v$ and join it to some set of vertices (say $Y^{'}$) in $Y \setminus \{v\}$ and some set of vertices (say $X^{'}$) in X.

Suppose $Y^{'} = Y \setminus \{v\}$ and $|X^{'}| = k_t$. If every vertex in $X^{'}$ has degree $r-1$, then $H \cong G$. If every vertex in $X^{'}$ has degree r, then $m - k_t = k_t$ and the resulting extension has no r-vertex, so it has no dacard isomorphic to $(r, G-u)$. We therefore assume that $X^{'}$ contains vertices of degree $r-1, r$ and that it contains no vertices of other degree. But then any r-vertex deleted dacard of the resulting extension must contain an $(r+1)$-vertex and hence such a dacard is not isomorphic to $G-u$.

Suppose $|Y'| = n-1+k_t-m$ and $X' = X$. Then every r-vertex deleted dacard must contain an $(r+1)$-vertex (because $k_t \leq m-1$) and hence it is not isomorphic to $G-u$.

Now we consider the remaining case that $\phi \neq X' \subsetneq X$ and $\phi \neq Y' \subsetneq Y \setminus \{v\}$. Then $|Y'| < n-2$, which implies $|X'| > k_t$ because the associated degree of $G-v$ is $n-1+k_t$. Since $G-v$ has exactly k_t vertices of degree $r-1$ (in X of it), it follows that X' must contain at least one r-vertex. But then this vertex will occur as an $(r+1)$-vertex in every r-vertex deleted dacard of the resulting extension H and so such a dacard is not isomorphic to $G-u$. Hence $drn(G) \leq 2$.

Case 2. $r = n-2$.
Now $n \geq 3$ and $k_t \leq m-1$. If $|Y_0|$ were at least two, then either r would be at most $n-3$ or G would have a complete vertex, giving a contradiction. Therefore $|Y_0| = 0$ or 1. Also if $|Y_0| = 0$, then, since $r = n-2$, it follows that $|Y_1| \leq 4$. If $|Y_1|$ were 3 or 4, then the order of G would be at most six. Thus, either $|Y_0| = 0$ and $|Y_1| \leq 2$, or else $|Y_0| = 1$.

Now proceeding as in Case 1 but with the two dacards $(n-1+k_t, G-v)$ and $(n-2, G-u)$, where $v \in Y_{k_t}$, $u \in X$ and u is nonadjacent to a vertex in Y_1, we will have $drn(G) \leq 2$ and by Theorem 1, $drn(G) = 2$.

3 Conclusion

For graphs with at least three vertices, knowing the degree of the deleted vertex is equivalent to knowing the total number of edges. A simple counting argument computes the size of the graph when its entire deck is known. So the dadeck gives the same information as the deck. However, the counting argument requires the entire deck, so an individual dacard gives more information than the corresponding card.

In the above sections, we have proved that the drn is at most 3 for a split graph G with regular independent set. There is a hope to complete a proof of $drn(G) \leq 3$ for all split graphs G. With reference to our results, it seems that the drn of bipartite graphs, with a regular independent partite set, is likely to be at most 3. However, extending this result to the family of all bipartite graphs needs intensive work as because reconstructibility of the family of all bipartite graphs remains open.

Acknowledgement. The work reported here is supported by the Research Project EMR/2016/000157 awarded to the second author by SERB, Government of India, New Delhi.

References

1. Anu, A., Monikandan, S.: Degree associated reconstruction number of certain connected graphs with unique end vertex and a vertex of degree $n-2$. Discrete Math. Algorithms Appl. **8**(4), 1650068 (2016). 13 p.
2. Anusha Devi, P., Monikandan, S.: Degree associated reconstruction number of graphs with regular pruned graph. Ars Combin. (to appear)
3. Anusha Devi, P., Monikandan, S.: Degree associated reconstruction parameters of total graphs. Contrib. Discrete Math. (to appear)
4. Anusha Devi, P., Monikandan, S.: Degree associated reconstruction number of connected digraphs with unique end vertex. Australas. J. Combin. **66**(3), 365–377 (2016)
5. Asciak, K.J., Francalanza, M.A., Lauri, J., Myrvold, W.: A survey of some open questions in reconstruction numbers. Ars Combin. **97**, 443–456 (2010)
6. Barrus, M.D., West, D.B.: Degree-associated reconstruction number of graphs. Discrete Math. **310**, 2600–2612 (2010)
7. Bondy, J.A.: A graph reconstructors manual, in surveys in combinatorics. In: Proceedings of the 13th British Combinatorial Conference. London Mathematical Society, Lecture Note Series, vol. 166, pp. 221–252 (1991)
8. Harary, F.: Graph Theory. Addison Wesley, Reading (1969)
9. Harary, F.: On the reconstruction of a graph from a collection of subgraphs. In: Fieldler, M. (ed.) Theory of Graphs and Its Applications, pp. 47–52. Academic Press, New York (1964)
10. Harary, F., Plantholt, M.: The graph reconstruction number. J. Graph Theory **9**, 451–454 (1985)
11. Kelly, P.J.: On isometric transformations. Ph.D. Thesis, University of Wisconsin Madison (1942)
12. Ma, M., Shi, H., Spinoza, H., West, D.B.: Degree-associated reconstruction parameters of complete multipartite graphs and their complements. Taiwanese J. Math. **19**(4), 1271–1284 (2015)
13. Monikandan, S., Sundar Raj, S.: Adversary degree associated reconstruction number. Discrete Math. Algorithms Appl. **7**(1), 1450069 (2015). 16 p.
14. Ramachandran, S.: On a new digraph reconstruction conjecture. J. Combin. Theory Ser. B. **31**, 143–149 (1981)
15. Ramachandran, S.: Degree associated reconstruction number of graphs and digraphs. Mano. Int. J. Math. Scis. **1**, 41–53 (2000)
16. Ramachandran, S.: Graph reconstruction-some new developments. AKCE J. Graphs Combin. **1**(1), 51–61 (2004)
17. Ramachandran, S., Monikandan, S.: Graph reconstruction conjecture: reductions using complement, connectivity and distance. Bull. Inst. Comb. Appl. **56**, 103–108 (2009)
18. Spinoza, H.: Degree-associated reconstruction parameters of some families of trees (2014, submitted)
19. Stockmeyer, P.K.: The falsity of the reconstruction conjecture for tournaments. J. Graph Theory **1**, 19–25 (1977)
20. Ulam, S.M.: A Collection of Mathematical Problems. Interscience Tracts in Pure and Applied Mathematics, vol. 8. Interscience Publishers, New York (1960)

Distance Antimagic Labelings of Graphs

N. Kamatchi[1(✉)], G.R. Vijayakumar[2], A. Ramalakshmi[1], S. Nilavarasi[1], and S. Arumugam[3]

[1] Department of Mathematics, Kamaraj College of Engineering and Technology, Virudhunagar 626001, Tamil Nadu, India
kamakrishna77@gmail.com

[2] Tata Institute of Fundamental Research, Mumbai, India
vijay@math.tifr.res.in

[3] Kalasalingam University, Anand Nagar, Krishnankoil 626126, Tamil Nadu, India
s.arumugam.klu@gmail.com

Abstract. Let $G = (V, E)$ be a graph of order n. Let $f : V(G) \rightarrow \{1, 2, \ldots, n\}$ be a bijection. For any vertex $v \in V$, the neighbor sum $\sum_{u \in N(v)} f(u)$ is called the weight of the vertex v and is denoted by $w(v)$. If $w(x) \neq w(y)$ for any two distinct vertices x and y, then f is called a distance antimagic labeling. A graph which admits a distance antimagic labeling is called a distance antimagic graph. If the weights form an arithmetic progression with first term a and common difference d, then the graph is called an (a, d)-distance antimagic graph.

In this paper we prove that the hypercube Q_n is an (a, d)-distance antimagic graph. Also, we present several families of disconnected distance antimagic graphs.

Keywords: (a, d)-distance antimagic graph · Distance antimagic graph

1 Introduction

By a graph $G = (V, E)$ we mean a finite, undirected graph with neither loops nor multiple edges. We further assume that G has no isolated vertices. The order $|V|$ and the size $|E|$ are denoted by n and m respectively. For graph theoretic terminology we refer to Chartrand and Lesniak [2].

A *distance magic labeling* of a graph G of order n is a bijection $f : V \rightarrow \{1, 2, \ldots, n\}$ with the property that there is a positive integer k such that $\sum_{y \in N(x)} f(y) = k$ for every $x \in V$. The constant k is called the *magic constant* of the labeling f.

The sum $\sum_{y \in N(x)} f(y)$ is called the *weight* of the vertex x and is denoted by $w(x)$.

Let G be a distance magic graph of order n with labeling f and magic constant k. Then $\sum_{u \in N_{G^c}(v)} f(u) = \frac{n(n+1)}{2} - k - f(v)$, and hence the set of all

S. Arumugam et al. (Eds.): ICTCSDM 2016, LNCS 10398, pp. 113–118, 2017.
DOI: 10.1007/978-3-319-64419-6_15

vertex weights in G^c is $\{\frac{n(n+1)}{2} - k - i : 1 \leq i \leq n\}$, which is an arithmetic progression with first term $a = \frac{n(n+1)}{2} - k - n$ and common difference $d = 1$.

Motivated by this observation, in [1] we introduced the following concept of (a, d)-distance antimagic graph.

Definition 1. [1] *A graph G is said to be (a, d)-distance antimagic if there exists a bijection $f : V \to \{1, 2, \ldots, n\}$ such that the set of all vertex weights is $\{a, a + d, a + 2d, ..., a + (n-1)d\}$ and any graph which admits such a labeling is called an (a, d)-distance antimagic graph.*

Thus the complement of every distance magic graph is an $(a, 1)$-distance antimagic graph.

We observe that if a graph G is (a, d)-distance antimagic with $d > 0$, then for any two distinct vertices u and v we have $w(u) \neq w(v)$. This observation naturally leads to the concept of distance antimagic labeling.

Definition 2. [3] *Let $G = (V, E)$ be a graph of order n. Let $f : V \to \{1, 2, \ldots, n\}$ be a bijection. If $w(x) \neq w(y)$ for any two distinct vertices x and y in V, then f is called a distance antimagic labeling. Any graph G which admits a distance antimagic labeling is called a distance antimagic graph.*

Definition 3. *The K_2-bistar graph $K_2(m, n)$ is the graph obtained by joining m copies of K_2 to a vertex of K_2 and n copies of K_2 to the other vertex of K_2.*

In this paper we prove that the hypercube Q_n is an (a, d)-distance antimagic graph. Also, we present several families of disconnected distance antimagic graphs.

2 Main Results

The following theorem gives an (a, d)-distance antimagic labeling of hypercubes.

Theorem 1. *For every $n \geq 3$, the hypercube Q_n is (a, d)-distance antimagic, where $a = 2^n + 2$ and $d = n - 2$. Moreover there exists an (a, d)-distance antimagic labeling $f_n : V(Q_n) \to \{1, 2, \ldots, 2^n\}$ such that if $f_n(v) = j$, then $w_{f_n}(v) = 2^n + 1 + (n-2)j, 1 \leq j \leq 2^n$.*

Proof. We prove this result by induction on n. For Q_3, the labeling f_3 given in Fig. 1 is a $(10, 1)$-distance antimagic labeling satisfying the condition that $w_{f_3}(j) = 9 + j = 2^n + 1 + j, 1 \leq j \leq 8$. We now assume that the theorem is true for Q_n. Let $f_n : V(Q_n) \to \{1, 2, 3, \ldots, 2^n\}$ be a $(2^n + 2, n - 2)$-distance antimagic labeling of Q_n such that if $f_n(v) = j$, then $w_{f_n}(v) = 2^n + 1 + j$ for all $j, 1 \leq j \leq 2^n$. Let $Q_n^{(1)}$ and $Q_n^{(2)}$ be two copies of Q_n in Q_{n+1}, with a perfect matching M consisting of edges joining a vertex of $Q_n^{(1)}$ with the corresponding vertex of $Q_n^{(2)}$. Now (See Fig. 2) define $f_{n+1} : V(Q_{n+1}) \to \{1, 2, \ldots, 2^{n+1}\}$ by

$$f_{n+1}(v) = \begin{cases} f_n(v) & \text{if } v \in V(Q_n^{(1)} \\ f_n(v_1) + 2^n & \text{if } v_1 \in V(Q_n^{(2)} \text{ and } vv_1 \in M \end{cases}$$

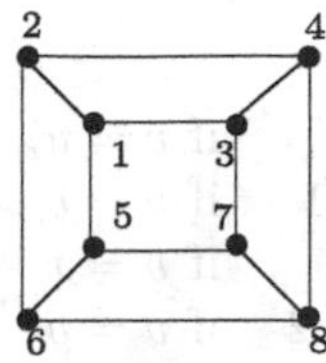

Fig. 1. Q_3 with $(10, 1)$-distance antimagic labeling

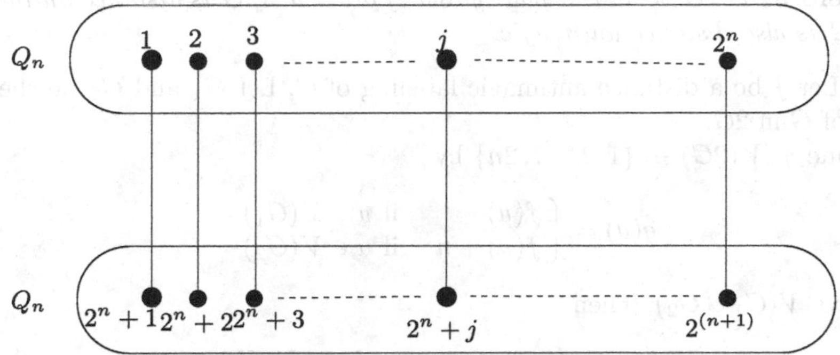

Fig. 2. Q_{n+1} with (a, d)-distance antimagic labeling

If $f_{n+1}(v) = j, 1 \leq j < 2^n$, then

$$\begin{aligned} w_{f_{n+1}}(v) &= w_{f_n}(v) + 2^n + j \\ &= 2^n + 1 + (n-2)j + 2^n + j \\ &= (2^{n+1} + 1) + (n-1)j. \end{aligned}$$

If $f_{n+1}(v_1) = j$, where $2^n + 1 \leq j \leq 2^{n+1}$ and $vv_1 \in M$, then

$$\begin{aligned} w_{f_{n+1}}(v_1) &= w_{f_{n+1}}(v) + n2^n + j \\ &= (1 + 2^n) + (n-2)j + n2^n + j \\ &= (1 + 2^{n+1}) + (n-1)(2^n + j). \end{aligned}$$

Thus, $w_f^{(n+1)}(j) = (1 + 2^{n+1}) + (n-1)j, j = 1, 2, 3, \ldots, 2^{n+1}$ and by induction the proof is complete.

Theorem 2. *The bistar $G = K_2(n, n)$ is distance antimagic.*

Proof. Let $V(G) = \{u_1, u_2, \ldots, u_n\} \cup \{v_1, v_2, \ldots, v_n\} \cup \{u, v\}$ and $E(G) = \{u_i u : 1 \leq i \leq n\} \cup \{v_i v : 1 \leq i \leq n\} \cup \{uv\}$. Define $f : V(G) \to \{1, 2, \ldots, 2n+2\}$, by

$$f(x) = \begin{cases} 2i & \text{if } x = u_i,\ i \leq i \leq n \\ 2i+1 & \text{if } x = v_i,\ i \leq i \leq n \\ 1 & \text{if } x = v \\ 2n+2 & \text{if } x = u \end{cases}$$

Then

$$w(x) = \begin{cases} 2i & \text{if } v = u_i,\ i \le i \le n \\ 2i+1 & \text{if } v = v_i,\ i \le i \le n \\ 1 & \text{if } v = x \\ 2n+2 & \text{if } v = y \end{cases}$$

Hence f is a distance antimagic labeling of G.

Theorem 3. *Let G be an r-regular graph of order n. If G is distance antimagic, then $2G$ is also distance antimagic.*

Proof. Let f be a distance antimagic labeling of G. Let G_1 and G_2 be the two copies of G in $2G$.

Define $g : V(2G) \to \{1, 2, \ldots, 2n\}$ by

$$g(u) = \begin{cases} f(u) & \text{if } u \in V(G_1) \\ f(u) + n & \text{if } u \in V(G_2) \end{cases}$$

Let $u, v \in V(G_1 \cup G_2)$. Then

$$w_g(u) = \begin{cases} w_f(u) & \text{if } u \in V(G_1) \\ w_f(u) + rn & \text{if } u \in V(G_2) \end{cases}$$

Hence it follows that $w_g(u) \neq w_g(v)$ if $u, v \in V(G_1)$ or $u, v \in V(G_2)$.

Now, let $u \in V(G_1)$ and $v \in V(G_2)$. Since $w_f(u) \neq w_f(v)$, without loss of generality we assume that $w_f(u) < w_f(v)$. Now, $w_g(u) = w_f(u) < w_f(v) < w_f(v) + rn = w_g(v)$. Thus g is a distance antimagic labeling of $2G$.

Theorem 4. *Let H be the graph obtained from the cycle C_3 by attaching a pendent vertex at one vertex. Let G be the union of r copies of H. Then G is distance antimagic.*

Proof. Let H_i be the ith copy of H in G. Let $V(H_i) = \{u_{i1}, u_{i2}, u_{i3}, u_{i4}\}$ and $E(H_i) = \{(u_{i1}, u_{i2}), (u_{i1}, u_{i3}), (u_{i2}, u_{i3}), (u_{i2}, u_{i4})\}$.

Define $f : V(G) \to \{1, 2, \ldots, 4r\}$, by

$$f(u_{ij}) = \begin{cases} 4(i-1)+1, & \text{if } j = 1 \\ 4(i-1)+2, & \text{if } j = 2 \\ 4(i-1)+3, & \text{if } j = 3 \\ 4(i-1)+4, & \text{if } j = 4 \end{cases}$$

where $1 \le i \le r$.

The vertex weights are given by

$$w(u_{ij}) = \begin{cases} 8i-3, & \text{if } j = 1 \\ 12i-4, & \text{if } j = 2 \\ 8i-5, & \text{if } j = 3 \\ 4i-2, & \text{if } j = 4 \end{cases}$$

Clearly the vertex weights are distinct.

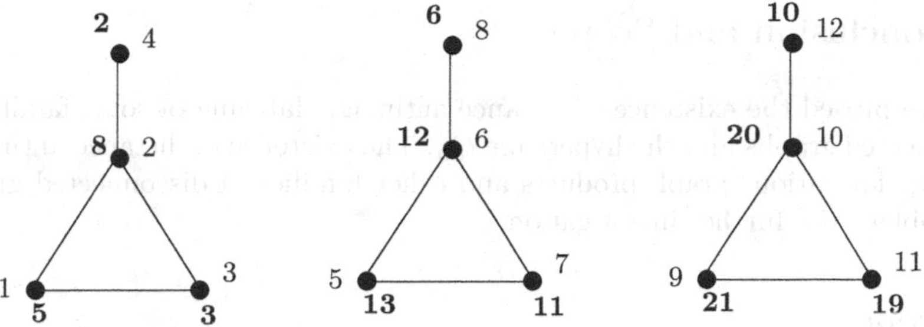

Fig. 3. Distance antimagic labeling of union of 3-pan

Example 1. The distance antimagic labeling of 3 copies of H is given in Fig. 3.

Theorem 5. *For $n = 2k+1$, let H_k be the graph obtained from the path $(u_1, u_2, \ldots, u_{2k+1})$ by adding the edges (u_i, u_{i+2}) where i is odd. Let G be the union of r copies of H_3 where $n \geq 1$. Then G is distance antimagic.*

Proof. Let G_i be the ith copy of H_3 in G, given in Fig. 4.

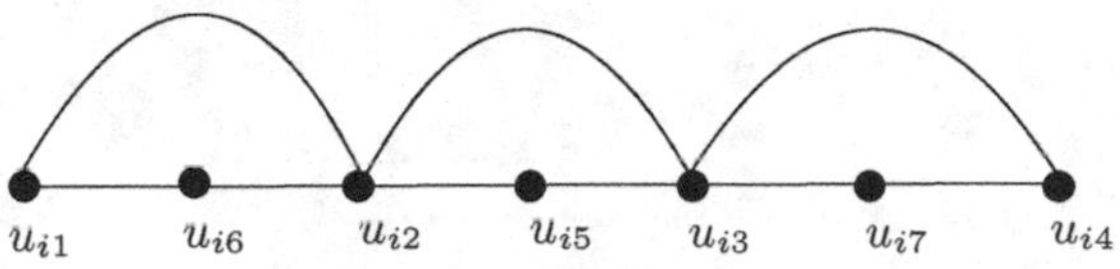

Fig. 4. The graph H_3

Define $f : V(G) \to \{1, 2, \ldots, 4r\}$ by

$$f(u_{ij}) = \begin{cases} 7(i-1)+1 & \text{if } j = 1 \\ 7(i-1)+2 & \text{if } j = 2 \\ 7(i-1)+3 & \text{if } j = 3 \\ 7(i-1)+4 & \text{if } j = 4 \\ 7(i-1)+5 & \text{if } j = 5 \\ 7(i-1)+6 & \text{if } j = 6 \\ 7(i-1)+7 & \text{if } j = 7 \end{cases}$$

The vertex weights are given by

$$w(u_{ij}) = \begin{cases} 14i-6 & \text{if } j = 1 \\ 28i-13 & \text{if } j = 2 \\ 28i-10 & \text{if } j = 3 \\ 14i-4 & \text{if } j = 4 \\ 14i-9 & \text{if } j = 5 \\ 14i-11 & \text{if } j = 6 \\ 14i-7 & \text{if } j = 7 \end{cases}$$

Clearly the vertex weights are distinct.

3 Conclusion and Scope

We have proved the existence of distance antimagic labeling of some families of disconnected graphs and the hypercube Q_n. The existence of distance antimagic labelings for various graph products and other families of disconnected graphs are problems for further investigation.

References

1. Arumugam, S., Kamatchi, N.: On (a, d)-distance antimagic graphs. Australas. J. Combin. **54**, 279–287 (2012)
2. Chartrand, G., Lesniak, L.: Graphs & Digraphs. Chapman and Hall, CRC, London (2005)
3. Kamatchi, N., Arumugam, S.: Distance antimagic graphs. J. Combin. Math. Combin. Comput. **84**, 61–67 (2013)

Probabilistic Conjunctive Grammar

K. Kanchan Devi([✉]) and S. Arumugam

National Centre for Advanced Research in Discrete Mathematics (n-CARDMATH), Kalasalingam University, Krishnankoil 626126, Tamil Nadu, India
kanchandevi.klu@gmail.com, s.arumugam.klu@gmail.com

Abstract. This paper extends conjunctive grammar to Probabilistic Conjunctive Grammar (PCG). This extension is motivated by the concept of probabilistic context free grammar which has many applications in the area of computational linguistics, computer science and bio-informatics. Our focus is to develop PCG for its application in linguistics and computer science. In bio-informatics stochastic conjunctive grammar has been defined to detect Pseudo knots in RNA.

Keywords: Conjunctive grammar · Probabilistic conjunctive grammar

1 Introduction

Conjunctive grammar, introduced in [1], is a context-free grammar augmented with an explicit set-theoretic intersection operation. Every rule in a conjunctive grammar is of the form

$$A \to \alpha_1 \& \cdots \& \alpha_n$$

where $n \geq 1$ and α_i are strings consisting of terminal and nonterminal symbols. Each of the above rules indicates that any string that can be generated from each α_i can be generated by A. Conjunctive grammar can express everything that ordinary context-free grammar can. An important property of conjunctive grammar is that the parse of a string generated by a grammar can be represented in the form of a tree with shared leaves, which generalizes ordinary context-free parse trees. By using this property we can define a probabilistic variant of conjunctive grammar as a generalization of probabilistic context-free grammar. A probabilistic context-free grammar [2,3] can be defined as a probability measure on a set of rooted trees. This measure is specified by a set of rules for evolving symbols known as non-terminals into sequences of non-terminals and terminals and by assigning probabilities to these rules. This gives a probability measure on the set of finite sequences of terminals. Applications of this grammar are seen in the area of bio-informatics, linguistics and computer science.

In computer science (computational linguistics), stochastic grammars have a longer tradition, and were studied and used mainly in the field of natural language processing [12,13]. Probabilistic methods have been employed in automatic speech recognition. Recognition of natural unrestricted speech requires a

S. Arumugam et al. (Eds.): ICTCSDM 2016, LNCS 10398, pp. 119–127, 2017.
DOI: 10.1007/978-3-319-64419-6_16

"language model" that attaches probabilities to the production of all possible strings of words [9]. In [10], two applications of the Inside-Outside algorithm to speech recognition tasks has been discussed. The first application involves the direct comparison of stochastic context-free grammars (SCFGs) with hidden Markov models (HMMs) in modeling isolated words. The second application investigates the role of SCFGs in representing the language model component of a speech recognizer.

Stochastic context free grammars (SCFGs) were introduced in bioinformatics for the purpose of modeling RNA secondary structure, the original early references being [4,5]. Stochastic context free grammars are applied to the problems of folding, aligning and modeling families of homologous RNA sequence [6]. The SCFG is used to differentiate the tRNA sequences from the other RNA sequences of similar length, to produce multiple alignments of large collections of tRNA sequences, and to determine structure of new tRNA sequences [5].

In Computer Science, the probabilistic context free grammars are used as a model for research in security and privacy. Using the PCFG word mangling rules are generated and then passwords are guessed to be used in password cracking [7]. PCFG models are used for understanding the choice of passwords by constructing password strength meter and password cracking utilities [8].

In bio-informatics, RNA pseudoknot prediction is done through stochastic conjunctive grammars [11]. Grammars work by rewriting non-terminals symbols using a set of production rules. Stochastic grammars can be trained to predict the most probable structure for an RNA sequence by assigning probabilities to each production rule.

In this paper we propose a probabilistic model of conjunctive grammar which can be used in computational linguistics, security and privacy in computer science. This model will be more effective than the probabilistic context-free model in the above areas of application since conjunctive grammar is a powerful extension of context-free grammar.

2 Context-Free Grammar (CFG) and Probabilistic Context-Free Grammar (PCFG)

Definition 2.1. *A CFG is a four tuple $G = (\Sigma, N, P, S)$, where Σ and N are disjoint finite non-empty sets of terminal and nonterminal symbols respectively, P is a finite set of rules of the form $A \to \beta$, where $A \in N$ and $\beta \in (\Sigma \cup N)^*$ and $S \in N$ is the start symbol.*

Definition 2.2. *A PCFG is a five tuple $G = (\Sigma, N, P, S, q)$, where Σ and N are disjoint finite non-empty sets of terminal and nonterminal symbols respectively, P is a finite set of rules of the form $A \to \beta$ where $A \in N$ and $\beta \in (\Sigma \cup N)^*$, $S \in N$ is the start symbol, each rule in P is augmented with a conditional probability assigned by a function q given as $q(A \to \beta)$ for every rule in P. For any $A \in N$ $\sum_{A \to \beta \in P} q(A \to \beta) = 1$. Also for a given parse tree T containing* *rules* $A_1 \to \beta_1,\ A_2 \to \beta_2, \cdots,\ A_n \to \beta_n$, *the probability of* T *under PCFG is* $p(T) = \prod_{i=1}^{n} q(A_i \to \beta_i)$.

Example 2.3. *Consider the PCFG,* $G = (\Sigma, N, P, S, q)$ *where* $N = \{S, A, B\}$, $\Sigma = \{a, b\}$, $S = \{S\}$ *and* P *and* q *are given below*

P	q
$S \to AB$	1.0
$A \to aA$	0.5
$A \to a$	0.5
$B \to bB$	0.7
$B \to b$	0.3

Then

$$\begin{aligned} S &\Rightarrow AB && (p = 1) \\ &\Rightarrow (aa)B && (p = 0.25) \\ &\Rightarrow (aa)(bB) && (p = 0.175) \\ &\Rightarrow (aa)(bb) && (p = 0.0525) \\ &\Rightarrow aabb && (p = 0.0525) \end{aligned}$$

Hence the probability of generating the sequence aabb is 0.0525.

3 Conjunctive Grammar (CG) and Probabilistic Conjunctive Grammar (PCG)

Definition 3.1. [1] *A conjunctive grammar is a quadruple* $G = (\Sigma,N,P,S)$, *in which* Σ *and* N *are disjoint finite non-empty sets of terminal and nonterminal symbols respectively;* P *is a finite set of rules of the form*

$$A \to \alpha_1 \& \cdots \& \alpha_n \ (\text{with } A \in N \text{ and } \alpha_1, \cdots, \alpha_n \in (\Sigma \cup N)^*) \qquad (*)$$

and $S \in N$ *is a nonterminal designated as the start symbol. For any rule of the form* $(*)$ *and any number* i $(1 \le i \le n)$, *an object* $A \to \alpha_i$ *is referred to as a conjunct.*

A rule $(*)$ means that the occurence of a noterminal symbol A can be replaced by $\alpha_1 \& \cdots \& \alpha_n$.

For any finite nonempty set Σ, let Σ^*denote the set of all words of finite length over Σ. The immediate derivability relation denoted by $\Rightarrow$ is defined as follows: Using a rule $A \to \alpha_1 \& \cdots \& \alpha_n \in P$, any occurence of a nonterminal symbol A in any term can be rewritten as $\cdots A \cdots \Rightarrow \cdots (\alpha_1 \& \cdots \& \alpha_n) \cdots$

A conjunction of several identical strings can be rewritten by one such string: For every $w \in \Sigma^*$, $\cdots (w \& \cdots \& w) \cdots \Rightarrow \cdots w \cdots$

Definition 3.2. *Let $G = (\Sigma, N, P, S)$ be a conjunctive grammar. The language generated by the term φ is $L_G(\varphi) = \{w \mid w \in \Sigma^*, \varphi \Rightarrow^* w\}$. The language generated by the grammar is $L(G) = L_G(S) = \{w \mid w \in \Sigma^*, S \Rightarrow^* w\}$.*

Any language represented as an intersection of finitely many context-free languages can be directly specified using conjunction for the start symbol.

As context-free grammar (CFG) has been extended to the probabilistic (stochastic) context-free grammar (PCFG), in the same way we extend the conjunctive grammar (CG) to its probabilistic form, namely Probabilistic Conjunctive Grammar (PCG).

Definition 3.3. *A Conjunctive grammar $G = (\Sigma, N, P, S, q)$, is called a probabilistic conjunctive grammar (PCG) if for each rule $A \rightarrow B$, a probability $q(A \rightarrow B)$ is assigned such that for any $A \in N$,*

$$\sum_{A \rightarrow \beta \in P} q(A \rightarrow \beta) = 1.$$

Also if T is a parse tree containing the rules $A_1 \rightarrow \beta_1,\ A_2 \rightarrow \beta_2, \cdots,\ A_n \rightarrow \beta_n$ then

$$p(T) = \prod_{i=1}^{n} q(A_i \rightarrow \beta_i).$$

Assigning probabilities is important when dealing with machine learning because a string can have many parse trees to derive it. These parse trees will possibly have very different structure but the application of probabilities helps us to select the most probable parse and the parsing will be faster by pruning off the low probability sub trees.

Definition 3.4. *The probabilistic conjunctive language $L(G_p)$ defined by a PCG is given as*

$$L(G_p) = \{(x, q(x)) \mid S \xrightarrow{q_i(x)} x, \quad for\ i = 1, \cdots, k, \quad x \in \Sigma^* \ and\ q(x) = \sum_{i=1}^{k} q_i(x)\}$$

where S is the start symbol and there are k distinctively different derivations of generating x from S.

Definition 3.5. *Let G_p be a probabilistic Conjunctive grammar. Then a probabilistic conjunctive grammar G_{p_1} is equivalent to G_p if and only if*

$$L(G_p) = L(G_{p_1}).$$

Example 3.6. *Consider the PCG $G_p = (\Sigma, N, P, S, q)$, where $N = \{S, A, B, C\}$, $\Sigma = \{a, b, \epsilon\}$, $S = \{S\}$ and P and q are below*

P	q
$S \to AB\&C$	1.0
$A \to aA$	0.4
$A \to a$	0.4
$A \to \epsilon$	0.2
$B \to bB$	0.7
$B \to \epsilon$	0.3
$C \to aC$	0.4
$C \to bC$	0.4
$C \to \epsilon$	0.2

Then

$$
\begin{aligned}
S &\Rightarrow AB\&C && (p = 1)\\
&\Rightarrow (aA)B\&C && (p = 0.4)\\
&\Rightarrow (a)B\&C && (p = 0.04)\\
&\Rightarrow (a)(bB)\&C && (p = 0.028)\\
&\Rightarrow (a)(b)\&C && (p = 0.0084)\\
&\Rightarrow (a)(b)\&(aC) && (p = 0.00336)\\
&\Rightarrow (a)(b)\&(a(bC)) && (p = 0.001344)\\
&\Rightarrow (a)(b)\&(a)(b) && (p = 0.0002688)\\
&\Rightarrow ab\ \&ab && (p = 0.0002688)\\
&\Rightarrow ab && (p = 0.0002688)
\end{aligned}
$$

3.1 Binary Normal Form of PCG

A binary normal form is a natural extension of Chomsky Normal Form for the conjunctive grammars.

Definition 3.7. *A PCG is said to be in binary normal form, if each rule in P is one of the following forms:*

(i) $A \to B_1C_1\& \cdots \&B_mC_m$, *where* $m \geq 1; A, B_i, C_i \in N$,
(ii) $A \to a$, *where* $A \in N, a \in \Sigma$,
(ii) $S \to \epsilon$, *if S does not appear in right parts of the production rules.*

Each rule r in P is augmented with a probability q(r) such that for any $A \in N$,

$$\sum_{r \in P} q(r) = 1.$$

The binary normal form theorem for CG was given in [1].

The following theorem on the binary normal form of PCG is analogous to the procedure used by Hoptcroft and Ullman [14] to prove the Chomsky Normal form for context-free grammars and [15] for normalized stochastic context free grammar.

Theorem 3.8. *Any PCG G_p is equivalent to a PCG G_{p_b} in which all the productions are in binary normal form.*

Proof. We consider two cases:

Case I: G is a PCG generating a language not containing ϵ.
We construct a PCG, G_1 which is equivalent to G such that there are no productions of the form of $A \xrightarrow{p_1} B$, where $A, B \in N$. Consider productions in G of the form $A \xrightarrow{p_1} B$ leading to a derivation chain

$$A \xrightarrow{p_1} B_1 \xrightarrow{p_2} B_2 \cdots \xrightarrow{p_m} B_m \xrightarrow{p_{m+1}} B \xrightarrow{p_{m+2}} w, \text{ where } w \notin N$$

We define a new production by $A \xrightarrow{p} w$ where $p = p_1 p_2 \cdots p_{m+2}$, provided that there are no loops among $A, B_1, \cdots, B_m, B \in N$. If there exists a loop between nonterminals A and B such that

$$A \xrightarrow{p_0} B, A \xrightarrow{p_i} \alpha_i, i = 1, \cdots, n \text{ and } B \xrightarrow{q_0} A, B \xrightarrow{q_j} \beta_j, j = 1, \cdots, m \quad (1)$$

then

$$p_0 + p_1 + \cdots + p_n = q_0 + q_1 + \cdots + q_m = 1.$$

Now we replace the set of productions in (1) by the following productions

$$\begin{aligned} A &\xrightarrow{r_i} \beta_i, i = 1, \cdots, m, \\ A &\xrightarrow{t_i} \alpha_i, i = 1, \cdots, n, \\ B &\xrightarrow{s_i} \alpha_i, i = 1, \cdots, n, \\ B &\xrightarrow{u_i} \beta_i, i = 1, \cdots, m, \end{aligned}$$

where

$$r_i = \frac{p_0 q_i}{1 - p_0 q_0}, i = 1, \cdots, m, \quad (2)$$

$$t_i = \frac{p_i}{1 - p_0 q_0}, i = 1, \cdots, n, \quad (3)$$

$$s_i = \frac{q_0 p_i}{1 - p_0 q_0}, i = 1, \cdots, n, \quad (4)$$

$$u_i = \frac{q_i}{1 - p_0 q_0}, i = 1, \cdots, m, \quad (5)$$

We prove one of the Eqs. (2), (3), (4) and (5). By applying the productions $A \xrightarrow{p_0} B$ and $B \xrightarrow{q_0} A$, an arbitrary number of times before replacing B by β_i with probability q_i, we can easily derive β_i from A,

$$A \xrightarrow{p_0} B \xrightarrow{q_i} \beta_i$$
$$A \xrightarrow{p_0} B \xrightarrow{q_0} A \xrightarrow{p_0} B \xrightarrow{q_i} \beta_i$$
$$A \xrightarrow{p_0} B \xrightarrow{q_0} A \xrightarrow{p_0} B \xrightarrow{q_0} A \xrightarrow{p_0} B \xrightarrow{q_i} \beta_i$$
$$\vdots$$

Therefore, β_i can be derived from A with probability r_i, which is defined as the sum of the probabilities of obtaining β_i from A by means of infinite number of derivations, such that

$$r_i = p_0q_i + p_0q_0p_0q_i + p_0q_0p_0q_0p_0q_i + \cdots = p_0q_i \sum_{n=0}^{\infty} (q_0p_0)^n = \frac{p_0q_i}{1 - p_0q_0}$$

Similarly the other equations can be proved.

In this way, loop between A and B is being eliminated. Hence we get an equivalent PCG G_1in which there are no productions of the form $A \xrightarrow{p} B, A, B \in N$.

We now construct a PCG G_2 equivalent to G_1 in which there are no productions of the form $A \xrightarrow{p} \alpha_1\alpha_2 \ldots \alpha_n$, $n \geq 2$ where $A \in N$ and $\alpha_i \in (\Sigma \cup N)^*$. Suppose that $\alpha_i \in \Sigma$ and let $\alpha_i$be a terminal symbol $'a'$. Then the α_i is replaced by a new non terminal B_i which is not appearing as the premise of any rule in G_1. Hence we get $A \xrightarrow{p} \alpha_1\alpha_2 \ldots B_i \ldots \alpha_n$ and $B_i \xrightarrow{1} a$. After repeating this procedure for all terminals in $\alpha_1 \ldots \alpha_n$ in all the production rules, we get a grammar G_2 in which all the productions are of the form

1. $A \xrightarrow{p} a$, where $A \in N$, $a \in \Sigma$, or
2. $A \xrightarrow{p} \alpha_1\alpha_2 \ldots \alpha_n$, $n \geq 2$, where $A, \alpha_i \in N$,

Clearly G_2 is equivalent to G_1.

Finally we construct a PCG G_3 equivalent to G_2 in which all the productions are of the form $A \xrightarrow{p} a$ or $A \xrightarrow{p} B_1C_1 \& B_2C_2 \& \cdots \& B_mC_m$, $A, B_i, C_i \in N$ and $a \in \Sigma$. Consider a typical production in G_2 of the form $A \xrightarrow{p} B_1B_2 \ldots B_m, m \geq 3$ and $A, B_i \in N$. We replace this production by the productions

$$A \xrightarrow{p} B_1D_1$$
$$D_1 \xrightarrow{1} B_2D_2$$
$$\vdots$$
$$D_{m-2} \xrightarrow{1} B_{m-1}B_m,$$

where $D's$ are the new nonterminals which do not appear as the premise of any production in G_2. After the elimination of this production from G_2, we get a grammar G_3 that is equivalent to G_2. Hence G_3 is in binary normal form and is equivalent to G.

Case II: G is a PCG generating a language containing ϵ.
We introduce a new start symbol S' and add, $S' \to \epsilon$ to P', and for each production $S \to C_1\&\ldots\&C_k \in P'$, a rule $S' \to C_1\&\ldots\&C_k$ added to P' as well.

Example 3.9. *Consider PCG $G_p = (\{s,t\},\{S,M,C\},P,S,q)$, where P with q are given below*

$$S \xrightarrow{0.8} Mt\&SC, \qquad S \xrightarrow{0.2} t$$
$$M \xrightarrow{0.4} tMss, \qquad M \xrightarrow{0.3} sM, \qquad M \xrightarrow{0.3} tss$$
$$C \xrightarrow{0.7} sC, \qquad C \xrightarrow{0.3} t.$$

The equivalent grammar to G_p in binary normal form is given by

$$G_{p_b} = (\{s,t\},\{A,B,C,D,S,M,N\},P,S,q)$$

where P with q are defined as:

$$S \xrightarrow{0.8} MB\&SC, \qquad S \xrightarrow{0.2} t$$
$$M \xrightarrow{0.4} ND, \qquad M \xrightarrow{0.3} AM, \qquad M \xrightarrow{0.3} BD$$
$$C \xrightarrow{0.7} AC, \qquad C \xrightarrow{0.3} t$$
$$N \xrightarrow{1} BM$$
$$D \xrightarrow{1} AA$$
$$A \xrightarrow{1} s$$
$$B \xrightarrow{1} t.$$

The derivation of the string "tsst" is given below

$$\begin{aligned}
S &\Rightarrow MB\&SC && (p = 0.8)\\
&\Rightarrow (BD)B\&SC && (p = 0.24)\\
&\Rightarrow ((t)D)B\&SC && (p = 0.24)\\
&\Rightarrow ((t)AA)B\&SC && (p = 0.24)\\
&\Rightarrow ((t)ss)B\&SC && (p = 0.24)\\
&\Rightarrow ((t)ss)t\&(SC) && (p = 0.24)\\
&\Rightarrow (tsst)\&(tC) && (p = 0.048)\\
&\Rightarrow (tsst)\&(t(AC)) && (p = 0.0336)\\
&\Rightarrow (tsst)\&(t(sC)) && (p = 0.0336)\\
&\Rightarrow (tsst)\&(ts(AC)) && (p = 0.02352)\\
&\Rightarrow (tsst)\&(ts(sC)) && (p = 0.02352)\\
&\Rightarrow (tsst)\&(ts(st)) && (p = 0.007056)\\
&\Rightarrow tsst && (p = 0.007056)
\end{aligned}$$

4 Conclusion

In this paper we have extended the concept of conjunctive grammar to probabilistic conjunctive grammar. Application of probabilistic conjunctive grammar to other areas will be reported in a subsequent paper.

Acknowledgments. The first author is thankful to the management of Kalasalingam University for providing fellowship.

References

1. Okhotin, A.: Conjunctive grammars. J. Automata Lang. Comb. **6**(4), 519–535 (2001)
2. Jelinek, F., Lafferty, J.D., Mercer, R.L.: Basic Methods of Probabilistic Context Free Grammars. Speech Recognition and Understanding. Springer, Heidelberg (1992)
3. Lari, K., Young, S.J.: The estimation of stochastic context-free grammars using the inside-outside algorithm. Comput. Speech Lang. **4**(1), 35–56 (1990)
4. Eddy, S.R., Durbin, R.: RNA sequence analysis using covariance models. Nucleic Acids Res. **22**(11), 2079–2088 (1994)
5. Sakakibara, Y., Brown, M., Hughey, R., Mian, I.S., Sjölander, K., Underwood, R.C., Haussler, D.: Stochastic context-free grammers for tRNA modeling. Nucleic Acids Res. **22**(23), 5112–5120 (1994)
6. Sakakibara, Y., Brown, M., Hughey, R., Mian, I.S.: The application of Stochastic Context-free Grammars to Folding, Aligning and Modeling Homologous RNA Sequences, Report, UC Santa Cruz (1993)
7. Weir, M., Aggarwal, S., De Medeiros, B., Glodek, B.: Password cracking using probabilistic context-free grammars. In: 30th IEEE Symposium on Security and Privacy, pp. 391–405. IEEE (2009)
8. Ma, J., Yang, W., Luo, M., Li, N.: A study of probabilistic password models. In: IEEE Symposium on Security and Privacy, pp. 689–704. IEEE (2014)
9. Bahl, L.R., Jelinek, F., Mercer, R.L.: A maximum likelihood approach to continuous speech recognition. IEEE Trans. Pattern Anal. Mach. Intell. **2**, 179–190 (1983)
10. Lari, K., Young, S.J.: Applications of stochastic context-free grammars using the inside-outside algorithm. Comput. Speech Lang. **5**(3), 237–257 (1991)
11. Zier-Vogel, R., Domaratzki, M.: RNA pseudoknot prediction through stochastic conjunctive grammars. In: Informal Proceedings of the Computability in Europe 2013, pp. 80–89 (2013)
12. Booth, T.L., Thompson, R.A.: Applying probability measures to abstract languages. IEEE Trans. Comput. **100**(5), 442–450 (1973)
13. Baker, J.K.: Trainable grammars for speech recognition. J. Acoust. Soc. Am. **65**(S1), S132 (1979)
14. Hopcroft, J.E., Ullman, J.D.: Formal Languages and their Relation to Automata. Addison-Wesley Publishing, Reading (1969)
15. Fuf, K.S.: Stochastic automata, stochastic languages and pattern recognition. J. Cybern. **1**(3), 31–49 (1971)

(1, 2)-Domination in Graphs

K. Kayathri(✉) and S. Vallirani

PG and Research Department of Mathematics, Thiagarajar College, Madurai 625 009, Tamilnadu, India
kayathrikanagavel@gmail.com, vallirani3@gmail.com

Abstract. A $(1,2)$-dominating set in a graph $G=(V,E)$ is a set having the property that for every vertex $v \in V-S$, there is at least one vertex in S at a distance 1 from v and a second vertex in S at a distance at most 2 from v. The $(1,2)-$domination number of G, denoted by $\gamma_{1,2}(G)$, is the minimum cardinality of a $(1,2)-$dominating set of G. In this paper, we have derived bounds of $\gamma_{1,2}$ in terms of the order and the maximum degree. For trees, we get the bounds in terms of the number of pendant vertices. We have also characterized the graphs G of order n, for which $\gamma_{1,2}(G)=n, n-1, n-2$.

Keywords: Domination · (1, 2)-dominating set

1 Introduction

Hedetniemi et al. [3] introduced the concept of $(1,k)$-domination in graphs. Let k be a positive integer. A subset S of vertices is called a $(1,k)$-dominating set in G if for every vertex $v \in V-S$, there are two distinct vertices $u, w \in S$ such that u is adjacent to v, and w is within distance k of v (*i.e.* $d_G(v,w) \leq k$). Hedetniemi et al. [4,5] examined $(1,k)$-domination along with the internal distances in $(1,k)$-dominating sets. Factor and Langley [1,2] studied $(1,2)$-domination of digraphs.

In this paper, we study $(1,2)$-domination in graphs. All our graphs are finite and simple.

2 Bounds of $\gamma_{1,2}$ in terms of Δ

We start with the following observations.

Observation 1. *For any two graphs G and H, $\gamma_{1,2}(G \cup H)=\gamma_{1,2}(G)+\gamma_{1,2}(H)$.*

Observation 2. *If H is a spanning supergraph of G, then $\gamma_{1,2}(H) \leq \gamma_{1,2}(G)$.*

Theorem 1. *If G is a graph of order $n \geq 4$ with $\Delta(G) \geq n-2$, then*

$$\gamma_{1,2}(G)=\begin{cases} 2 & \text{if } G \text{ is connected} \\ 3 & \text{if } G \text{ is disconnected.} \end{cases}$$

S. Arumugam et al. (Eds.): ICTCSDM 2016, LNCS 10398, pp. 128–133, 2017.
DOI: 10.1007/978-3-319-64419-6_17

Proof. When $\Delta(G) = n - 1$, let u be a full-degree vertex; and v be any other vertex in G. Then $\{u, v\}$ is a $(1,2)$-dominating set and so $\gamma_{1,2}$(G) =2.

When $\Delta(G) = n - 2$, let u be a vertex of degree $n - 2$; and v be the vertex which is not adjacent to u.

Case 1. G is connected.
Let w be a neighbour of v. Then $\{u, w\}$ is a $(1,2)$-dominating set and so $\gamma_{1,2}(G) = 2$.

Case 2. G is disconnected.
Then v is an isolated vertex. Let S be any $(1,2)$-dominating set of G. Since every isolated vertex must lie in S, $\gamma_{1,2}(G) \geq 3$. Clearly $\{u, v, x\}$ is a $(1,2)$-dominating set for every $x \in N(u)$ and $\gamma_{1,2}(G) = 3$.

Corollary 1. $\gamma_{1,2}(G) = 2$ *for the graphs* $G = K_n, K_{1,n}, W_n, F_n$ *and* $H + K_1$ *where* H *is any graph.*

Theorem 2. *Let* G *be a connected graph of order* $n \geq 5$ *with* $2 \leq \Delta(G) \leq n-3$. *Then* $\gamma_{1,2}(G) \leq n - \Delta(G)$.

Proof. Let G be a connected graph with the given hypothesis. Let $\Delta(G) = n-1-k$. Then $2 \leq k \leq n-3$ and $n - \Delta(G) = k+1$. Let $V(G) = \{u, v_i | 1 \leq i \leq n-1\}$, where u is a vertex of degree $\Delta(G)$, and $N(u) = \{v_{k+1}, v_{k+2}, ..., v_{n-1}\}$. Then $V(G) = N[u] \cup V_1$, where $V_1 = \{v_1, v_2, ..., v_k\}$. Since G is connected, at least one vertex in V_1 has a neighbour in $N(u)$.

Case 1. Every vertex in V_1 has some neighbour in $N(u)$.
Without loss of generality, assume that v_i is adjacent to v_{j_i} in $N(u)$, for $1 \leq i \leq k$. The vertices $v_{j_1}, v_{j_2}, ..., v_{j_k}$ need not be distinct. Let $V_2 = \{v_{j_i} | 1 \leq i \leq k\} \subseteq N(u)$. Let $S = \{u, v_{j_1}, v_{j_2}, ..., v_{j_k}\} (= V_2 \cup \{u\})$.

Every vertex $v_i \in N(u) - S$ is adjacent to u and at a distance at most 2 from v_{j_1}. Every $v_i \in V_1$ is adjacent to v_{j_i} and at a distance 2 from u. Hence S is a $(1,2)$-dominating set and so $\gamma_{1,2}(G) \leq k+1$.

Case 2. Some vertex in V_1 has no neighbour in $N(u)$.
Without loss of generality, let $V_1^{'} = \{v_1, v_2, ..., v_r\} \subseteq V_1$ be the set of vertices that have no neighbours in $N(u)$. Let $V_1^{''} = V_1 - V_1^{'} = \{v_{r+1}, v_{r+2}, ..., v_k\}$. Then $V(G) = N[u] \cup V_1^{'} \cup V_1^{''}$. Since G is connected, at least one vertex in $V_1^{'}$ is adjacent to some vertex in $V_1^{''}$. Without loss of generality, let v_1 be adjacent to v_{r+1}. Without loss of generality, assume that v_i is adjacent to v_{j_i} in $N(u)$, for $r+1 \leq i \leq k$. The vertices $v_{j_{r+1}}, v_{j_{r+2}}, ..., v_{j_k}$ need not be distinct. Let $V_2 = \{v_{j_i} | r+1 \leq i \leq k\} \subseteq N(u)$.

Let $S = \{u, v_{j_{r+1}}, v_{j_{r+2}}, ..., v_{j_k}, v_1, v_2, ..., v_r\} (= V_2 \cup V_1^{'} \cup \{u\})$. Every $v_i \in N(u) - V_2$ is adjacent to u and at a distance at most 2 from $v_{j_{r+1}}$. Every $v_i \in V_1^{''}$ is adjacent to v_{j_i} and at a distance 2 from u. Hence S is a $(1,2)$-dominating set and so $\gamma_{1,2}(G) \leq k+1$.

A *wounded spider* is the graph formed by subdividing at most $n-1$ of the edges of a star $K_{1,n}$ for $n \geq 2$. Let $WS_{n,t}$ denote the wounded spider formed by subdividing t edges of $K_{1,n}$, $1 \leq t \leq n-1$.

Corollary 2. $\gamma_{1,2}(WS_{n,t}) = t + 1$.

Proof. Let $V[WS_{n,t}] = \{u, v_1, v_2, ..., v_n, v_1', v_2', ..., v_t'\}$ and $E[WS_{n,t}] = \{uv_j, v_iv_i' | 1 \leq j \leq n, 1 \leq i \leq t\}$. Let S be any $(1, 2)$-dominating set of $WS_{n,t}$. For $1 \leq i \leq t$, to dominate v_i, either $v_i \in S$ or $v_i' \in S$. Moreover, for $t+1 \leq j \leq n$, to dominate v_j, *either* $u \in S$ *or* $v_j \in S$. Therefore, $|S| \geq t + 1$.

Note that $t = n - \Delta - 1$. When $t \geq 2$, then $\Delta(WS_{n,t}) \leq n - 3$; and by Theorem 7, $\gamma_{1,2}(WS_{n,t}) \leq t + 1$. Hence $\gamma_{1,2}(WS_{n,t}) = t + 1$.

When $t = 1$, then $\Delta(WS_{n,t}) \geq n - 2$; and by Theorem 1, $\gamma_{1,2}(WS_{n,t}) = 2$.

3 Composition of Two Graphs

Theorem 3. *Let G be a non-trivial connected graph. Then for any graph H, $\gamma_{1,2}(GoH) = |V(G)|$.*

Proof. Let $V(G) = \{v_1, v_2, ..., v_n\}$ and $V(H) = \{u_1, u_2, ..., u_s\}$. Let $H_1, H_2, ..., H_n$ denote the copies of H, where every vertex of H_i is adjacent to v_i, $1 \leq i \leq n$. Let $V(H_i) = \{u_1^i, u_2^i, ..., u_s^i\}$. Let S be any $(1, 2)$-dominating set of GoH. Since there is no adjacency between the vertices in H_i and H_j for $i \neq j$, every u_r^i in H_i is adjacent to either v_i or u_k^i, where $u_k^i \in N[u_r^i]$. Hence for each i, $1 \leq i \leq n$, to dominate $V(H_i)$, we need at least one vertex in S. Hence $\gamma_{1,2}(GoH) \geq n$. Let $S_1 = \{v_1, v_2, ..., v_n\}$. For every $u_r^i, 1 \leq i \leq n, 1 \leq r \leq s, d_{GoH}(u_r^i, v_i) = 1$ *and* $d_{GoH}(u_r^i, v_j) = 2$ for every $v_j \in N_G(v_i)$. Hence S_1 is a $(1, 2)$-dominating set and so $\gamma_{1,2}(GoH) = n$.

Corollary 3. *Let G be any graph having t isolates. Then for any graph H, $\gamma_{1,2}(GoH) = |V(G)| + t$, where $t \geq 0$.*

Proof. Let $G_1, G_2, ..., G_k$ be the components of G.

Then $\gamma_{1,2}(GoH) = \sum_{i=1}^{k} \gamma_{1,2}(G_ioH)$.

Case 1. $t = 0$.
Since each G_i is connected, by Theorem 3, $\gamma_{1,2}(G_ioH) = |V(G_i)|$. Hence $\gamma_{1,2}(GoH) = |V(G)|$.

Case 2: $t \neq 0$.
Without loss of generality, let $G_1, G_2, ..., G_t$ denote the components of order 1. Then G_ioH has a full - degree vertex; and so by Theorem 1, $\gamma_{1,2}(G_ioH) = 2$, for $1 \leq i \leq t$. By Theorem 3, $\gamma_{1,2}(G_ioH) = |V(G_i)|$, for $t + 1 \leq i \leq k$. Thus, we get the result.

4 Some Characterizations

Theorem 4. *Let G be a connected graph of order $n \geq 2$. Then $\gamma_{1,2}(G) = n$ if and only if $n = 2$.*

Proof. When $G = K_2$, the result is obvious. Conversely, suppose that $n \neq 2$.

Claim. $\gamma_{1,2}(G) < n$.

We prove this result by induction on n.

When $n = 3$, a set of any two vertices of G is a $(1,2)$-dominating set of G and so $\gamma_{1,2}(G) = 2 < n$.

Assume the result for $n = k$ with $k \geq 3$.

Next, let G be a connected graph of order $n = k+1$. Let v be a vertex that is not a cut vertex in G. Then $G - v$ is connected, and of order $n - 1 = k$. By the induction hypothesis, $G - v$ has a $(1,2)$-dominating set S with $|S| < k$. (i.e.) $|S| \leq k - 1$.

Let u be a neighbour of v.

Case 1. $u \in S$.
Since G is connected, $n \geq 3$ and v is not a cut-vertex in G, u has another neighbour (say) w. Then $S \cup \{w\}$ is a $(1,2)$-dominating set in G and so $\gamma_{1,2}(G) \leq k < n$.

Case 2. $u \notin S$.
Since S is a $(1,2)$-dominating set in $G - v$, there exists a vertex $w \in S$ that is adjacent to u. Then $S \cup \{u\}$ is a $(1,2)$-dominating set in G and so $\gamma_{1,2}(G) \leq k < n$.

Thus, by induction, the result follows.

Theorem 5. *Let G be a connected graph of order $n \geq 3$. Then $\gamma_{1,2}(G) = n - 1$ iff $n = 3$. i.e. $\gamma_{1,2}(G) = n - 1$ iff $G = P_3$ or K_3.*

Proof. When $n = 3$, a set of any two vertices of G is a $(1,2)$-dominating set of G and so $\gamma_{1,2}(G) = 2 = n - 1$. Conversely, suppose that $n \neq 3$.

Claim. $\gamma_{1,2}(G) < n - 1$.

We shall prove this result by induction on n.

When $n = 4$, since G is connected, $\Delta(G) \geq 2$. Now, any two adjacent vertices form a (1, 2)-dominating set and so $\gamma_{1,2}(G) = 2 < n - 1$.

Assume the result for $n = k$ with $k \geq 4$.

Next, let G be a connected graph of order $n = k + 1$. The rest of the proof is similar to the proof of Theorem 4.

Corollary 4. *Let G be any graph of order n. Then*

(i) $\gamma_{1,2}(G) = n$ *iff* $G = sK_1 \cup \frac{n-s}{2} K_2$, *with* $0 \leq s \leq n$.
(ii) $\gamma_{1,2}(G) = n - 1$ *iff* $G = sK_1 \cup \frac{n-s-3}{2} K_2 \cup H$, *where* $H \cong P_3$ *or* K_3, *with* $0 \leq s \leq n - 3$.

Theorem 6. *Let G be a connected graph of order $n \geq 4$. Then $\gamma_{1,2}(G) = n - 2$ iff $G = P_5$ or G is of order 4.*

Proof. If $G = P_5$ or G is of order 4, it is easy to verify that $\gamma_{1,2}(G) = n-2$. Conversely, suppose that

$$\gamma_{1,2}(G) = n - 2. \tag{1}$$

Let $n = 5$. If $\Delta(G) \geq 3 (= n-2)$, then $\gamma_{1,2}(G) = 2$ (by Theorem 1), contradicting (1). If $\Delta(G) = 2$, then G is either P_5 or C_5; but $\gamma_{1,2}(C_5) = 2$, and so $G = P_5$. Now, let $n \geq 6$.

Claim. $\gamma_{1,2}(G) < n-2$.

We shall prove this result by induction on n.

When $n = 6$, if $\Delta(G) \geq n-2$, then $\gamma_{1,2}(G) = 2$ (by Theorem 1); if $3 \leq \Delta(G) \leq n-3$, then $\gamma_{1,2}(G) \leq n-3$ (by Theorem 7); if $\Delta(G) = 2$, then G is either P_6 or C_6 and $\gamma_{1,2}(G) \leq 3$; and in all these cases, we get a contradiction to (1).

Assume the result for $n = k$ with $k \geq 6$. Next, let G be a connected graph of order $n = k+1$. The rest of the proof is similar to the proof of Theorem 4.

Corollary 5. *For any graph G of order n, $\gamma_{1,2}(G) = n-2$ iff G is one of the following graphs:*

(i) $G = sK_1 \cup \frac{n-6-s}{2} K_2 \cup H$, *where* $H = 2P_3, 2K_3$ *or* $P_3 \cup K_3$, *with* $0 \leq s \leq n-6$.
(ii) $G = sK_1 \cup P_5 \cup \frac{n-5-s}{2} K_2$, *with* $0 \leq s \leq n-5$.
(ii) $G = sK_1 \cup \frac{n-4-s}{2} K_2 \cup H$ *where H is a connected graph of order 4, with* $0 \leq s \leq n-4$.

5 Trees

Theorem 7. *Let T be a tree of order $n \geq 2$. Then $\gamma_{1,2}(T) = 2$ if and only if T is a Star or Double Star.*

Proof. Suppose that $\gamma_{1,2}(T) = 2$. Let $S = \{u, v\}$ be a $(1,2)$-dominating set of T. Then every vertex in T is adjacent with either u or v. Hence $V(T) = N[u] \cup N[v]$. Then for every $x \in N(u)$ and $y \in V(T)$, $d(x,y) \leq d(x,u) + d(u,y) \leq 3$; similarly, for every $x \in N(v)$ and $y \in V(T)$, $d(x,y) \leq 3$; for every $x \in V - \{u,v\}$, $d(u,x) + d(x,v) \leq 3$; and so $d(u,v) \leq 3$. Hence $diam(T) \leq 3$; and so T is a Star or a Double Star $D_{r,\, s}$ (where $r + s = n-2$). Converse is obvious.

Theorem 7 deals with the trees of diameter 2 and 3. The next result deals with trees of diameter ≥ 4.

Theorem 8. *Let T be a tree of order n with r pendant vertices. Then*

(i) $3 \leq \gamma_{1,2}(T) \leq n-r$, *if* $diam(T) \geq 5$
(ii) $\gamma_{1,2}(T) = n-r$, *if* $diam(T) = 3$ *or* 4.

Proof. Let $diam(T) \geq 3$. Let V_1 denote the set of all pendant vertices in T. Then $|V - V_1| \geq 2$ and $V - V_1$ is a $(1,2)$-dominating set; and so

$$\gamma_{1,2}(T) \leq n - r. \tag{2}$$

Using Theorem 7, $\gamma_{1,2}(T) \geq 3$; and so (i) follows.

When $diam(T) = 3$, T is a double star; and by Theorem 7, $\gamma_{1,2}(T) = 2 = n - r$. When $diam(T) = 4$, $diam(T - V_1) = 2$; and so $T - V_1$ is $K_{1,\ n-r-1}$, where $n - r - 1 \geq 2$. Let $V(K_{1,\ n-r-1}) = \{u, u_1, u_2, ..., u_{n-r-1}\}$. For $1 \leq j \leq d_T(u_i) - 1$, let v_{i_j} denote a pendant vertex adjacent to u_i. For $1 \leq t \leq d_T(u) - n - r - 1$, let w_t denote a pendant vertex adjacent to u. (If $d_T(u) = n - r - 1$, then there is no w_t's).

By (2), $\gamma_{1,2}(T) \leq n - r$. Assume the contrary that $\gamma_{1,2}(T) \neq n - r$. Then there is a $(1, 2)$-dominating set S_1 of cardinality $n - r - 1$.

If $S_1 = \{u_1, u_2, ..., u_{n-r-1}\}$, then there is no vertex at a distance 2 from v_{i_j}, for $1 \leq i \leq n - r - 1$ and $1 \leq j \leq d_T(u_i) - 1$, which is a contradiction.

Then $u_i \notin S_1$, for some i, $1 \leq i \leq n - r - 1$. Without loss of generality, let $u_1, u_2, ..., u_k \notin S_1$ and $u_{k+1}, u_{k+2}, ..., u_{n-r-1} \in S_1$, where $1 \leq k \leq n - r - 1$. For $1 \leq i \leq k$, $u_i \notin S_1$; and so all v_{i_j}'s must lie in S_1. But $|S_1| = n - r - 1$. Hence it follows that, $d(u_i) = 2$ for $i = 1, 2, 3, ..., k$, and $S_1 = \{v_{1_1}, v_{2_1}, ..., v_{k_1}, u_{k+1}, u_{k+2}, ..., u_{n-r-1}\}$.

Case 1. $k = n - r - 1$.
Now $S_1 = \{v_{1_1}, v_{2_1}, ..., v_{(n-r-1)_1}\}$; and so u is not $(1, 2)$-dominated by S_1, a contradiction.

Case 2. $k < n - r - 1$.
Now there is no vertex in S_1 at a distance at most 2 from v_{s_j}, $k+1 \leq s \leq n-r-1$, a contradiction.
Hence $\gamma_{1,2}(T) = n - r$.

References

1. Factor, K.A.S., Langley, L.J.: An introduction to (1, 2)-domination graphs. Congr. Numer. **199**, 33–38 (2009)
2. Factor, K.A.S., Langley, L.J.: A characterization of connected (1, 2)- domination graphs of tournaments. AKCE Int. J. Graphs Comb. **8**(1), 51–62 (2011)
3. Hedetniemi, S.M., Hedetniemi, S.T., Rall, D.F., Knisely, J.: Secondary domination in graphs. AKCE Int. J. Graphs Comb. **5**(2), 103–115 (2008)
4. Hedetniemi, J.T., Hedetniemi, K.D., Hedetniemi, S.M., Hedetniemi, S.T.: Secondary and internal distances of sets in graphs. AKCE Int. J. Graphs Comb. **6**(2), 239–266 (2009)
5. Hedetniemi, J.T., Hedetniemi, K.D., Hedetniemi, S.M., Hedetniemi, S.T.: Secondary and internal distances of sets in graphs II. AKCE Int. J. Graphs Comb. **9**(1), 85–113 (2012)

New Construction on SD-Prime Cordial Labeling

A. Lourdusamy and F. Patrick(✉)

Department of Mathematics, St. Xavier's College (Autonomous), Palayamkottai 627002, India
lourdusamy15@gmail.com, patrick881990@gmail.com

Abstract. Given a bijection $f : V(G) \rightarrow \{1, 2, \cdots, |V(G)|\}$, we associate two integers $S = f(u) + f(v)$ and $D = |f(u) - f(v)|$ with every edge uv in $E(G)$. The labeling f induces an edge labeling $f' : E(G) \rightarrow \{0, 1\}$ such that for any edge uv in $E(G)$, $f'(uv) = 1$ if $gcd(S, D) = 1$, and $f'(uv) = 0$ otherwise. Let $e_{f'}(i)$ be the number of edges labeled with $i \in \{0, 1\}$. We say f is SD-prime cordial labeling if $|e_{f'}(0) - e_{f'}(1)| \leq 1$. Moreover G is SD-prime cordial if it admits SD-prime cordial labeling. In this paper, we investigate some new construction of SD-prime cordial graph.

Keywords: SD-prime labeling · SD-prime cordial labeling · SD-prime cordial graph

1 Introduction

All graphs considered here are simple, finite, connected and undirected. For all other standard terminology and notations we follow Harary [3]. A labeling of a graph is a map that carries the graph elements to the set of numbers, usually to the set of non-negative or positive integers. If the domain is the set of vertices the labeling is called vertex labeling. If the domain is the set of edges then the labeling is called edge labeling. If the labels are assigned to both vertices and edges then the labeling is called total labeling. For all detailed survey of graph labeling we refer Gallian [2]. In [5,6], Lau and Shiu have introduced the concepts SD-prime labeling. In [4], Lau et al. have introduced SD-prime cordial labeling and they discussed SD-prime cordial labeling for some standard graphs. Lourdusamy et al. [7] proved that splitting graph of star and bistar, shadow graph of star and bistar, degree splitting graph of star and bistar, subdivision of star and bistar, square graph of bistar and path, $K_{1,3} * K_{1,n}$, closed helm, gear graph, flower graph, total graph of path and cycle, the graph obtained by duplication of each vertex of path and cycle by an edge, quadrilateral snake, alternative triangular snake, triangular ladder, $P_n \odot K_1$, $C_n \odot K_1$, jewel graph and $K_2 + mK_1$ admit SD-prime cordial labeling. In this paper, we discussed some new construction of graphs on SD-prime cordial labeling concerning star, path and fan related graph.

S. Arumugam et al. (Eds.): ICTCSDM 2016, LNCS 10398, pp. 134–143, 2017.
DOI: 10.1007/978-3-319-64419-6_18

Given a bijection $f : V(G) \to \{1, 2, \cdots, |V(G)|\}$, we associate two integers $S = f(u) + f(v)$ and $D = |f(u) - f(v)|$ with every edge uv in E.

Definition 1. [5] *A bijection $f : V(G) \to \{1, 2, \cdots, |V(G)|\}$ induces an edge labeling $f' : E(G) \to \{0, 1\}$ such that for any edge uv in G, $f'(uv) = 1$ if $gcd(S, D) = 1$, and $f'(uv) = 0$ otherwise. We say f is SD-prime labeling if $f'(uv) = 1$ for all $uv \in E(G)$. Moreover, G is SD-prime if it admits SD-prime labeling.*

Definition 2. [4] *A bijection $f : V(G) \to \{1, 2, \cdots, |V(G)|\}$ induces an edge labeling $f' : E(G) \to \{0, 1\}$ such that for any edge uv in G, $f'(uv) = 1$ if $gcd(S, D) = 1$, and $f'(uv) = 0$ otherwise. The labeling f is called SD-prime cordial labeling if $|e_{f'}(0) - e_{f'}(1)| \leq 1$. We say that G is SD-prime cordial if it admits SD-prime cordial labeling.*

Definition 3. [1] *If $G_1(p_1, q_1)$ and $G_2(p_2, q_2)$ are two connected graphs, $G_1 \hat{o} G_2$ is obtained by superimpose any selected vertex of G_2 on any selected vertex of G_1. The resultant graph $G = G_1 \hat{o} G_2$ consists of $p_1 + p_2 - 1$ vertices and $q_1 + q_2$ edges.*

Definition 4. *The join of two graphs G_1 and G_2 is denoted by $G_1 + G_2$ and whose vertex set is $V(G_1 + G_2) = V(G_1) \bigcup V(G_2)$ and edge set is $E(G_1 + G_2) = E(G_1) \bigcup E(G_2) \bigcup \{uv : u \in V(G_1), v \in V(G_2)\}$.*

Definition 5. *The graph $F_n = P_n + K_1$ is called a fan.*

2 Main Results

Lemma 1. *Let f be a SD-prime cordial labeling of a graph G of order p and size q. Let $w \in V(G)$ be such that $f(w) = 2$. The graph obtained by identifying a vertex w in G and a vertex of degree n in $K_{1,n}$ admits a SD-prime cordial labeling if one of the following conditions holds:*

1. *n is even,*
2. *n is odd and q is even,*
3. *n is odd, q is odd, p is odd and $e_{f'}(1) = \frac{q+1}{2}$,*
4. *n is odd, q is odd, p is even and $e_{f'}(1) = \frac{q-1}{2}$.*

Proof. Let f be a SD-prime cordial labeling of a graph G of order p and size q. That is the vertices of g are labeled with numbers $\{1, 2, \cdots, p\}$ and $|e_{f'}(1) - e_{f'}(0)| \leq 1$.

Let $w \in V(G)$ be such that $f(w) = 2$. Let us denote by H the graph obtained by identifying a vertex w in G and a vertex of degree n in $K_{1,n}$.

We define a vertex labeling g of H such that

$$g(v) = f(v),\ v \in V(G);$$
$$g(x_i) = p + i,\ i = 1, 2, \cdots, n.$$

Thus for the induced edge labeling we get

$$g'(uv) = f'(vu), \quad v \in V(G);$$

and for $i = 1, 2, \cdots, n$

$$g'(wx_i) = 0 \text{ if } (p \text{ is odd and } i \text{ is odd}) \text{ or } (p \text{ is even and } i \text{ is even});$$
$$g'(wx_i) = 1 \text{ if } (p \text{ is odd and } i \text{ is even}) \text{ or } (p \text{ is even and } i \text{ is odd}).$$

Let us denote by $e^*_{g'}(k)$ number of edges wx_i labeled with k, where $k = 0, 1$. Then

$$\begin{aligned}|e_{g'}(1) - e_{g'}(0)| &= \left|(e_{f'}(1) - e^*_{g'}(1)) - (e_{f'}(0) - e^*_{g'}(0))\right| \\ &= \left|e_{f'}(1) - e_{f'}(0) + e^*_{g'}(1) - e^*_{g'}(0)\right|.\end{aligned}$$

Thus, if n is even and q is even then

$$|e_{g'}(1) - e_{g'}(0)| = \left|\frac{q}{2} - \frac{q}{2} + \frac{n}{2} - \frac{n}{2}\right| = 0;$$

if n is even, q is odd and $e_{f'}(1) = \frac{q+1}{2}$ then

$$|e_{g'}(1) - e_{g'}(0)| = \left|\frac{q+1}{2} - \frac{q-1}{2} + \frac{n}{2} - \frac{n}{2}\right| = 1;$$

if n is even, q is odd and $e_{f'}(1) = \frac{q-1}{2}$ then

$$|e_{g'}(1) - e_{g'}(0)| = \left|\frac{q-1}{2} - \frac{q+1}{2} + \frac{n}{2} - \frac{n}{2}\right| = 1;$$

if n is odd, q is even and p is even then

$$|e_{g'}(1) - e_{g'}(0)| = \left|\frac{q}{2} - \frac{q}{2} + \frac{n+1}{2} - \frac{n-1}{2}\right| = 1;$$

if n is odd, q is even and p is odd then

$$|e_{g'}(1) - e_{g'}(0)| = \left|\frac{q}{2} - \frac{q}{2} + \frac{n-1}{2} - \frac{n+1}{2}\right| = 1;$$

if n is odd, q is odd, p is odd and $e_{f'}(1) = \frac{q+1}{2}$ then

$$|e_{g'}(1) - e_{g'}(0)| = \left|\frac{q+1}{2} - \frac{q-1}{2} + \frac{n-1}{2} - \frac{n+1}{2}\right| = 0;$$

if n is odd, q is odd, p is even and $e_{f'}(1) = \frac{q-1}{2}$ then

$$|e_{g'}(1) - e_{g'}(0)| = \left|\frac{q-1}{2} - \frac{q+1}{2} + \frac{n+1}{2} - \frac{n-1}{2}\right| = 0.$$

Lemma 2. *Let f be a SD-prime cordial labeling of a graph G of order p and size q. Let $w \in V(G)$ be such that $f(w) = 1$. The graph obtained by identifying a vertex w in G and a vertex of degree n in $K_{1,n}$ admits a SD-prime cordial labeling if one of the following conditions holds:*

1. *n is even,*
2. *n is odd and q is even,*
3. *n is odd, q is odd, p is even and $e_{f'}(1) = \frac{q+1}{2}$,*
4. *n is odd, q is odd, p is odd and $e_{f'}(1) = \frac{q-1}{2}$.*

Proof. The proof is analogous to that of Lemma 1.

Theorem 1. *Let f be a SD-prime cordial labeling of a graph G of order p and size q. Let $w \in V(G)$ be such that $f(w) = 2^s$, where $s = 0, 1, 2, \cdots$. The graph obtained by identifying a vertex w in G and a vertex of degree n in $K_{1,n}$ admits a SD-prime cordial labeling.*

Proof. The proof follows from Lemmas 1 and 2.

Lemma 3. *Let f be a SD-prime cordial labeling of a graph G of order p and size q. Let $w \in V(G)$ be such that $f(w) = 2$. The graph obtained by identifying a vertex w in G and a vertex of degree 1 in P_n admits a SD-prime cordial labeling if one of the following conditions holds:*

1. *n is odd,*
2. *n is even and q is even,*
3. *n is even, q is odd, p is odd and $e_{f'}(1) = \frac{q+1}{2}$,*
4. *n is even, q is odd, p is even and $e_{f'}(1) = \frac{q-1}{2}$.*

Proof. Let f be a SD-prime cordial labeling of a graph G of order p and size q. That is the vertices of g are labeled with numbers $\{1, 2, \cdots, p\}$ and $|e_{f'}(1) - e_{f'}(0)| \leq 1$.

Let $w \in V(G)$ be such that $f(w) = 2$. Let us denote by H the graph obtained by identifying a vertex w in G and a vertex of degree 1 in P_n.
We define a vertex labeling g of H such that

$$g(v) = f(v), \quad v \in V(G);$$

If p is odd and n is odd,

$$g(x_i) = \begin{cases} p+i-1 & \text{if } i \equiv 2,3 \ (mod\ 4) \text{ and } 2 \leq i \leq n \\ p+i & \text{if } i \equiv 0 \ (mod\ 4) \text{ and } 2 \leq i \leq n \\ p+i-2 & \text{if } i \equiv 1 \ (mod\ 4) \text{ and } 2 \leq i \leq n; \end{cases}$$

If p is even and n is odd,

$$g(x_i) = \begin{cases} p+i & \text{if } i \equiv 2 \ (mod\ 4) \text{ and } 2 \leq i \leq n \\ p+i-2 & \text{if } i \equiv 3 \ (mod\ 4) \text{ and } 2 \leq i \leq n \\ p+i-1 & \text{if } i \equiv 0,1 \ (mod\ 4) \text{ and } 2 \leq i \leq n; \end{cases}$$

If n is even,

$$g(x_i) = \begin{cases} p+i-1 & \text{if } i \equiv 2,1 \ (mod\ 4) \text{ and } 2 \le i \le n \\ p+i & \text{if } i \equiv 3 \ (mod\ 4) \text{ and } 2 \le i \le n \\ p+i-2 & \text{if } i \equiv 0 \ (mod\ 4) \text{ and } 2 \le i \le n. \end{cases}$$

Thus for the induced edge labeling we get

$$\begin{aligned} g'(uv) &= f'(vu),\ v \in V(G); \\ g'(wx_2) &= 0 \text{ if } (n \text{ is odd}) \text{ or } (p \text{ is odd and } n \text{ is even}); \\ g'(wx_2) &= 1 \text{ if } (p \text{ is even and } n \text{ is even}); \\ g'(x_{2i-1}x_{2i}) &= 0 \text{ if } n \text{ is odd and } 2 \le i \le \frac{n-1}{2}; \\ g'(x_{2i-1}x_{2i}) &= 1 \text{ if } n \text{ is even and } 2 \le i \le \frac{n}{2}; \\ g'(x_{2i}x_{2i+1}) &= 1 \text{ if } n \text{ is odd and } 1 \le i \le \frac{n-1}{2}; \\ g'(x_{2i}x_{2i+1}) &= 0 \text{ if } n \text{ is even and } 1 \le i \le \frac{n-2}{2}. \end{aligned}$$

Let us denote by $e^*_{g'}(k)$ number of edges wx_2, x_ix_{i+1} labeled with k, where $k = 0, 1$.
Then

$$\begin{aligned} |e_{g'}(1) - e_{g'}(0)| &= \left|(e_{f'}(1) - e^*_{g'}(1)) - (e_{f'}(0) - e^*_{g'}(0))\right| \\ &= \left|e_{f'}(1) - e_{f'}(0) + e^*_{g'}(1) - e^*_{g'}(0)\right|. \end{aligned}$$

Thus, if n is odd and q is even then

$$|e_{g'}(1) - e_{g'}(0)| = \left|\frac{q}{2} - \frac{q}{2} + \frac{n-1}{2} - \frac{n-1}{2}\right| = 0;$$

if n is odd, q is odd and $e_{f'}(1) = \frac{q+1}{2}$ then

$$|e_{g'}(1) - e_{g'}(0)| = \left|\frac{q+1}{2} - \frac{q-1}{2} + \frac{n-1}{2} - \frac{n-1}{2}\right| = 1;$$

if n is odd, q is odd and $e_{f'}(1) = \frac{q-1}{2}$ then

$$|e_{g'}(1) - e_{g'}(0)| = \left|\frac{q-1}{2} - \frac{q+1}{2} + \frac{n-1}{2} - \frac{n-1}{2}\right| = 1;$$

if n is even, q is even and p is odd then

$$|e_{g'}(1) - e_{g'}(0)| = \left|\frac{q}{2} - \frac{q}{2} + \frac{n-2}{2} - \frac{n}{2}\right| = 1;$$

if n is even, q is even and p is even then

$$|e_{g'}(1) - e_{g'}(0)| = \left|\frac{q}{2} - \frac{q}{2} + \frac{n}{2} - \frac{n-2}{2}\right| = 1;$$

if n is even, q is odd, p is odd and $e_{f'}(1) = \frac{q+1}{2}$ then

$$|e_{g'}(1) - e_{g'}(0)| = \left|\frac{q+1}{2} - \frac{q-1}{2} + \frac{n-2}{2} - \frac{n}{2}\right| = 1;$$

if n is even, q is odd, p is even and $e_{f'}(1) = \frac{q-1}{2}$ then

$$|e_{g'}(1) - e_{g'}(0)| = \left|\frac{q-1}{2} - \frac{q+1}{2} + \frac{n}{2} - \frac{n-2}{2}\right| = 1.$$

Lemma 4. *Let f be a SD-prime cordial labeling of a graph G of order p and size q. Let $w \in V(G)$ be such that $f(w) = 1$. The graph obtained by identifying a vertex w in G and a vertex of degree 1 in P_n admits a SD-prime cordial labeling if one of the following conditions holds:*

1. *n is odd,*
2. *n is even and q is even,*
3. *n is even, q is odd, p is even and $e_{f'}(1) = \frac{q+1}{2}$,*
4. *n is even, q is odd, p is odd and $e_{f'}(1) = \frac{q-1}{2}$.*

Proof. The proof is analogous to that of Lemma 3.

Theorem 2. *Let f be a SD-prime cordial labeling of a graph G of order p and size q. Let $w \in V(G)$ be such that $f(w) = 2^s$, where $s = 0, 1, 2, \cdots$. The graph obtained by identifying a vertex w in G and a vertex of degree 1 in P_n admits a SD-prime cordial labeling.*

Proof. The proof follows from Lemmas 3 and 4.

Lemma 5. *Let f be a SD-prime cordial labeling of a graph G of order p and size q. Let $w \in V(G)$ be such that $f(w) = 2$. The graph obtained by identifying a vertex w in G and a vertex of degree n in F_n admits a SD-prime cordial labeling if one of the following conditions holds:*

1. *n is even and q is even,*
2. *n is even, q is odd, p is even and $e_{f'}(1) = \frac{q-1}{2}$,*
3. *n is odd and q is even,*
4. *n is odd, q is odd, p is odd and $e_{f'}(1) = \frac{q+1}{2}$,*
5. *n is odd, q is odd, p is even and $e_{f'}(1) = \frac{q-1}{2}$.*

Proof. Let f be a SD-prime cordial labeling of a graph G of order p and size q. That is the vertices of g are labeled with numbers $\{1, 2, \cdots, p\}$ and $|e_{f'}(1) - e_{f'}(0)| \leq 1$.

Let $w \in V(G)$ be such that $f(w) = 2$. Let us denote by H the graph obtained by identifying a vertex w in G and a vertex of degree n in F_n.
We define a vertex labeling g of H such that

$$g(v) = f(v), \quad v \in V(G);$$

If n is even,

$$g(x_i) = \begin{cases} p+i & \text{if } i \equiv 1,2 \ (mod\ 4) \text{ and } 1 \leq i \leq n \\ p+i+1 & \text{if } i \equiv 3 \ (mod\ 4) \text{ and } 1 \leq i \leq n \\ p+i-1 & \text{if } i \equiv 0 \ (mod\ 4) \text{ and } 1 \leq i \leq n; \end{cases}$$

If n is odd,

$$g(x_i) = \begin{cases} p+i & \text{if } i \equiv 1,0 \ (mod\ 4) \text{ and } 1 \leq i \leq n \\ p+i+1 & \text{if } i \equiv 2 \ (mod\ 4) \text{ and } 1 \leq i \leq n \\ p+i-1 & \text{if } i \equiv 3 \ (mod\ 4) \text{ and } 1 \leq i \leq n. \end{cases}$$

Thus for the induced edge labeling we get

$$g'(uv) = f'(vu), \quad v \in V(G);$$

for $1 \leq i \leq n$,

$$\begin{aligned} g'(wx_i) = 0 \quad & \text{if } (p \text{ is odd, } n \text{ is even and } i \equiv 1,0 \ (mod\ 4)) \text{ or} \\ & (p \text{ is even, } n \text{ is even and } i \equiv 2,3 \ (mod\ 4)) \text{ or} \\ & (p \text{ is odd, } n \text{ is odd and } i \equiv 1,2 \ (mod\ 4)) \text{ or} \\ & (p \text{ is even, } n \text{ is odd and } i \equiv 3,0 \ (mod\ 4)); \end{aligned}$$

$$\begin{aligned} g'(wx_i) = 1 \quad & \text{if } (p \text{ is odd, } n \text{ is even and } i \equiv 2,3 \ (mod\ 4)) \text{ or} \\ & (p \text{ is even, } n \text{ is even and } i \equiv 1,0 \ (mod\ 4)) \text{ or} \\ & (p \text{ is odd, } n \text{ is odd and } i \equiv 3,0 \ (mod\ 4)) \text{ or} \\ & (p \text{ is even, } n \text{ is odd and } i \equiv 1,2 \ (mod\ 4)); \end{aligned}$$

for $1 \leq i \leq n-1$,

$$g'(x_i x_{i+1}) = \begin{cases} 1 & \text{if } (n \text{ is even and } i \text{ is odd) or } (n \text{ is odd and } i \text{ is even}) \\ 0 & \text{if } (n \text{ is even and } i \text{ is even) or } (n \text{ is odd and } i \text{ is odd}). \end{cases}$$

Let us denote by $e^*_{g'}(k)$ number of edges $wx_i, x_i x_{i+1}$ labeled with k, where $k = 0, 1$.

Then

$$\begin{aligned}|e_{g'}(1) - e_{g'}(0)| &= \left|(e_{f'}(1) - e^*_{g'}(1)) - (e_{f'}(0) - e^*_{g'}(0))\right| \\ &= \left|e_{f'}(1) - e_{f'}(0) + e^*_{g'}(1) - e^*_{g'}(0)\right|.\end{aligned}$$

Thus, if n is even and q is even then

$$|e_{g'}(1) - e_{g'}(0)| = |\frac{q}{2} - \frac{q}{2} + n - (n-1)| = 1;$$

if n is even, q is odd and $e_{f'}(1) = \frac{q-1}{2}$ then

$$|e_{g'}(1) - e_{g'}(0)| = |\frac{q-1}{2} - \frac{q+1}{2} + n - (n-1)| = 0;$$

if n is odd, q is even and p is odd then

$$|e_{g'}(1) - e_{g'}(0)| = |\frac{q}{2} - \frac{q}{2} + (n-1) - n| = 1;$$

if n is odd, q is even and p is even then

$$|e_{g'}(1) - e_{g'}(0)| = |\frac{q}{2} - \frac{q}{2} + n - (n-1)| = 1;$$

if n is odd, q is odd, p is odd and $e_{f'}(1) = \frac{q+1}{2}$ then

$$|e_{g'}(1) - e_{g'}(0)| = |\frac{q+1}{2} - \frac{q-1}{2} + (n-1) - n| = 0;$$

if n is odd, q is odd, p is even and $e_{f'}(1) = \frac{q-1}{2}$ then

$$|e_{g'}(1) - e_{g'}(0)| = |\frac{q-1}{2} - \frac{q+1}{2} + n - (n-1)| = 0.$$

Lemma 6. *Let f be a SD-prime cordial labeling of a graph G of order p and size q. Let $w \in V(G)$ be such that $f(w) = 1$. The graph obtained by identifying a vertex w in G and a vertex of degree n in F_n admits a SD-prime cordial labeling if one of the following conditions holds:*

1. *n is even and q is even,*
2. *n is even, q is odd, p is even and $e_{f'}(1) = \frac{q-1}{2}$,*
3. *n is odd and q is even,*
4. *n is odd, q is odd, p is even and $e_{f'}(1) = \frac{q+1}{2}$,*
5. *n is odd, q is odd, p is odd and $e_{f'}(1) = \frac{q-1}{2}$.*

Proof. Let f be a SD-prime cordial labeling of a graph G of order p and size q. That is the vertices of g are labeled with numbers $\{1, 2, \cdots, p\}$ and

$$|e_{f'}(1) - e_{f'}(0)| \leq 1.$$

Let $w \in V(G)$ be such that $f(w) = 1$. Let us denote by H the graph obtained by identifying a vertex w in G and a vertex of degree n in F_n.
We define a vertex labeling g of H such that

$$g(v) = f(v), \quad v \in V(G);$$

If n is even,

$$g(x_i) = \begin{cases} p+i & \text{if } i \equiv 1,2 \ (mod\ 4) \text{ and } 1 \le i \le n \\ p+i+1 & \text{if } i \equiv 3 \ (mod\ 4) \text{ and } 1 \le i \le n \\ p+i-1 & \text{if } i \equiv 0 \ (mod\ 4) \text{ and } 1 \le i \le n; \end{cases}$$

If n is odd,

$$g(x_i) = \begin{cases} p+i & \text{if } i = 1 \\ p+i+1 & \text{if } i = 2 \\ p+i-1 & \text{if } i = 3 \\ p+i & \text{if } i \equiv 1,2 \ (mod\ 4) \text{ and } 4 \le i \le n \\ p+i+1 & \text{if } i \equiv 3 \ (mod\ 4) \text{ and } 4 \le i \le n \\ p+i-1 & \text{if } i \equiv 0 \ (mod\ 4) \text{ and } 4 \le i \le n. \end{cases}$$

Using the above labeling and similar method of Lemma 5, one can easily verify that the graph obtained by identifying a vertex w in G and a vertex of degree n in F_n admits a SD-prime cordial labeling.

Theorem 3. *Let f be a SD-prime cordial labeling of a graph G of order p and size q. Let $w \in V(G)$ be such that $f(w) = 2^s$, where $s = 0, 1, 2, \cdots$. The graph obtained by identifying a vertex w in G and a vertex of degree n in F_n admits a SD-prime cordial labeling.*

Proof. The proof follows from Lemmas 5 and 6.

Conjecture 1. Let f be a SD-prime cordial labeling of a graph G_1 of order p and size q. Let $w \in V(G_1)$ be such that $f(w) = 2^s$, where $s = 0, 1, 2, \cdots$. The graph obtained by identifying a vertex w in G_1 and any one of the vertices in any graph G_2 admits a SD-prime cordial labeling.

References

1. Baskar Babujee, J., Babitha, S.: New constructions of edge bimagic graphs from magic graphs. Appl. Mathe. **2**, 1393–1396 (2011)
2. Gallian, J.A.: A dynamic survey of graph labeling. Electronic J. Combin. 18, #DS6 (2015)
3. Harary, F.: Graph Theory. Addison-wesley, Reading (1972)
4. Lau, G.C., Chu, H.H., Suhadak, N., Foo, F.Y., Ng, H.K.: On SD-prime cordial graphs. Int. J. Pure Appl. Mathe. **106**(4), 1017–1028 (2016)

5. Lau, G.C., Shiu, W.C.: On SD-prime labeling of graphs. Utilitas Math. (2014, accepted)
6. Lau, G.C., Shiu, W.C., Ng, H.K., Ng, C.D., Jeyanthi, P.: Further results on SD-prime labeling. J. Combin. Math. Combin. Comput. **98**, 151–170 (2016)
7. Lourdusamy, A., Patrick, F.: Some results on SD-prime labeling of graphs. Proyecciones. **36**(3) (2017)

Dominator Coloring of Generalized Petersen Graphs

J. Maria Jeyaseeli(✉), Nazanin Movarraei, and S. Arumugam

National Centre for Advanced Research in Discrete Mathematics (n-CARDMATH), Kalasalingam University, Anand Nagar, Krishnankoil 626126, Tamil Nadu, India
mariyajeyaseeli@gmail.com, nazanin.movarraei@gmail.com, s.arumugam.klu@gmail.com

Abstract. A vertex coloring $\mathcal{C} = \{V_1, V_2, \ldots, V_k\}$ of a graph G is called a dominator coloring of G if every vertex v of G is adjacent to all the vertices of at least one color class V_i. The dominator chromatic number $\chi_d(G)$ is the minimum number of colors required for a dominator coloring. In this paper we determine the dominator chromatic number of the generalized Petersen graph $P(n, k)$ where $1 \leq k \leq 3$.

Keywords: Dominator coloring · Dominator chromatic number · Generalized petersen graph

1 Introduction

By a graph $G = (V, E)$, we mean a finite undirected graph with neither loops nor multiple edges. The order $|V|$ and the size $|E|$ of G are denoted by n and m respectively. For graph theoretic terminology we refer to Chartrand and Lesniak [4].

Graph coloring and domination are two major areas in graph theory that have been well studied. The concept of dominator coloring which was introduced by Hedetniemi et al. [10] has flavour of both these concepts.

Gera et al. [8] also studied this concept. Several results on dominator colorings are given in [1,5,8,9]. Algorithmic aspects of dominator colorings problem have been investigated in Arumugam et al. [2].

Let $G = (V, E)$ be graph and let $v \in V$. The open neighborhood $N(v)$ and the closed neighborhood $N[v]$ are defined by $N(v) = \{u \in V : uv \in E\}$ and $N[v] = N(v) \cup \{v\}$. A subset S of V is called a dominating set of G if $N(v) \cap S \neq \emptyset$ for all $v \in V - S$. The minimum cardinality of a dominating set of G is called the domination number of G and is denoted by $\gamma(G)$. For an excellent treatment of fundamentals of domination we refer to [11].

A vertex coloring of G is an assignment of colors to the vertices of G such that adjacent vertices receive distinct colors. The minimum number of colors required for a coloring of G is called the chromatic number of G and is denoted by $\chi(G)$.

S. Arumugam et al. (Eds.): ICTCSDM 2016, LNCS 10398, pp. 144–151, 2017.
DOI: 10.1007/978-3-319-64419-6_19

Definition 1.1 [8,10]. *Let $G = (V, E)$ be a graph, $S \subseteq V$ and let $v \in V$. We say that v dominates S if v is adjacent to all the vertices in S. A vertex coloring $\mathcal{C} = \{V_1, V_2, \ldots, V_k\}$ is called a dominator coloring of G if every vertex v dominates at least one color class V_i. The dominator chromatic number $\chi_d(G)$ is the minimum number of colors required for a dominator coloring of G. A dominator coloring of G using χ_d colors is called a χ_d-coloring of G.*

Observation 1.2. *Let $\mathcal{C}$ be a χ_d-coloring of G. Since any closed neighborhood $N[v]$ contains a color class from $\mathcal{C}$, it follows that $\chi_d(G) \geq k$ where k is the maximum number of disjoint closed neighborhoods in G. Further, if $\chi_1 = \chi(H)$ where H is the subgraph of G induced by the set of vertices of G not covered by the above k color classes, then $\chi_d(G) \geq k + \chi_1$.*

Theorem 1.3 [9]. *For any graph G, we have $\max\{\chi(G), \gamma(G)\} \leq \chi_d(G) \leq \chi(G) + \gamma(G)$. In particular, if G is bipartite, then $\gamma(G) \leq \chi_d(G) \leq \gamma(G) + 2$.*

Definition 1.4. *The Cartesian product $G \square H$ of two graphs G and H is the graph with $V(G \square H) = V(G) \times V(H)$ and $E(G \square H) = \{(g_1, h_1)(g_2, h_2) : g_1 = g_2$ and $h_1 h_2 \in E(H)$ or $h_1 = h_2$ and $g_1 g_2 \in E(G)\}$.*

Definition 1.5 [13]. *Let n and k be positive integers with $n \geq 3$ and $k \leq n - 1$. The generalized Petersen graph $P(n, k)$ is the graph with $V(P(n, k)) = \{u_1, u_2, u_3, \ldots, u_n\} \cup \{v_1, v_2, v_3, \ldots, v_n\}$ and $E(P(n, k)) = \{v_i v_{i+1} : 1 \leq i \leq n\} \cup \{v_i u_i : 1 \leq i \leq n\} \cup \{u_i u_{i+k} : 1 \leq i \leq n\}$ where addition in the suffix is modulo n.*

Domination in generalized Petersen graphs have been investigated in several papers [3,6,7].

In this paper we determine the dominator chromatic number of the generalized Petersen graph $P(n, k)$ where $1 \leq k \leq 3$.

We need the following theroems.

Theorem 1.6. *The generalized Petersen graph $P(n, k)$ is bipartite if and only if n is even and k is odd.*

Theorem 1.7 [12]. *Let $G = P(n, 1), n \geq 3$. Then*

$$\gamma(G) = \begin{cases} \frac{n}{2} + 1 \; if \; n \equiv 2 (mod\ 4) \\ \lceil \frac{n}{2} \rceil \quad otherwise. \end{cases}$$

Theorem 1.8 [12]. *Let $G = P(n, 2), n \geq 5$. Then $\gamma(G) = \lceil \frac{3n}{5} \rceil$.*

Theorem 1.9 [12]. *Let $G = P(n, 3), n \geq 7$. Then*

$$\gamma(G) = \begin{cases} \frac{n}{2} + 1 & n \equiv 2 (mod\ 4) \\ \lceil \frac{n}{2} \rceil & n \equiv 1, 0 (mod\ 4) or n = 11 \\ \lceil \frac{n}{2} \rceil + 1 & n \equiv 3 (mod\ 4), n \neq 11. \end{cases}$$

2 Main Results

The generalized Petersen graph $P(n,1)$ is isomorphic to the Cartesian product $C_n \square K_2$. We now proceed to determine $\chi_d(P(n,1))$. We start with the following lemma.

Lemma 2.1. *Let $G = P_n \square K_2$ and let $(u_1, u_2, \ldots, u_n)$ and $(v_1, v_2, \ldots, v_n)$ be the two copies of P_n in G with $u_i v_i \in E(G)$. Let $G_1 = G - \{u_1, v_n\}$, if $n \equiv 1(mod\ 4)$ and $G_2 = G - \{u_1, u_n\}$, if $n \equiv 3(mod\ 4)$. Then $\chi_d(G_1) = \chi_d(G_2) = \lfloor \frac{n}{2} \rfloor + 2$.*

Proof. Let $\Im_1 = \{N[u_{4i}] : 1 \leq i \leq \frac{n-1}{4}\} \cup \{N[v_{4i-2}] : 1 \leq i \leq \frac{n-1}{4}\}$ if $n \equiv 1(mod\ 4)$ and let $\Im_2 = \{N[u_{4i}] : 1 \leq i \leq \frac{n-3}{4}\} \cup \{N[v_{4i-2}] : 1 \leq i \leq \frac{n-3}{4} + 1\}$ if $n \equiv 3(mod\ 4)$. Then both $\Im_1$ and $\Im_2$ are two sets of disjoint closed neighbourhoods in G_1 and G_2 respectively with $|\Im_1| = |\Im_2| = \lfloor \frac{n}{2} \rfloor$. It follows from Observation 1.2 that $\chi_d(G_1) \geq \lfloor \frac{n}{2} \rfloor + 2$ and $\chi_d(G_2) \geq \lfloor \frac{n}{2} \rfloor + 2$.

Now, let $\mathcal{C}_1 = \{\{u_{4i}\} : 1 \leq i \leq \frac{n-1}{4}\} \cup \{\{v_{4i-2}\} : 1 \leq i \leq \frac{n-1}{4}\} \cup \{\{v_1, v_3, \ldots, v_{n-2}, u_2, u_6, \ldots, u_{n-3}\}\} \cup \{\{v_4, v_8, \ldots, v_{n-1}, u_3, u_5, \ldots, u_n\}\}$ if $n \equiv 1(mod\ 4)$ and let $\mathcal{C}_2 = \{\{u_{4i}\} : 1 \leq i \leq \frac{n-3}{4}\} \cup \{\{v_{4i-2}\} : 1 \leq i \leq \frac{n-3}{4} + 1\} \cup \{\{v_1, v_3, \ldots v_n, u_2, u_6, \ldots, u_{n-1}\}\} \cup \{\{v_4, v_8, \ldots, v_{n-3}, u_3, u_5, \ldots, u_{n-2}\}\}$ if $n \equiv 3(mod\ 4)$. Clearly $\mathcal{C}_i$ is a dominator coloring of G_i, for i = 1,2 and $|\mathcal{C}_1| = |\mathcal{C}_2| = \lfloor \frac{n}{2} \rfloor + 2$.

Thus $\chi_d(G_1) = \chi_d(G_2) = \lfloor \frac{n}{2} \rfloor + 2$.

Corollary 2.2. *If $\mathcal{C}$ is a dominator coloring of G_1 or G_2 with $\{v_1\} \in \mathcal{C}$, then $|\mathcal{C}| \geq \lfloor \frac{n}{2} \rfloor + 3$.*

Proof. Since $\Im_1$ and $\Im_2$ defined in Lemma 2.1 are $\lfloor \frac{n}{2} \rfloor$ disjoint closed neighbourhoods of $G_1 - v_1$ and $G_2 - v_1$ respectively, and $\mathcal{C} - \{\{v_1\}\}$ is a dominator coloring of $G_1 - v_1$ or $G_2 - v_1$, we have $|\mathcal{C} - \{v_1\}| \geq \lfloor \frac{n}{2} \rfloor + 2$. Hence $|\mathcal{C}| \geq \lfloor \frac{n}{2} \rfloor + 3$.

Corollary 2.3. *Let $\mathcal{C}$ be a χ_d-coloring of G_1 or G_2. Then for each closed neighbourhood in $\Im_1$ or $\Im_2$, its central vertex is a color class in $\mathcal{C}$.*

Proof. Since v_1 is a pendent vertex of G_1 and G_2, either $\{v_1\}$ or $\{v_2\}$ is a color class in $\mathcal{C}$. Also, since $|\mathcal{C}| = \lfloor \frac{n}{2} \rfloor + 2$, it follows from Corollary 2.2 that $\{v_1\} \notin \mathcal{C}$. Hence $\{v_2\} \in \mathcal{C}$. Let $H_1 = \langle N[v_2] \rangle$. Now let $\mathcal{C}_1 = \{C \cap (V(G_1) - V(H_1)) : C \in \mathcal{C}$ and $C \neq \{v_2\}\}$. Clearly $\mathcal{C}_1$ is a dominator coloring of $G - V(H_1)$ and $|\mathcal{C}_1| = \lfloor \frac{n}{2} \rfloor + 1$. Since u_3 is a pendent vertex of $V(G_1) - V(H_1)$, it follows that $\{u_3\}$ or $\{u_4\}$ is a color class in $\mathcal{C}_1$. By Corollary 2.2, $\{u_3\} \notin \mathcal{C}_1$ and hence $\{u_4\} \in \mathcal{C}_1$. Thus $\{u_4\} \in \mathcal{C}$. By a similar argument the central vertex of each closed neighbourhood in $\Im_1$ or $\Im_2$ is a color class in $\mathcal{C}$.

Theorem 1. *For the generalized Petersen graph $G = P(n,1) = C_n \square K_2$, we have*

$$\chi_d(G) = \begin{cases} 3 & if\ n = 3 \\ \lceil \frac{n}{2} \rceil + 3 & if\ n \equiv 2(mod\ 4) \\ \lceil \frac{n}{2} \rceil + 2 & otherwise. \end{cases}$$

Proof. The result is obvious if $n = 3$. Now, let $n = 4k + j$ where $k \geq 1$ and $0 \leq j \leq 3$. Let $(v_1, v_2, \ldots, v_n, v_1)$ and $(u_1, u_2, \ldots, u_n, u_1)$ be the two copies of C_n in G and let $u_iv_i \in E(G)$. Let $\Im = \{N[v_{4i-3}] : 1 \leq i \leq k\} \cup \{N[u_{4i-1}] : 1 \leq i \leq k\}$. Then $\Im$ is a family of $2k$ disjoint closed neighbourhoods in G and it follows from Observation 1.2 that $\chi_d(G) \geq 2k + 2$. To prove the reverse inequality we consider the following cases:

Case 1: $j = 0$.
In this case $\Im$ covers all the vertices of G and $\mathcal{C} = \{\{v_{4i-3}\} : 1 \leq i \leq k\} \cup \{\{u_{4i-1}\} : 1 \leq i \leq k\} \cup \{\{v_2, v_4, \ldots, v_n, u_1, u_5, \ldots, u_{n-3}\}\} \cup \{\{v_3, v_7, \ldots, v_{n-1}, u_2, u_4, \ldots, u_n\}$ is a dominator coloring of G. Thus $\chi_d(G) \leq |\mathcal{C}| = 2k + 2$. Hence, $\chi_d(G) = 2k + 2 = \lceil \frac{n}{2} \rceil + 2$.

Case 2: $j = 1$.
In this case, $\Im$ is a collection of $2k$ disjoint closed neighbourhoods in G and the vertices u_n and v_{n-1} are not covered by these neighbourhoods. Now, by Corollary 2.3 in any dominator coloring of G using $2k + 2$ colors, the central vertices of the closed neighbourhoods are sigleton color classes and the vertices u_n and v_{n-2} do not dominate any of the above color classes.

So, $\chi_d(G) \geq 2k + 3 = \lceil \frac{n}{2} \rceil + 2$. Hence, $\chi_d(G) = 2k + 3 = \lceil \frac{n}{2} \rceil + 2$.

Case 3: $j = 2$.
In this case, $\mathcal{C} = \{\{v_{4i-3}\} : 1 \leq i \leq k\} \cup \{\{u_{4i-1}\} : 1 \leq i \leq k\} \cup \{v_2, v_4, \ldots, v_{n-2}, v_n, u_1, u_5, \ldots, u_{n-5}\} \cup \{v_3, v_7, \ldots, v_{n-3}, u_2, u_4, \ldots, u_{n-2}, u_n\} \cup \{u_{n-1}\} \cup \{v_{n-1}\}$ is a dominator coloring of G. Thus $\chi_d(G) \leq |\mathcal{C}| = 2k + 4 = \lceil \frac{n}{2} \rceil + 3$. On the other hand, let $\mathcal{C}$ be a dominator coloring of G. Let $\mathcal{D} = \{C \in \mathcal{C} : C$ is dominated by at least one vertex in $G\}$. Now C is dominated by four vertices of G if and only if $|C| = 1$. Since $|V(G)| = 8k + 4$, it follows that $|\mathcal{D}| \geq 2k + 1$ and $|\mathcal{D}| = 2k + 1$ if and only if $|C| = 1$ for all $C \in \mathcal{D}$ and there exist $2k+1$ disjoint closed neighbourhoods in G. However the number of disjoint closed neighbourhoods in G is 2k and hence it follows that $|\mathcal{D}| \geq 2k + 2$. Thus $|\mathcal{C}| \geq 2k + 4 = \lceil \frac{n}{2} \rceil + 3$. Hence $\chi_d(G) = |\mathcal{C}| = 2k + 4 = \lceil \frac{n}{2} \rceil + 3$.

Case 4: $j = 3$.
In this case, $N[v_{4k+1}] \cup \Im$ is a collection of $2k+1$ disjoint closed neighbourhoods in G and the vertices u_{n-1} and u_n are not covered by these neighbourhoods. Now, by Corollary 2.3 in any dominator coloring of G using $2k + 3$ colors, the central vertices of the closed neighbourhoods are sigleton color classes and the vertices u_{n-1} and u_n do not dominate any of the above color classes.

So, we need atleast one more color and we have $\chi_d(G) \geq 2k + 4 = \lceil \frac{n}{2} \rceil + 2$. Hence, $\chi_d(G) = 2k + 4 = \lceil \frac{n}{2} \rceil + 2$.

Lemma 2.4. *Let G be the graph of order $2n$ with $V(G) = \{u_1, u_2, \ldots, u_n\} \cup \{v_1, v_2, \ldots, v_n\}$ given in Fig. 1 and let $n \equiv 0 (mod\ 5)$ and $n = 5k$ where $k \geq 1$. Then $\chi_d(G) = 3k + 2$. Further if $\mathcal{C}$ is any χ_d-coloring of G, then $\{u_i\} \in \mathcal{C}$ for all $i \equiv 2$ or $4 (mod\ 5)$ and $\{v_i\} \in \mathcal{C}$ for all $i \equiv 3 (mod\ 5)$.*

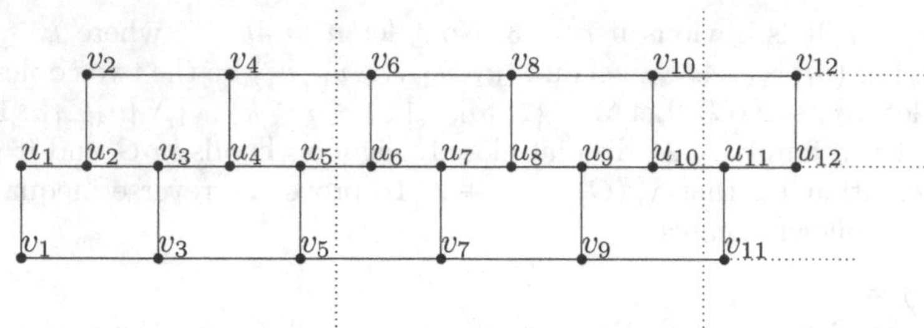

Fig. 1. Graph of Lemma 2.4.

Proof. Let $\mathcal{C}_1 = \{\{v_{5i-2}\} : 1 \leq i \leq k\} \cup \{\{u_{5i-3}\} : 1 \leq i \leq k\} \cup \{\{v_{5i-1}\} : 1 \leq i \leq k\}$. Since $V(G) - S$ where S is the set of vertices not covered by $\mathcal{C}_1$ is bipartite, $V(G) - S$ can be colored with two color classes V_1 and V_2. Hence $\mathcal{C} = \mathcal{C}_1 \cup \{V_1, V_2\}$ is a dominator coloring of G and $\chi_d(G) \leq |\mathcal{C}| = 3k + 2$.

Now let H_i be the subgraph of G induced by the set $S_i = \{u_{5i-4}, \ldots, u_{5i}, v_{5i-4}, \ldots, v_{5i}\}$, where $1 \leq i \leq k$. Clearly, $\{S_i : 1 \leq i \leq k\}$ forms a partition of $V(G)$. Now, let $\mathcal{C}$ be any χ_d-coloring of G and let $\Im = \{D \in \mathcal{C} : D$ is dominated by a vertex $v\}$. Clearly $|D| \leq 3$ for all $D \in \Im$. If $|D| = 2$ or 3, then exactly one vertex of V dominates the color class D. Also if $|D| = 1$ and $D = \{v\}$, then $deg\, v+1$ vertices dominate the color class D. Let r be the number of color classes D in $\Im$ with $|D| \geq 2$ and let s be the number of color classes D in $\Im$ with $|D| = 1$. If we choose one vertex from each color class in $\Im$, then the set of all choosen vertices is a dominating set of G. Hence $|\Im| = r + s \geq \gamma(G) = 3k$. Further at most $3r + s$ vertices are colored by the color classes in $\Im$ and we need at least two colors for coloring the remaining vertices. Hence $\chi_d(G) \geq |\Im| + 2 \geq 3k + 2$. Thus $\chi_d(G) = 3k + 2$.

We now claim that

$$\Im \cap V(H_i) = \{\{v_{5i-2}\}, \{u_{5i-3}\}, \{u_{5i-1}\}\} \text{ for all } i, 1 \leq i \leq k \tag{1}$$

The proof is by induction on k. Let $k = 1$. Then $G = H_1$ and $\chi_d(G) = 5$. If $|D| \geq 2$ for some $D \in \Im$, then exactly one vertex dominates D and for the remaining nine vertices we need at least three color classes for domination. Hence $|\mathcal{C}| \geq 6$ which is a contradiction. Thus $|D| = 1$ for each $D \in \Im \cap V(H_1)$. If $D = \{v_2\}$ or $\{u_1\}$ or $\{v_1\}$ or $\{v_5\}$ or $\{u_5\}$ or $\{u_4\}$, then exactly two vertices dominate D and for the remaining seven vertices we need at least three color classes for domination. If $D = \{u_3\}$ then three vertices dominate D and for the remaining vertices we need at least three classes for domination. Thus in all cases $|\mathcal{C}| \geq 6$, giving a contradiction. Hence (1) follows when $k = 1$. We now assume that the result is true for $k - 1$ and let $H = H_1 \cup H_2 \cup \cdots \cup H_{k-1}$. By our assumption no vertex of H_k dominates a color class in H. Hence there exist three color classes in H_k such that every vertex of H_k dominates one of these color classes and these three color classes are necessarily $\{v_{5k-2}\}, \{u_{5k-3}\}$ and $\{u_{5k-1}\}$.

Theorem 2.5. *Let $G = P(n, 2)$ with $n \geq 5$. Then $\chi_d(G) = \lceil \frac{3n}{5} \rceil + 2$.*

Proof. Let $V(G) = \{v_1, v_2, \ldots, v_n\} \cup \{u_1, u_2, \ldots, u_n\}$ and let $u_iv_i \in E(G)$. Let $n = 5k+j$ where $k \geq 1$ and $0 \leq j \leq 4$. Let $\mathcal{C}_1 = \{\{v_{5i-3}\} : 0 \leq i \leq \frac{n}{5}\} \cup \{\{u_{5i-2}\} : 1 \leq i \leq \frac{n}{5}\} \cup \{\{v_{5i-1}\} : 0 \leq i \leq \frac{n}{5}\}\}$.

If $j = 0$, let $V_1 = \{v_1, v_3, v_6, v_8, \ldots, v_{n-4}, v_{n-2}, u_4, u_5, u_9, u_{10}, \ldots, u_{n-1}, u_n\}$ and $V_2 = \{u_1, u_2, u_6, u_7, \ldots, u_{n-4}, u_{n-3}, v_5, v_{10}, \ldots, v_n\}$ and $\mathcal{C} = \mathcal{C}_1 \cup \{V_1, V_2\}$.

If $j = 1$, let $V_1 = \{v_1, v_3, v_6, v_8, \ldots, v_{n-3}, u_4, u_5, u_9, u_{10}, \ldots, u_{n-2}, u_{n-1}\}$ and $V_2 = \{u_1, u_2, u_6, u_7, \ldots, u_{n-5}, u_{n-4}, v_5, v_{10}, \ldots, v_{n-1}\}$ and $\mathcal{C} = \mathcal{C}_1 \cup \{V_1, V_2\} \cup \{u_n\} \cup \{v_n\}$.

If $j = 2$, let $V_1 = \{v_1, v_3, v_6, v_8, \ldots, v_{n-4}, v_{n-1}, u_4, u_5, u_9, u_{10}, \ldots, u_{n-3}, u_{n-2}\}$ and $V_2 = \{u_1, u_2, u_6, u_7, \ldots, u_{n-6}, u_{n-5}, v_5, v_{10}, \ldots, v_{n-2}, v_n\}$ and $\mathcal{C} = \mathcal{C}_1 \cup \{V_1, V_2\} \cup \{u_{n-1}\} \cup \{u_n\}$.

If $j = 3$, let $V_1 = \{v_3, v_6, v_8, \ldots, v_{n-5}, v_{n-2}, u_1, u_4, u_5, u_9, u_{10}, \ldots, u_{n-4}, u_{n-3}\}$ and $V_2 = \{u_2, u_6, u_7, \ldots, u_{n-2}, u_{n-1}, v_1, v_5, v_{10}, \ldots, v_{n-3}\}$ and $\mathcal{C} = \mathcal{C}_1 \cup \{V_1, V_2\} \cup \{v_{n-1}\} \cup \{u_n\}$.

If $j = 4$, let $V_1 = \{v_1, v_3, v_6, v_8, \ldots, v_{n-3}, v_{n-1}, u_4, u_5, u_9, u_{10}, \ldots, u_{n-5}, u_{n-4}, u_n\}$ and $V_2 = \{u_1, u_2, u_6, u_7, \ldots, u_{n-3}, u_{n-2}, v_5, v_{10}, \ldots, v_{n-4}\}$ and $\mathcal{C} = \mathcal{C}_1 \cup \{V_1, V_2\} \cup \{v_n\} \cup \{u_{n-1}\} \cup \{v_{n-2}\}$.

Clearly $\mathcal{C}$ is a dominator coloring of G. Thus $\chi_d(G) \leq \lceil \frac{3n}{5} \rceil + 2$. We now prove the reverse inequality. Let $S_i = \{u_{5i-4}, \ldots, u_{5i}, v_{5i-4}, \ldots, v_{5i}\}, 1 \leq i \leq k$ and let H_i be the subgraph of G induced by S_i. We consider the following cases.

Case 1. $j = 0$

In this case S_i covers all the vertices of G and by Lemma 2.4, each H_i contains at least three color classes. Since we need at least two colors for the remaining vertices, $\chi_d(G) \geq 3 \lfloor \frac{n}{5} \rfloor + 2 = \lceil \frac{3n}{5} \rceil + 2$.

Case 2. $j = 1$

In this case two vertices u_i and v_i are not covered by S_i. By Lemma 2.4 each H_i has at least three color classes such that each vertex of H_i dominates one of these color classes. Further we need at least one color class which is to be dominated by the vertices not covered by H_i. Hence $\chi_d(G) \geq 3 \lfloor \frac{n}{5} \rfloor + 3 = \lceil \frac{3n}{5} \rceil + 2$.

Case 3. $j = 2$

In this case four vertices are not covered by S_i. It follows from Lemma 2.4 that $\chi_d(G) \geq 3 \lfloor \frac{n}{5} \rfloor + 4 = \lceil \frac{3n}{5} \rceil + 2$.

Case 4. $j = 3$

In this case six vertices are not covered by S_i. It follows from Lemma 2.4 that $\chi_d(G) \geq 3 \lfloor \frac{n}{5} \rfloor + 4 = \lceil \frac{3n}{5} \rceil + 2$.

Case 5. $j = 4$

In this case eight vertices are not covered by S_i. It follows from Lemma 2.4 that $\chi_d(G) \geq 3 \lfloor \frac{n}{5} \rfloor + 5 = \lceil \frac{3n}{5} \rceil + 2$.

Theorem 2.6. *Let $G = P(n,3)$. Then*

$$\chi_d(G) = \begin{cases} \lceil \frac{n}{2} \rceil + 2 & \text{if } n \equiv 0 (mod 4) \text{ or } n \equiv 1 (mod 4) \text{ or } n = 11 \\ \lceil \frac{n}{2} \rceil + 3 & \text{otherwise.} \end{cases}$$

Proof. Let $n = 4k + j$ where $k \geq 1$ and $0 \leq j \leq 3$. Let $V(G) = \{v_1, v_2, \ldots, v_n\} \cup \{u_1, u_2, \ldots, u_n\}$ and let $u_i v_i \in E(G)$. Let $\mathcal{C}_1 = \{\{v_{4i-2}\} : 1 \leq i \leq k\} \cup \{\{u_{4i}\} : 1 \leq i \leq k\}$.

If $j = 1$ or $n = 11$, let $V_1 = \{v_1, v_3, \ldots, v_{n-4}, v_{n-1}, u_2, u_6, \ldots, u_{n-3}, u_{n-2}\}$ and $V_2 = \{u_1, u_3, \ldots, u_{n-8}, u_{n-6}, u_{n-4}, v_4, v_8, \ldots, v_{n-5}, v_{n-2}, v_n\}$ and $\mathcal{C} = \mathcal{C}_1 \cup \{V_1, V_2\} \cup \{u_n\}$.

If $j = 3$ and $n \neq 11$, let $V_1 = \{v_1, v_3, \ldots, v_{n-6}, v_{n-3}, u_2, u_6, \ldots, u_{n-5}, u_{n-4}, u_n\}$ and $V_2 = \{u_1, u_3, \ldots, u_{n-6}, u_{n-1}, v_4, v_8, \ldots, v_{n-7}, v_{n-4}, v_{n-2}, v_n\}$ and $\mathcal{C} = \mathcal{C}_1 \cup \{V_1, V_2\} \cup \{v_2\} \cup \{u_{n-2}\}$.

Clearly $\mathcal{C}$ is a dominator coloring of G. Thus

$$\chi_d(G) \leq \begin{cases} \lceil \frac{n}{2} \rceil + 2 & \text{if } n \equiv 1 (mod\ 4) \text{ or } n = 11 \\ \lceil \frac{n}{2} \rceil + 3 & \text{if } n \equiv 3 (mod\ 4), n \neq 11 \end{cases}$$

If $j = 0$ or $j = 2$, it follows from Theorem 1.6 that G is bipartite. Hence by Theorems 1.3 and 1.9 we have

$$\chi_d(G)] \leq \begin{cases} \lceil \frac{n}{2} \rceil + 2 & \text{if } n \equiv 0 (mod\ 4) \\ \lceil \frac{n}{2} \rceil + 3 & \text{if } n \equiv 2 (mod\ 4) \end{cases}$$

To prove the reverse inequality, we consider the following cases.

Case 1. $j = 0$
In this case there exist $2k$ disjoint closed neighbourhoods in G. Hence $\chi_d(G) \geq 2k + 2 = \frac{n}{2} + 2$.

Case 2. $j = 1$
In this case there exist $2k-1$ disjoint closed neighbourhoods in G and six vertices are not covered by these neighbourhoods. Now, let $\mathcal{C}$ be a χ_d-coloring of G. Let $\Im = \{C \in \mathcal{C} : C$ is dominated by at least one vertex in $G\}$. Since each color class $C \in \Im$ is dominated by at most four vertices in G and $|V(G)| = 8k + 2$, it follows that $|\Im| \geq 2k + 1$ Hence $\chi_d(G) \geq 2k + 3 = \lceil \frac{n}{2} \rceil + 2$.

Case 3. $j = 2$
In this case there exist $2k$ disjoint closed neighbourhoods in G and four vertices are not covered by these neighbourhoods. Since $|V(G)| = 8k + 4$ it follows that $|\Im| \geq 2k + 1$ and $|\Im| = 2k + 1$ if and only if $|C| = 1$ for all $C \in \Im$ and there exist $2k + 1$ disjoint closed neighbourhoods in G. However the maximum number of disjoint closed neighbourhoods in G is $2k$ and hence it follows that $|\Im| \geq 2k + 2$. Hence $\chi_d(G) \geq 2k + 4 = \lceil \frac{n}{2} \rceil + 3$.

Case 4. $j = 3$
In this case there exist $2k+1$ disjoint closed neighbourhoods in G and two vertices are not covered by these neighbourhoods. The corresponding $2k + 1$ color classes dominate all the vertices of these neighbourhoods if and only if the center vertices of each neighbourhood is a singleton color and the two uncovered vertices do not dominate any of these color classes. Hence, $\chi_d(G) \geq 2k + 5 = \lceil \frac{n}{2} \rceil + 3$.

3 Conclusion

In this paper we have determined the value of $\chi_d(G)$ when $G = P(n, k)$ where $1 \leq k \leq 3$. The problem remains open for $P(n, k)$ where $k \geq 4$ and results for these cases will be reported in a subsequent paper.

Acknowledgments. The first two authors are thankful to the management of Kalasalingam University for providing fellowship.

References

1. Arumugam, S., Bagga, J., Raja Chandrasekar, K.: On dominator coloring in graphs. Proc. Indian Acad. Sci J. **122**, 561–571 (2012)
2. Arumugam, S., Chandrasekar, K.R., Misra, N., Philip, G., Saurabh, S.: Algorithmic aspects of dominator colorings in graphs. In: Iliopoulos, C.S., Smyth, W.F. (eds.) IWOCA 2011. LNCS, vol. 7056, pp. 19–30. Springer, Heidelberg (2011). doi:10.1007/978-3-642-25011-8_2
3. Behzad, A., Behzad, M., Prager, C.E.: On the domination number of the generalized Petersen graphs. Discrete Math. **308**, 603–610 (2008)
4. Chartrand, G., Lesniak, L.: Graphs and Digraphs. Chapman and Hall, London (2005)
5. Chellali, M., Maffray, F.: Dominator coloring in some classes of graphs. Graphs Comb. **28**, 97–107 (2012)
6. Fu, X., Yang, Y., Jiangh, B.: On the domination number of generalized Petersen graph $P(n, 3)$.. Ars Combin. **84**, 73–83 (2007)
7. Fu, X., Yang, Y., Jiangh, B.: On the domination number of generalized Petersen graph $P(n, 2)$. Discrete Math. (to appear)
8. Gera, R., Rasmussen, C., Horton, S.: Dominator coloring and safe clique partitions. Congr. Numer. **181**, 19–32 (2006)
9. Gera, R.M.: On dominator coloring in graphs. Graph Theory Notes N.Y. **LII**, 25–30 (2007)
10. Hedetniemi, S.M., Hedetniemi, S.T., Mcrae, A.A., Blour, J.R.S.: Dominator coloring of graphs. (2006, preprint)
11. Haynes, T.W., Hedetniemi, S.T., Slater, P.J.: Fundamentals of Domination in Graphs. Marcel Dekker, New York (1998)
12. Javad Ebrahimi, B., Jahanbakht, N., Mahmodian, E.S.: Vertex domination of generalized Petersen graphs. Discrete Math. **309**, 4355–4361 (2009)
13. Watkins, M.: A theorem on Tait coloring with an application to the generalized Petersen graph. J. Combin. Theory **6**, 152–164 (1969)

Super Edge-Antimagic Gracefulness of Disconnected Graphs

G. Marimuthu(✉) and P. Krishnaveni(✉)

Department of Mathematics, The Madura College,
Madurai 625011, Tamil Nadu, India
yellowmuthu@yahoo.com, krishnaswetha82@gmail.com

Abstract. For a graph $G = (V, E)$, a bijection g from $V(G) \cup E(G)$ into $\{1, 2, \ldots, |V(G)| + |E(G)|\}$ is called (a, d)-edge-antimagic graceful labeling of G if the edge-weights $w(xy) = |g(x) + g(y) - g(xy)|, xy \in E(G)$, form an arithmetic progression starting from a and having a common difference d. An (a, d)-edge-antimagic graceful labeling is called super (a, d)-edge-antimagic graceful if $g(V(G)) = \{1, 2, \ldots, |V(G)|\}$. We study super (a, d)-edge-antimagic graceful labelings of disconnected graphs, mC_n, mK_n and mP_n.

Keywords: Edge-antimagic graceful labeling · Super edge-antimagic graceful labeling

2010 Mathematical Subject Classification Number: 05C

1 Introduction

We consider finite undirected nontrivial graphs with neither loops nor multiple edges. We denote by $V(G)$ and $E(G)$ the set of vertices and the set of edges of a graph G, respectively. Let $|V(G)| = p$ and $|E(G)| = q$ be the order and size of G respectively. General references for graph-theoretic notions are [2,21].

A labeling of a graph is any function that carries some set of graph elements to numbers. Kotzig and Rosa [12,13] introduced the concept of edge-magic labeling. For more information on edge-magic and super edge-magic labelings, see [8].

Hartsfield and Ringel [9] introduced the concept of antimagic labeling. An antimagic labeling of a (p, q) graph G as a bijection f from $E(G)$ to the set $\{1, 2, \ldots, q\}$ such that the sums of label of the edges incident with each vertex $v \in V(G)$ are distinct. The concept of (a, d)-edge-antimagic total labeling was introduced by Simanjuntak et al. in [19]. This labeling is the extension of the notion of edge-magic labeling, see [12,13].

For a graph $G = (V, E)$, a bijection g from $V(G) \cup E(G)$ into $\{1, 2, \ldots, |V(G)| + |E(G)|\}$ is called an (a, d)-edge-antimagic total labeling of G if the edge-weights $w(xy) = g(x) + g(y) + g(xy), xy \in E(G)$, form an arithmetic progression starting from a and having a common difference d. The $(a, 0)$-edge-antimagic total labeling is usually called edge-magic total lableing [7,12,13]. An (a, d)-edge

S. Arumugam et al. (Eds.): ICTCSDM 2016, LNCS 10398, pp. 152–155, 2017.
DOI: 10.1007/978-3-319-64419-6_20

antimagic total labeling is called super if the smallest possible labels appear on the vertices.

All cycles and paths have a (a, d)-edge antimagic total labeling for some values of a and d, see [19]. In [1], Baca et al. proved the (a, d)-edge-antimagic properties of certain classes of graphs. Ivanco and Luckanicova [10] described some constructions of super edge-magic total (super $(a, 0)$-edge-antimagic total) labelings for disconnected graphs, namely, $nC_k \cup mP_k$ and $K_{1,m} \cup K_{1,n}$. Super edge-amtimagic total labelings of mK_n are given in [5]. Super (a, d)-edge-antimagic labelings for $P_n \cup P_{n+1}, nP_2 \cup P_n$ and $nP_2 \cup P_{n+2}$ have been described by Sudarsana et al. in [20].

In [6], Dafik et al. proved super edge-antimagicness of a disjoint union of m copies of C_n and m copies of P_n. For most recent research in the subject, refer to [3,4,11,14,16–18].

A (p, q) graph G is called edge magic graceful if there exists a bijection $g : V(G) \cup E(G) \to \{1, 2, \ldots, p + q\}$ such that $|g(x) + g(y) - g(xy)| = k$, a constant for any edge xy of G. The graph G is said to be super edge magic graceful if further $g(V(G)) = \{1, 2, \ldots, p\}$.

In [15] Marimuthu et al. presented some properties of super edge magic graceful graphs and proved that some classes of graphs are super edge magic graceful.

An (a, d)-edge-antimagic graceful labeling is defined as a bijection from $V(G) \cup E(G)$ to $\{1, 2, 3, \ldots, p + q\}$ so that the set of edge-weights of all edges in G is equal to $\{a, a + d, a + 2d, \ldots, a + (q - 1)d\}$, for two integers $a \geq 0$ and $d \geq 0$.

An (a, d)-edge-antimagic graceful labeling g is called super (a, d)-edge-antimagic graceful if $g(V(G)) = \{1, 2, \ldots, p\}$. A graph G is called (a, d)-edge-antimagic graceful or super (a, d)-edge-antimagic graceful if there exists an (a, d)-edge-antimagic graceful or a super (a, d)-edge-antimagic graceful labeling of G.

In this paper we study super (a, d)-edge-antimagic graceful labelings of certain classes of graphs including mC_n, mK_n and mP_n.

2 Main Results

We first consider super edge-antimagic gracefulness of mC_n, disjoint union of m copies of C_n, where $m > 1$. Let $V(mC_n) = \{x_i^j : 1 \leq i \leq n, 1 \leq j \leq m\}$ and $E(mC_n) = \{x_i^j x_{i+1}^j : 1 \leq i \leq n - 1, 1 \leq j \leq m\} \cup \{x_n^j x_1^j : 1 \leq j \leq m\}$.

Theorem 1. *The graph mC_n has a super $(0, 1)$ -edge-antimagic graceful labeling for every $m \geq 2$ and $n \geq 3$.*

Proof. Define $f : V(mC_n) \cup E(mC_n) \to \{1, 2, \ldots, 2mn\}$ as follows:

$$f(x_i^j) = i + (j - 1)n, \qquad if \qquad 1 \leq i \leq n, 1 \leq j \leq m$$

$$f(x_i^j x_{i+1}^j) = mn + j + i + (n - 1)(j - 1), \quad 1 \leq i \leq n - 1, 1 \leq j \leq m$$

$$f(x_n^j x_1^j) = (m + 1)n - (n - 1) + n(j - 1) \quad if \quad 1 \leq j \leq m.$$

The edge-weights of mC_n, under the labeling f are given by $W = W_f^1 \cup W_f^2$ where

$W_f^1 = \{W_f^1(x_i^j x_{i+1}^j) = mn - i - n(j-1) : 1 \leq i \leq n-1 \quad and \quad 1 \leq j \leq m\}$ and $W_f^2 = \{W_f^2(x_n^j x_1^j) = mn - nj : \quad 1 \leq j \leq m\}$.

Clearly $W = \{0, 1, 2, \ldots, mn-1\}$. Thus f is a super $(0,1)$-edge-antimagic graceful labeling.

Theorem 2. *The graph mK_n, where $m \geq 2$ and $n \geq 2$ is super $(a,1)$-edge-antimagic graceful.*

Proof. Let $\{x_1^j, x_2^j, \ldots, x_n^j\}$ be the vertex set of the jth copies of K_n where $1 \leq j \leq m$. Define $f : V(mK_n) \cup E(mK_n) \rightarrow \{1, 2, \ldots, \frac{mn(n+1)}{2}\}$ as follows.

$$f(x_i^j) = m(n-i+1) - j + 1 \quad for \quad 1 \leq i \leq n \quad and \quad 1 \leq j \leq m$$

$$f(x_i^j x_{i+k}^j) = m(n(k+1) - \frac{k(k+1)}{2} + 1 - i) + 1 - j \quad for \quad 1 \leq i \leq n-1,$$

$$1 \leq k \leq n-i \quad and \quad 1 \leq j \leq m.$$

The edge-weight of $x_i^j x_{i+k}^j$, is given by

$$w(x_i^j x_{i+k}^j) = m(n(k-1) - \frac{k(k+1)}{2} + (k-1) + i) - 1 + j \text{ where}$$
$$1 \leq i \leq n-1, \quad 1 \leq k \leq n-i \quad and \quad 1 \leq j \leq m.$$

It can be easily verified that the set of edge-weights is $\{0, 1, 2, \ldots, \frac{mn(n-1)}{2} - 1\}$. Hence f is a super $(a,1)$-edge-antimagic graceful labeling of mK_n.

We now consider mP_n, disjoint union of m copies of P_n. $V(mP_n) = \{x_i^j : 1 \leq i \leq n, 1 \leq j \leq m\}$ and $E(mP_n) = \{x_i^j x_{i+1}^j : 1 \leq i \leq n-1, 1 \leq j \leq m\}$.

Theorem 3. *The graph mP_n where $m \geq 2$ and $n \geq 2$ is super $(a,1)$-edge-antimagic graceful.*

Proof. Define $f : V(mP_n) \cup E(mP_n) \rightarrow \{1, 2, \ldots, m(2n-1)\}$ by $f(x_i^j) = j + (i-1)m$ and $f(x_i^j x_{i+1}^j) = nm + j + (i-1)m$ where $1 \leq i \leq n-1$ and $1 \leq j \leq m$.

The set of edge-weights consists of the consecutive integers $\{0, 1, 2, \ldots, (m(n-1) - 1\}$ and hence f is a super $(a,1)$-edge-antimagic graceful labeling of mP_n.

References

1. Baca, M., Lin, Y., Miller, M., Youssef, M.Z.: Edge-antimagic graphs. Discrete Math. **307**, 1232–1244 (2007)
2. Baca, M., Miller, M.: Super Edge-Antimgic Graphs: A Wealth of Problems and Some Solutions. Brown Walker Press, Boca Raton (2008)

3. Baca, M., Barrientos, C.: Graceful and edge antimagic labelings. Ars Combin. **96**, 505–513 (2010)
4. Baca, M., Baskoro, E.T., Simanjuntak, R., Sugeng, K.A.: Super edge-antimagic labelings of the generalized Petersen graph $P\left(n, \frac{n-1}{2}\right)$. Util. Math. **70**, 119–127 (2006)
5. Baca, M., Barrientos, C.: On super edge-antimagic total labelings of mK_n. Discrete Math. **308**, 5032–5037 (2008)
6. Dafik, D., Miller, M., Ryan, J., Baca, M.: On super (a, d)-edge-antimagic total labeling of disconnected graphs. Discrete Math. **309**, 4909–4915 (2009)
7. Enomoto, H., Llado, A.S., Nakamigawa, T., Ringel, G.: Super edge-magic graphs. SUT J. Math. **34**, 105–109 (1998)
8. Gallian, J.A.: A dynamic survey of graph labeling. Electron. J. Combin. # DS6, 1–408 (2016)
9. Hartsfield, N., Ringel, G.: Pearls in Graph Theory. Academic Press, Boston (1990)
10. Ivanco, J., Luckanicova, I.: On edge-magic disconnected graphs. SUT J. Math. **38**, 175–184 (2002)
11. Javaid, M., Bhatti, A.A.: On super (a, d)-edge-antimagic total labeling of subdivided stars. Ars Combin. **105**, 503–512 (2012)
12. Kotzig, A., Rosa, A.: Magic Valuations of Complete Graphs. CRM Publisher, New York (1972)
13. Kotzig, A., Rosa, A.: Magic valuations of finite graphs. Canadian Math. Bull. **13**, 451–461 (1970)
14. Lee, M.J.: On super $(a, 1)$-edge-antimagic total labelings of cartesian product graphs. J. Discrete Math. Sci. Cryptogr. **16**(2–3), 117–124 (2013)
15. Marimuthu, G., Balakrishnan, M.: Super edge magic graceful graphs. Inform. Sci. **287**, 140–151 (2014)
16. Rahmawati, S., Silaban, D.R., Miller, M., Baca, M.: Constuction of new larger (a, d)-edge-antimagic vertex graphs by using adjacency matrices. Australian J. Combin. **56**, 257–272 (2013)
17. Roshini Leely Pushpam, P., Saibulla, A.: On super (a, d)-edge-antimagic total labeling of certain families of graphs. Discuss. Math. Graph Theory. **32**(3), 535–543 (2012)
18. Roshini Leely Pushpam, P., Saibulla, A.: Super (a, d)-edge-antimagic total labelings of some classes of graphs. SUT J. Math. **48**(1), 1–12 (2012)
19. Simanjuntak, R., Bertault, F., Miller, M.: Two new (a, d)-antimagic graph labelings. In: Proceedings of 11th Australian Workshop of Combinatorial Algorithm, pp. 179–189 (2000)
20. Sudarsana, I.W., Ismaimuza, D., Baskoro, E.T., Assiyatun, H.: On super (a, d)-edge-antimagic total labeling of disconnected graphs. J. Combin. Math. Combin. Comput. **55**, 149–158 (2005)
21. West, D.B.: An Introduction to Graph Theory. Prentice Hall, Engelwood Cliffs (1996)

Mixed Noise Elimination and Data Hiding for Secure Data Transmission

P. Mohanakrishnan[1], K. Suthendran[2](✉), S. Arumugam[2], and T. Panneerselvam[2]

[1] Department of Computer Science, Axis College of Engineering, Thirssur, Kerala, India
Mohanp2248@gmail.com

[2] Kalasalingam University, Anand Nagar, Krishnankoil 626126, Tamil Nadu, India
{k.suthendran,s.arumugam}@klu.ac.in,tpsphc@gmail.com

Abstract. In image processing, mixed noise elimination from the image is a difficult task since the noise distribution usually does not have a parametric model. The Additive White Gaussian Noise (AWGN) together with impulse noise (IN) is one typical example of mixed noise. Most of the noise removal methods detect the locations of impulse noise pixels and then removes mixed noise. The presence of strong mixed noise leads to unwanted artifacts and to solve this issue a weighted encoding with sparse nonlocal regularization (WESNR) method is available and it removes mixed noise by soft impulse detection through weighted encoding. In this work, WESNR is used to eliminate mixed noise. Reversible Data Hiding (RDH) technique is used to encrypt denoised image and hides data for the purpose of secure communication. Experimental results showed that the proposed method can attain real reversibility after data extraction without affecting image quality.

Keywords: Noise estimation · AWGN · IN · WESNR · RDH

1 Introduction

A major portion of information received by a human being from the environment is visual. Hence processing visual information by computer has been drawing a very significant attention of researchers over the last few decades. Removal of noise is an essential and challenging operation in image processing. Before performing any process in an image, it must be first restored because images may be corrupted by noise during image acquisition and transmission. Nature of noise removal depends on the type of noise which has corrupted the image. To remove mixed noise there is a novel method which consists of two stages, namely detection of noise in image and elimination of noise from the image. The effect of this process will be minimal and different type of noise in an image can increase the strength of adaptive Gaussian noise and impulse noise [1]. The aim of this

S. Arumugam et al. (Eds.): ICTCSDM 2016, LNCS 10398, pp. 156–164, 2017.
DOI: 10.1007/978-3-319-64419-6_21

work is to introduce effective weighted encoding with sparse nonlocal regularization for mixed noise removal. The chosen noise removal technique [2] is widely used in areas such as medical imagery, military imagery and law forensics. The image denoising technique can hide denoised images and data [3,7]. Denoising has a broad spectrum of applications and it covers wide area of image processing applications. The survey report [4] indicates a method of restoration of images corrupted by Gaussian and uniform impulsive noise. The proposed blind inpainting algorithm removes the noise in iteration basis and results in better performance. The review article [5] presents two-phase approach of deblurring images corrupted by impulse plus Gaussian noise. Restoration of blurred images corrupted with impulse noise is a difficult problem. The image deblurring from noisy data is a fundamental problem in image processing and in real applications, practical systems can sometimes suffer from few or more pixels, called outliers that are much noisier than others. The developed method outperforms them by at least 2 to 6 dB in PSNR and gives satisfactory results even if noise level is high. The paper [6] deals with Switching Bilateral Filter (SBF) and Texture/Noise Detector for universal noise removal. The operation was carried out in two stages, namely detection followed by filtering. For detection, sorted quadrant median vector (SQMV) scheme was used with features like edge or texture information. The proposed SQMV was mainly used to edge/texture detection, noise detection and switching bilateral filter. The edge detector, obtains a reference median value of noise detection, which detects impulse and Gaussian noise and both detectors were based on robust estimators of SQMV. Many noise removal algorithms, such as bilateral filtering, tendancy to treat impulse noise as edge pixels end up with unsatisfactory results. In order to process impulse pixels and edge pixels differently, two detectors were introduced based on SQMV in a pixel of neighborhood. With regard to impulse detection rate and classification rate, the noise detector shows a good performance in identifying noise even in mixed noise models. In most of the noise model cases, proposed filter outperforms other filters, both in PSNR and visually. In [8] patch-based approach for removing mixed Gaussian-impulse noise was explained and it was addressing the problem of image restoration which have been affected by a mixture of Gaussian and impulse noise. During the last fifteen years, great deal of image processing techniques have been developed in order to take advantage of self-similarities of images. The idea of restoration methods is simple and nice, since it relies on assumption that a given patch can be found almost identically in different places. The patch can be restored in two steps. In the first step corresponding patches in an image are identified and trade-off between discriminative power and robustness to noise for each patch is measured. In the second step the real underlying patch behind the damaged versions is estimated [9]. Further investigation has led to the conclusion that the patch based approach is efficient in removing mixtures of Gaussian and impulse noises. The obtained result was widely studied in the particular case of gaussian noise as well extension to impulse degradations of both the similarity measure between patches and statistical estimator of original patches. The papers [10–12] deal with Reversible Data Hiding in Encrypted

Image. In [12], a novel reversible data hiding scheme for encrypted image was proposed. The scheme consists of three phases, namely, image encryption, data embedding and data extraction /image recovery. The data of original image are entirely encrypted by a stream cipher and a receiver can decrypt this encrypted image containing embedded data, by using encryption key and the decrypted version is similar to original image. According to data-hiding key, with the aid of the spatial correlation in natural image, the embedded data can be correctly extracted and original image can perfectly recovered. Although the knowledge of encryption key can be used to obtain a decrypted image and detect presence of hidden data using LSB-steganalytic methods and without data-hiding key, it is still impossible to extract the additional data and recovery of original image. To ensure correct data-extraction and perfect image recovery, the block side length has to be a big value like 32, or error correction mechanism has to be introduced before data hiding to protect additional data with a cost of payload reduction. Existing mixed noise removal methods [13] are based on detection and it involves two sequential steps, namely detection of IN pixels and removal of noise. The two phase strategy will become less effective when the AWGN or IN is very strong. The major difficulty of removal of IN and AWGN mixed noise lies in complex distribution of mixed noise and it has heavy tail so that it cannot be readily characterized by parametric model. In this connection the present study focuses effective encoding based method for removal of mixed noise, namely, a weighted encoding with sparse nonlocal regularization (WESNR). There is no explicit impulse pixel detection in WESNR, and each noise-corrupted patch is encoded over a pre-learned dictionary to remove IN and AWGN simultaneously in a soft impulse pixel detection manner. In WESNR, the mixed noise is suppressed by weighting encoding residual so that the final encoding residual tends to follow Gaussian distribution. The weighted encoding and sparse nonlocal regularization are unified into a variation framework and it removes the mixed noise. By using this technique it is possible to have communication between sender and receiver using denoised image and data which can be hidden in an image. So, our proposed method has more security on images as well data.

2 Module Description

Proposed system includes Pre-processing, filtering, denoising transmitter side and receiver side.

2.1 Pre-processing

In order to achieve the effective impact of mixed noise technique, as a preprocessing step we add both additive noise and impulse noise to image. Addition of additive white gaussian noise is done by adding a random value to image pixel. Further impulse noise is added to image. There are two types of impulse noise, namely, random valued impulse noise and salt and pepper impulse noise. The impulse noise is added in an image by adding a function of Matlab.

2.2 Filtering

The adaptive median filtering was used for filtering purpose. The adaptive median filter performs spatial processing to determine an image pixel and it has been affected by impulse noise. The adaptive median filter classifies pixels as noise by comparing each pixel in image to its surrounding neighbor pixels. The size of neighborhood is adjustable. A pixel that is different from a majority of its neighbors, as well as being not structurally aligned with those similar pixels is labeled as impulse noise. These noise pixels were replaced by median and neighborhood value of pixels and they have to pass noise labeling test.

2.3 Denoising

Five high-quality images were used (independent of test images used in this paper) to train PCA dictionaries. A number of 876,359 patches (size: 7×7) are extracted from the five images and they are clustered into 200 clusters by using K-means clustering algorithm. For each cluster, a compact local PCA dictionary is learnt. Meanwhile, centroid of each cluster is calculated. For a given image patch, the euclidian distance between it and the centroid of each cluster is computed, and the PCA dictionary associated with its closest cluster is chosen to encode the given patch. In addition, final denoising results are not sensitive to training images used in PCA dictionary learning.

2.4 Transmitter Side

After denoising, enough space is reserved on original image and the image is converted into its encrypted version with encryption key. Then the data to be embedded in the image is fixed in the image and embedded data is again locked with other encrypted key. In addition the original content was unknown to receiver end. Data-hider compresses least significant bits (LSB) of encrypted image using a data-hiding key to create a sparse space to accommodate additional data. Since data embedding only affects LSB, a decryption with encryption key can result in an image similar to original image.

2.5 Receiver Side

Data extraction and image recovery takes place at receiver side. Using data hiding key the receiver can extract the data and original image by using encryption key. In this connection there are mainly two cases:

- Extracting Data from Encrypted Images: The database manager gets data hiding key, can decrypt LSB-planes and extract additional data by directly reading decrypted version. The database manager can update information through LSB replacement and encrypt updated information according to data hiding key all over again.
- Extracting Data from Decrypted Images: The user wants to decrypt image and for this it is necessary to extract the data from decrypted image.

3 Proposed System Process Flow Diagram

The selected random image is added with addictive white gaussian noise, salt and pepper impulse noise and the random image is turned into noisy image (Figs. 1 and 2). The noisy image is passed through adaptive median filtering and these noise pixels are then replaced by median pixel value of the pixels in the neighborhood that have passed noise labeling test. The noisy image is passed through adaptive filter image and it is divided into patches of denoising. The divided patches are compared with patches and stored in dictionary. Iteratively patches which are obtained from image and dictionary are combined by removing noise in image Fig. 2.

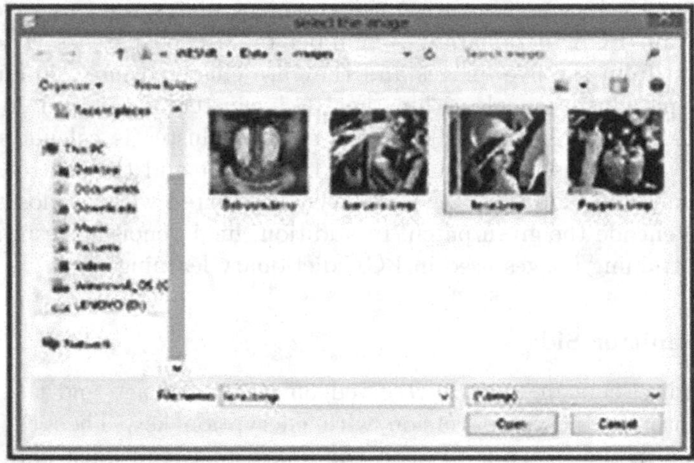

Fig. 1. The selected image of "lena.bmp"

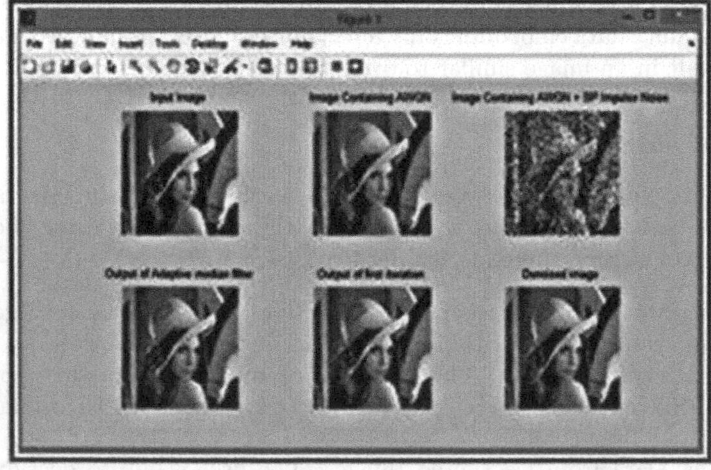

Fig. 2. Input image, noisy image and denoised image.

The encryption key was given to encrypt the image and it was divided into three main steps, namely, image partition, self reversible embedding and image encryption. The encrypted image is used for data hiding, so that the sender can provide data to be hidden and final PSNR and total elapsed time completion for of the process. Further the data is given to sender with a data hiding key to access hidden data and image. From these noise free images around 876,359 patches (size: 7 × 7) are extracted and they are clustered into 200 clusters by using the K-means clustering algorithm (Fig. 3). Using PCA 200 clusters are saved as dictionary and we search for similar patches in dictionary. The first output is encrypted image with data hidden in the image. In the receiver side the same data hidden key provided to sender will be given to open hidden data.

The next stage encryption key is used to get encrypted image and final image was denoised image of sender side (Fig. 4).

The whole processing of the system is shown in Fig. 5.

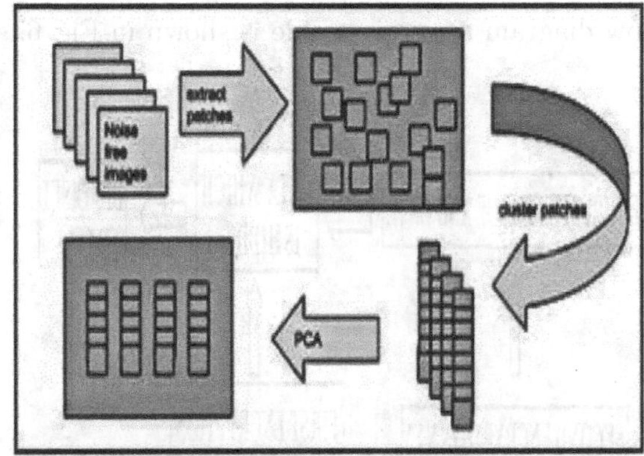

Fig. 3. The flow diagram of dictionary used in process of denoising

Fig. 4. The final output of denoising image

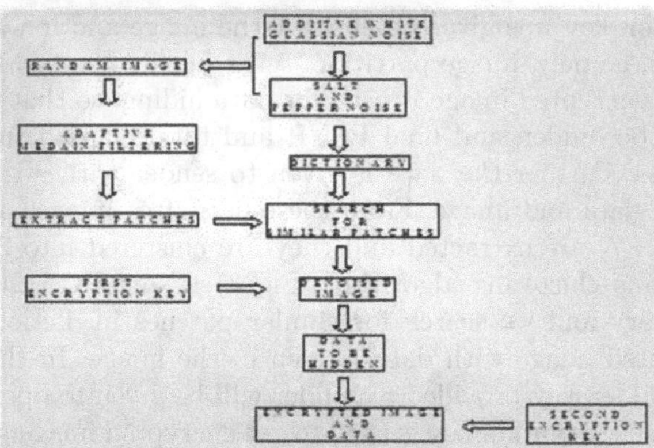

Fig. 5. The flow diagram of transmitting denoised image

The process flow diagram at receiver side is shown in Fig. 6.

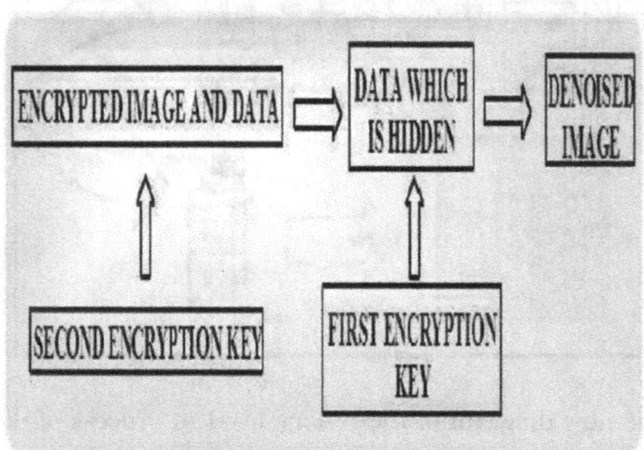

Fig. 6. The flow diagram of receiver side

In the receiver side it will authenticate both encryption key and datahiding key so as to ensure security of image as well as data. If authentication fails then error message is displayed.

3.1 Results and Discussion

Noise elimination from an image is an important task and there is a need to find whether local image variations are due to color, appearance, or brightness from an image itself, or because of noise. The focus of this work is on the development of algorithms for removal of mixed noise from digital images in spatial domain. There are many types of noises present in images such as impulse noise, adaptive white gaussian noise, short noise, quantization noise, film grain and among these one or more are coupled together to form a mixed noise. The additive white gaussian noise (AWGN) coupled with impulse noise (IN) is one typical kind of mixed noise. A scheme for the removal of mixed noises by weighted encoding technique is proposed. The mixture of IN and AWGN, is a difficult problem of denoising and this is because of different properties as well types of noises. The proposed methodology focuses not only on image noise removal but also on encryption of image and hiding data for the purpose of secure communication. Here we used a Reversible Data Hiding (RDH) technique for encrypting image and through this method the original image can be recovered and thus it is very easy for data hider to reversibly embed data in encrypted image. This method is useful to achieve real reversibility, data extraction and error free image recovery. The observed experimental results showed that the WESNR method achieves leading mixed noise removal performance in terms of both quantitative measures and visual quality. The experimental results were highlighting that the PSNR, quality of image and elapsed time are interconnected and the proposed system is showing high PSNR value. Experiment carried out in our proposed system indicates that the quality of image increases as PSNR value increases. In the proposed system a range of 0–25 AWGN values and 0.3–0.5 SP values were used. From the obtained results it was clear that the noise intensity of image increases by increasing the value of AWGN and SP. Hence the conclusion is PSNR, intensity of noise and total time taken to remove noise all are linked, and details are included in Table 1.

Table 1. PSNR values with different amount of noise

AWGN	S.P	Initial PSNR	Final PSNR	Elapsed time
0	0.3	10.659	35.119	67.245
	0.5	8.443	33.753	83.9410
05	0.3	10.654	34.763	67.339
	0.5	8.438	33.277	66.430
10	0.3	10.630	33.304	65.698
	0.5	8.431	32.2634	66.207
15	0.3	10.546	31.884	86.34
	0.5	8.409	30.929	86.58
20	0.3	10.427	30.957	86.155
	0.5	8.352	29.980	86.31
25	0.3	10.350	30.126	106.021
	0.5	8.307	28.906	105.42

4 Conclusion

In any communication system the noise occurrence is unavoidable. However, by using filters noise can be removed. When the noise presence is strong we need other techniques to remove the same. One kind of noise is called mixed noise. The removal this noise from natural images is a challenging task as well noise distribution usually does not have a parametric model and has a heavy tail. Using weighted encoding technique, noise is removed simultaneously and this technique does it without detecting the type of noise in image. Moreover, the proposed system has taken the image with mixed noise like additive white gaussian noise and impulse noise. The proposed system not only removes noises but also encrypts image and data to be hidden in the image.

References

1. Rani, S.U., Jayamma, R.: Reversible records whacking in encrypted images by reserving possibility before encryption. Int. J. Comput. Sci. Mobile Comput. **3**, 194–200 (2014)
2. Verma, R., Ali, D.J.: A comparative study of various types of image noise and efficient noise removal techniques. Int. J. Adv. Res. Comput. Sci. Softw. Eng. **3**, 617–622 (2013)
3. Motwani, M.C., Gadiya, M.C., Motwani, R.C., Harris, F.C.: Survey of image denoising techniques. In: Proceedings of GSPX, pp. 27–30 (2004)
4. Yan, M.: Restoration of images corrupted by impulse noise and mixed Gaussian impulse noise using blind inpainting. SIAM J. Imaging Sci. **6**, 1227–1245 (2013)
5. Cai, J.-F., Chan, R.H., Nikolova, M.: Two-phase approach for deblurring images corrupted by impulse plus Gaussian noise. Inverse Prob. Imaging **2**, 187–204 (2008)
6. Lin, C.-H., Tsai, J.-S., Chiu, C.-T.: Switching bilateral filter with a texture/noise detector for universal noise removal. IEEE Trans. Image Process. **19**, 2307–2320 (2010)
7. Liu, J., Tai, X.-C., Huang, H., Huan, Z.: A weighted dictionary learning model for denoising images corrupted by mixed noise. IEEE Trans. Image Process. **22**, 1108–1120 (2013)
8. Delon, J., Desolneux, A.: A patch-based approach for removing impulse or mixed gaussian-impulse noise. SIAM J. Imaging Sci. **6**, 1140–1174 (2013)
9. Johnson, M., Ishwar, P., Prabhakaran, V., Schonberg, D., Ramchandran, K.: On compressing encrypted data. IEEE Trans. Signal Process. **52**, 2992–3006 (2004)
10. Ni, Z., Shi, Y.-Q., Ansari, N., Su, W.: Reversible data hiding. IEEE Trans. Circuits Syst. Video Technol. **16**, 354–362 (2006)
11. Nosrati, M., Karimi, R., Hariri, M.: Reversible data hiding: principles, techniques, and recent studies. World Appl. Program. **2**, 349–353 (2012)
12. Ma, K., Zhang, W., Zhao, X., Yu, N., Li, F.: Reversible data hiding in encrypted images by reserving room before encryption. IEEE Trans. Inf. Forensics Secur. **8**, 553–562 (2013)
13. Jiang, J., Zhang, L., Yang, J.: Mixed noise removal by weighted encoding with sparse nonlocal regularization. IEEE Trans. Image Process. **23**, 2651–2662 (2014)

Analysis of Particle Swarm Optimization Based 2D FIR Filter for Reduction of Additive and Multiplicative Noise in Images

V. Muneeswaran[1(✉)] and M. Pallikonda Rajasekaran[2]

[1] Department of Instrumentation and Control Engineering, Kalasalingam University, Krishnankoil 626126, Tamil Nadu, India
munees.klu@gmail.com

[2] Department of Electronics and Communication Engineering, Kalasalingam University, Krishnankoil 626126, Tamil Nadu, India
m.p.raja@klu.ac.in

Abstract. Noise in digital images is the major cause of severe artifacts. Filter design for denoising applications can also be addressed with optimization techniques as conventional filters incur in this. Exploration and Exploitation capability features of the Meta Heuristic Optimization Techniques make them applicable to noise reduction in digital images. An increasing number of Meta Heuristic Optimization algorithms make it suitable for designing FIR filters. In the proposed method, Particle Swarm Optimization, a global optimizer algorithm was used in calculating the appropriate coefficients for 2D FIR Filter. The proposed filter was applied to standard test images for testing its noise suppression capability. Indicators of performance, such as Peak signal to noise ratio (PSNR) values and Structural Content (SC) were used in accessing the efficiency of the proposed method and to the adaptability of the method for removing different noise types. Thus a brief comparison for noise suppression in digital images with both multiplicative and additive noise types using PSO optimized 2D FIR filter is addressed in this paper.

Keywords: 2D FIR filter · Particle Swarm Optimization · Peak signal to noise ratio · Meta Heuristic Optimization · Structural Content

1 Introduction

Noise in digital images is unavoidable in all imaging modalities as the instrumentation facilities and the environmental factors in which the images captured interfere with the internal attributes of the image [4]. All noise types either multiplicative or additive in nature conduce for the degradation of the image almost in all cases. Due to the prevalence of modern imaging facilities it becomes essential to limit or remove the noise signals present in it. In this study the efficiency of the noise elimination scheme using a two dimensional Finite Impulse Response filter based on Particle Swarm Optimization algorithm together with a median

S. Arumugam et al. (Eds.): ICTCSDM 2016, LNCS 10398, pp. 165–174, 2017.
DOI: 10.1007/978-3-319-64419-6_22

filter for removing additive and multiplicative noise present in the image is studied and the ability to recover its noiseless form is discussed. The preliminary step to identify the noise present in the image is to analyze the histogram. The basic difference between the additive and multiplicative noise is that, assume a variable $x(t)$ following a stochastic differential equation. If the corresponding random term in the stochastic differential equation of a variable $x(t)$ does not reckon on the state of the system $x(t)$, we call it additive noise. If the random term in the stochastic differential equation depends on the state of the system $x(t)$, then the noise is assumed to be multiplicative in nature. The goal for a denoising filter consists of suppressing the noise while preserving all the useful features such as edges and textural features. In conventional filters the removal of additive and multiplicative noise will result in the blurring and distorted features in the filtered image. The use of population based optimization techniques eliminates the need of performing local statistics and diffusion based methods that was computationally high [9,13]. Optimization techniques do not have the need to have prior knowledge about the amount of noise present.

2 Previous Works

Hitherto more number of studies have been performed on image denoising in the literature. Denoising process in wavelet domain and frequency domain requires optimal threshold and cut-off frequency as their basic components. A few important and recent notable works in the denoising field is discussed in this section. Ratha Jeyalakshmi and Ramar used to modify the morphological image filtering algorithm with arbitrary structuring elements for speckle reduction [10]. Andria and his team produced a denoising scheme using simlet 5 mother function, which are filtered with linear phase. It also involves processing of horizontal, vertical, diagonal and approximation denoised images [2]. Behrenbruch reviewed filtering approaches on the post processing scenario that clarifies the misconception in filtering techniques [4]. Vikrant Bhateja and his team modified the diffusion equation of Perona and Malik by replacing the diffusion coefficient with a non-linear function of coefficient of variations. Noise reduction is achieved in his work by summing up the weighted Laplacian images [16]. In another work Vikrant Bhateja and his research group suppressed the noise content by processing the non-homogenous regions with the application of modified average filtering templates on it [17]. Team of members headed by Nagashettappa Biradar combined fuzzy filters with triangular membership function and conventional SRAD filter in homomorphic domain and non-homomorphic domain for noise reduction [5]. Fatma Latifoglu used artificial bee colony optimization algorithm for determining the optimal co-efficients of the 2D FIR filter [11]. Gupta used soft thresholding process and multiscale decomposition for denoising that is computationally hard [7].

3 Two Dimensional FIR Filter

Two dimensional FIR filter is used for image processing in various applications. Two dimensional Finite Impulse Response filter was always characterized by their filter coefficients $h(m, n)$ [6,8]. The frequency response of the 2D Filter is therefore given by Eq. (1).

$$H(\omega_u, \omega_v) = \sum_{m=-M}^{M} \sum_{n=-N}^{N} h(m, n) e^{-j(m\omega_u + n\omega_v)} \tag{1}$$

where $\omega_u = \frac{2\Pi u}{M}$ and $\omega_v = \frac{2\Pi v}{N}$.

In this equation $h(m, n)$ were the filter coefficients which will be found iteratively with the help of optimization algorithm. The stability condition of a two dimensional filter is given in the following Eq. (2).

$$\sum_{m=-x}^{X} \sum_{n=-x}^{X} |h(m, n)| < X \tag{2}$$

where X is the number of elements from the origin in the mask size of the filter coefficients. In this study the effect of filter coefficients with appropriate zero locations in the complex plane produced using particle swarm optimization algorithm is analyzed for suppression of Gaussian and speckle noise present in images.

4 Median Filter

Median filter is one of the conventional filters that is extensively used in the spatial filtering process due to its non-linear property. It is widely used in image processing algorithm with the intention of noise reduction and in pre-processing [12]. The process of median filtering is accomplished by placing median of a window as a value instead of its original value. While calculating the median value the following procedure is adopted. All the values in the mask will be sorted in numerical order and the middle value in the sorted order will be considered as the value to be replaced. Thus the property of median filter is achieved.

5 Additive and Multiplicative Noise in Images

5.1 Additive Noise

Gaussian Noise. A probability density function (PDF) of the Gaussian noise will resemble the normal distribution. Thus the noise value are Gaussian distributive in nature. The prime cause of Gaussian noise in images occurs during capture e.g. noise due to improper illumination and/or due to abrupt changes in temperature, and/or during transmission e.g. noise of electronic circuit. Most commonly Gaussian noise can be suppressed using a spatial filtering approach, despite the smoothing of image, an unwanted outcome may end up in the blurring of edges and details as they will be processed in the task of blocking high frequencies. Traditional spatial filtering approach for noise reduction comprises:

mean filtering technique, median filtering technique and Gaussian smoothing technique for a random variable z its probability density function P is given by Eq. (3).

$$P_G(Z) = \frac{1}{\sigma\sqrt{2\pi}} e^{-\frac{(z-\mu)^2}{2\sigma^2}} \tag{3}$$

where Z is the gray level present in the image, μ and σ are the mean and standard deviation respectively.

Salt and Pepper Noise. An image is considered to be getting exposed to salt and pepper noise only if it has random occurrences of white and black pixels. It is observed that over heated imaging components may cause salt and pepper noise.

5.2 Multiplicative Noise

Speckle Noise. Images that are formed with coherent energy sources and imaging systems impose a serious threat of speckle noise. It is often termed as dominant multiplicative noise. Removing speckle noise becomes harder as its intensity varies with the image intensity [14]. Speckle noise in rare cases may contain useful texture information. As speckle noise is multiplicative in nature it is modeled only with the random value multiplications as given in Eq. (4).

$$J = I + n * I \tag{4}$$

where J is the speckle affected image, I is noiseless input image and n is the noisy image of variance v.

6 Particle Swarm Optimization

Optimization is the process of finding the best available values from the input values [13]. Particle Swarm Optimization is a mathematical modelling of social behavior of certain animals within their team. Particle Swarm Optimization is often preferred for its robustness in finding the global best location of particles [1]. For a iteration l the velocity of the particle i is calculated by sum of global best solution g_{best}, its current best value p_{best} and its current velocity v^l. Considering $v_i^{l=0} = 0$ the new velocity vector is calculated by the Eq. (5).

$$v_i^l + 1 = W * v_i^l + \alpha * C_1 * [g_{best}^i - x_i^l] + \beta * C_2 * [p_{best}^i - x_i^l] \tag{5}$$

The tradeoff between p_{best} and g_{best} is controlled by W the inertial weight parameter. The relative attraction between p_{best} and g_{best} is indicated by C_1 and C_2. α and β are random values between 0 and 1. The new position is calculated as

$$x_i^{l+1} = x_i^l + v_i^{l+1} \tag{6}$$

The range of v_i lies between $[v_{min}, v_{max}]$. When the new position is calculated the particle will shift to it and at the last iteration the g_{best} becomes the optimal solution found.

7 Design Formulation

In the given scheme two image signals were used on the input side such as the noiseless image $I_{org}(n)$ and $I_{noisy}(n)$ is the noisy image contaminated by either additive and multiplicative noise. The 2D FIR filter system with optimization using Particle Swarm Optimization together with median filter will produce the denoised image. The objective of the optimization process is to reduce the Mean Square Error value that results as a difference between noisy image and 2D FIR filter output [11] as shown in Fig. 1.

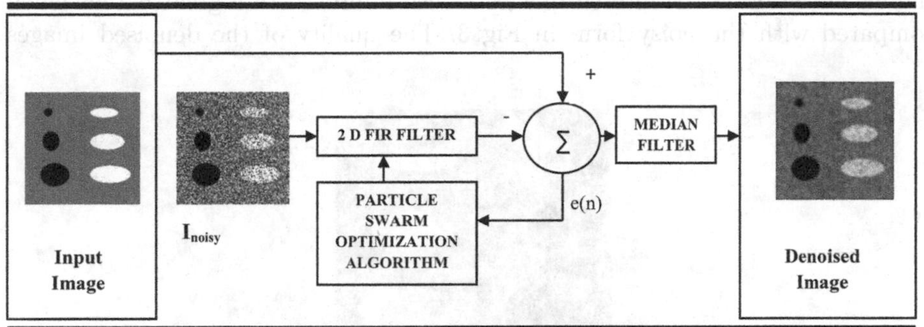

Fig. 1. Proposed denoising scheme.

Coefficients of filter were adjusted by minimization of the Mean Square Error value between filter output and $I_{org}(n)$ and is given as

$$MSE = \frac{1}{KL} \sum_{k=0}^{K-1} \sum_{l=0}^{L-1} [I_{org}(k,l) - I_{Noisy}(k,l)]^2 \tag{7}$$

In this proposed methodology, Particle Swarm Optimization is used in finding optimal coefficients. The steps of PSO based 2D-FIR are given below

1. Set the number of population(window size of filter), learning parameter (C1, C2).
2. Generate the swarm with the condition given in Eq. (2).
3. Update the variables p_{best} and g_{best} at the current iteration based upon the fitness function The fitness function for this problem is given is given in Eq. (7).
4. Generate new p_{best} and g_{best} (values) with fitness values and compute the velocity and position using Eq. (5).
5. Look up for the termination condition and repeat steps 3–5 till the optimum value of g_{best} is reached or upto the termination condition (Number of iterations is set as 100 in this case).

8 Results and Discussion

The simulated test image, as shown in Fig. 2 in its JPG format with 128 × 28 pixels was used in the experiments. For its noisy version the standard image is corrupted with additive and multiplicative noise at different noise level and it is shown in Fig. 3. Consciously degrading an image with noise will allow us to validate the effectiveness of an image denoising operator to noise and assess its performance as shown in Tables 1, 3 and 4.

It can be seen that from Fig. 4 the amount of the additive and multiplicative noise is reduced on the application of the proposed denoising scheme and the visualization of the filtered images is also improved to a great extent when compared with the noisy form in Fig. 3. The quality of the denoised images

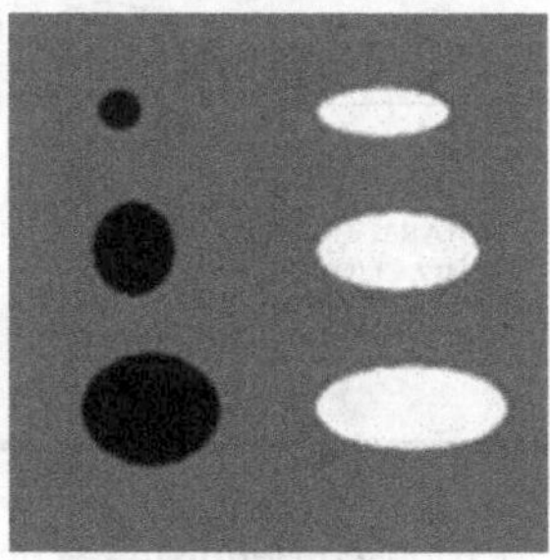

Fig. 2. Simulated image

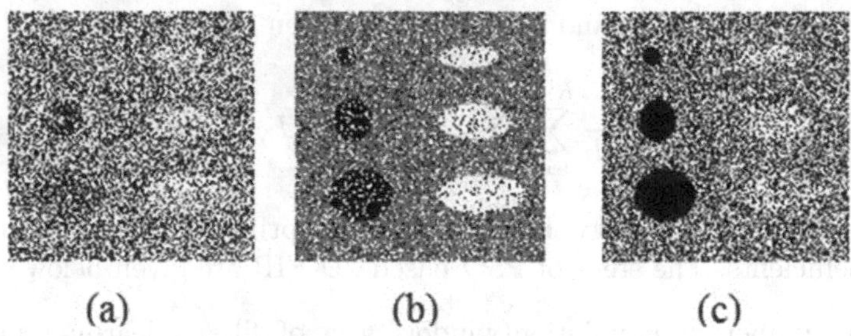

Fig. 3. Noised form of simulated image

Table 1. 3 × 3 Mask

a_{00}	a_{01}	a_{02}
0.0621	0.1024	0.0616
a_{10}	a_{11}	a_{12}
0.1021	0.2736	0.1026
a_{20}	a_{21}	a_{22}
0.0614	0.1022	0.0621

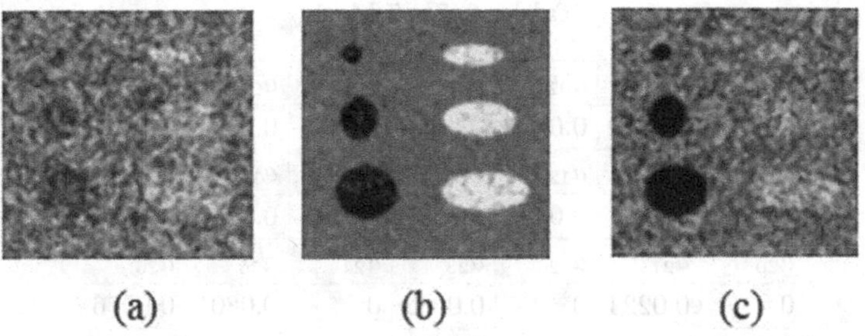

Fig. 4. Denoised form of simulated image

Table 2. Quality metrics

Metrics	Gaussian noise			S & P noise			Speckle noise		
	Mask size			Mask size			Mask size		
	3×3	5×5	7×7	3×3	5×5	7×7	3×3	5×5	7×7
PSNR	13.48	12.23	10.34	17.46	17.23	15.03	15.63	15.11	13.54
MSE	0.022	0.026	0.030	0.010	0.010	0.012	0.014	0.016	0.024
SC	1.703	1.792	1.890	0.932	0.972	1.239	1.375	1.484	1.643
ENL	6.994	6.843	6.544	9.223	9.094	8.564	7.512	7.012	6.843

Table 3. 5×5 Mask

a_{00}	a_{01}	a_{02}	a_{03}	a_{04}
0.0238	0.0281	0.0137	0.0092	0.0784
a_{10}	a_{11}	a_{12}	a_{13}	a_{14}
0	0.0390	0.0268	0.0153	0.0602
a_{20}	a_{21}	a_{22}	a_{23}	a_{24}
0.0569	0.0609	0.1260	0.0885	0.0429
a_{30}	a_{31}	a_{32}	a_{33}	a_{34}
0.0635	0.0497	0.0782	0.0582	0.0403
a_{40}	a_{41}	a_{42}	a_{43}	a_{44}
0.0079	0.0373	0.0510	0	0.0119

were evaluated by standard metrics such as Peak Signal to Noise ratio (PSNR) [8], Mean square error (MSE) [3], Structural Content (SC) [15] and Equivalent Number of Looks (ENL) [18]. It is evident that there is very less blurring in the filtered image with 3×3 window mask and the value of quality metrics for different window size supports the aforementioned fact. The limit on window size reduces the computational complexity as well as effect of blurring in the resultant images. The obtained results are above compromising level even at high noise densities without much iterative application of the filtering algorithm. It is seen

Table 4. 7 × 7 Mask

a_{00}	a_{01}	a_{02}	a_{03}	a_{04}	a_{05}	a_{06}
0.0525	0.0122	0.0497	0.0516	0.0136	0.0391	0
a_{10}	a_{11}	a_{12}	a_{13}	a_{14}	a_{15}	a_{16}
0.0399	0.0685	0.0524	0.1058	0.0748	0.0767	0
a_{20}	a_{21}	a_{22}	a_{23}	a_{24}	a_{25}	a_{26}
0	0.0224	0	0.0082	0	0.0801	0.0076
a_{30}	a_{31}	a_{32}	a_{33}	a_{34}	a_{35}	a_{36}
0	0.1031	0.0916	0.0076	0.0431	0.0325	0.0007
a_{40}	a_{41}	a_{42}	a_{43}	a_{44}	a_{45}	a_{46}
0.0713	0.0282	0	0.0545	0.0430	0.0061	0.0120
a_{50}	a_{51}	a_{52}	a_{53}	a_{54}	a_{55}	a_{56}
0	0.0397	0.0048	0	0.0884	0.0456	0

from Table 2, that the coefficients for 3 × 3 mask performs well in the following hierarchy of removing noise, it efficiently removes salt and pepper noise in the images, whereas the ability to remove the speckle content and salt and pepper noise content in the images was relatively low. The effect of increasing window size is also clearly illustrated in Table 2. It is clearly seen with the augumentation in window size the quality of the image decreases which is clearly illustrated in Table 2. Thus the 3 × 3 mask can be preferred for denoising applications such as Ultrasound Images, SAR Images etc.

9 Conclusion

We have proposed a optimization based filtering technique for image denoising process. In this proposed denoising technique, the qualitative and quantitative aspect of filtering are discussed. It is clearly realized that the filter coefficients produced by the mask size 3 is more suited in purging salt and pepper noise, whereas the same values when applied for removing speckle noise and Gaussian noise performs relatively low. The performance of the optimization based filtering technique was illustrated with efficient quality indicators. Hence the proposed optimization technique based FIR filter can be used for noise elimination process. Future work includes the development of noise elimination schemes using different meta-heuristic optimization algorithms. Filter coefficients that can suppress both the additive and multiplicative noise can also be a future work.

Acknowledgments. The author would like to thank the management of Kalasalingam University for providing financial assistance under the University Research Fellowship. Also we thank the Department of Electronics and Communication Engineering of Kalasalingam University, Tamil Nadu, India for permitting to use the computational facilities available in Centre for Research in Signal Processing and VLSI

Design which was setup with the support of the Department of Science and Technology (DST), New Delhi under FIST Program in 2013 (Reference No: SR/FST/ETI-336/2013 dated November 2013).

References

1. Aggarwal, A., Rawat, T.K., Upadhyay, D.K.: Design of optimal digital FIR filters using evolutionary and swarm optimization techniques. AEU - Int. J. Electron. Commun. **70**(4), 373–385 (2016)
2. Andria, G., Attivissimo, F., Cavone, G., Giaquinto, N., Lanzolla, A.M.L.: Linear filtering of 2-D wavelet coefficients for denoising ultrasound medical images. Meas. J. Int. Meas. Confed. **45**, 1792–1800 (2012)
3. Arunprasath, T., Pallikonda Rajasekaran, M., Kannan, S.: ANFIS-EM approach for PET brain image reconstruction. Int. J. Imaging Syst. Technol. **25**, 1–6 (2015)
4. Behrenbruch, C.P., Petroudi, S., Bond, S., Declerck, J.D., Leong, F.J., Brady, J.M.: Image filtering techniques for medical image post-processing: an overview. Br. J. Radiol. **77**, S126–S132 (2004)
5. Biradar, N., Dewal, M.L., Rohit, M.K.: A novel hybrid homomorphic fuzzy filter for speckle noise reduction. Biomed. Eng. Lett. **4**, 176–185 (2014)
6. Chandra, A., Chattopadhyay, S.: A new strategy of image denoising using multiplier-less FIR filter designed with the aid of differential evolution algorithm. Multimed. Tools Appl. **75**, 1079–1098 (2016)
7. Gupta, S., Chauhan, R.C., Sexana, S.C.: Wavelet-based statistical approach for speckle reduction in medical ultrasound images. Med. Biol. Eng. Comput. **42**, 189–192 (2004)
8. Hua, J., Kuang, W., Gao, Z., Meng, L., Xu, Z.: Image denoising using 2-D FIR filters designed with DEPSO. Multimed. Tools Appl. **69**, 157–169 (2014)
9. Jain, P., Tyagi, V.: A survey of edge-preserving image denoising methods. Inf. Syst. Front. **18**, 159–170 (2016)
10. Jeyalakshmi, T.R., Ramar, K.: A modified method for speckle noise removal in ultrasound medical images. Int. J. Comput. Electr. Eng. **2**(1), 54–58 (2010)
11. Latifoglu, F.: A novel approach to speckle noise filtering based on Artificial Bee Colony algorithm: an ultrasound image application. Comput. Methods Programs Biomed. **111**(3), 561–569 (2013)
12. Maity, A., Pattanaik, A., Sagnika, S., Pani, S.: A comparative study on approaches to speckle noise reduction in images. In: Proceedings of the 1st International Conference on Computational Intelligence and Networks, CINE, pp. 148–155 (2015)
13. Muneeswaran, V., Pallikonda Rajasekaran, M.: Performance evaluation of radial basis function networks based on tree seed algorithm. In: Proceeding of the 2016 International Conference of Circuit Power and Computing Technologies, pp. 1–4. IEEE Explore (2016)
14. Park, J., Kang, J.B., Chang, J.H., Yoo, Y.: Speckle reduction techniques in medical ultrasound imaging. Biomed. Eng. Lett. **4**, 32–40 (2014)
15. Rosa, R., Monteiro, F.C.: Performance analysis of speckle ultrasound image filtering. Comput. Methods Biomech. Biomed. Eng. Imaging Vis. **4**, 193–201 (2014)
16. Bhateja, V., Singh, G., Srivastava, A.: A novel weighted diffusion filtering approach for speckle suppression in ultrasound images. In: Satapathy, S., Udgata, S., Biswal, B. (eds.) FICTA 2013. AISC, vol. 247, pp. 459–466. Springer, Cham (2014). doi:10.1007/978-3-319-02931-3_52

17. Bhateja, V., Srivastava, A., Singh, G., Singh, J.: A modified speckle suppression algorithm for breast ultrasound images using directional filters. In: Satapathy, S., Avadhani, P., Udgata, S., Lakshminarayana, S. (eds.) ICT and Critical Infrastructure: Proceedings of the 48th Annual Convention of Computer Society of India- Vol II. AISC, vol. 249, pp. 219–226. Springer, Cham (2014). doi:10.1007/978-3-319-03095-1_24
18. Subrahmanyam, G.R.K.S., Rajagopalan, A.N., Aravind, R.: A recursive filter for despeckling SAR images. IEEE Trans. Image Process. **17**, 1969–1974 (2008)

Irregularity Strength of Corona of Two Graphs

K. Muthu Guru Packiam[1], T. Manimaran[2(✉)], and A. Thuraiswamy[3]

[1] Department of Mathematics, Rajah Serfoji Government College, Thanjavur 613005, Tamilnadu, India
gurupackiam@yahoo.com

[2] Department of Mathematics, Ramco Institute of Technology, Rajapalayam 626117, Tamilnadu, India
tm_maran@yahoo.com

[3] Department of Mathematics, Kalasalingam University, Anand Nagar, Krishnankoil 626126, Tamilnadu, India
thuraiswamy@yahoo.com

Abstract. Let $G = (V, E)$ be a connected graph of order $n \geq 3$. Let $f : E \rightarrow \{1, 2, ..., k\}$ be a function and let the weight of a vertex v be defined by $\omega(v) = \sum_{v \in V} f(v)$. Then f is called an irregular labeling if all the vertex weights are distinct. The irregularity strength $s(G)$ is the smallest positive integer k such that there is an irregular labeling $f : E \rightarrow \{1, 2, ..., k\}$. In this paper we determine the irregularity strength of corona $G \circ H$ where $G = P_n$ or C_n and $H = mK_1$ or K_2 or K_3.

Keywords: Irregular labeling · Irregularity strength · Corona

1 Introduction

Let $G = (V, E)$ be a connected graph of order $n \geq 3$. Let $f : E \rightarrow \{1, 2, ..., k\}$ be a function and let the weight of a vertex v be defined by $\omega(v) = \sum_{v \in V} f(v)$. Then f is called an irregular labeling if all the vertex weights are distinct. The irregularity strength $s(G)$ is the smallest positive integer k such that there is an irregular labeling $f : E \rightarrow \{1, 2, ..., k\}$. The irregularity strength of a graph was introduced by Chartrand et al. [4]. Further they proved the following lower bound for $s(G)$.

Proposition 1. *If G is a connected graph of order at least 3 containing p_i vertices of degree i, for some positive integer i, then $s(G) \geq \frac{p_i - 1}{i} + 1$.*

Aigner and Triesh [1] proved that $\mathrm{s}(G) \leq n - 1$ if G is a connected graph of order n, and $\mathrm{s}(G) \leq n + 1$ otherwise. Nierhoff [11] refined their method and showed that $\mathrm{s}(G) \leq n - 1$ for all graphs with finite irregularity strength, except for K_3. This bound is tight, for example star $K_{1,n}$. Faudree and Lehel [6] showed that if G is d-regular and $d \geq 2$, then $\lceil \frac{n+d-1}{d} \rceil \leq \mathrm{s}(G) \leq \lceil \frac{n}{2} \rceil + 9$, and they conjectured that $\mathrm{s}(G) \leq \lceil \frac{n}{d} \rceil + c$ for some constant c. Przybylo in [12] proved

S. Arumugam et al. (Eds.): ICTCSDM 2016, LNCS 10398, pp. 175–181, 2017.
DOI: 10.1007/978-3-319-64419-6_23

that s(G) $\leq 16\frac{n}{d}+6$. Kalkowski et al. [9] showed that $s(G)\leq 6\frac{n}{\delta}+6$, where δ is the minimum degree of G. Currently Majerski and Przybylo [10] proved that $s(G)\leq (4+o(1))\frac{n}{\delta}+4$ for graphs with minimum degree $\delta\geq\sqrt{n}\ln n$. Other interesting results on the irregularity strength can be found in [2,3,5–8].

Theorem 1 [8]**.** *If $G=H\circ K_2$, where H is graph with $p\geq 3$ vertices such that $\delta(H)\geq 2$, then $s(G)=p+1$.*

2 Main Results

Definition 1. *The corona $G_1\circ G_2$ of two graphs G_1 and G_2 is the graph G obtained by taking one copy G_1 which has n vertices and n copies of G_2 and then joining i^{th} vertex of G_1 to every vertex in the ith copy of G_2.*

Theorem 2. *Let P_n be the path on n vertices. Then $s(P_n\circ mK_1)=mn$.*

Proof. Let $G=P_n\circ mK_1$. Let $V(G)=\{u_i:1\leq i\leq n\}\cup\{v_i^j:1\leq i\leq n, 1\leq j\leq m\}$ and $E(G)=\{u_iu_{i+1}:1\leq i\leq n-1\}\cup\{u_iv_i^j:1\leq i\leq n, 1\leq j\leq m\}$. Define $f:E(G)\to\{1,2,3,...,mn\}$ as follows.

$$f(u_iu_{i+1})=mn \text{ for } 1\leq i\leq n-1 \text{ and}$$
$$f(u_iv_i^j)=n(j-1)+i \text{ for } 1\leq i\leq n, 1\leq j\leq m.$$

Hence weights of the vertices of G are

$$\omega(v_i^j)=n(j-1)+i \text{ for all } i \text{ and } j,$$
$$\omega(u_i)=\begin{cases}\frac{nm(m+1)}{2}+im & \text{if } i=1,n\\ \frac{nm(m+3)}{2}+im & \text{if } 2\leq i\leq n-1.\end{cases}$$

Since the weights of all vertices of G are distinct and f is an irregular labeling, $s(G)\leq mn$. By Proposition 1, $s(G)\geq mn$. Hence $s(G)=mn$.

Theorem 3. *Let C_n be the cycle on n vertices. Then $s(C_n\circ mK_1)=mn$.*

Proof. Let $G=C_n\circ mK_1$. Let $V(G)=\{u_i:1\leq i\leq n\}\cup\{v_i^j:1\leq i\leq n, 1\leq j\leq m\}$ and $E(G)=\{e_i=u_iu_{i+1}:1\leq i\leq n-1\}\cup\{e_n=u_nu_1\}\cup\{u_iv_i^j:1\leq i\leq n, 1\leq j\leq m\}$.
Define $f:E(G)\to\{1,2,3,...,mn\}$ as follows.

$$f(e_i)=mn \text{ for } 1\leq i\leq n \text{ and}$$
$$f(u_iv_i^j)=n(j-1)+i \text{ for } 1\leq i\leq n, 1\leq j\leq m.$$

Hence weights of the vertices are given by

$$\omega(v_i^j)=n(j-1)+i \text{ for all } i \text{ and } j \text{ and}$$
$$\omega(u_i)=\frac{nm(m+3)}{2}+im \text{ for } 1\leq i\leq n.$$

Since the weights of all vertices of G are distinct and f is an irregular labeling, $s(G)\leq mn$. By Proposition 1, $s(G)\geq mn$. Hence $s(G)=mn$.

Theorem 4. *Let P_n be the path on n vertices. Then $s(P_n \circ K_2) = n+1$.*

Proof. Let $G = P_n \circ K_2$. Let $V(G) = \{u_i : 1 \le i \le n\} \cup \{v_i^j : 1 \le i \le n, j = 1,2\}$ and $E(G) = \{u_iu_{i+1} : 1 \le i \le n-1\} \cup \{u_iv_i^j : 1 \le i \le n, j = 1,2\} \cup \{v_i^1v_i^2 : 1 \le i \le n\}$.
Define $f : E(G) \to \{1,2,3,...,n+1\}$ as follows.

$$f(u_iu_{i+1}) = n+1 \text{ for } 2 \le i \le n-2,$$
$$f(u_iv_i^1) = i \text{ for } 1 \le i \le n,$$
$$f(u_iv_i^2) = i+1 \text{ for } 1 \le i \le n,$$
$$f(v_i^1v_i^2) = \begin{cases} i & \text{if } 1 \le i \le \lceil \frac{n+1}{2} \rceil \\ i+1 & \text{if } \lceil \frac{n+1}{2} \rceil < i \le n, \end{cases}$$

for $n \ne 5$,

$$f(u_1u_2) = f(u_{n-1}u_n) = \begin{cases} n & \text{if } n \text{ is odd} \\ n+1 & \text{if } n \text{ is even,} \end{cases}$$

for $n = 5$, $f(u_1u_2) = 5, f(u_4u_5) = 6$.
Hence weights of the vertices are given by

$$\omega(v_i^1) = \begin{cases} 2i & \text{if } 1 \le i \le \lceil \frac{n+1}{2} \rceil \\ 2i+1 & \text{if } \lceil \frac{n+1}{2} \rceil < i \le n, \end{cases}$$
$$\omega(v_i^2) = \begin{cases} 2i+1 & \text{if } 1 \le i \le \lceil \frac{n+1}{2} \rceil \\ 2i+2 & \text{if } \lceil \frac{n+1}{2} \rceil < i \le n, \end{cases}$$
$$\omega(u_1) = \begin{cases} n+3 & \text{if } n \text{ is odd} \\ n+4 & \text{if } n \text{ is even.} \end{cases}$$

For $n = 3$, $\omega(u_2) = 11$.
For $n > 3$,

$$\omega(u_2) = \begin{cases} 2n+6 & \text{if } n \text{ is odd} \\ 2n+7 & \text{if } n \text{ is even.} \end{cases}$$

For $n > 3$,

$$\omega(u_{n-1}) = \begin{cases} 4n & \text{if } n \text{ is odd and } n \ne 5 \\ 4n+1 & \text{if } n \text{ is even } or n = 5. \end{cases}$$

For $n \ge 2$,

$$\omega(u_n) = \begin{cases} 3n+1 & \text{if } n \text{ is odd and } n \ne 5 \\ 3n+2 & \text{if } n \text{ is even or } n = 5 \end{cases}$$

and $\omega(u_i) = 2(n+i)+3$ for $2 < i < n-1$.

Since the weights of all vertices of G are distinct and f is an irregular labeling, $s(G) \le n+1$. By Proposition 1, $s(G) \ge n+1$. Hence $s(G) = n+1$.

Theorem 5. *Let C_n be the cycle on n vertices. Then $s(C_n \circ K_2) = n + 1$.*

Proof. By Theorem 1, $s(C_n \circ K_2) = n + 1$.

Theorem 6. *Let P_n be the path on n vertices. Then $s(P_n \circ K_3) = n + 1$.*

Proof. Let $G = P_n \circ K_3$. Let $V(G) = \{u_i : 1 \leq i \leq n\} \cup \{v_i^j : 1 \leq i \leq n, j = 1, 2, 3\}$ and $E(G) = \{u_i u_{i+1} : 1 \leq i \leq n-1\} \cup \{u_i v_i^j : 1 \leq i \leq n, j = 1, 2, 3\} \cup \{v_i^1 v_i^2, v_i^2 v_i^3, v_i^3 v_i^1 : 1 \leq i \leq n\}$.
Define $f : E(G) \to \{1, 2, 3, ..., n+1\}$ as follows.
For $n = 2$,

$$f(u_1v_1^1) = f(v_1^1v_1^2) = f(v_1^2v_1^3) = f(v_1^3v_1^1) = f(u_2v_2^1) = 1,$$
$$f(u_1v_1^2) = f(v_2^1v_2^2) = f(v_2^2v_2^3) = 2,$$
$$f(u_1u_2) = f(u_1v_1^3) = f(u_2v_2^2) = f(u_2v_2^3) = f(v_2^3v_2^1) = 3.$$

For $n = 3$,

$$f(u_1v_1^1) = f(v_1^1v_1^2) = f(v_1^2v_1^3) = f(v_1^3v_1^1) = 1,$$
$$f(u_1v_1^2) = f(u_2v_2^1) = f(v_2^1v_2^2) = f(v_2^2v_2^3) = f(v_2^3v_2^1) = f(u_3v_3^1) = 2,$$
$$f(u_1u_2) = f(u_1v_1^3) = f(u_2v_2^2) = f(u_3v_3^2) = 3,$$
$$f(u_2u_3) = f(u_2v_2^3) = f(u_3v_3^3) = f(v_3^1v_3^2) = f(v_3^2v_3^3) = f(v_3^3v_3^1) = 4.$$

For $n = 4$,

$$f(u_1v_1^1) = f(v_1^1v_1^2) = f(v_1^2v_1^3) = f(v_1^3v_1^1) = 1,$$
$$f(u_1v_1^2) = f(u_2v_2^1) = f(v_2^1v_2^2) = f(v_2^2v_2^3) = f(v_2^3v_2^1) = f(u_3v_3^1) = 2,$$
$$f(u_1u_2) = f(u_1v_1^3) = f(u_2v_2^2) = f(u_3v_3^2) = f(u_4v_4^1) = 3,$$
$$f(u_2u_3) = f(u_2v_2^3) = f(u_3v_3^3) = f(v_3^1v_3^2) = f(v_3^2v_3^3) = f(v_3^3v_3^1) = f(u_4v_4^2) = 4,$$
$$f(u_3u_4) = f(u_4v_4^3) = f(v_4^1v_4^2) = f(v_4^2v_4^3) = f(v_4^3v_4^1) = 5.$$

For $n \geq 5$,

$$f(v_i^1v_i^2) = f(v_i^2v_i^3) = f(v_i^3v_i^1) = \begin{cases} 1 & \text{if } 1 \leq i \leq \lfloor \frac{n+1}{3} \rfloor \\ i+1 & \text{if } \lfloor \frac{n+1}{3} \rfloor < i \leq n, \end{cases}$$

$$f(u_1v_1^j) = j \text{ for } j = 1, 2, 3,$$

$$f(u_iv_i^1) = \begin{cases} i+2 & \text{if } 2 \leq i \leq \lfloor \frac{n+1}{3} \rfloor \\ i-1 & \text{if } \lfloor \frac{n+1}{3} \rfloor < i \leq n, \end{cases}$$

$$f(u_iv_i^2) = \begin{cases} \lfloor \frac{n+4}{3} \rfloor + i & \text{if } 2 \leq i \leq \lfloor \frac{n+1}{3} \rfloor \\ i & \text{if } \lfloor \frac{n+1}{3} \rfloor < i \leq n, \end{cases}$$

$$f(u_iv_i^3) = \begin{cases} 2\lfloor \frac{n+1}{3} \rfloor + i & \text{if } 2 \leq i \leq \lfloor \frac{n+1}{3} \rfloor \\ i+1 & \text{if } \lfloor \frac{n+1}{3} \rfloor < i \leq n, \end{cases}$$

$$f(u_1u_2) = 3(\lfloor \frac{n+1}{3} \rfloor - 1) \text{ for all } n \geq 5 \text{ and}$$

if $n \equiv 0(mod\ 3)$, then

$$f(u_i u_{i+1}) = \begin{cases} n & \text{if } 2 \le i \le \lfloor \frac{n+1}{3} \rfloor \\ n+1 & \text{if } \lfloor \frac{n+1}{3} \rfloor < i \le n-1, \end{cases}$$

if $n \equiv 1(mod\ 3)$, then

$$f(u_i u_{i+1}) = \begin{cases} n+1 & \text{if } 2 \le i \le \lfloor \frac{n+1}{3} \rfloor + 1 \\ n & \text{if } \lfloor \frac{n+1}{3} \rfloor + 1 < i \le n-2 \\ n-1 & \text{if } i = n-1 \text{ and } n \ne 7 \\ n & \text{if } i = n-1 \text{ and } n = 7, \end{cases}$$

if $n \equiv 2(mod\ 3)$, then

$$f(u_i u_{i+1}) = \begin{cases} n+1 & \text{if } 2 \le i \le \lfloor \frac{n+1}{3} \rfloor \\ n & \text{if } \lfloor \frac{n+1}{3} \rfloor + 1 < i \le n-1 \\ n+1 & \text{if } i = \lfloor \frac{n+1}{3} \rfloor + 1 \text{ and } n = 5 \\ n & \text{if } i = \lfloor \frac{n+1}{3} \rfloor + 1 \text{ and } n \ne 5. \end{cases}$$

Clearly the weights of the vertices of G are distinct for $n = 2, 3, 4$.

For $n \ge 5$ the weights of the vertices of G are given below.

$$\omega(v_1^j) = j + 2 \text{ for } j = 1, 2, 3,$$

$$\omega(v_i^1) = \begin{cases} i+4 & \text{if } 2 \le i \le \lfloor \frac{n+1}{3} \rfloor \\ 3i+1 & \text{if } \lfloor \frac{n+1}{3} \rfloor < i \le n, \end{cases}$$

$$\omega(v_i^2) = \begin{cases} \lfloor \frac{n+4}{3} \rfloor + i + 2 & \text{if } 2 \le i \le \lfloor \frac{n+1}{3} \rfloor \\ 3i+2 & \text{if } \lfloor \frac{n+1}{3} \rfloor < i \le n, \end{cases}$$

$$\omega(v_i^3) = \begin{cases} 2\lfloor \frac{n+1}{3} \rfloor + i + 2 & \text{if } 2 \le i \le \lfloor \frac{n+1}{3} \rfloor \\ 3i+3 & \text{if } \lfloor \frac{n+1}{3} \rfloor < i \le n, \end{cases}$$

$$\omega(u_1) = 3\left(\left\lfloor \frac{n+1}{3} \right\rfloor + 1\right),$$

$$\omega(u_2) = \begin{cases} 5\lfloor \frac{n+1}{3} \rfloor + \lfloor \frac{n+4}{3} \rfloor + n + 5 & \text{if } n \equiv 0(mod\ 3) \\ 5\lfloor \frac{n+1}{3} \rfloor + \lfloor \frac{n+4}{3} \rfloor + n + 6 & \text{if } n \equiv 1, 2(mod\ 3), \end{cases}$$

$$\omega(u_n) = \begin{cases} 4n+1 & \text{if } n \equiv 0(mod\ 3) \\ 4n-1 & \text{if } n \equiv 1(mod\ 3) \text{ and } n \ne 7 \\ 4n & \text{if } n \equiv 2(mod\ 3) \text{ and if } n = 7, \end{cases}$$

if $n = 0(mod\ 3)$, then

$$\omega(u_i) = \begin{cases} 2\lfloor \frac{n+1}{3} \rfloor + \lfloor \frac{n+4}{3} \rfloor + 2n + 3i + 2 & \text{if } 2 < i \le \lfloor \frac{n+1}{3} \rfloor \\ 2n + 3i + 1 & \text{if } i = \lfloor \frac{n+1}{3} \rfloor + 1 \\ 2n + 3i + 2 & \text{if } \lfloor \frac{n+1}{3} \rfloor + 1 < i \le n-1, \end{cases}$$

if $n \equiv 1(mod\ 3)$, then

$$\omega(u_i)=\begin{cases}2\lfloor\frac{n+1}{3}\rfloor+\lfloor\frac{n+4}{3}\rfloor+2n+3i+4 & \text{if } 2<i\leq\lfloor\frac{n+1}{3}\rfloor\\ 2n+3i+2 & \text{if } i=\lfloor\frac{n+1}{3}\rfloor+1\\ 2n+3i+1 & \text{if } i=\lfloor\frac{n+1}{3}\rfloor+2\\ 2n+3i & \text{if } \lfloor\frac{n+1}{3}\rfloor+2<i\leq n-2\\ 2n+3i-1 & \text{if } i=n-1 \text{ and } n\neq 7\\ 2n+3i & \text{if } i=n-1 \text{ and } n=7,\end{cases}$$

if $n \equiv 2(mod\ 3)$, then

$$\omega(u_i)=\begin{cases}2\lfloor\frac{n+1}{3}\rfloor+\lfloor\frac{n+4}{3}\rfloor+2n+3i+4 & \text{if } 2<i\leq\lfloor\frac{n+1}{3}\rfloor\\ 2n+3i+2 & \text{if } i=\lfloor\frac{n+1}{3}\rfloor+1 \text{ and } n=5\\ 2n+3i+1 & \text{if } i=\lfloor\frac{n+1}{3}\rfloor+1 \text{ and } n\neq 5\\ 2n+3i+1 & \text{if } i=\lfloor\frac{n+1}{3}\rfloor+2 \text{ and } n=5\\ 2n+3i & \text{if } i=\lfloor\frac{n+1}{3}\rfloor+2 \text{ and } n\neq 5\\ 2n+3i & \text{if } \lfloor\frac{n+1}{3}\rfloor+2<i\leq n-1.\end{cases}$$

Since the weights of all vertices of G are distinct and f is an irregular labeling, $s(G) \leq n+1$. By Proposition 1, $s(G) \geq n+1$. Hence $s(G) = n+1$.

Theorem 7. *Let C_n be the cycle on n vertices. Then $s(C_n \circ K_3) = n+1$.*

Proof. Let $G = C_n \circ K_3$. Let $V(G) = \{u_i : 1 \leq i \leq n\} \cup \{v_i^j : 1 \leq i \leq n, j = 1,2,3\}$ and $E(G) = \{e_i = u_iu_{i+1} : 1 \leq i \leq n-1\} \cup \{e_n = u_nu_1\} \cup \{u_iv_i^j : 1 \leq i \leq n, j = 1,2,3\} \cup \{v_i^1v_i^2, v_i^2v_i^3, v_i^3v_i^1 : 1 \leq i \leq n\}$.
Define $f : E(G) \to \{1,2,3,...,n+1\}$ as follows.

$$f(u_iv_i^2) = f(v_i^2v_i^3) = f(e_i) = n+1 \text{ for } 1 \leq i \leq n,$$
$$f(u_iv_i^1) = f(v_i^3v_i^1) = 1 \text{ for } 1 \leq i \leq n,$$
$$f(u_iv_i^3) = f(v_i^1v_i^2) = i \text{ for } 1 \leq i \leq n.$$

The weights of the vertices of G are given by

$$\omega(v_i^1) = i+2 \text{ for } 1 \leq i \leq n,$$
$$\omega(v_i^2) = 2n+i+2 \text{ for } 1 \leq i \leq n,$$
$$\omega(v_i^3) = n+i+2 \text{ for } 1 \leq i \leq n,$$
$$\omega(u_i) = 3n+i+4 \text{ for } 1 \leq i \leq n,$$

Since the weights of all vertices of G are distinct and f is an irregular labeling, $s(G) \leq n+1$. By Proposition 1, $s(G) \geq n+1$. Hence $s(G) = n+1$.

3 Conclusion and Scope

In this paper we have determined the irregularity strength of $P_n \circ K_m$ and $C_n \circ K_m$ when $1 \leq m \leq 3$. The problem remains open for $m \geq 4$.

References

1. Aigner, M., Triesch, E.: Irregular assignments of trees and forests. SIAM J. Discrete Math. **3**, 439–449 (1990)
2. Anholcer, M., Palmer, C.: Irregular labellings of circulant graphs. Discrete Math. **312**, 3461–3466 (2012)
3. Bohman, T., Kravitz, D.: On the irregularity strength of trees. J. Graph Theory **45**, 241–254 (2004)
4. Chartrand, G., Jacobson, M.S., Lehel, J., Oellermann, O.R., Ruiz, S., Saba, F.: Irregular networks. Congr. Numer. **64**, 187–192 (1988)
5. Ebert, G., Hemmeter, J., Lazebnik, F., Wolder, A.: Irregularity strength for certain graphs. Congr. Numer. **71**, 39–52 (1990)
6. Faudree, R.J., Schelp, R.H., Jacobson, M.S., Lehel, J.: Irregular networks, regular graphs and integer matrices with distinct row and column sums. Discrete Math. **76**, 223–240 (1989)
7. Frieze, A., Gould, R.J., Karonski, M., Pfender, F.: On graph irregularity strength. J. Graph Theory **41**, 120–137 (2002)
8. Jinnah, M.I., Santhosh Kumar, K.R.: Irregularity strength of corona product of two graphs. In: Proceedings of National Seminar on Algebra, Analysis and Discrete Mathematics, pp. 191–199. University of Kerala, Thiruvananthapuram, Kerala (2012)
9. Kalkowski, M., Karonski, M., Pfender, F.: A new upper bound for the irregularity strength of graphs. SIAM J. Discrete Math. **25**(3), 1319–1321 (2011)
10. Majerski, P., Przybylo, J.: On the irregularity strength of dense graphs. SIAM J. Discrete Math. **28**(1), 197–205 (2014)
11. Nierhoff, T.: A tight bound on the irregularity strength of graphs. SIAM J. Discrete Math. **13**, 313–323 (2000)
12. Przybylo, J.: Irregularity strength of regular graphs. Electron. J. Combin. 15 (2008). R82

1-Distant Irregularity Strength of Graphs

K. Muthu Guru Packiam[1], T. Manimaran[2(✉)], and A. Thuraiswamy[3]

[1] Department of Mathematics, Rajah Serfoji Government College, Thanjavur 613005, Tamil Nadu, India
gurupackiam@yahoo.com

[2] Department of Mathematics, Ramco Institute of Technology, Rajapalayam 626117, Tamil Nadu, India
tm_maran@yahoo.com

[3] Department of Mathematics, Kalasalingam University, Anand Nagar, Krishnankoil 626126, Tamil Nadu, India
thuraiswamy@yahoo.com

Abstract. Let $G = (V, E)$ be a connected graph of order $n \geq 3$. Let $f : E \rightarrow \{1, 2, ..., k\}$ be a function and let the weight of a vertex v be defined by $\omega(v) = \sum\limits_{v \in V} f(v)$. Then f is called an irregular labeling if all the vertex weights are distinct. The irregularity strength $s(G)$ is the smallest positive integer k such that there is an irregular labeling $f : E \rightarrow \{1, 2, ..., k\}$. In this paper we prove that for some families of graphs, irregularity strength and r-distant irregularity strength are equal. Further exact value of 1-distant irregularity strength of some classes of graphs are determined.

Keywords: Irregular labeling · Irregularity strength · 1-Distant irregularity strength

1 Introduction

Let $G = (V, E)$ be a connected graph of order $n \geq 3$. Let $f : E \rightarrow \{1, 2, ..., k\}$ be a function and let the weight of a vertex v be defined by $\omega(v) = \sum\limits_{v \in V} f(v)$. Then f is called an irregular labeling if all the vertex weights are distinct. The irregularity strength $s(G)$ is the smallest positive integer k such that there is an irregular labeling $f : E \rightarrow \{1, 2, ..., k\}$.

The irregularity strength of a graph was introduced by Chartrand et al. [1], and the irregularity strength of many graphs were determined in [1], e.g., $s(K_n) = 3, n \geq 3$. Further the irregularity strength of a graph was studied by numerous authors see [2–4,6,7]. Ebert et al. [2] proved that

$$s(W_n) = \begin{cases} \lceil \frac{n+1}{3} \rceil & \text{if } n \geq 6, \\ \lceil \frac{n+1}{3} \rceil + 1 & \text{if } n = 4 \text{ or } 5. \end{cases}$$

Kathiresan et al. [7] proved that $s(F_n) = n+1$. Two more characteristics, namely total vertex irregularity strength and total edge irregularity strength were introduced by Baca et al. [5]. Also the characteristic neighbour-distinguishing k-total

S. Arumugam et al. (Eds.): ICTCSDM 2016, LNCS 10398, pp. 182–190, 2017.
DOI: 10.1007/978-3-319-64419-6_24

weighting was introduced by Przybylo et al. [8]. The bounds for k was found by many authors. The bound was improved to $k = 5$ by Kalkowski et al. [9]. The r-distant irregularity strength of a graph was introduced by Przybylo [10]. In this paper we prove that for some families of graphs, irregularity strength and r-distant irregularity strength are equal. Further exact value of 1-distant irregularity strength of some classes of graphs are determined.

Theorem 1 [3]. *If G is an d-regular graph of order n with $d \geq \frac{n}{2}$, then $s(G) \leq \lceil \frac{n}{2} \rceil + 1$.*

Theorem 2 [3]. $s(K_{m,n}) = 3$ *if* $1 < \frac{n}{2} \leq m < n$.

Definition 1. *Let $G = (V, E)$ be a graph. A labeling $f : E(G) \rightarrow \{1, 2, 3, ..., k\}$ is called r-distant irregular, if for every $v \in V$, the weights of the vertices in $N_r[v]$ (the set of all vertices which are at distance less than or equal to r from v) are pairwise distinct, where the weight of the vertex is the sum of the labels of the edges which are incident with that vertex. The minimum k for which there exists an r-distant irregular labeling of G, is called r-distant irregularity strength $s_r(G)$ of the graph G.*

Proposition 1. *For any graph G, distance irregularity strength chain $s_1(G) \leq s_2(G) \leq ... \leq s_{diam(G)}(G) = s(G)$ holds.*

Proof. For any vertex v in G, $N_1[v] \subseteq N_2[v] \subseteq ... \subseteq N_{diam(G)}[v] = V(G)$. Therefore $s_1(G) \leq s_2(G) \leq ... \leq s_{diam(G)}(G) = s(G)$.

Proposition 2. *If G is a graph of order n and $\Delta(G) = n-1$, then $s_r(G) = s(G)$ for all r.*

Proof. Let $\deg v = n - 1$. Then $N_r[v] = V(G)$ for all r. Therefore the weights of all vertices in G are distinct. Hence, $s_r(G) = s(G)$ for all r.

Corollary 1. $s_r(K_n) = s(K_n) = 3$ *for all* $n \geq 3$.

Corollary 2. *Let $K_{1,n}$ be the star with n pendant vertices. Then $s_r(K_{1,n}) = s(K_{1,n}) = n$.*

Corollary 3. *Let $W_n, n \geq 3$ be the wheel on n vertices. Then*

$$s_r(W_n) = s(W_n) = \begin{cases} \lceil \frac{n+1}{3} \rceil & if \;\; n \geq 6, \\ \lceil \frac{n+1}{3} \rceil + 1 & if \;\; n = 4 \; or \; 5. \end{cases}$$

Corollary 4. *Let $F_n, n \geq 2$ be the friendship graph on $2n + 1$ vertices. Then $s_r(F_n) = s(F_n) = n + 1$.*

Converse of the Proposition 2 is not true. That is if $s_r(G) = s(G)$ for all r, then $\Delta(G)$ need not be equal to $n - 1$. For example, in the following graph G, $s_r(G) = s(G)$ for all r, but there is no vertex of degree $n - 1$ (Fig. 1).

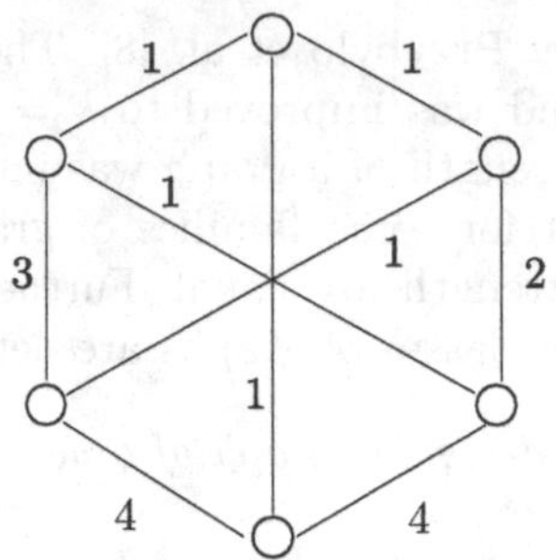

Fig. 1. A graph with $s_r(G) = s(G)$ and $\Delta(G) < n - 1$.

2 1-Distant Irregularity Strength of Certain Families of Graphs

Theorem 3. *Let $P_n, n > 2$ be the path on n vertices. Then*

$$s_1(P_n) = \begin{cases} 2 \ if \ \ n \leq 5, \\ 3 \ if \ \ n > 5. \end{cases}$$

Proof. Let $P_n = (v_1, e_1, v_2, e_2, ..., e_{n-1}, v_n)$.
Define $f : E(P_n) \to \{1, 2, 3\}$ is as follows.
For $P_3 : f(e_1) = 1, f(e_2) = 2$.
For $P_4 : f(e_1) = 1, f(e_2) = 2, f(e_3) = 2$.
For $P_5 : f(e_1) = 1, f(e_2) = 1, f(e_3) = 2, f(e_4) = 2$.
For $n > 5$,

$$f(e_i) = \begin{cases} 1 & \text{if } i \equiv 1 (mod\ 3), \\ 2 & \text{if } i \equiv 2 (mod\ 3), \\ 3 & \text{if } i \equiv 0 (mod\ 3), \text{ for } 1 \leq i < n - 4, \end{cases}$$

$$f(e_{n-3}) = \begin{cases} 2 & \text{if } n \equiv 2 (mod\ 3), \\ 3 & \text{if } n \not\equiv 2 (mod\ 3), \end{cases}$$

$$f(e_{n-2}) = \begin{cases} 3 & \text{if } n \equiv 2 (mod\ 3), \\ 1 & \text{if } n \not\equiv 2 (mod\ 3), \end{cases}$$

and $f(e_{n-1}) = 1$.

Since the weights of all vertices of P_n are distinct and f is an 1-distant irregular labeling,

$$s_1(P_n) = \begin{cases} 2 & \text{if } \ n \leq 5, \\ 3 & \text{if } \ n > 5. \end{cases}$$

Theorem 4. *Let C_n be the cycle on n vertices. Then $s_1(C_n) = 3$.*

Proof. Let $C_n = (v_1, e_1, v_2, e_2, v_3, ..., v_n, e_n, v_1)$.
Define $f : E(C_n) \to \{1, 2, 3\}$ as follows.
If $n \not\equiv 2(mod\ 3)$, then

$$f(e_i) = \begin{cases} 1 & \text{if } i \equiv 1(mod\ 3), \\ 2 & \text{if } i \equiv 2(mod\ 3), \\ 3 & \text{if } i \equiv 0(mod\ 3). \end{cases}$$

If $n \equiv 2(mod\ 3)$, then for $1 \le i \le n-2$,

$$f(e_i) = \begin{cases} 1 & \text{if } i \equiv 1(mod\ 3), \\ 2 & \text{if } i \equiv 2(mod\ 3), \\ 3 & \text{if } i \equiv 0(mod\ 3), \end{cases}$$

and $f(e_{n-1}) = 3, f(e_n) = 1$.
Since the weights of all vertices of C_n are distinct and f is an 1-distant irregular labeling, $s_1(C_n) = 3$.

Theorem 5. *Let G be a d-regular graph on n vertices. Then $s_1(G) = s(G) \le \lceil \frac{n}{2} \rceil + 1$ if $d \ge \frac{n}{2}$.*

Proof. Since $d \ge \frac{n}{2}$, for every vertex v in G, $N_1[v]$ must contain at least $\frac{n}{2} + 1$ vertices. Hence for any two arbitrary vertices v_i and v_j, $N_1[v_i] \cap N_1[v_j] \ne \phi$. Let $v \in N_1[v_i] \cap N_1[v_j]$, then $v_i, v_j \in N_1[v]$. Hence weights of v_i, v_j are distinct for all i and j. Thus, $s_1(G) = s(G)$. By Theorem 1, $s_1(G) \le \lceil \frac{n}{2} \rceil + 1$.

Theorem 6. *Let $K_{m,n}$ be the complete bipartite graph. Then $s_1(K_{m,n}) = s(K_{m,n})$ for all m and n. Also $s_1(K_{m,n}) = 3$ if $1 < \frac{n}{2} \le m < n$.*

Proof. Let $K_{m,n}$ be the complete bipartite graph with bipartition (V_1, V_2) and let $|V_1| = m, |V_2| = n$. Since every vertex of V_1 is adjacent to all the vertices of $V_2, V_2 \subset N_1[x]$ for all x in V_1. Therefore the weights of all vertices in $V_2 \cup \{x\}$ are distinct for all x in V_1. Similarly the weights of all vertices in $V_1 \cup \{y\}$ are distinct for all y in V_2. Hence the weights of all vertices of $K_{m,n}$ are distinct, and so $s_1(K_{m,n}) = s(K_{m,n})$ for all m and n. By Theorem 2, $s_1(K_{m,n}) = 3$ if $1 < \frac{n}{2} \le m < n$.

Theorem 7. *Let $G = P_n \square K_2$, where $\square$ denotes the Cartesian product. Then $s_1(G) = 3$.*

Proof. Construct the graph G by joining the two paths $P_n : v_1e_1v_2e_2v_3...v_{n-1}e_{n-1}v_n$ and $P'_n : v'_1e'_1v'_2e'_2v'_3...v'_{n-1}e'_{n-1}v'_n$ with the edges $v_iv'_i, i = 1, 2, 3, ..., n$.
Define $f : E(G) \to \{1, 2, 3\}$ as follows.
An optimal 1-distant irregular labelings for $n = 2, 3$ are shown in Fig. 2.

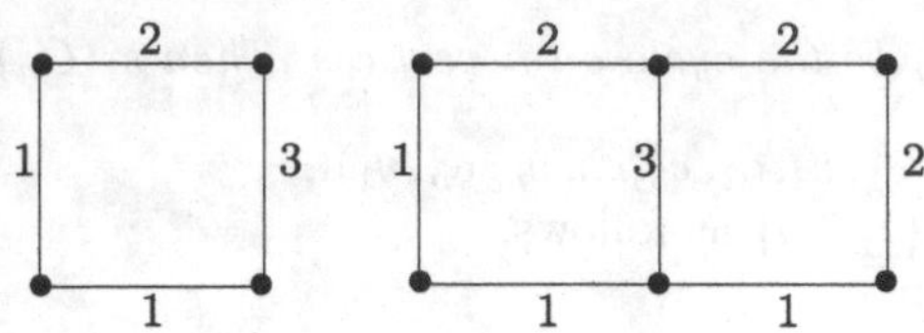

Fig. 2. Optimal 1-distant irregular labelings for $n = 2, 3$

For $n > 3$,

$$f(e_i) = \begin{cases} 1 & \text{if } i \equiv 1(mod\ 2), \\ 2 & \text{if } i \equiv 2(mod\ 4), \\ 3 & \text{if } i \equiv 0(mod\ 4), \text{ for } 1 \leq i < n-2, \end{cases}$$

$$f(e_{n-2}) = \begin{cases} 2 & \text{if } n \equiv 0, 3(mod\ 4), \\ 3 & \text{if } n \equiv 1, 2(mod\ 4), \end{cases}$$

$$f(e_{n-1}) = \begin{cases} 1 & \text{if } n \not\equiv 1(mod\ 4), \\ 2 & \text{if } n \equiv 1(mod\ 4), \end{cases}$$

$$f(e'_i) = \begin{cases} 1 & \text{if } i \equiv 0(mod\ 4), \\ 2 & \text{if } i \not\equiv 0(mod\ 4), \text{ for } 1 \leq i \leq n-1, \end{cases}$$

$$f(v_iv'_i) = \begin{cases} 1 & \text{if } i \equiv 0(mod\ 2), \\ 3 & \text{if } i \equiv 1(mod\ 2), \text{ for } 2 \leq i \leq n-1, \end{cases}$$

and $f(v_1v'_1) = f(v_nv'_n) = 1$.

Since the weights of all vertices of G are distinct and f is an 1-distant irregular labeling, $s_1(G) = 3$.

Theorem 8. *Let $G = C_n \square K_2, n \geq 3$. Then $s_1(G) = 3$.*

Proof. The graph G consists of two cycles $C_n = (v_1, e_1, v_2, e_2, ..., v_n, e_n, v_1)$ and $C'_n = (v'_1, e'_1, v'_2, e'_2, ..., v'_n, e'_n, v'_1)$ with the edges $v_iv'_i, i = 1, 2, 3, ..., n$.

Define $f : E(G) \rightarrow \{1, 2, 3\}$ as follows.

For $n = 3$,

$$f(e_1) = f(e_2) = f(e_3) = f(v_1v'_1) = 1,$$
$$f(e'_1) = f(e'_2) = f(e'_3) = f(v_3v'_3) = 3 \text{ and } f(v_2v'_2) = 2.$$

For $n = 5$,

$$f(e_1) = f(e_5) = f(e'_2) = f(e'_3) = 1,$$
$$f(v_iv'_i) = 1 \text{ for } 1 \leq i \leq 5,$$
$$f(e_2) = f(e'_4) = 2, f(e_3) = f(e_4) = f(e'_1) = f(e'_5) = 3.$$

For $n = 4$ or $n \geq 6$,

$$f(e_i) = \begin{cases} 1 & \text{if } i \equiv 0, 1 (mod\ 4), \\ 2 & \text{if } i \equiv 2, 3 (mod\ 4), \text{ for } 1 \leq i \leq n-3, \end{cases}$$

$$f(e_{n-1}) = f(e_{n-2}) = \begin{cases} 2 & \text{if } n \equiv 0 (mod\ 4), \\ 1 & \text{if } n \not\equiv 0 (mod\ 4), \end{cases}$$

and $f(e_n) = 1$.

$$f(e'_1) = \begin{cases} 3 & \text{if } n \equiv 0 (mod\ 4), \\ 2 & \text{if } n \not\equiv 0 (mod\ 4), \end{cases}$$

$$f(e'_i) = \begin{cases} 2 & \text{if } i \equiv 2 (mod\ 4), \\ 3 & \text{if } i \not\equiv 2 (mod\ 4), \text{ for } 2 \leq i \leq n-4, \end{cases}$$

$$f(e'_{n-2}) = \begin{cases} 2 & \text{if } n \equiv 0 (mod\ 4), \\ 3 & \text{if } n \not\equiv 0 (mod\ 4), \end{cases}$$

$$f(e'_{n-1}) = \begin{cases} 2 & \text{if } n \equiv 1, 2 (mod\ 4), \\ 3 & \text{if } n \equiv 0, 3 (mod\ 4), \end{cases}$$

and $f(e'_n) = f(e'_{n-3}) = 3$.

$$f(v_i v'_i) = \begin{cases} 1 & \text{if } i \equiv 1, 2 (mod\ 4), \\ 3 & \text{if } i \equiv 0, 3 (mod\ 4), \text{ for } 1 \leq i \leq n-4, \end{cases}$$

$$f(v_{n-3} v'_{n-3}) = \begin{cases} 1 & \text{if } n \equiv 0 (mod\ 4), \\ 2 & \text{if } n \equiv 1 (mod\ 4), \\ 3 & \text{if } n \equiv 2, 3 (mod\ 4), \end{cases}$$

$$f(v_{n-2} v'_{n-2}) = \begin{cases} 1 & \text{if } n \equiv 0, 3 (mod\ 4), \\ 3 & \text{if } n \equiv 1, 2 (mod\ 4), \end{cases}$$

$$f(v_{n-1} v'_{n-1}) = \begin{cases} 3 & \text{if } n \equiv 0 (mod\ 4), \\ 2 & \text{if } n \not\equiv 0 (mod\ 4), \end{cases}$$

and $f(v_n v'_n) = 3$.

Since the weights of all vertices of G are distinct and f is an 1-distant irregular labeling, $s_1(G) = 3$.

Theorem 9. *Let $T_n, n \geq 2$ be the fan graph obtained by joining each vertex of the path P_n to a vertex x by an edge. Then*

$$s_1(T_n) = s(T_n) = \begin{cases} 3 & if \quad n = 2, \\ \lceil \frac{n+1}{3} \rceil & if \quad n > 2. \end{cases}$$

Proof. Let $V(T_n) = \{v_i : 1 \leq i \leq n\} \cup \{x\}$ and $E(T_n) = \{e_i = v_i v_{i+1} : 1 \leq i \leq n-1\} \cup \{x v_i : 1 \leq i \leq n\}$. Since degree of x is n, by Proposition 2, $s_1(T_n) = s(T_n)$. Now define $f : E(T_n) \to \{1, 2, 3, ..., \lceil \frac{n+1}{3} \rceil\}$ as follows.

For $n = 2, 3$ the labelings are shown in Fig. 3.

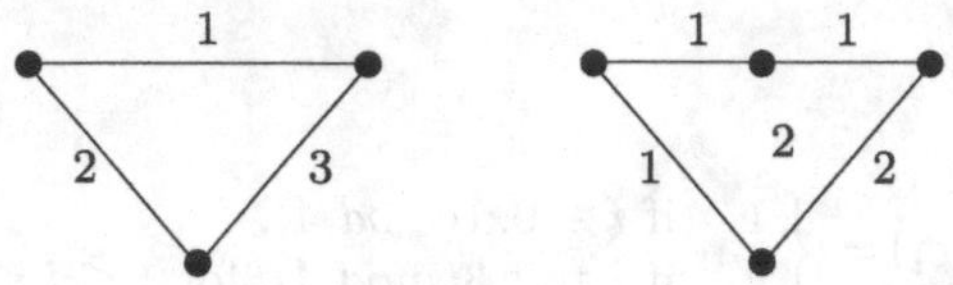

Fig. 3. 1-Distant irregular labelings for $n = 2, 3$

For $n > 3$, define

$$f(e_i) = \left\lceil \frac{i+1}{3} \right\rceil \quad \text{for } 1 \le i \le n-1,$$
$$f(xv_1) = 1,$$
$$f(xv_i) = \left\lceil \frac{i-1}{3} \right\rceil \quad \text{for } 2 \le i \le \left\lfloor \frac{2n-4}{3} \right\rfloor \text{ and } i = n,$$
$$f(xv_i) = \left\lceil \frac{i+2}{3} \right\rceil \quad \text{for } \left\lfloor \frac{2n-4}{3} \right\rfloor < i < n.$$

By the above labeling we have $s_1(T_n) \le \lceil \frac{n+1}{3} \rceil$, for $n > 2$. Since the weights of $v_1, v_2, v_3, ..., v_n$ and x are distinct, the maximum weight of the vertices $v_1, v_2, v_3, ..., v_n$ is at least $n + 1$. Since degree of these vertices are 2 or 3, at least one of the label is greater than or equal to $\frac{n+1}{3}$. Therefore $s_1(T_n) \ge \lceil \frac{n+1}{3} \rceil$ for $n > 2$. Hence $s_1(T_n) = \lceil \frac{n+1}{3} \rceil$ for $n > 2$. Further we cannot label T_2 with fewer than 3 labels. Hence we have

$$s_1(T_n) = \begin{cases} 3 & \text{if n} = 2, \\ \lceil \frac{n+1}{3} \rceil & \text{if n} > 2. \end{cases}$$

Definition 2. *A caterpillar is a graph derived from a path by hanging any number of leaves from the vertices of the path.*

Theorem 10. *Let $G = P_r \circ \overline{K_k}$, where $\circ$ denotes corona. Then*

$$s_1(G) = \begin{cases} 2 \ if \ \ k = 1, \\ 2 \ if \ \ k = 2 \ and \ r \le 5, \\ 3 \ if \ \ k = 2 \ and \ r > 5, \\ k \ if \ \ k > 2. \end{cases}$$

Proof. Let $V(G) = \{c_i : 1 \le i \le r\} \cup \bigcup_{i=1}^{r} \{x_i^j : 1 \le j \le k\}$ and $E(G) = \{c_i c_{i+1} : 1 \le i \le r-1\} \cup \bigcup_{i=1}^{r} \{c_i x_i^j : 1 \le j \le k\}$.

Define $f : E(G) \rightarrow \{1, 2, 3, ..., k\}$ as follows.

Case 1. $k = 1$.

If $r = 3$, then $f(c_1c_2) = f(c_1x_1^1) = f(c_2x_2^1) = f(c_3x_3^1) = 1, f(c_2c_3) = 2$. If $r > 3$, then for $1 \le i \le r-1$,

$$f(c_ic_{i+1}) = f(c_ix_i^1) = \begin{cases} 1 & \text{if } i \equiv 1, 2 (mod\ 4), \\ 2 & \text{if } i \equiv 0, 3 (mod\ 4), \end{cases}$$

and

$$f(c_r x_r^1) = \begin{cases} 1 & \text{if } r \not\equiv 0(mod\ 4), \\ 2 & \text{if } r \equiv 0(mod\ 4). \end{cases}$$

Case 2. $k = 2$ and $r \leq 5$.
Assign label $f(c_i x_i^1) = 1, f(c_i x_i^2) = 2$ for $1 \leq i \leq r$
If $r = 3$, then $f(c_1c_2) = 1, f(c_2c_3) = 2$.
If $r = 4$, then $f(c_1c_2) = 1, f(c_2c_3) = f(c_3c_4) = 2$.
If $r = 5$, then $f(c_1c_2) = f(c_2c_3) = 1, f(c_3c_4) = f(c_4c_5) = 2$.

Case 3. $k = 2, r > 5$ or $k > 2$.
Assign label $f(c_i x_i^j) = j$ for $1 \leq i \leq r, 1 \leq j \leq k$.
If $r \equiv 1(mod\ 3)$, then for $1 \leq i \leq r - 4$,

$$f(c_i c_{i+1}) = \begin{cases} 1 & \text{if } i \equiv 1(mod\ 3), \\ 2 & \text{if } i \equiv 2(mod\ 3), \\ 3 & \text{if } i \equiv 0(mod\ 3), \end{cases}$$

and $f(c_{r-3}c_{r-2}) = 3, f(c_{r-2}c_{r-1}) = f(c_{r-1}c_r) = 1$.
If $r \not\equiv 1(mod\ 3)$, then for $1 \leq i \leq r - 1$,

$$f(c_i c_{i+1}) = \begin{cases} 1 & \text{if } i \equiv 1(mod\ 3), \\ 2 & \text{if } i \equiv 2(mod\ 3), \\ 3 & \text{if } i \equiv 0(mod\ 3). \end{cases}$$

Since the weights of all vertices of G are distinct and f is an 1-distant irregular labeling,

$$s_1(G) = \begin{cases} 2 & \text{if } k = 1, \\ 2 & \text{if } k = 2 \text{ and } r \leq 5, \\ 3 & \text{if } k = 2 \text{ and } r > 5, \\ k & \text{if } k > 2. \end{cases}$$

3 Conclusion and Scope

The determination of r-distant irregularity strength of path, cycle, ladder graph, fan graph and caterpillar for $r \geq 2$ are still open.

References

1. Chartrand, G., Jacobson, M.S., Lehel, J., Oellermann, O.R., Ruiz, S., Saba, F.: Irregular networks. Congr. Numer. **64**, 187–192 (1988)
2. Ebert, G., Hemmeter, J., Lazebnik, F., Wolder, A.: Irregularity strength for certain graphs. Congr. Numer. **71**, 39–52 (1990)
3. Faudree, R.J., Schelp, R.H., Jacobson, M.S., Lehel, J.: Irregular networks, regular graphs and integer matrices with distinct row and column sums. Discrete Math. **76**, 223–240 (1989)

4. Przybylo, J.: Irregularity strength of regular graphs. Electron. J. Combin. **15**, R82 (2008)
5. Baca, M., Jendrol, S., Miller, M., Ryan, J.: On irregular total labeling. Discrete Math. **307**, 1378–1388 (2007)
6. Kathiresan, K.M., Muthu Guru Packiam, K.: Change in irregularity strength by an edge. J. Combin. Math. Combin. Comput. **64**, 49–64 (2008)
7. Kathiresan, K.M., Muthu Guru Packiam, K.: A study on stable, positive and negative edges with respect to irregularity strength of a graph. Ars Combin. **103**, 479–489 (2012)
8. Przybylo, J., Wozniak, M.: On a 1, 2 conjecture. Discrete Math. Theor. Comput. Sci. **12**(1), 101–108 (2010)
9. Kalkowski, M., Karonski, M., Pfender, F.: Vertex-coloring edge-weightings: towards the 1-2-3-conjecture. J. Combin. Theory Ser. B. **100**, 347–349 (2010)
10. Przybylo, J.: Distant irregularity strength of graphs. Discrete Math. **313**(24), 2875–2880 (2013)

Cloud Data Security Based on Data Partitions and Multiple Encryptions

B. Muthulakshmi(✉) and M. Venkatesulu

Kalasalingam University, Anand Nagar, Krishnankoil 626126, Tamil Nadu, India
selvamayil2010@gmail.com, venkatesulum2000@gmail.com

Abstract. The benefits of cloud computing are measured services, unlimited and automatic access and release of resources. But, data security issues such as stealing credentials, unauthorized exposure of data may scare businesses away. Many popular security techniques try to protect data in cloud. Still, various state-of-art techniques are emerging as new types of attacks are being exposed. This paper proposes a scheme that is based on partitioning the plain text into different data parts, generating multiple cipher texts for each part and each cipher text is stored in different centers of cloud storage. Decryption of a data part is done with all cipher texts of each data part and a private key. By obtaining all plain texts data parts by this method, the original text is delivered to authorized user while preserving confidentiality and integrity.

Keywords: Cloud computing · Cloud data security · Data partition · Multiple cipher texts · Confidentiality · Integrity

1 Introduction

Cloud computing is a combination of technologies such as Service Oriented Architecture (SOA), virtualization and networking. The Cloud is classified into three types such as Software as a Service (SaaS), Platform as a Service (PaaS) and Infrastructure as a Service (IaaS) based on the service it offers and into four types namely, Private Cloud, Public Cloud, Community Cloud and Hybrid Cloud based on deployment. Maddineni and Ragi emphasized that essential characteristic, for example flexibility, measured services continue to enticing ventures to receive Cloud computing [13]. But security issues, particularly data security issues backtrack to adopt this state of art technique. Some of the real time situations and some proposed solutions for data security issues are presented below. Now-a-days, data theft and cyber breach are every day events around the world. Yahoo Inc. reported one of the largest cyber-attacks recently. This situation emphasized the importance of precautionary measures to be taken by customers of Cloud and in turn the Cloud Service Provider (CSP) by tightening the control over their customers data, by screening the employees now and then and checking the background of users who access the data. Sun et al. proposed that the major issues in Cloud are resource security, resource administration, resource

S. Arumugam et al. (Eds.): ICTCSDM 2016, LNCS 10398, pp. 191–196, 2017.
DOI: 10.1007/978-3-319-64419-6_25

monitoring and data security [7]. Hashizume et al. and Grobaver et al. recorded distinctive analysis of data security issues in SaaS, PaaS and IaaS, for example data leakage due to multitenancy, data theft due to flaws in application in Cloud and legal and compliance issues due to exposure of data to various regions of the world [4,5]. Tumulak suggested that amazing Cloud security solution has to encompass data lockdown, access policies and security intelligence [14]. Yakonbov et al. discussed some precise cryptographic techniques such as homomorphic encryption, verifiable computation and multi party computation which are some modes of security in Cloud [6]. Esporito et al. and Paillier explained Paillier homomorphic crypto system which create more cipher texts for a single plaintext and cipher texts that are randomly chosen used for decryption and not only this, Pairing-Based, Identity-Based encryption techniques, ABE, Digital signature and secure routing were also discussed [2,11]. Gomathisankaran et al. described HORNs, a scheme in which Residue Number System is used to obtain homomorphic encryption on data shares, a way of protecting data in cloud [3]. The authors addressed data security issues of confidentiality, integrity and cloud collusion and propose some realistic solutions too. Zhou et al. discussed how RBE could be applied for effective consumer revocation and RBAC is utilized to segregate data to public or private cloud depending upon the level of sensitivity [9]. Shaikh and Sasikumar studied data classification as a method of attaining cloud data security [6]. Wang et al. and Balu, Kuppusamy focused key escrow and inexpressiveness of attribute as problems in making use of CP-ABE scheme for Cloud data security, but they furnished a way to the above problems through enforcing an improved key issuing protocol and weighted attribute procedure which ensures the security proof of data confidentiality and privacy [1,8]. Thilakanathan et al. pointed out varieties of assaults on the Cloud, motives of malicious customers and the way ABE and Proxy Re-encryption facilitates secure and confidential data sharing in the Cloud [10].

2 Proposed Scheme

Security methods presented in the above section are supposed to preserve confidentiality, integrity or authenticate the access to rights for information, applications and other resources or to preclude the malicious insiders of the organization from having entry to access to customers data. Though, the above security procedures possess some merits and demerits, they are supposed to redress some vulnerabilities and attacks as aforementioned. This paper proposes yet another novel data/information protection technique based on partitioning of data/information, prime numbers and a secret key.

2.1 Algorithm for Encryption

1. Let M be the information/data, for example a file to be secured.
2. Partition M by reading the file into $M_1, M_2, \ldots, M_n$ where n is the number of data parts.

3. Generate $2n$ prime numbers where n is the number of data parts. Number of prime numbers to be generated depends on the number of cipher texts to be produced for each part M_i. With a pair of prime numbers 4 encrypted parts will be produced for each data part and with 4 prime numbers 8 cipher texts will be formed.
4. Arrange $2n$ prime numbers into n number of pairs such as $[P_i, Q_i]$where P_i and Q_i are two consecutive prime numbers. Here Q_i is nothing but P_{i+1}.
5. Define $P_i^c = 2^p - P_i, Q_i^c = 2^p - Q_i$, where p is the size of the prime number.
6. Generate a set of very large integers $D_1, D_2, \ldots, D_n$ and split each D_i into 4 parts d_1, d_2, d_3 *and* d_4. This partition may depend on the number of prime numbers to be taken to encrypt each data part. In this scheme, for a pair of primes and their prime complements, four partitions are required.
7. Define cipher texts C_{i1} to C_{i4} for each M_i
$f_1\ (M_i, P_i, Q_i, d_1)$,
$f_2\ (M_i, P_i, Q_i^c, d_2)$,
$f_3\ (M_i, P_i^c, Q_i, d_3)$ and
$f_4\ (M_i, P_i^c, Q_i^c, d_4)$
8. Each cipher text C_{ij} is intended to be stored in different folders of a Cloud storage.

2.2 Algorithm for Decryption

1. Get $M_i = C_{i1} + C_{i2} + C_{i3} + C_{i4} - D_i$ and omit the extra zeros.
2. Get the plaintext M, by changing each M_i into original format from the format at the time of partitioning.

3 Implementation and Testing

This scheme has been developed in Java. The results of the scheme have been tested with a system of processor Intel(R) Core (TM)2 Duo CPU, 4 GB of RAM and Windows7 as the platform. A JPEG file has been taken as an input. Partition the file by changing the format and produce multiple cipher texts for all data parts. Create as many number of folders in Cloud storage as the number of cipher texts for each data part. First set of cipher texts is stored in one of the folder of Cloud storage. Successive sets of cipher texts are stored different folders of Cloud storage.

Figure 1 is the image that was taken as an input to test our implementation. The size of the Fig. 1 image is 21.1 KB, which is a JPEG image of .jpg extension. The image format of Fig. 1 is converted to byte array. Figure 2 shows the screenshot of the converted byte array format.

The byte array format of the image is changed to a series of integers and the obtained series is given below.

−2555936; 1067590; 1229324289;16777217;65536; −2424700; 591367; 303174165; 320017174; 353703704; 387454744; 387389210; 387258904; 404231960;404231962;

Fig. 1. Image taken for testing

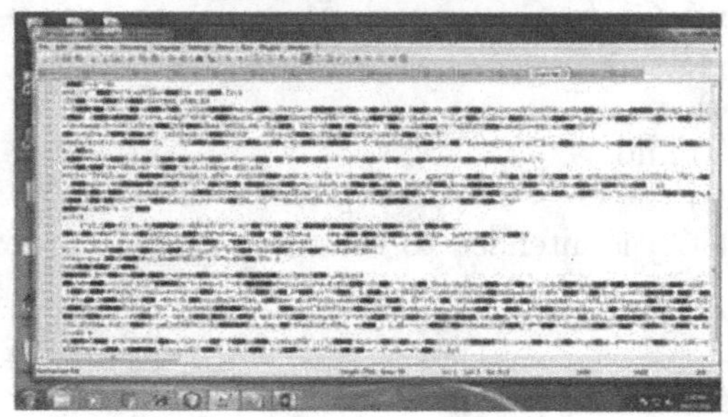

Fig. 2. Byte array form of the input image.

489168920; 488972055; 404828449; 657009454; 774772767; 859321132; 925379886; 721488394; 168693006; 454037530; 757407014; 758066479;757935405 (repeated 10 times); 791489837; 757989312; 1116160; −520036093; 19005442; 285278993; 19005442; 285278993; 33539072; 452984834;50397441; 0; 0; 1029; 33752576; 17301444.

To encrypt the given data/information, a set of prime numbers has been generated. Again for encryption, complement of each prime number is required. Complement of a prime can be calculated using the formula 2^p- (a prime number).

To enhance the security of data a set of keys consisting of very large integers are to be generated. Each large integer is to be split into parts. Figure 3 shows the generation of very large integers. These will act as a key to decryption. The data owner will give this set of keys to authorized users.

The first set of split of very large integers is used to create first set of cipher texts. Similarly, the three other sets of split of very large integers take part to create second, third and fourth sets of cipher texts. The four sets of cipher texts thus generated are stored in the successive folders of Cloud storage. The proposed algorithm decrypts the encrypted shares successfully (Fig. 4).

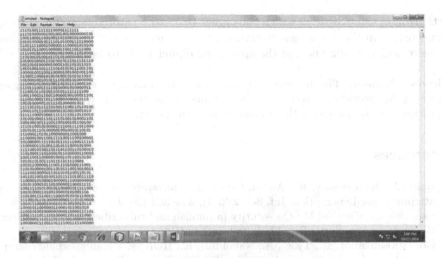

Fig. 3. Generation of large integers which acts as keys

Fig. 4. Image created by decrypted set of integers

4 Conclusion

Bring Your Own Cloud (BYOC) is now becoming the trend of an IT field. But, data security issues such as data leakage, data loss, unauthorized access and modification of data are challenging the adoption of cloud technology by government agencies and business enterprises. The scheme proposed in this paper preserves the privacy of data owner by ensuring confidentiality and integrity. The data to be secured is partitioned and encrypted and each encrypted form is supposed to be stored in various data centers of Cloud storage entirely in different format. So, the proposed design preserves the confidentiality. As long as, the data in data centers of Cloud remain unchanged, decryption is possible. So, the above scheme checks the integrity of data too.

5 Future Work

In future, our plan is to improve the implementation by encrypting files of larger size and of all formats and to do crypt-analysis of the proposed design.

Furthermore, the CSP with which the cipher texts to be stored is assumed to be trust-worthy. In the worst case scenario, CSP can act as a malicious insider. So, our next design would encrypt the encrypted cipher texts to be stored in cloud.

Acknowledgment. The first author is thankful to the management of Kalasalingam University for providing fellowship. The authors thank the referee for the valuable suggestions which resulted in the present form of the paper.

References

1. Balu, A., Kuppusamy, K.: An expressive and provably secure ciphertext-policy attribute based encryption. Inf. Sci. **276**(4), 354–362 (2014)
2. Esporito, C., Ciauyie, M.: On security in publish and subscribe services: a survey. IEEE. Commun. Surv. Tut. **17**(2), 966–997 (2015)
3. Gomathisankaran, M., Tyagi, A., Namuduri, K.: HORNs: a homomorphic encryption scheme for cloud computing using Residue Number System. In: 5th Annual Conference on Information Sciences and Systems, Baltimore, MD, pp. 1–5 (2011)
4. Grobaver, B., Walloschek, T., Stocker, E.: Understanding cloud computing vulnerabilities. IEEE Secur. Priv. **9**(2), 50–57 (2011)
5. Hashizume, K., Rosado, D.G., Fernandez Medina, E., Fernandez, E.B.: An analysis of security issues for cloud computing. J. Internet Serv. Appl. 4(1) (2013). 13 pages
6. Shaikh, R., Sasikumar, M.: Data classification for achieving security in cloud computing. Procedia Comput. Sci. **45**, 493–498 (2015). Science Direct
7. Sun, Y., Zhang, J., Xiong, Y., Zhu, G.: Data security and privacy in cloud computing. Int. J. Distrib. Sens. N. (2014). Article ID 190903, 9 pages, doi:10.1155/2014/190903
8. Wang, S., Liang, K., Lin, J.K., Chen, J., Yu, J., Xie, W.: Attribute based data sharing scheme revisited in cloud computing. IEEE T. Inf. Foren. Sec. **11**(8), 1661–1673 (2016)
9. Zhou, L., Varadharajan, V., Hitchens, M.: Achieving secure role based access on encrypted data in cloud storage. IEEE T. Inf. Foren. Sec. **8**(12), 1947–1960 (2014)
10. Thilakanathan, D., Chen, S., Nepal, S., Calvo, R.A.: Secure Data Sharing in the Cloud. In: Nepal, S., Pathan, M. (eds.) Security, Privacy and Trust in Cloud Systems. Springer, Heidelberg (2014). doi:10.1007/978-3-642-38586-5_2
11. Paillier, P.: Public-key crypto systems based on composite degree repository classes. In: Proceedings 17th International Conference Theory Application Cryptography Technology, pp. 223–238 (1999)
12. Yakonbov, S., Gadepally, V., Schear, N., Shen, E., Yerukhimovich, A.: A survey of cryptographic approaches to securing big-data analytics in the cloud. In: High Performance Extreme Computing Conference (HPEC), pp. 1–6. IEEE (2014)
13. Maddineni, V.S.K., Ragi, S.: Security techniques for protecting data in cloud computing. A Master thesis, School of Computing, Blekinge Institute of Technology, Sweden (2011)
14. Tumulak, D.: Data Security in the Cloud. Vormetric, Gartner Inc. (2012)

Upper Majority Domination Number of a Graph

A. Muthuselvi(✉) and S. Arumugam

National Centre for Advanced Research in Discrete Mathematics (n-CARDMATH), Kalasalingam University, Anand Nagar, Krishnankoil 626126, Tamil Nadu, India
amuthuselvi90@gmail.com, s.arumugam.klu@gmail.com

Abstract. A majority dominating function of a graph $G = (V, E)$ is a function $g : V \to \{-1, 1\}$ such that $g(N[v]) \geq 1$, for at least half of the vertices $v \in V$, where $N[v]$ is the closed neighborhood of v and $g(N[v]) = \sum\limits_{u \in N[v]} g(u)$. The weight of g is defined by $g(V) = \sum\limits_{v \in V} g(v)$. The majority domination number $\gamma_{maj}(G)$ is the minimum weight of a majority dominating function of G. The maximum cardinality of a minimal majority dominating function of G is called the upper majority domination number of G and is denoted by $\Gamma_{maj}(G)$. In this paper we initiate a study of this parameter and present several basic results.

Keywords: Majority domination · Majority domination number · Upper majority domination number

1 Introduction

By a graph $G = (V, E)$ we mean a finite undirected graph with neither loops nor multiple edges. The order $|V|$ and the size $|E|$ are denoted by n and m respectively. For graph theoretic terminology we refer to Chartrand and Lesniak [7]. One of the major areas within graph theory is the study of domination and related subset problems. For an excellent treatment of fundamentals of domination we refer to the book by Haynes et al. [5]. Survey of general advanced topics in domination is given in the book edited by Haynes et al. [6]. Several domination related concepts have been formulated in terms of functions satisfying the condition that the closed neighborhood sum of any vertex is at least one. For a survey of results on several topics dealing with domination related functions we refer to Chaps. 1, 2, 3, 4 and 5 of [6]. Dunbar et al. [2] introduced the concept of minus dominating function.

Let $G = (V, E)$ be any graph and let $v \in V$. The open neighborhood $N(v)$ of v is the set of vertices adjacent to v. The closed neighborhood $N[v]$ of v is $N[v] = N(v) \cup \{v\}$. For a set S of vertices, we define $N(S) = \bigcup\limits_{v \in S} N(v)$ and $N[S] = N(S) \cup S$. A subset S of V is called a dominating set of G if $N[S] = V$. The domination number $\gamma(G)$ of a graph G is the minimum cardinality of a dominating set in G. For any real valued function $g : V \to R$ and $S \subseteq V$, let $g(S) = \sum\limits_{u \in S} g(u)$. A minus dominating function is defined in [2] as a function

S. Arumugam et al. (Eds.): ICTCSDM 2016, LNCS 10398, pp. 197–202, 2017.
DOI: 10.1007/978-3-319-64419-6_26

$g : V \to \{-1, 0, 1\}$ such that $g(N[v]) \geq 1$ for all $v \in V$. A minus dominating function is minimal if and only if for every vertex $v \in V$ with $g(v) \geq 0$, there exists a vertex $u \in N[v]$ with $g(N[u]) = 1$. The minus domination number of G is defined by $\gamma^-(G)$ =min$\{g(V) : g$ is a minus dominating function of $G\}$. The concept of signed dominating function was introduced by Dunbar et al. [3]. A function $g : V \to \{-1, 1\}$ is called a signed dominating function of G for every vertex $v \in V$, $g(N[v]) \geq 1$. The signed domination number $\gamma_s(G)$ =min$\{g(V) : g$ is a signed dominating function of $G\}$. Broere et al. [1] introduced the concept of majority domination in graphs.

Definition 1.1. *A function $g : V \to \{-1, 1\}$ is called a majority dominating function if $g(N[v]) \geq 1$ for at least half of the vertices $v \in V$. The majority domination number of graph G is $\gamma_{maj}(G)$ =min$\{g(V) : g$ is a majority dominating function of $G\}$.*

Definition 1.2. *A majority dominating function g of G is called a minimal majority dominating function, if there does not exist a majority dominating function $h : V \to \{-1, 1\}$ such that $h \neq g$ and $h(v) \leq g(v)$ for every $v \in V$.*

Theorem 1.3 [1]. *For any integer $n \geq 1$,*

$$\gamma_{maj}(K_n) = \begin{cases} 1 & \text{if } n \text{ is odd} \\ 2 & \text{if } n \text{ is even.} \end{cases}$$

Theorem 1.4 [1]. *For any integer $n \geq 2$,*

$$\gamma_{maj}(K_{1,n}) = \begin{cases} 1 & \text{if } n \text{ is even} \\ 2 & \text{if } n \text{ is odd.} \end{cases}$$

Theorem 1.5 [1]. *For $t \geq s \geq 2$,*

$$\gamma_{maj}(K_{s,t}) = \begin{cases} 2 - t & \text{if } s \text{ is even} \\ 3 - t & \text{if } s \text{ is odd.} \end{cases}$$

Proposition 1.6 [1]. *A majority dominating function g of a graph G is minimal only if for every vertex $v \in V$ with $g(v) = 1$, there exists a vertex $u \in N[v]$ with $g(N[u]) \in \{1, 2\}$.*

Lemma 1.7 [4]. *A signed dominating function g of a graph G is minimal if and only if for every vertex $v \in V$ with $g(v) = 1$, there exists a vertex $u \in N[v]$ with $g(N[u]) \in \{1, 2\}$.*

Definition 1.8 [4]. *The upper signed domination number of a graph G, denoted by $\Gamma_s(G)$, is defined as $\Gamma_s(G) = max\{g(V) : g$ is a minimal signed dominating function on $G\}$.*

2 Upper Majority Domination

Throughout this paper we use the following notation.

Let $g : V \to \{-1, 1\}$ be a majority dominating function of G. Let

$$\begin{aligned} P &= \{v \in V : g(v) = 1\}, \\ M &= \{v \in V : g(v) = -1\} \text{ and} \\ S &= \{v \in V : g(N[v]) \geq 1\}. \end{aligned}$$

Since g is a majority dominating function, $|S| \geq \frac{n}{2}$. For any $v \in P$, let $s(v) = \{w \in N[v] : g(N[w]) = 1 \text{ or } 2\}$. Let $k(v) = |s(v)|$ and $k = \min_{v \in P} k(v)$.

Theorem 2.1. *A majority dominating function g of a graph G is minimal if and only if $|S| - k < \lceil \frac{n}{2} \rceil$.*

Proof. Suppose $|S| - k < \lceil \frac{n}{2} \rceil$. Let $f : V \to \{-1, 1\}$ be a function such that $f < g$. Hence there exists $v \in V$ such that $g(v) = 1$ and $f(v) = -1$. Let $S_1 = \{u \in V : f(N[u]) \geq 1\}$. Then $|S_1| \leq |S| - k < \lceil \frac{n}{2} \rceil$. Hence f is not a majority dominating function of G. Thus g is a minimal majority dominating function of G.

Conversely, let g be a minimal majority dominating function of G and let $g(v) = 1$. Define $h : V \to \{-1, 1\}$ by $h(v) = -1$ and $h(w) = g(w)$, for all $w \neq v$. Let $S_2 = \{v \in V : h(N[v]) \geq 1\}$. Now for any $w \in N[v]$ with $g(N[w]) = 2$ or 1, we have $f(N[w]) = 0$ or -1. Hence $|S_2| = |S| - k < \lceil \frac{n}{2} \rceil$.

Definition 2.2. *The upper majority domination number $\Gamma_{maj}(G)$ of G is defined by $\Gamma_{maj}(G) = max\{g(V) : g$ is a minimal majority dominating function of $G\}$.*

It follows from the definition that $\gamma_{maj}(G) \leq \Gamma_{maj}(G)$. We now proceed to determine the upper majority domination number of some standard graphs, which in turn gives several class of graphs for which $\gamma_{maj}(G) = \Gamma_{maj}(G)$.

Theorem 2.3. *For any integer $n \geq 2$,*

$$\Gamma_{maj}(K_n) = \begin{cases} 1 & \text{if } n \text{ is odd} \\ 2 & \text{if } n \text{ is even.} \end{cases}$$

Proof. Let $V(K_n) = \{v_1, v_2, \ldots, v_n\}$. Any function $g : V \to \{-1, 1\}$ is a majority dominating function of K_n if and only if $|P| \geq \lceil \frac{n+1}{2} \rceil$. Further if $|P| > \lceil \frac{n+1}{2} \rceil$, then $g(N[v_i]) \geq 3$ for all $v_i \in V$. Hence $k(v_i) = 0$ for all $v_i \in V$, so that $k = 0$. Also $|S| = n$ and hence $|S| - k > \lceil \frac{n}{2} \rceil$. Thus it follows from Theorem 2.1 that g is not a minimal majority dominating function of K_n. Hence g is a minimal majority dominating function of K_n if and only if $|P| = \lceil \frac{n+1}{2} \rceil$. Therefore for any minimal dominating function of K_n, $g(V) = \begin{cases} 1 & \text{if } n \text{ is odd} \\ 2 & \text{if } n \text{ is even} \end{cases}$. Hence

$$\Gamma_{maj}(K_n) = \begin{cases} 1 & \text{if } n \text{ is odd} \\ 2 & \text{if } n \text{ is even.} \end{cases}$$

Corollary 2.4. $\gamma_{maj}(K_n) = \Gamma_{maj}(K_n)$.

Proof. Follows from Theorems 1.3 and 2.3.

Theorem 2.5. *For any integer $n \geq 2$,*

$$\Gamma_{maj}(K_{1,n-1}) = \begin{cases} 1 & \text{if } n \text{ is odd} \\ 2 & \text{if } n \text{ is even.} \end{cases}$$

Proof. Let $G = K_{1,n-1}$. Let $V(G) = \{v, v_1, v_2, \ldots, v_{n-1}\}$ and $d(v) = n$. Let $g : V \to \{-1, 1\}$ be a majority dominating function of G. If $g(v) = -1$, then $g(N[v_i]) \leq 0$ for all i, $1 \leq i \leq n-1$, which is a contradiction. Hence $g(v) = 1$. Now $g(N[v_i]) \geq 1$ if and only if $g(v_i) = 1$ for atleast $\lceil \frac{n-1}{2} \rceil$ vertices v_i. Thus $v \in P$ and $|P| \geq \lceil \frac{n-1}{2} \rceil + 1$. Also if $|P| > \lceil \frac{n-1}{2} \rceil + 1$, then $g(N[v]) \geq 2$, for all $v \in P$. Thus $k = 1$ and $|S| > \lceil \frac{n-1}{2} \rceil + 1$. Hence $|S| - k > \lceil \frac{n}{2} \rceil$ and by Theorem 2.1, g is not a minimal majority dominating function of G. Therefore g is a minimal majority dominating function of G if and only if $|P| = \lceil \frac{n+1}{2} \rceil$ and hence $\Gamma_{maj}(K_{1,n-1}) = g(V) = \begin{cases} 1 & \text{if } n \text{ is odd} \\ 2 & \text{if } n \text{ is even.} \end{cases}$

Corollary 2.6. $\gamma_{maj}(K_{1,n-1}) = \Gamma_{maj}(K_{1,n-1})$.

Proof. Follows from Theorems 1.3 and 2.5.

Theorem 2.7. *Let G be the complete bipartite graph $K_{m,n}$ with $2 \leq m \leq n$. Then*

$$\Gamma_{maj}(G) = \begin{cases} 0 & \text{if } m \text{ and } n \text{ are even} \\ 2 & \text{if } m \text{ and } n \text{ are odd} \\ 1 & \text{otherwise.} \end{cases}$$

Proof. Let $U = \{u_1, u_2, \ldots, u_m\}$ and $W = \{w_1, w_2, \ldots, w_n\}$ be the bipartition of G. Let $g : V \to \{-1, 1\}$ be any function. Let $P_U = \{u_i \in U : g(u_i) = 1\}$, $P_W = \{w_i \in W : g(w_i) = 1\}$, $S_U = \{u_i \in U : g(N[u_i]) \geq 1\}$ and $S_W = \{w_i \in W : g(N[w_i]) \geq 1\}$.

$$\text{Then } S_U = \begin{cases} U & \text{if } |P_W| \geq \lceil \frac{n}{2} \rceil + 1 \\ P_U & \text{if } |P_W| = \lceil \frac{n}{2} \rceil \\ \emptyset & \text{if } |P_W| < \lceil \frac{n}{2} \rceil. \end{cases}$$

$$\text{Similarly } S_W = \begin{cases} W & \text{if } |P_U| \geq \lceil \frac{m}{2} \rceil + 1 \\ P_W & \text{if } |P_U| = \lceil \frac{m}{2} \rceil \\ \emptyset & \text{if } |P_U| < \lceil \frac{m}{2} \rceil. \end{cases}$$

Hence it follows that g is a minimal majority dominating function if one of the following holds:

(i) $|P_U| = \lceil \frac{m}{2} \rceil$ and $|P_W| = \lceil \frac{n}{2} \rceil$.
(ii) $|P_U| = \lceil \frac{m}{2} \rceil$ and $|P_W| \geq \lceil \frac{n}{2} \rceil + 1$.
(iii) $|P_U| \geq \lceil \frac{m}{2} \rceil + 1$ and $|P_W| = \lceil \frac{n}{2} \rceil$.
(iv) $|P_U| \geq \lceil \frac{m}{2} \rceil + 1$ and $|P_W| \geq \lceil \frac{n}{2} \rceil + 1$.

(v) $|P_U| \geq \lceil \frac{m}{2} \rceil + 1$ and $|P_W| < \lceil \frac{n}{2} \rceil$.
(vi) $|P_U| < \lceil \frac{m}{2} \rceil$ and $|P_W| \geq \lceil \frac{n}{2} \rceil + 1$ and $m = n$.

Further g is a minimal majority dominating function if and only if either (i) holds or (v) holds with $|P_U| = \lceil \frac{m}{2} \rceil + 1$ and $|P_W| = 0$ or (vi) holds with $|P_W| = \lceil \frac{n}{2} \rceil + 1$ and $|P_U| = 0$. Hence g satisfying (i) gives a minmal majority dominating function with $g(V) = \Gamma_{maj}(G) = \begin{cases} 0 & \text{if } m \text{ and } n \text{ are even} \\ 2 & \text{if } m \text{ and } n \text{ are odd} \\ 1 & \text{otherwise.} \end{cases}$

Theorem 2.8. *Let G denote the friendship graph with t triangles (u_0, u_i, u_{i+t}, u_0), where $1 \leq i \leq t$. Then $\Gamma_{maj}(G) = 1$.*

Proof. Clearly $n = |V(G)| = 2t + 1$ and hence $\lceil \frac{n}{2} \rceil = t + 1$.

Define $g : V \to \{-1, 1\}$ as follows.

$$g(u_i) = g(u_{i+t}) = 1 \text{ if } 1 \leq i \leq \lceil \frac{t}{2} \rceil$$

$$g(u_i) = 1 \text{ if } t \text{ is even and } i = \lceil \frac{t}{2} \rceil + 1$$

$$g(v) = -1 \text{ for all the remaining vertices.}$$

$$\text{Then } |S| = \begin{cases} t+1 & \text{if } t \text{ is even} \\ t+2 & \text{if } t \text{ is odd} \end{cases}$$

$$\text{and } k = \begin{cases} 1 & \text{if } t \text{ is even} \\ 3 & \text{if } t \text{ is odd.} \end{cases}$$

Clearly $|S| - k < \lceil \frac{n}{2} \rceil$ and hence g is a minimal majority dominating function on G. Thus $\Gamma_{maj} \geq \sum_{u_i \in V} g(u_i) = 1$. Now, let $f : V \to \{-1, 1\}$ be any minimal majority dominating function on G with $\sum_{u_i \in V} f(u_i) = \Gamma_{maj}(G)$. Hence $|S| \geq t+1$ and $k > 0$.

Now, suppose $f(N[u_0]) = \Gamma_{maj}(G) \geq 3$. Then $|P| \geq t+2$. Since $f(N[u_0]) \geq 3$ and $k > 0$, we have $f(u_0) = -1$. Hence $f(N[u_i]) = 1$ for all $u_i \in P$. Thus $|S| \geq t + 3$ and $k \leq 2$. Now $|S| - k > t + 1$, which is a contradiction. Hence $\Gamma_{maj}(G) \leq 1$. Thus $\Gamma_{maj}(G) = 1$.

3 Conclusion and Scope

In this paper we have introduced the concept of upper majority domination number of a graph. Results connecting upper majority domination number and upper signed domination number may be reported in a subsequent paper.

Acknowledgments. The first author is thankful to the management of Kalasalingam University for providing fellowship.

References

1. Broere, I., Hattingh, J.H., Henning, M.A., McRae, A.A.: Majority dominating functions in graphs. Discret. Math. **138**, 125–135 (1995)
2. Dunbar, J., Hedetniemi, S., Henning, M.A., McRae, A.: Minus domination in graphs. Discret. Math. **199**, 35–47 (1999)
3. Dunbar, J.E., Hedetniemi, S., Henning, M.A., Slater, P.J.: Signed domination in graphs. In: Graph Theory Combinatorics and Algorithms, vols. 1 and 2, pp. 311–321, Wiley, New York (1995)
4. Tang, H., Chen, Y.: Note on upper signed domination number. Discret. Math. **308**, 3416–3419 (2008)
5. Haynes, T.W., Hedetniemi, S.T., Slater, P.J.: Fundamentals of Domination in Graphs. Marcel Dekker Inc., New York (1998)
6. Haynes, T.W., Hedetniemi, S.T., Slater, P.J.: Domination in Graphs-Advanced Topics. Marcel Dekker Inc., New York (1998)
7. Chartrand, G., Lesniak, L.: Graphs & Digraphs, 4th edn. Chapman & Hall/CRC, Boca Raton (2005)

Super $(a,3)$-edge Antimagic Total Labeling for Union of Two Stars

M. Nalliah[1(✉)] and S. Arumugam[2]

[1] Department of Mathematics, School of Advanced Sciences, VIT University, Vellore 632 014, India
nalliah.moviri@vit.ac.in

[2] National Centre for Advanced Research in Discrete Mathematics (n-CARDMATH), Kalasalingam University, Anand Nagar, Krishnankoil 626126, Tamil Nadu, India
arumugam.klu@gmail.com

Abstract. An *(a,d)-edge antimagic total labeling* of a (p,q)-graph G is bijection $f : V \cup E \to \{1, 2, 3, \ldots, p+q\}$ with the property that the edge-weights $w(uv) = f(u) + f(v) + f(uv)$ where $uv \in E(G)$ form an arithmetic progression $a, a+d, \ldots, a+(q-1)d$, where $a > 0$ and $d \geq 0$ are two fixed integers. If such a labeling exists, then G is called an *(a,d)-edge antimagic total graph.* If further the vertex labels are the integers $\{1, 2, 3, \ldots, p\}$, then f is called a *super (a,d)-edge antimagic total labeling* of G ((a,d)-SEAMT labeling) and a graph which admits such a labeling is called a *super (a,d)-edge antimagic total graph* ((a,d)-SEAMT graph). If $d = 0$, then the graph G is called a *super edge-magic graph.* In this paper we investigate the existence of super $(a,3)$-edge antimagic total labelings for union of two stars.

Keywords: Total labeling · Antimagic total labeling · Super antimagic · Total labeling

1 Introduction

By a graph $G = (V, E)$ we mean a finite, undirected graph with neither loops nor multiple edges. The order and size of G are denoted by n and m respectively. For graph theoretic terminology we refer to Chartrand and Lesniak [3].

An *(a,d)-edge antimagic total labeling* of a (p,q)-graph G is bijection $f : V \cup E \to \{1, 2, 3, \ldots, p+q\}$ with the property that the edge-weights $w(uv) = f(u) + f(v) + f(uv)$ where $uv \in E(G)$ form an arithmetic progression $a, a + d, \ldots, a+(q-1)d$, where $a > 0$ and $d \geq 0$ are two fixed integers. If such a labeling exists, then G is called an *(a,d)-edge antimagic total graph.* If further

S. Arumugam—Also at School of Electrical Engineering and Computer Science, The University of Newcastle, NSW 2308, Australia; Department of Computer Science, Liverpool Hope University, Liverpool, UK; Department of Computer Science, Ball State University, USA.

S. Arumugam et al. (Eds.): ICTCSDM 2016, LNCS 10398, pp. 203–211, 2017.
DOI: 10.1007/978-3-319-64419-6_27

the vertex labels are the integers $\{1, 2, 3, \ldots, p\}$, then f is called a *super (a,d)-edge antimagic total labeling* of G ((a, d)-SEAMT labeling) and a graph which admits such a labeling is called a *super (a,d)-edge antimagic total graph* ((a, d)-SEAMT graph). If $d = 0$ then the graph G is called a *super edge-magic graph.*

Let $f : V \cup E \to \{1, 2, 3, \ldots, p+q\}$ be an (a, d)-edge antimagic total labeling of a graph $G = (V, E)$. Then $W = \{w(uv) : w(uv) = f(u) + f(v) + f(uv), uv \in E(G)\} = \{a, a+d, \ldots, a+(q-1)d\}$. In the computation of the edge-weights of G each edge label is used once and each label of the vertex v is used $deg(v)$ times. Thus the following equation holds.

$$\sum_{v \in V(G)} deg(v) f(v) + \sum_{e \in E(G)} f(e) = \sum_{e \in E(G)} w(e) \tag{1}$$

This equation was first observed by Bača and Youssef [2], which we repeatedly use.

We now present some basic results on super (a, d)-edge antimagic total labeling of graphs.

Theorem 1.1 [5]. *Let G be a (a, d)-SEAMT graph with p vertices and q edges. Let f be an (a, d)-SEAMT labeling of G. Then the labeling $\overline{f}$ defined by*

$$\overline{f}(v) = p + 1 - f(v) \text{ for all } v \in V \text{ and}$$
$$\overline{f}(e) = 2p + q + 1 - f(e) \text{ for all } e \in E$$

is a $(4p + q + 3 - a - (q-1)d, d)$-SEAMT labeling of G.

Super edge-antimagicness of disjoint union of two stars have been investigated by Dafik et al. [4]. Let $G = K_{1,m} \cup K_{1,n}$ with $m \geq n$. Dafik et al. [4] have observed that if $G = K_{1,m} \cup K_{1,n}$ admits an (a, d)-SEAMT labeling, then $d \leq 3 + \frac{2}{m+n-1}$. Hence if $m+n \geq 4$, then $d \leq 3$. They have proved the existence of (a, d)-SEAMT labelings with $d \leq 2$. If $m + n$ is odd, $m \geq n \geq 2$ and m is a multiple of $n + 1$, then G has a $(a, 1)$-SEAMT labeling. Further if $m \geq n \geq 2$ and m is a multiple of $n+1$, then G has an $(a, 2)$-SEAMT labeling. They have also proved the existence of $(4m+6, 1)$-SEAMT labeling and $(2m+7, 3)$-SEAMT labeling for $K_{1,m} \cup K_{1,n}$, where $m = n \geq 2$. They posed the following problem.

Problem 1.2 [4]. *For the graph $K_{1,m} \cup K_{1,n}$, $m > n \geq 2$ determine if there is an $(a, 3)$-SEAMT labeling.*

In [1] Arumugam and Nalliah proved that the graph $K_{1,m} \cup K_{1,n}$, $m \geq n+2$, has no $(a, 3)$-SEAMT labeling except when $n = 2$ and $m = 4$ and hence the Problem 1.2 reduces to the following:

Problem 1.3 [1]. *Determine the values of n for which the graph $K_{1,n+1} \cup K_{1,n}$ admits an $(a, 3)$-SEAMT labeling.*

It has also been proved in [1] that the graph $K_{1,n+1} \cup K_{1,n}$ admits an $(a, 3)$-SEAMT labeling if $1 \leq n \leq 16$, $n \neq 5, 8, 11, 14$. Further $a = 2n + 8$ if $n \neq 2$ and $a = 13$ when $n = 2$.

In this paper we continue with the investigation of Problem 1.3 and present further results on the existence and nonexistence of $(a, 3)$-SEAMT labelings.

2 Main Results

Theorem 2.1. *The graph $G = K_{1,6} \cup K_{1,5}$ has no $(a, 3)$-SEAMT labeling.*

Proof. The order and size of G are given by $p = 13$ and $q = 11$. Let $V(G) = \{c_1, u_1, u_2, \ldots, u_6\} \cup \{c_2, v_1, v_2, \ldots, v_5\}$ where c_1 and c_2 are respectively the central vertices of $K_{1,6}$ and $K_{1,5}$. Suppose there exists an $(a, 3)$-SEAMT labeling $f : V \cup E \to \{1, 2, 3, \ldots, 24\}$ for G. Let $f(c_1) = i_1$ and $f(c_2) = i_2$. Since $deg\ c_1 = 6$, $deg\ c_2 = 5$ and $deg\ v = 1$ for all $v \in V - \{c_1, c_2\}$, Eq. (1) gives

$$a = \frac{5i_1 + 4i_2 + 135}{11} \tag{2}$$

Since the minimum possible edge-weight is 17, we have $a \geq 17$. Also the maximum possible edge-weight is at most 48, which implies that $a \leq 19$. Thus $17 \leq a \leq 19$. Let e_1 and e_2 denote the edges in G with minimum and maximum weight respectively.

Case 1. $a = 17$.
Substituting the value a in Eq. (2), we get,

$$5i_1 + 4i_2 = 52 \tag{3}$$

Suppose e_2 is incident with c_1. Then $f(c_1) = i_1 \in \{13, 12, 11, 10\}$ and $f(c_2) = i_2 \in \{1, 2\}$. For any of these values of i_1 and i_2 the value of $5i_1+4i_2$ is greater than 52, which is a contradiction. A similar contradiction arises, if $f(c_1) = i_1 \in \{1, 2\}$ and $f(c_2) = i_2 \in \{13, 12, 11, 10\}$.

Case 2. $a = 19$.
If f is a $(19, 3)$-SEAMT labeling of G, then by Theorem 1.1, $\overline{f}$ is a $(17, 3)$-SEAMT labeling, which does not exist by Case 1.

Case 3. $a = 18$.
Substituting the value a in Eq. (2), we get,

$$5i_1 + 4i_2 = 63 \tag{4}$$

Suppose e_2 is incident with c_1. Then $f(c_1) = i_1 \in \{13, 12, 11\}$ and $f(c_2) = i_2 \in \{1, 2, 3\}$. Hence Eq. (4) is satisfied only when $i_1 = 11$ and $i_2 = 2$. Similarly if e_1 is incident with c_2, Eq. (4) is satisfied only when $i_1 = 3$ and $i_2 = 12$.

Subcase i. $i_1 = 11$ and $i_2 = 2$.
Then $f(c_1) = 11$ and $f(c_2) = 2$. Also the set of edge-weights is given by $W = \{18, 21, 24, 27, 30, 33, 36, 39, 42, 45, 48\}$. Let $e_1 = c_2v_1$ and $e_2 = c_1u_6$ be the edges with $w(e_1) = 18$ and $w(e_2) = 48$. Then $f(e_1) = 15$, $f(v_1) = 1$ $f(c_2) = 2$ and $f(e_2) = 24, f(u_6) = 13$, $f(c_1) = 11$. Now for any edge e of $K_{1,5}$ we have $w(e) \leq$

2+12+23 = 37. Also for any edge e of $K_{1,6}$ we have $w(e) \geq 11+3+14 = 28$. Hence the set W_1 of all edge-weights of $K_{1,5}$ is a subset of $\{18, 21, 24, 27, 30, 33, 36\}$. Also the set W_2 of all edge-weights of $K_{1,6}$ is a subset of $\{30, 33, 36, 39, 42, 45, 48\}$. Hence we may assume without loss of generality that if $e_3 = c_1v_5$, then $w(e_3) \in \{30, 33, 36\}$.

If $w(e_3) = 30$ then we have $21 + 24 + 27 + 30 = \sum_{i=2}^{5}[f(v_i) + f(e_i)] + 8$ and hence $\sum_{i=2}^{5}[f(v_i) + f(e_i)] = 94$. Now, we need four vertex labels of $K_{1,5}$ from the set $A = \{3, 4, 5, 6, 7, 8, 9, 10, 12\}$ and four edge labels of $K_{1,5}$ from the set $B = \{14, 16, 17, 18, 19, 20, 21, 22, 23\}$ such that the sum of these four vertex labels and four edge labels is 94. There exist 19 possible such sets of 8 elements, denoted by $C_i = A_i \cup B_i$, where $A_i \subseteq A$ and $B_i \subseteq B, 1 \leq i \leq 19$, which are given below.

$$
\begin{aligned}
C_1 &= \{3, 4, 5, 6, 16, 17, 20, 23\}\\
C_2 &= \{3, 4, 5, 7, 14, 18, 20, 23\}\\
C_3 &= \{3, 4, 5, 8, 14, 17, 20, 23\}\\
C_4 &= \{3, 4, 6, 9, 14, 16, 19, 23\}\\
C_5 &= \{3, 4, 6, 9, 16, 17, 19, 20\}\\
C_6 &= \{3, 4, 9, 10, 14, 16, 18, 20\}\\
C_7 &= \{3, 5, 6, 7, 14, 16, 20, 23\}\\
C_8 &= \{3, 5, 6, 7, 14, 17, 19, 23\}\\
C_9 &= \{3, 5, 7, 10, 14, 17, 18, 20\}\\
C_{10} &= \{3, 5, 7, 9, 14, 16, 17, 23\}\\
C_{11} &= \{3, 5, 7, 9, 14, 17, 19, 20\}\\
C_{12} &= \{3, 6, 7, 9, 14, 16, 19, 20\}\\
C_{13} &= \{3, 6, 7, 10, 14, 16, 18, 20\}\\
C_{14} &= \{3, 6, 9, 10, 14, 16, 17, 19\}\\
C_{15} &= \{4, 5, 6, 7, 14, 17, 18, 23\}\\
C_{16} &= \{4, 5, 6, 9, 14, 16, 17, 23\}\\
C_{17} &= \{3, 7, 8, 9, 14, 16, 17, 20\}\\
C_{18} &= \{4, 5, 6, 7, 14, 17, 20, 21\} \text{ and}\\
C_{19} &= \{4, 5, 6, 10, 14, 17, 18, 20\}.
\end{aligned}
$$

Also we have $33 + 36 + 39 + 42 + 45 = \sum_{i=1}^{5}[f(u_i) + f(e_i)] + 55$ and hence $\sum_{i=1}^{5}[f(u_i) + f(e_i)] = 140$. Now, we need five vertex labels of $K_{1,6}$ from the set $A' = \{1, 2, \ldots, 14\} - \{\{1, 2, 11, 13\} \cup A_i, 1 \leq i \leq 19\}$. Further we need five edge labels of $K_{1,6}$ from the set $B' = \{14, 15, \ldots, 24\} - \{\{15, 24\} \cup B_i, 1 \leq i \leq 19\}$ such that the sum of these five vertex labels and five edge labels is 140. There exist

19 possible such sets of 10 elements, denoted by $D_i = A'_i \cup B'_i$ where $A'_i \subseteq A'$ and $B_i \subseteq B', 1 \leq i \leq 19$, which are given below.

$$D_1 = \{7, 8, 9, 10, 12, 14, 18, 19, 21, 22\}$$
$$D_2 = \{6, 8, 9, 10, 12, 16, 17, 19, 21, 22\}$$
$$D_3 = \{6, 7, 9, 10, 12, 16, 18, 19, 21, 22\}$$
$$D_4 = \{5, 7, 8, 10, 12, 17, 18, 20, 21, 22\}$$
$$D_5 = \{5, 7, 8, 10, 12, 14, 18, 21, 22, 23\}$$
$$D_6 = \{5, 6, 7, 8, 12, 17, 19, 21, 22, 23\}$$
$$D_7 = \{4, 8, 9, 10, 12, 17, 18, 19, 21, 22\}$$
$$D_8 = \{4, 8, 9, 10, 12, 16, 18, 20, 21, 22\}$$
$$D_9 = \{4, 6, 8, 9, 12, 16, 19, 21, 22, 23\}$$
$$D_{10} = \{4, 6, 8, 10, 12, 18, 19, 20, 21, 22\}$$
$$D_{11} = \{4, 6, 8, 10, 12, 16, 18, 21, 22, 23\}$$
$$D_{12} = \{4, 5, 8, 10, 12, 17, 18, 21, 22, 23\}$$
$$D_{13} = \{4, 5, 8, 9, 12, 17, 19, 21, 22, 23\}$$
$$D_{14} = \{4, 5, 7, 8, 12, 18, 20, 21, 22, 23\}$$
$$D_{15} = \{3, 8, 9, 10, 12, 16, 19, 20, 21, 22\}$$
$$D_{16} = \{3, 7, 8, 10, 12, 18, 19, 20, 21, 22\}$$
$$D_{17} = \{4, 5, 6, 10, 12, 18, 19, 21, 22, 23\}$$
$$D_{18} = \{3, 8, 9, 10, 12, 16, 18, 19, 22, 23\} \text{ and}$$
$$D_{19} = \{3, 7, 8, 9, 12, 16, 19, 21, 22, 23\}.$$

Let $S_i = \{C_1 \cup D_i, 1 \leq i \leq 19\}$. If S_1 is used for getting the edge-weights $21, 24, 33, 36, 39, 42$ and 45, then there is no edge with weight 27 or 30, which is a contradiction. A similar contradiction arises for $S_i, 2 \leq i \leq 19$.

If $w(e_3) = 33$, then we have $21 + 24 + 27 + 33 = \sum_{i=2}^{5} [f(v_i) + f(e_i)] + 8$ and hence $\sum_{i=2}^{5} [f(v_i) + f(e_i)] = 97$. Now, we need four vertex labels of $K_{1,5}$ from the set $A = \{3, 4, 5, 6, 7, 8, 9, 10, 12\}$ and four edge labels of $K_{1,5}$ from the set $B = \{14, 16, 17, 18, 19, 20, 21, 22, 23\}$ such that the sum of these four vertex labels and four edge labels is 97. There exist 4 possible such sets of 8 elements, denoted by $C_i = A_i \cup B_i$, where $A_i \subseteq A$ and $B_i \subseteq B, 1 \leq i \leq 4$, which are given below.

$$C_1 = \{3, 4, 6, 8, 16, 17, 20, 23\}$$
$$C_2 = \{4, 5, 6, 8, 14, 17, 20, 23\}$$
$$C_3 = \{3, 4, 6, 10, 16, 17, 20, 21\} \text{ and}$$
$$C_4 = \{4, 5, 6, 10, 14, 17, 20, 21\}.$$

Also we have $30 + 36 + 39 + 42 + 45 = \sum_{i=1}^{5}[f(u_i) + f(e_i)] + 55$ and hence $\sum_{i=1}^{5}[f(u_i) + f(e_i)] = 137$. Now, we need five vertex labels of $K_{1,6}$ from the set $A' = \{1, 2, \ldots, 14\} - \{\{1, 2, 11, 13\} \cup A_i, 1 \leq i \leq 4\}$ and five edge labels of $K_{1,6}$ from the set $B' = \{14, 15, \ldots, 24\} - \{\{15, 24\} \cup B_i, 1 \leq i \leq 4\}$ such that the sum of these five vertex labels and five edge labels is 137. There exist 4 possible such sets of 10 elements, denoted by $D_i = A'_i \cup B'_i$ where $A'_i \subseteq A'$ and $B'_i \subseteq B', 1 \leq i \leq 4$ which are given below.

$$\begin{aligned}
D_1 &= \{5, 7, 9, 10, 12, 14, 18, 19, 21, 22\}\\
D_2 &= \{3, 7, 9, 10, 12, 16, 18, 19, 21, 22\}\\
D_3 &= \{5, 7, 8, 9, 12, 14, 18, 19, 22, 23\} \text{ and}\\
D_4 &= \{3, 7, 8, 9, 12, 16, 18, 19, 22, 23\}.
\end{aligned}$$

Let $S_i = \{C_i \cup D_i, 1 \leq i \leq 4\}$. If S_1 is used for getting the edge-weights 21, 24, 27, 30, 36, 39, 42 and 45, then there is no edge with weight 33. If S_2 is used for getting the edge-weights 21, 24, 30, 36, 39, 42 and 45, then there is no edge with weight 27 or 33. If S_3 or S_4 is used for getting the edge-weights 21, 30, 36, 39, 42 and 45, then there is no edge with weights 24, 27 and 33.

If $w(e_3) = 36$, then we have $\sum_{i=2}^{5}[f(v_i) + f(e_i)] = 100$. Hence there is no possible set of 8 elements of weights $\{21, 24, 27, 36\}$ for $K_{1,5}$. Also, we have $\sum_{i=1}^{5}[f(u_i) + f(e_i)] = 134$ and hence there is no possible set of 10 elements of weights $\{30, 33, 39, 42, 45\}$ for $K_{1,6}$. Thus the case $i_1 = 11$ and $i_2 = 2$ leads to a contradiction in all possibilities.

A similar argument can be used to prove that $i_1 = 3$ and $i_2 = 4$ leads to a contradiction. Hence G does not admit an $(a, 3)$-SEAMT labeling.

Theorem 2.2. *If* $n = 3(2^{r+3} - 3)$, $r \geq 1$, *then* $G = K_{1,n+1} \cup K_{1,n}$ *admits an* $(2n + 8, 3)$*-SEAMT labeling.*

Proof. Let $n = 3(2^{r+3} - 3)$, $r \geq 1$. Let $k = 2^r - 1, I_1 = \frac{3k+1}{2}$ and $I_{s+1} = \frac{3k+1-\sum_{i=1}^{s} 2^i}{2^{s+1}}$ where $1 \leq s \leq r - 1$.

We define a bijection $g : V \cup E \to \{1, 2, 3, \ldots, 4n + 4\}$ as follows:

$$\begin{aligned}
g(c_2) &= 2,\\
g(c_1) &= 2n + 1,\\
g(v_i) &= 2i - 1, 1 \leq i \leq n,\\
g(u_i) &= 2i + 2, 1 \leq i \leq n,\\
g(u_{n+1}) &= 2n + 3,\\
g(c_2 v_i) &= 2n + 4 + i, 1 \leq i \leq n,\\
g(c_1 u_{n+1}) &= 4n + 4,
\end{aligned}$$

$$
\begin{aligned}
g(c_1 u_{n+1-i}) &= 4n+3-i, 1 \le i \le \frac{n-1}{2},\\
g\left(c_1 u_{\frac{n+1}{2}}\right) &= 2n+4,\\
g(c_1 u_i) &= 3n+5+i, 1 \le i \le \frac{n-3}{4},\\
g\left(c_1 u_{\frac{n+1}{4}}\right) &= 4n+3,\\
g\left(c_1 u_{\frac{n+1}{4}+3i-2}\right) &= 3n+4+\frac{n-3}{4}+3i, \quad if\ 1 \le i \le \frac{n-15}{24},\\
g\left(c_1 u_{\frac{n+1}{4}+3i-1}\right) &= 3n+8+\frac{n-3}{4}+3i, \quad if\ 1 \le i \le \frac{n-15}{24},\\
g\left(c_1 u_{\frac{n+1}{4}+3i}\right) &= 3n+3+\frac{n-3}{4}+3i, \quad if\ 1 \le i \le \frac{n-15}{24}\\
g\left(c_1 u_{3\left(\frac{n+1}{8}\right)-1}\right) &= 3n+7+\frac{n-3}{4}+\frac{n-15}{8},\\
g\left(c_1 u_{3\left(\frac{n+1}{8}\right)}\right) &= 3n+5,\\
g\left(c_1 u_{3\left(\frac{n+1}{8}\right)+1}\right) &= 27\left(\frac{n-15}{8}\right)+54,\\
g\left(c_1 u_{3\left(\frac{n+1}{8}\right)+1+I_1}\right) &= 78k+56,\\
g\left(c_1 u_{3\left(\frac{n+1}{8}\right)+I_1}\right) &= 81k+56+\frac{3k+1}{2},\\
g\left(c_1 u_{3\left(\frac{n+1}{8}\right)+1+3i-2}\right) &= 81k+55+3i,\\
g\left(c_1 u_{3\left(\frac{n+1}{8}\right)+1+3i-1}\right) &= 81k+59+3i,\\
g\left(c_1 u_{3\left(\frac{n+1}{8}\right)+1+3i}\right) &= 81k+54+3i, \text{where } 1 \le i \le \frac{k-1}{2}.
\end{aligned}
$$

For $1 \le s \le r-1$.

$$
\begin{aligned}
g\left(c_1 u_{3\left(\frac{n+1}{8}\right)+1+\sum_{t=1}^{s}(I_t+1)}\right) &= 81k+54+\sum_{t=1}^{s}(I_t+1),\\
g\left(c_1 u_{3\left(\frac{n+1}{8}\right)+\sum_{t=1}^{s+1}(I_t+1)}\right) &= 81k+59+\sum_{t=1}^{s}(I_t+1)-2(I_{s+1}+1),\\
g\left(c_1 u_{3\left(\frac{n+1}{8}\right)-1+\sum_{t=1}^{s+1}(I_t+1)}\right) &= 81k+55+\sum_{t=1}^{s+1}(I_t+1),\\
g\left(c_1 u_{3\left(\frac{n+1}{8}\right)+1+\sum_{t=1}^{s}(I_t+1)+3i-2}\right) &= 81k+55+3i+\sum_{t=1}^{s}(I_t+1),
\end{aligned}
$$

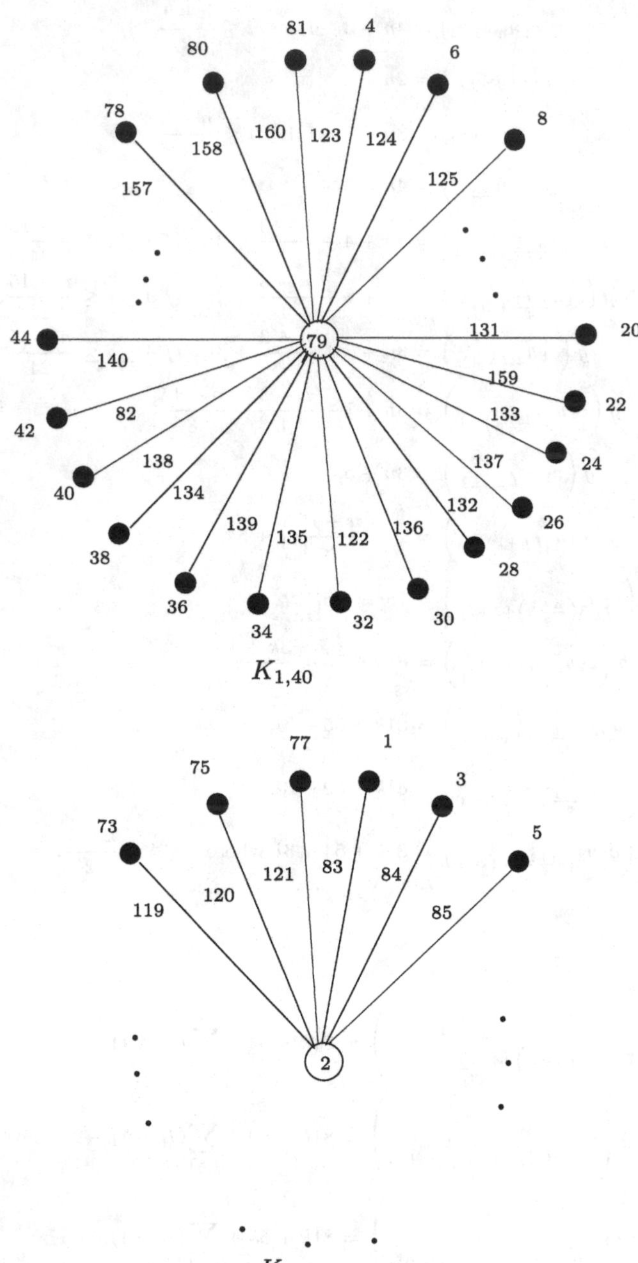

Fig. 1. A (86, 3)-SEAMT labeling of $K_{1,40} \cup K_{1,39}$.

$$g\left(c_1 u_{3\left(\frac{n+1}{8}\right)+1+\sum_{t=1}^{s}(I_t+1)+3i-1}\right) = 81k + 59 + 3i + \sum_{t=1}^{s}(I_t + 1),$$

$$g\left(c_1 u_{3\left(\frac{n+1}{8}\right)+1+\sum_{t=1}^{s}(I_t+1)+3i}\right) = 81k + 54 + 3i + \sum_{t=1}^{s}(I_t + 1),$$

$$\text{where } 1 \leq i \leq \frac{I_{s+1} - 2}{3}$$

$$g\left(c_1 u_{3\left(\frac{n+1}{8}\right)+1+\sum_{t=1}^{1\leq s\leq r}(I_t+1)}\right) = 84k + 54.$$

It can be verified that the set of edge weights induced by g is an arithmetic progresssion with $a = 2n + 8$ and $d = 3$ and we omit the details.

Example 2.3. A $(86, 3)$-SEAMT labeling of $K_{1,40} \cup K_{1,39}$ is given in Fig. 1.

3 Conclusion and Scope

The problem of determining whether the union of two stars $K_{1,n+1} \cup K_{1,n}$ admits an $(a, 3)$-SEAMT labeling for the remaining values of n not covered in the above two theorems is still open.

References

1. Arumugam, S., Nalliah, M.: Super $(a$, 3)-edge-antimagic total labelings for union of two stars. Util. Math. (To appear)
2. Bača, M., Youssef, M.Z.: Further results on antimagic graph labelings. Australas. J. Combin. **38**, 163–172 (2007)
3. Chartrand, G., Lesniak, L.: Graphs and Digraphs, 5th edn. CRC Press, Boca Raton (2005)
4. Miller, M., Ryan, J., Bača, M.: Antimagic labeling of the union of two stars. Australas. J. Combin. **42**, 35–44 (2008)
5. Sudarsana, I.W., Ismaimuza, D., Baskoro, T., Assiyatun, H.: On super (a, d)-edge-antimagic total labeling of disconnected graphs. J. Combin. Math. Combin. Comput. **55**, 149–158 (2005)

On Determinant of Laplacian Matrix and Signless Laplacian Matrix of a Simple Graph

Olayiwola Babarinsa(✉) and Hailiza Kamarulhaili

School of Mathematical Sciences, Universiti Sains Malaysia, USM, 11800 George Town, Penang, Malaysia
babs3in1@googlemail.com

Abstract. In a simple graph, Laplacian matrix and signless Laplacian matrix are derived from both adjacency matrix and degree matrix. Although, determinant of Laplacian matrix is always zero, yet we express it using only the adjacency matrix and square of its adjacency matrix. Likewise, we express the determinant of signless Laplacian matrix using only the diagonal elements provided the signless Laplacian matrix is equal to the square of its adjacency matrix.

Keywords: Laplacian matrix · Signless Laplacian matrix · Adjacency matrix · Determinant

1 Introduction

A graph $G = (V, E)$ consists of a finite set of vertices V and a set of edges E [1]. Let us denote the vertex set and the edge set of a graph G as $V(G)$ and $E(G)$ respectively. A simple graph is without loops or multiple edges. An unweighted graph is a weighted graph in which each of the edges has weight 1 [2].

Let G be a simple graph with finite number of vertices $V(G) = \{1, ..., n\}$. The adjacency matrix A of G is the $n \times n$ matrix where a_{ij} is given by

$$a_{ij} = \begin{cases} 1 & \text{if there is an edge from } v_i \text{ to } v_j \\ 0 & \text{otherwise.} \end{cases}$$

We denote $det(A)$ to be the determinant of the adjacency matrix and is given by

$$det(A) = \sum_{i=1}^{n} (-1)^{e_i} 2^{c_i}$$

where e_i is the number of even components of G and c_i is the number of components of G containing more than two points, consisting of a single undirected cycle. The adjacency matrix is a (0,1)-matrix for the case of a finite simple graph with zeros on its diagonal. If the graph is undirected, the adjacency matrix

S. Arumugam et al. (Eds.): ICTCSDM 2016, LNCS 10398, pp. 212–217, 2017.
DOI: 10.1007/978-3-319-64419-6_28

is symmetric which has a complete set of real eigenvalues and an orthogonal eigenvector basis such that $\sum \lambda_i = tr(A) = 0$. The adjacency matrix can also determine if the graph is connected or not. In algebraic graph theory, adjacency matrix is useful when the nonzero elements are replaced with algebraic variables [3,12].

Let D be the Degree matrix of graph G. Then D is a diagonal matrix (valency matrix) [4], where

$$d_{ij} = \begin{cases} deg(v_i) & \text{if } i = j \\ 0 & \text{otherwise.} \end{cases}$$

The adjacency matrix and degree matrix play the role in expressing Laplacian matrix and signless Laplacian matrix. However, we are interested in expressing the determinant of the matrices without the degree matrix using solely the adjacency matrix and its square.

The remaining part of this paper is developed and presented as follows: Sect. 2 provides the basic knowledge of Laplacian matrix and signless Laplacian matrix, their representations and uses. In Sect. 3, we give the proposition about determinant of Laplacian matrix, conjecture of the determinant of signless Laplacian matrix and as well raise a question.

2 Laplacian Matrix and Signless Laplacian Matrix

Let G be a graph with adjacency matrix $A(G)$ and let $D(G)$ be the $n \times n$ diagonal matrix of G. Denote the Laplacian matrix of G as $L(G)$ and the Signless Laplacian matrix as $Q(G)$. Then Laplacian matrix of G is

$$L(G) = D(G) - A(G).$$

Hence, $(i,j)^{th}$ entry $L_{i,j}$ is given by

$$L_{i,j} = \begin{cases} deg(v_i) & \text{if } i = j \\ -1 & \text{if } i \neq j \text{ and } v_i \text{ is adjacent to } v_j \\ 0 & \text{otherwise.} \end{cases}$$

The Signless Laplacian matrix of G is

$$L(G) = D(G) + A(G).$$

Hence $(i,j)^{th}$ entry $Q_{i,j}$ is given by

$$Q_{i,j} = \begin{cases} deg(v_i) & \text{if } i = j \\ 1 & \text{if } i \neq j \text{ and } v_i \text{ is adjacent to } v_j \\ 0 & \text{otherwise.} \end{cases}$$

where $deg(v_i)$ is degree of the vertex i [9].

A standout amongst the majority imperative and helpful grid representations of a graph is the combinatorial graph Laplacian matrix or Kirchhoff matrix or admittance matrix. The spectral properties of the Laplacian matrix is used to examine numerous combinatorial properties of a graph. Further the number of spanning trees of a simple graph is equal to any cofactor of the Laplacian matrix [8,19]. The set of non-negative eigenvalues of a graph is the spectrum of the graph. Among all eigenvalues of Laplacian of a graph, the algebraic connectivity of a graph is the second smallest eigenvalue of L, $\mu = \mu(G)$. The eigenvectors corresponding to μ are called Fiedler vectors of the graph G and the smallest nonzero eigenvalue of L is called the Fiedler value or Spectral gap. Thus if the connectivity is different from zero then the graph is connected [7]. The Laplacian is useful in the study of Markov chain, spectral clustering, graph partitioning, random walks, coordination of multi-agent systems, formation control and synchronization, see [10,11,13,17,20].

Laplacian matrix is a class of Z- matrices, particularly M-matrix - diagonally dominant. The Laplacian matrix is positive semi-definite. Also row sum of a matrix is the sum of all the elements in a row and the column sum is defined similarly. Obviously, Laplacian matrix is symmetric with zero row and column sums. The multiplicity of the eigenvalue 0 is the number of connected components of the graph. The determinant of a matrix that is obtained by deleting any single row and any column of the Laplace matrix is, except possibly for the sign, the number of spanning trees of the corresponding graph [15].

For regular graphs the existing theory of spectral of the adjacency matrix and of the Laplacian matrix is exactly assigned to the signless Laplacian [5]. The signless Laplacian matrix appears to be suitable for studying graph properties. Also signless Laplacian matrix is a positive semi-definite matrix [16]. The trace of the signless Laplacian matrix is equal to the sum of vertex degrees of G [6].

3 Determinant of Laplacian Matrix and Signless Laplacian Matrix

In Laplacian matrix and signless Laplacian matrix, degree matrix and adjacency matrix are needed. The link to express determinant of Laplacian matrix using only the adjacency matrix together with the square of the adjacency matrix and without the degree matrix is now fully discussed. We therefore explain the link as follows: We observe that

$$diag(L) = diag(A^2) = diag(D).$$

If $i \neq j$, then the entry in position (i, j) of A^2 is the number of common neighbours of v_i and v_j. The matrix multiplication puts into position (i, j) the product of row i and column j; that is $\sum_{k=1}^{n} a_{i,k}a_{k,j}$. When G is simple, the entries in A are 1 or 0, depending on whether the corresponding vertices are adjacent. Hence

$a_{i,k}a_{k,j} = 1$, if v_k is a common neighbour of v_i and v_j; otherwise, the contribution is 0. Thus the number of contributions of 1 to entry (i, j) is the number of common neighbours of v_i and v_j [18]. If $i = j$ the degree $deg(v_i)$ of a vertex sums the $(0, 1)$ in each row and represented in a degree matrix D. Therefore,

$$\begin{aligned} diag(D) &= diag[A^2(\sum_{k=1}^{n} a_{i,k}a_{k,j})] \\ &= \sum_{k=1}^{n}(a_{i,i}^2 + a_{i,k}a_{k,j}) \end{aligned}$$

The diagonal $deg(v_i)$ of Laplacian matrix can hence be replaced to have elements of L. This has successfully linked the diagonal of square of adjacency matrix with diagonal of the Laplacian matrix, thereby eliminating the need of degree matrix to evaluate the determinant of Laplacian matrix. Though A^2 cannot be used to express the entries in Laplacian matrix solely but at least it contains part of the information of Laplacian matrix. For an unweighted adjacency matrix of simple graph, the determinant of A^2 is always equal to square of determinant of A [14].

Proposition 1. *Let L and A be Laplacian matrix and adjacency matrix respectively. Then*

$$det(L) = (-1)^{det(A)}[det(A)]^2 - (-1)^{det(A^2)}det(A^2)$$

Proof. Let $det(A) = x$, thus $det(A^2) = x^2$ for $x \in \mathbb{Z}$. Then

$$\begin{aligned} det(L) &= (-1)^x x^2 - (-1)^{x^2} x^2 \\ &= x^2[(-1)^x - (-1)^{x^2}] \end{aligned}$$

Vividly, if the value of x is odd then

$$det(L) = x^2[-1 + 1] = 0$$

Also, if the value of x is even we have

$$det(L) = x^2[1 - 1] = 0.$$

Irrespective of the value of x the determinant of the Laplacian matrix is still zero.

Example 1. Given adjacency matrix $A = \begin{bmatrix} 0\,1\,1 \\ 1\,0\,1 \\ 1\,1\,0 \end{bmatrix}$ of a simple graph.

We need to verify the diagonal of L, A^2 and D are the same, where

$$D = \begin{bmatrix} 2\,0\,0 \\ 0\,2\,0 \\ 0\,0\,2 \end{bmatrix}$$

$$L = D - A = \begin{bmatrix} 2 & 0 & 0 \\ 0 & 2 & 0 \\ 0 & 0 & 2 \end{bmatrix} - \begin{bmatrix} 0 & 1 & 1 \\ 1 & 0 & 1 \\ 1 & 1 & 0 \end{bmatrix} = \begin{bmatrix} 2 & -1 & -1 \\ -1 & 2 & -1 \\ -1 & -1 & 2 \end{bmatrix}$$

$$A^2 = \begin{bmatrix} 0 & 1 & 1 \\ 1 & 0 & 1 \\ 1 & 1 & 0 \end{bmatrix} \times \begin{bmatrix} 0 & 1 & 1 \\ 1 & 0 & 1 \\ 1 & 1 & 0 \end{bmatrix} = \begin{bmatrix} 2 & 1 & 1 \\ 1 & 2 & 1 \\ 1 & 1 & 2 \end{bmatrix}$$

Since $diag[L(2,2,2)] = diag[A^2(2,2,2)] = diag[D(2,2,2)]$. Now

$$\begin{aligned} det(L) &= (-1)^{det(A)}[det(A)]^2 - (-1)^{det(A^2)}det(A^2) \\ &= (-1)^2 2^2 - (-1)^4 4 \\ &= 0 \end{aligned}$$

The square of adjacency matrix for the type of signless Laplacian matrix considered here need to have a constant scalar in its diagonal, hence the diagonal matrix D is in fact a scalar matrix.

Conjecture 1. Let Q and A be signless Laplacian matrix and adjacency matrix respectively. Then

$$det(Q) = (d-1)d^{n-1}$$

provided $Q = A^2$, where d is the element in diagonal entry of A^2.

Example 2. Given adjacency matrix $A = \begin{bmatrix} 0 & 1 & 1 \\ 1 & 0 & 1 \\ 1 & 1 & 0 \end{bmatrix}$ of a simple graph.

Then

$$Q = D + A = \begin{bmatrix} 2 & 0 & 0 \\ 0 & 2 & 0 \\ 0 & 0 & 2 \end{bmatrix} + \begin{bmatrix} 0 & 1 & 1 \\ 1 & 0 & 1 \\ 1 & 1 & 0 \end{bmatrix} = \begin{bmatrix} 2 & 1 & 1 \\ 1 & 2 & 1 \\ 1 & 1 & 2 \end{bmatrix}$$

and

$$A^2 = \begin{bmatrix} 0 & 1 & 1 \\ 1 & 0 & 1 \\ 1 & 1 & 0 \end{bmatrix} \times \begin{bmatrix} 0 & 1 & 1 \\ 1 & 0 & 1 \\ 1 & 1 & 0 \end{bmatrix} = \begin{bmatrix} 2 & 1 & 1 \\ 1 & 2 & 1 \\ 1 & 1 & 2 \end{bmatrix}$$

Now $d = 2$ and $det(Q) = (d-1)d^{n-1} = (2-1)2^{3-1} = 4$.

We now pose the following question which, of course, mainly arise from the Conjecture (1).

Question 1. Is there any $n \times n$ signless Laplacian matrix of a simple graph which is equal to the square of its adjacency matrix with vertices $n > 3$?

4 Conclusion

We have given a conjecture about the determinant of Signless Laplacian matrix as well as a question. We have shown the importance of square of adjacency matrix and further studies may unveil the need for square of adjacency matrix in determinant of signless Laplacian matrix.

References

1. Bapat, R., Kirkland, S.J., Neumann, M.: On distance matrices and laplacians. Linear Algebra Appl. **401**, 193–209 (2005)
2. Bapat, R.B.: On minors of the compound matrix of a laplacian. Linear Algebra Appl. **439**(11), 3378–3386 (2013)
3. Biggs, N.: Algebraic Graph Theory. Cambridge University Press, Cambridge (1993)
4. Chung, F., Lu, L., Vu, V.: Spectra of random graphs with given expected degrees. P. Natl Acad. Sci. **100**(11), 6313–6318 (2003)
5. Cvetkovic, D.: New theorems for signless laplacians eigenvalues. Bull. Acad. Serbe Sci. Arts Cl. Sci. Math. Natur. Sci. Math **137**(33), 131–146 (2008)
6. Cvetkovic, D., Rowlinson, P., Simic, S.K.: Signless laplacians of finite graphs. Linear Algebra Appl. **423**(1), 155–171 (2007)
7. De Abreu, N.M.M.: Old and new results on algebraic connectivity of graphs. Linear Algebra Appl. **423**(1), 53–73 (2007)
8. Godsil, C.D., Royle, G.: Algebraic Graph Theory. Springer, New York (2001)
9. Goldberg, F., Kirkland, S.: On the sign patterns of the smallest signless laplacian eigenvector. Linear Algebra Appl. **443**, 66–85 (2014)
10. Gutman, I., Lee, S., Chu, C., Luo, Y.: Chemical applications of the laplacian spectrum of molecular graphs: studies of the wiener number. Indian J. Chem. A. **33**, 603–608 (1994)
11. Mesbahi, M., Egerstedt, M.: Graph Theoretic Methods in Multiagent Networks. Princeton University Press, Princeton (2010)
12. Mohar, B., Alavi, Y., Chartrand, G., Oellermann, O.: The laplacian spectrum of graphs. Graph Theory Comb. Appl. **2**, 871–898 (1991)
13. Spielman, D.A.: Algorithms, graph theory, and linear equations in laplacian matrices. Proc. Int. Congr. Mathe. **4**, 2698–2722 (2010)
14. Tam, B., Huang, P.: Nonnegative square roots of matrices. Linear Algebra Appl. **498**, 404–440 (2016)
15. Teufl, E., Wagner, S.: Determinant identities for laplace matrices. Linear Algebra Appl. **432**(1), 441–457 (2010)
16. Van Dam, E.R., Haemers, W.H.: Which graphs are determined by their spectrum? Linear Algebra Appl. **373**, 241–272 (2003)
17. Von Luxburg, U.: A tutorial on spectral clustering. Stat. Comput. **17**(4), 395–416 (2007)
18. West, D.B.: Introduction to Graph Theory. N.J. Prentice Hall, Upper Saddle River (2001)
19. Zelazo, D., Burger, M.: On the definiteness of the weighted laplacian and its connection to effective resistance. In: 53rd IEEE Conference on Decision and Control, pp. 2895–2900 (2014)
20. Zhang, X.: The laplacian eigenvalues of graphs: a survey. Linear Algebra Research Advances p. arXiv preprint (2011). arXiv:1111.2897

On the Complexity of Minimum Cardinality Maximal Uniquely Restricted Matching in Graphs

B.S. Panda[1(✉)] and Arti Pandey[2]

[1] Department of Mathematics, Indian Institute of Technology Delhi, Hauz Khas, New Delhi 110016, India
bspanda@maths.iitd.ac.in

[2] Department of Mathematics, Indian Institute of Technology Ropar, Nangal Road, Rupnagar 140001, Punjab, India
arti@iitrpr.ac.in

Abstract. For a graph $G = (V, E)$, a set $M \subseteq E$ is called a *matching* in G if no two edges in M share a common vertex. A matching M in G is called an *uniquely restricted matching* in G if there is no other matching of the same cardinality in the graph induced on the vertices saturated by M. An uniquely restricted matching M is called *maximal* if M is not properly contained in any uniquely restricted matching of G. The minimum maximal uniquely restricted matching (`Min-UR-Matching`) problem is the problem of finding a minimum cardinality maximal uniquely restricted matching. In this paper, we initiate the study of the `Min-UR-Matching` problem. We prove that the decision version of the `Min-UR-Matching` problem is NP-complete for general graphs. In particular, this answers an open question posed by Hedetniemi [AKCE J. Graphs. Combin. 3(1)(2006) 1–37] regarding the complexity of the `Min-UR-Matching` problem. We also prove that this problem remains NP-complete for bipartite graphs with maximum degree 7. Next, we show that the `Min-UR-Matching` for bipartite graphs cannot be approximated within a factor of $n^{1-\epsilon}$ for any positive constant $\epsilon > 0$ unless $P = NP$. Finally, we prove that the `Min-UR-Matching` problem is linear-time solvable for chain graphs, a subclass of bipartite graphs.

Keywords: Matching · Uniquely restricted matching · Bipartite graphs · Graph algorithm · NP-complete

1 Introduction

Let $G = (V, E)$ be a graph. For a set $S \subseteq V$ of the graph $G = (V, E)$, the subgraph of G *induced* by S is defined as $G[S] = (S, E_S)$, where $E_S = \{xy \in E | x, y \in S\}$. A set of edges $M \subseteq E$ is a *matching* in G if no two edges of M are incident on a common vertex. Vertices incident to the edges of a matching M

The work was done when the second author (Arti Pandey) was in IIIT Guwahati.

S. Arumugam et al. (Eds.): ICTCSDM 2016, LNCS 10398, pp. 218–227, 2017.
DOI: 10.1007/978-3-319-64419-6_29

are *saturated* by M. A matching M in G is called *uniquely restricted matching*, if there is no other matching of the same cardinality in $G[S]$, where S is the set of vertices saturated by M. The concept of uniquely restricted matching was first introduced by Golumbic et al. in 2001 [2]. They also gave a characterization of uniquely restricted matching on arbitrary graphs using alternating cycles. A matching M in a graph G is uniquely restricted if and only if G does not contain an alternating cycle with respect to M [2].

A uniquely restricted matching M is *maximal* if no other uniquely restricted matching in G contains M. A maximal uniquely restricted matching of maximum cardinality is known as *maximum maximal uniquely restricted matching.* We also call it *maximum uniquely restricted matching.* The maximum uniquely restricted matching is well studied in literature, see [2,5–8].

A maximal uniquely restricted matching of minimum cardinality is called *minimum maximal uniquely restricted matching.* The Minimum maximal uniquely restricted matching (`Min-UR-Matching`) problem is to find a minimum maximal uniquely restricted matching in a given graph. In this paper, we initiate the study of the `Min-UR-Matching` problem. The *minimum maximal uniquely restricted matching number* of G is the cardinality of a minimum maximal uniquely restricted matching in G, and is denoted by $\mu'_r(G)$. A minimum maximal uniquely restricted matching is a maximal uniquely restricted matching that contains $\mu'_r(G)$ edges. The minimum maximal uniquely restricted matching problem and its decision version are defined as follows:

<u>`Min-UR-Matching` problem (MUMP)</u>

Instance: A graph $G = (V, E)$.
Solution: A maximal uniquely restricted matching M in G.
Measure: Cardinality of the set M.

<u>`Min-UR-Matching-Decision` problem(MUMDP)</u>

Instance: A graph $G = (V, E)$ and a positive integer $k \leq |E|$.
Question: Does there exist a maximal uniquely restricted matching M in G such that $|M| \leq k$?

The motivation for studying minimum maximal uniquely restricted matching comes from the minimum maximal matching problem. Note that maximum matching problem can be solved in polynomial time. However, the minimum maximal matching problem is NP-hard [1]. Here we show that the `Min-UR-Matching-Decision` problem is NP-complete. This also answers a question posed by Hedetniemi [3]. Also, the minimum maximal matching problem admits a 2-approximate solution. But here we show an interesting result that the `Min-UR-Matching` problem is hard to approximate even for bipartite graphs.

The rest of the paper is organized as follows. In Sect. 2, some pertinent definitions and some preliminary results are discussed. In Sect. 3, we show that the `Min-UR-Matching-Decision` problem is NP-complete for general graphs and even for bipartite graphs with maximum degree 7. In Sect. 4 we prove that the `Min-UR-Matching` problem for bipartite graphs is hard to approximate within a factor of $n^{1-\epsilon}$ for any positive constant $\epsilon > 0$ unless $P = NP$. In Sect. 5, we present a linear-time algorithm to solve the `Min-UR-Matching` problem in chain graphs, a subclass of bipartite graphs. Finally, Sect. 6 concludes the paper.

2 Preliminaries

Let $G = (V, E)$ be a simple undirected graph with a vertex set V and an edge set E. The vertex set and edge set of a graph G may also be denoted by $V(G)$ and $E(G)$, respectively. For $v \in V$, the sets $N_G(v) = \{u \in V | uv \in E\}$ and $N_G[v] = N_G(v) \cup \{v\}$ denote the *open neighborhood* and the *closed neighborhood* of v in G, respectively. The *degree* of a vertex $v \in V$, denoted by $d_G(v)$, is the number of neighbors of v, that is, $d_G(v) = |N_G(v)|$. The *minimum degree* and *maximum degree* of a graph G is defined by $\delta(G) = \min_{v \in V} d_G(v)$ and $\Delta(G) = \max_{v \in V} d_G(v)$, respectively. A graph G is said to be *complete* if every pair of distinct vertices of G are adjacent in G, and a complete graph on n vertices is denoted by K_n. A graph $G = (V, E)$ is said to be *bipartite* if $V(G)$ can be partitioned into two disjoint sets X and Y such that every edge of G joins a vertex in X to another vertex in Y, and such a partition (X, Y) of V is called a *bipartition*. A bipartite graph with bipartition (X, Y) of V is denoted by $G = (X, Y, E)$. A bipartite graph $G = (X, Y, E)$ is *complete* if each vertex of X is adjacent to all the vertices of Y. A complete bipartite graph $G = (X, Y, E)$ with $|X| = p$ and $|Y| = q$ is denoted by $K_{p,q}$. Let n and m denote the number of vertices and number of edges of G, respectively. In this paper, we only consider connected graphs with at least two vertices.

A matching M in G is said to be *maximal* if there is no other matching in G which properly contains M. A maximum maximal matching (maximum matching) is a maximal matching of maximum cardinality, and a minimum maximal matching is a maximal matching of minimum cardinality. Let $\mu(G)$ denote the cardinality of a maximum matching in G, and $\rho(G)$ denote the cardinality of a minimum maximal matching in G. Similarly, let $\mu_r(G)$ denote the cardinality of maximum uniquely restricted matching in G, and $\mu'_r(G)$ denote the cardinality of minimum maximal uniquely restricted matching in G. Note that the difference between the cardinality of minimum maximal matching and the cardinality of minimum maximal uniquely restricted matching can be arbitrary. For example,

$$\rho(K_n) = \lfloor \tfrac{n}{2} \rfloor \qquad \mu'_r(K_n) = 1$$
$$\rho(K_{n,n}) = n \qquad \mu'_r(K_{n,n}) = 1.$$

3 NP-completeness Results

In this section we first prove that the **Min-UR-Matching-Decision** problem is NP-complete for general graphs and then prove that it remains NP-complete for bipartite graphs as well as for bipartite graphs with maximum degree at most 7.

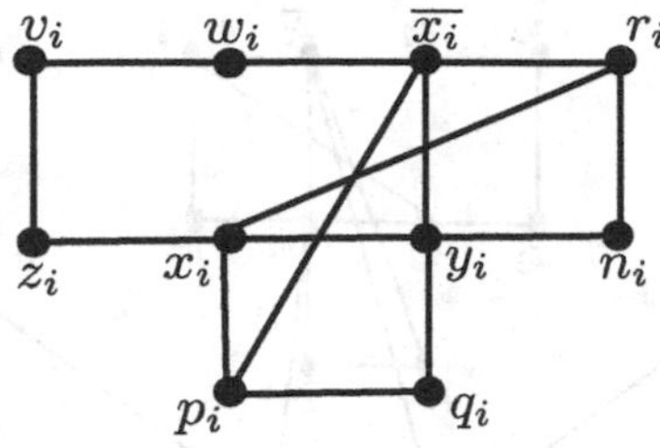

Fig. 1. The subgraph F_i.

We prove the NP-completeness of the **Min-UR-Matching-Decision** problem for graphs by proposing a polynomial time reduction from 3-SAT, a well known NP-complete problem.

3-SAT

c_j —— d_j

Fig. 2. The subgraph H_j.

Instance: A collection $C = \{c_1, c_2, \ldots, c_m\}$ of clauses over a set $X = \{x_1, x_2, \ldots, x_p\}$ of boolean variables such that $|c_j| = 3$ and $c_j = l_{j,1} \vee l_{j,2} \vee l_{j,3}$, $l_{j,i} \in X \cup \overline{X}$, for $j = 1, 2, \ldots, m$, $i = 1, 2, 3$, where $\overline{X} = \{\overline{x} : x \in X\}$.

Question: Is there a truth assignment for X that satisfies all the clauses in C?

Theorem 1. *The Min-UR-Matching-Decision problem is NP-complete for graphs.*

Proof. Clearly, the Min-UR-Matching-Decision problem is in NP. To show the hardness, we give a polynomial time reduction from 3-SAT. Let (C, X), where $C = \{c_1, c_2, \ldots, c_m\}$ and $X = \{x_1, x_2, \ldots, x_p\}$ be an instance of 3-SAT. We construct the graph $G_{(C,X)}$ in the following way:

- For each variable x_i we construct the graph F_i as follows. First take a cycle on six vertices, $C^i = (x_i, y_i, \overline{x_i}, w_i, v_i, z_i, x_i)$. Take four more vertices r_i, n_i, p_i and q_i and add the edges: $r_i n_i$, $\overline{x_i} r_i$, $y_i n_i$, $r_i x_i$, $p_i \overline{x_i}$, $p_i x_i$, $p_i q_i$, $q_i y_i$. The graph F_i is shown in Fig. 1.
- For each clause c_j, we construct a graph H_j on two vertices c_j and d_j with the unique edge $c_j d_j$, where c_j is called the clause vertex. The graph H_j is shown in Fig. 2.
- If q, r, s are the indices of literals in clause c_j, then join d_j with y_q, y_r and y_s.
- clause vertex c_j is connected to three literal vertices corresponding to literals in clause c_j.

The graph $G_{(C,X)}$ associated with the instance (C, X) of 3-SAT, where $C = \{c_1 = (x_1 \vee \overline{x_2} \vee x_3)\}$ and $X = \{x_1, x_2, x_3\}$ is shown in Fig. 3.

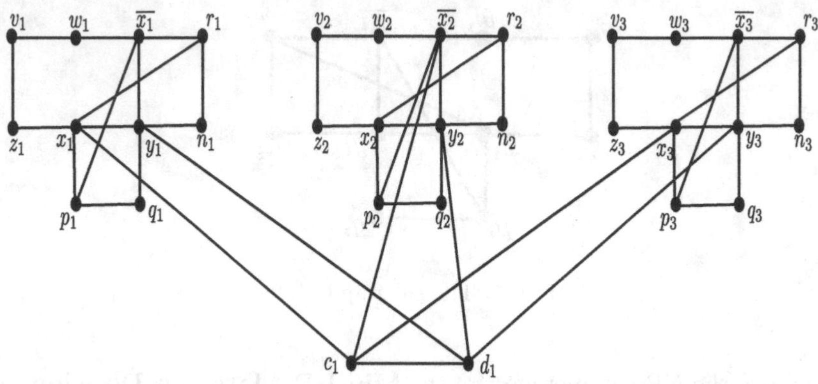

Fig. 3. The graph $G_{(C,X)}$ for (C, X), where $C = \{c_1 = (x_1 \vee \overline{x_2} \vee x_3)\}$ and $X = \{x_1, x_2, x_3\}$.

It is easy to see that $G_{(C,X)}$ can be constructed in polynomial time in $m = |C|$ and $p = |X|$. To complete the proof, it now suffices to show that the following claim is true.

Claim. There exists a satisfying truth assignment for C if and only if $G_{(C,X)}$ has a maximal uniquely restricted matching of size $k = 2p$.

Proof. First, suppose that there exists a truth assignment satisfying C. We construct a maximal uniquely restricted matching M in G as follows. If x_i is assigned the value 1, then include the edges $x_i y_i$ and $\overline{x_i} w_i$ in M; otherwise include the edges $\overline{x_i} y_i$ and $z_i x_i$ in M. Clearly M is a maximal uniquely restricted matching of size $2p$.

Conversely, suppose that M is a maximal uniquely restricted matching of size $2p$. Note that each maximal uniquely restricted matching in G contains at least two edges of each subgraph F_i. Since $|M| = 2p$, we have $|M \cap F_i| = 2$ for each $i, 1 \leq i \leq p$ and M contains no other edges of G.

Since $c_j d_j \notin M$, either c_j is adjacent to some x_i and $x_i y_i \in M$ or c_j is adjacent to some $\overline{x_i}$ and $\overline{x_i} y_i \in M$. The edge corresponding to a literal x_i(respectively, $\overline{x_i}$) is $x_i y_i$(respectively, $\overline{x_i} y_i$). M cannot contain both $x_i y_i$ and $\overline{x_i} y_i$. We can define a satisfying truth assignment to C setting a literal to be true if and only if the corresponding edge is in M. This completes the proof of our claim.

Note that the constructed graph $G_{(C,X)}$ is clearly a bipartite graph and (V_1, V_2) is a bipartition of its vertex set, where V_1 and V_2 are:

$$V_1 = \{x_i, \overline{x_i}, n_i, q_i : i = 1, 2, \ldots, p\} \cup \{d_j : j = 1, 2, \ldots, m\}$$
$$V_2 = \{w_i, z_i, p_i, r_i, y_i : i = 1, 2, \ldots, p\} \cup \{c_j : j = 1, 2, \ldots, m\}$$

Hence we have the following corollary.

Corollary 1. *The Min-UR-Matching-Decision problem is NP-complete for bipartite graphs.*

Also, note that the modified version of 3-SAT, 3-SAT in which every clause contains at most three literals and every variable appears, negated or not, at most three times, is also NP-complete. If we take an instance (C, X) of the modified 3-SAT and construct the graph $G_{(C,X)}$, then $deg(x) \leq 7$ for all $x \in V(G_{(C,X)})$. So we have the following corollary to the above theorem.

Corollary 2. *The Min-UR-Matching-Decision problem is also NP-complete for bipartite graphs with maximum degree* 7.

4 Hardness of Approximating Min-UR-Matching

In this section, we show that the Min-UR-Matching problem is hard to approximate even for bipartite graphs.

First we construct a bipartite graph from an instance of 3-SAT according to the following construction rule:

Construction Rule R: For given an instance (C, X) of 3-SAT with a set C of m clauses and a set X of p variables, and for a given integer $t \geq 2$, we can construct a bipartite graph $G_{(C,X),t}$ on $10p+tp(p+2m)$ vertices in the following way:

We construct the graph $G = G_{(C,X),t}$ with vertex set $C' \cup D \cup X' \cup Y \cup R \cup R' \cup P \cup P' \cup Z \cup W \cup M'$, where $X' = \{x_i, \overline{x_i}, 1 \leq i \leq p\}$, $Z = \{z_i, \overline{z_i} : 1 \leq i \leq p\}$,
$R = \{r_i, 1 \leq i \leq p\}$, $R' = \{\overline{r_i}, 1 \leq i \leq p\}$,
$P = \{p_i, 1 \leq i \leq p\}$, $P' = \{\overline{p_i}, 1 \leq i \leq p\}$,
$Y = \{y_i, 1 \leq i \leq p\}$,
$W = \{w_i, 1 \leq i \leq p\}$,
$C' = \{c_{j,k}, 1 \leq j \leq m, 1 \leq k \leq tp\}$,
$D = \{d_{j,k}, 1 \leq j \leq m, 1 \leq k \leq tp\}$,
$M' = \{m_{i,k}, 1 \leq i \leq p, 1 \leq k \leq tp\}$
are disjoint sets. The edges of $G = G_{(C,X),t}$ are such that:

- The set $X' \cup Y \cup Z \cup W$ induces p cycles $C^i = (x_i, y_i, \overline{x_i}, \overline{z_i}, w_i, z_i, x_i)$, each of length 6.
- The set $C' \cup D$ induces a matching $\{c_{j,k}d_{j,k}, 1 \leq j \leq m, 1 \leq k \leq tp\}$.
- For each vertex y_i, introduce tp edges $y_i m_{i,k}$, $k = 1, 2, \ldots, tp$ between Y and M in G.
- In addition, take the edges $x_i p_i$, $p_i\overline{p_i}$, $\overline{p_i}y_i$, $y_i\overline{r_i}$, $\overline{r_i}r_i$, $r_i\overline{x_i}$, for each $i = 1, 2, \ldots, p$.
- For each clause $c_j = l_j^1 \vee l_j^2 \vee l_j^3$, introduce three edges $c_{j,k}l_j^1$, $c_{j,k}l_j^2$, $c_{j,k}l_j^3$, $k = 1, 2, \ldots, tp$, between C' and X' in G.
- For each clause c_j, introduce three edges $d_{j,k}y_l$, $d_{j,k}y_q$, $d_{j,k}y_r$, $k = 1, 2, \ldots, tp$, (where l, q and r are the indices of literals present in the clause c_j) between D and Y in G.

The graph $G_{(C,X),t}$ associated with the instance (C, X) of 3-SAT, where $C = \{c_1 = (x_1 \vee x_2 \vee \overline{x_3})\}$, $X = \{x_1, x_2, x_3\}$ and $t = 2$ is shown in Fig. 4.

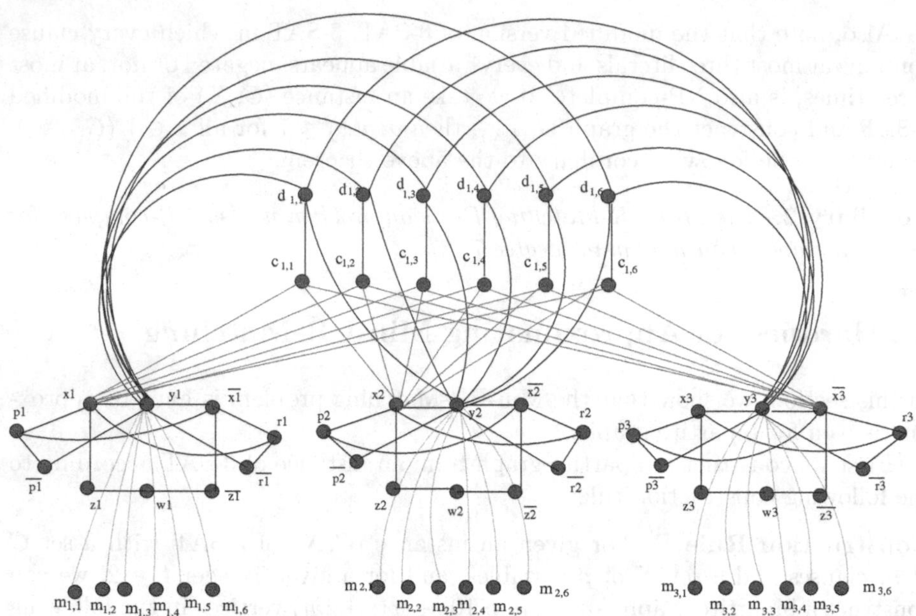

Fig. 4. The graph $G_{(C,X),t}$ corresponding to the instance (C, X), where $C = \{c_1 = (x_1 \vee x_2 \vee \overline{x_3})\}$, $X = \{x_1, x_2, x_3\}$ and $t = 2$.

Note that $G_{(C,X),t}$ is a bipartite graph as $(C' \cup Y \cup R \cup P \cup Z, D \cup X' \cup P' \cup R' \cup W)$ is a bipartition of $V(G_{(C,X),t})$.

Lemma 1. *For each instance (C, X) of 3-SAT with a set C of m clauses and a set X of p variables, and for each integer $t \geq 2$, there exists a bipartite graph $G = G_{(C,X),t}$ on $10p + tp(p + 2m)$ vertices such that the following property holds for the minimum maximal uniquely restricted matching number:*

$$\mu_r'(G) = \begin{cases} \leq 2p, & \textit{if } C \textit{ is satisfiable} \\ > tp, & \textit{if } C \textit{ is not satisfiable} \end{cases}$$

Proof. Let $C = \{c_1, c_2, \ldots, c_m\}$ and $X = \{x_1, x_2, \ldots.., x_p\}$ be an instance of 3-SAT, and let $t \geq 2$ be an integer. We construct the graph $G = G_{(C,X),t}$ according to the **Construction Rule *R*.**

Now, by assuming that there exists a truth assignment satisfying C, we will construct a uniquely restricted matching M consisting of $2p$ edges. So, $\mu_r'(G) \leq 2p$ in this case.

On the other hand, by assuming that C is not satisfiable. We will show that $\mu_r'(G) > tp$. The details of the proof have been deferred to the longer version of the paper.

We are now in a position to present the main theorem.

Theorem 2. *The Min-UR-Matching problem for bipartite graphs cannot be approximated within a factor of $n^{1-\epsilon}$ for any constant $\epsilon > 0$ unless $P = NP$, where n denotes the number of vertices in the input graph.*

Proof. Give the constant $\epsilon > 0$, we define $k = max\{2, \lceil 3/\epsilon \rceil\}$. Given an instance (C, X) of 3-SAT with $|C| = m$ and $|X| = p$, we set $t = 2p^{k-2}$. Now we construct the graph $G = G_{(C,X),t}$ using the **Construction Rule *R*.**

Claim. Approximating $\mu_r'(G)$ for $G = G_{(C,X),t}$ within a factor of p^{k-2} is NP-hard.

Proof. The proof of the claim have been deferred to the longer version of the paper. □

Now we estimate $t = 2p^{k-2}$ in terms of $n = |V(G)| = 10p + 2p^{k-1}(2m + p)$. We may assume that $p \geq 16$ and $p \geq m$. Indeed, 3-SAT remains NP-complete under these additional restrictions. We have $n \geq p^k$. Using the assumption $p \geq m$, we obtain

$$p^{k-2} = \frac{n - 10p}{2(p^2 + 2mp)} \geq \frac{n - 10p}{6p^2} \geq \frac{n - 10p}{6n^{2/k}} \geq \frac{n^{1-2/k}}{16} \geq n^{1-3/k}$$

According to the above claim, approximating $\mu_r'(G)$ within a factor of $n^{1-\epsilon}$ is NP-hard, since $p^{k-2} \geq n^{1-3/k} \geq n^{1-\epsilon}$

5 Chain Graphs

We have already seen that the Min-UR-Matching problem is even hard to approximate for bipartite graphs. In this section, we show that the Min-UR-Matching problem can be solved in linear time for chain graphs, a subclass of bipartite graphs.

A bipartite graph $G = (X, Y, E)$ is called a *chain graph* if the neighborhoods of the vertices of X form a *chain*, that is, the vertices of X can be linearly ordered, say $x_1, x_2, \ldots, x_p$, such that $N_G(x_1) \subseteq N_G(x_2) \subseteq \ldots \subseteq N_G(x_p)$. If $G = (X, Y, E)$ is a chain graph, then the neighborhoods of the vertices of Y also form a chain [10]. An ordering $\alpha = (x_1, x_2, \ldots, x_p, y_1, y_2, \ldots, y_q)$ of $X \cup Y$ is called a *chain ordering* if $N_G(x_1) \subseteq N_G(x_2) \subseteq \cdots \subseteq N_G(x_p)$ and $N_G(y_1) \supseteq N_G(y_2) \supseteq \cdots \supseteq N_G(y_q)$. It is well known that every chain graph admits a chain ordering [4,10].

Theorem 3. *Let $G = (X, Y, E)$ be a connected chain graph and $\alpha = (x_1, x_2, \ldots, x_p, y_1, y_2, \ldots, y_q)$ is chain ordering of $X \cup Y$. Then $M = \{x_p y_1\}$ is a maximal uniquely restricted matching in G. Furthermore, M is also a minimum maximal uniquely restricted matching in G, and $\mu_r'(G) = 1$.*

Proof. The proof is easy and hence is omitted.

Since, a chain ordering of a chain graph $G = (X, Y, E)$ can be computed in linear time [9], we have the following result as the corollary of the above theorem.

Corollary 3. *A minimum maximal uniquely restricted matching of a chain graph can be computed in $O(n + m)$ time.*

Note that for a chain graph G, $\rho(G) = \mu_r'(G) = 1$, but $\mu(G)$ and $\mu_r(G)$ can be arbitrary large. Consider a chain graph $G^* = (X, Y, E)$ with $|X| = |Y| = p$, and a chain ordering $\alpha = (x_1, x_2, \ldots, x_p, y_1, y_2, \ldots, y_p)$. Also assume that $|N_{G^*}(x_1)| = 1$ and $|N_{G^*}(x_i) \setminus N_{G^*}(x_{i-1})| = 1$ for each i, $2 \leq i \leq p$. Then $\mu(G^*) = \mu_r(G^*) = p$. Even the difference between $\mu(G)$ and $\mu_r(G)$ can also be arbitrary large. Consider a complete bipartite graph $K_{n,n}$. Clearly $K_{n,n}$ is also a chain graph, and $\mu(K_{n,n}) = n$, but $\mu_r(K_{n,n}) = 1$.

6 Conclusion

In this paper, we initiated the study of the Min-UR-Matching problem. We proved that the Min-UR-Matching-Decision problem is NP-complete for general graphs, and it remains NP-complete even for bipartite graphs with maximum degree 7. We have also shown that the Min-UR-Matching problem for bipartite graphs is hard to approximate within a factor of $n^{1-\epsilon}$ for any positive constant $\epsilon > 0$ unless $P = NP$. Finally, we proved that the Min-UR-Matching problem is linear-time solvable for chain graphs, a subclass of bipartite graphs. It will be an interesting problem to study the complexity status of the Min-UR-Matching problem for other important subclasses of bipartite graphs, for example bipartite permutations graphs, convex bipartite graphs, and chordal bipartite graphs.

References

1. Garey, M.R., Johnson, D.S.: Computers and Interactability: A Guide to the Theory of NP-Completeness. W.H. Freeman and Co., San Francisco, New York (1979)
2. Golumbic, M.C., Hirst, T., Lewenstein, M.: Uniquely restricted matchings. Algorithmica **31**, 139–154 (2001)
3. Hedetniemi, S.T.: Unsolved algorithmic problem on trees. AKCE J. Graphs. Combin. **3**, 1–37 (2006)
4. Kloks, T., Kratsch, D., Müller, H.: Bandwidth of chain graphs. Inform. Process. Lett. **68**(6), 313–315 (1998)
5. Levit, V.E., Mandrescu, E.: Local maximum stable sets in bipartite graphs with uniquely restricted maximum matchings, Stability in graphs and related topics. Discrete Appl. Math. **132**, 163–174 (2003)
6. Levit, V.E., Mandrescu, E.: On unicyclic graphs with uniquely restricted maximum matchings. Graphs Comb. **29**, 1867–1879 (2013)
7. Levit, V.E., Mandrescu, E.: Triangle-free graphs with uniquely restricted maximum matchings and their corresponding greedoids. Discrete Appl. Math. **155**, 2414–2425 (2007)
8. Mishra, S.: On the maximum uniquely restricted matching for bipartite graphs. Electron. Notes Discrete Math. **37**, 345–350 (2011)

9. Uehara, R., Uno, Y.: Efficient algorithms for the longest path problem. In: Fleischer, R., Trippen, G. (eds.) ISAAC 2004. LNCS, vol. 3341, pp. 871–883. Springer, Heidelberg (2004). doi:10.1007/978-3-540-30551-4_74
10. Yannakakis, M.: Node-and edge-deletion NP-complete problems. In: Conference Record of the Tenth Annual (ACM) Symposium on Theory of Computing (San Diego, Calif., 1978), pp. 253–264. ACM, New York (1978)

Partial Grundy Coloring in Some Subclasses of Bipartite Graphs and Chordal Graphs

B.S. Panda and Shaily Verma(✉)

Computer Science and Application Group, Department of Mathematics,
Indian Institute of Technology Delhi, Hauz Khas, New Delhi 110016, India
bspanda@maths.iitd.ac.in, shailyverma048@gmail.com

Abstract. A proper k-coloring of $G = (V, E)$ is an assignment of k colors to vertices of G such that no two adjacent vertices receive the same color. A proper k-coloring of a graph $G = (V, E)$ partitions V into independent sets or color classes $V_1, V_2, \ldots, V_k$. A vertex $v \in V_i$ is a Grundy vertex if it is adjacent to at least one vertex in each color class V_j for every $j < i$. A coloring is a partial Grundy coloring if every color class has at least one Grundy vertex in it and the partial Grundy number, $\delta\Gamma(G)$ of a graph G is the maximum number of colors used in a partial Grundy coloring. Given a graph G and an integer $k(1 \leq k \leq n)$, the Partial Grundy Number Decision problem is to decide whether $\delta\Gamma(G) \geq k$. It is known that the Partial Grundy Number Decision problem is NP-complete for bipartite graphs. In this paper, we strengthen this result by proving that this problem remains NP-complete even for perfect elimination bipartite graphs, a proper subclass of bipartite graphs. On the positive side, we propose a linear time algorithm to determine the partial Grundy number of a chain graph, a proper subclass of perfect elimination bipartite graphs. It is also known that the Partial Grundy Number Decision problem is NP-complete for (disconnected) chordal graphs. We strengthen this result by proving that the Partial Grundy Number Decision problem remains NP-complete even for (connected) doubly chordal graphs, a proper subclass of chordal graphs. On the positive side, we propose a linear time algorithm to determine the partial Grundy number of split graphs, a well known subclass of chordal graphs.

Keywords: Partial Grundy coloring · Perfect elimination bipartite graphs · NP-completeness · Polynomial time algorithms

1 Introduction

A proper k-coloring of $G = (V, E)$ is an assignment of k colors to vertices of G such that no two adjacent vertices receive the same color. A k-coloring of a graph $G = (V, E)$ partitions the vertex set V into k independent sets or color classes. A vertex $v \in V_i$ is a Grundy vertex if it is adjacent to at least one vertex in each color class V_j for every $j < i$. A coloring is a partial Grundy coloring if every color class has at least one Grundy vertex in it and the partial Grundy number, $\delta\Gamma(G)$

S. Arumugam et al. (Eds.): ICTCSDM 2016, LNCS 10398, pp. 228–237, 2017.
DOI: 10.1007/978-3-319-64419-6_30

of a graph G is the maximum number of colors used in a partial Grundy coloring. A Grundy k-coloring is a k-coloring if every vertex is a Grundy vertex. The Grundy number, denoted as $\Gamma(G)$, is the largest integer k such that there exist a Grundy k-coloring of G. Grundy coloring was introduced by Christen and Selkow [3] in 1979. Recently, the concept of partial Grundy coloring was introduced by Erdös et al. in 2003 [5]. Note that every Grundy coloring is a partial Grundy coloring and so for any graph G, $\chi(G) \leq \Gamma(G) \leq \delta\Gamma(G) \leq \Delta(G) + 1$. Recently in 2015, Balakrishnan and Kavaskar [1] proved that the partial Grundy coloring admits an interpolation theorem similar to the one for the Grundy coloring, that is, for any graph G and any integer k, $\chi(G) \leq k \leq \delta\Gamma(G)$, there exists a partial Grundy coloring of G using k colors.

The PARTIAL GRUNDY NUMBER DECISION PROBLEM is stated as follows.
PARTIAL GRUNDY NUMBER DECISION PROBLEM:
Instance: A graph $G = (V, E)$ and a positive integer k
Question: Does G have a partial Grundy coloring with at least k colors?

From a computational point of view, Zhengnan Shi et al. [9] in 2005 showed that the PARTIAL GRUNDY NUMBER DECISION PROBLEM is NP-complete for chordal graphs and bipartite graphs. They have also given a linear time algorithm to determine the partial Grundy number of trees [9]. Very recently, Effantin et al. [4] proved that given a fixed integer k, it can be checked in polynomial time whether $\delta\Gamma(G) \geq k$. In this paper, we study the computational complexity of PARTIAL GRUNDY NUMBER DECISION PROBLEM for some subclasses of bipartite graphs and chordal graphs.

The contributions of the paper are summarized as below.

1. We prove that the PARTIAL GRUNDY NUMBER DECISION PROBLEM remains NP-complete for perfect elimination bipartite graphs.
2. We propose a linear time algorithm to compute the Partial Grundy number of chain graphs, a subclass of perfect elimination bipartite graphs.
3. We show that the PARTIAL GRUNDY NUMBER DECISION PROBLEM remains NP-complete for doubly chordal graphs.
4. We show that partial Grundy number of a split graph can be computed in linear time.

2 Preliminaries

All the graphs considered in this paper are simple and undirected. For a graph $G = (V, E)$, the sets $N(v) = \{u \in V(G) | uv \in E\}$ and $N[v] = N(v) \cup \{v\}$ denote the *open neighborhood* and *closed neighborhood* of a vertex v, respectively. For a set $S \subseteq V$, the sets $N(S) = \cup_{u \in S} N(u)$ and $N[S] = N(S) \cup S$ denote the *open neighborhood* and the *closed neighborhood* of S, respectively. The *degree* of a vertex v is $|N(v)|$ and is denoted by $d(v)$. If $d(v) = 1$, then v is called a *pendant vertex*. An edge incident on some pendant vertex is called a *pendant edge*. A set $I \subseteq V$ is called an *independent set* if no two vertices in I are adjacent in G. A set $C \subseteq V$ is called a *clique* if there is an edge between any pair of vertices in S. The size of the clique S is $|S|$, that is the number of vertices in S. The *clique number* of G, denoted as $\omega(G)$, is the largest size of a clique present in G. For $S \subseteq V$, let $G[S]$ denote the subgraph induced by G on S. A graph $G = (V, E)$ is said to be *bipartite* if $V(G)$ can be partitioned into two disjoint sets X and Y such that every edge of G joins a vertex in X to another vertex in Y. Such a partition (X, Y) of V is called a *bipartition.* A bipartite graph with bipartition (X, Y) of V is denoted by $G = (X, Y, E)$.

3 Partial Grundy Coloring in Perfect Elimination Bipartite Graph

It is known that PARTIAL GRUNDY NUMBER DECISION PROBLEM is NP-complete for bipartite graphs [9]. In this section, we prove that PARTIAL GRUNDY NUMBER DECISION PROBLEM remains NP-complete even for perfect elimination bipartite graphs, a proper subclass of bipartite graphs.

Let $G = (X, Y, E)$ be a bipartite graph. An edge $e = xy$ is called a *bisimplicial edge* if $G[N(x) \cup N(y)]$ is a complete bipartite graph. Let $\sigma = (x_1y_1, x_2y_2, \ldots, x_ky_k)$ be a sequence of pairwise nonadjacent edges of G. Denote $S_j = \{x_1, x_2, \ldots, x_j\} \cup \{y_1, y_2, \ldots, y_j\}$ and let $S_0 = \emptyset$. Then σ is said to be a *perfect edge elimination scheme* for G if each edge $x_{j+1}y_{j+1}$ is bisimplicial in $G[(X \cup Y) \setminus S_j]$ for j, $0 \leq j \leq k-1$ and $G[(X \cup Y) \setminus S_k]$ has no edge. A graph for which there exists a perfect edge elimination scheme is a *perfect elimination bipartite graph* [8].

Theorem 1. *It is NP-complete to decide whether $\delta\Gamma(G) = \Delta(G)+1$ for perfect elimination bipartite graphs.*

Proof. Given a proper coloring f of G, it can be checked in polynomial time whether f is partial Grundy $(\Delta(G)+1)$-coloring of G. Hence PARTIAL GRUNDY NUMBER DECISION PROBLEM is in NP. To show that it is NP-complete, we present a reduction from the NP-complete problem, 3-colorability of a graph with degree bounded by 4 [7], which asks whether a graph G of maximum degree at most 4 can be properly colored using at most 3 colors. Let $G = (V, E)$ be an arbitrary graph of maximum degree bounded by 4 with n vertices and m edges. We construct the graph G' from G as follows:

1. Construct the vertex-edge incidence graph $I(G)$ of G, that is, the bipartite graph with vertex set $V(I) = V(G) \cup E(G)$ and edge set $E(I) = \{ve|$ where edge e is incident on the vertex v in $G\}$.
2. Add a vertex s_i and add an edge e_is_i for each $i, 1 \leq i \leq m$.
3. For vertex e_m, add the vertices $u_1, u_2, \ldots, u_{m-1}$ and make them adjacent with e_m.
4. For each vertex $e_i, 2 \leq i \leq m-1$, add the edge e_iu_j for all j, $m+1-i \leq j \leq m-1$.
5. Add a vertex w and add an edge between w and each e_i, where $1 \leq i \leq m$.
6. Add a $P_4 = x_1, y_1, x_2, y_2$ and make w adjacent with x_1 and x_2 and add two new vertices, x_3 and x_4 and add the pendant edges wx_3 and wx_4.

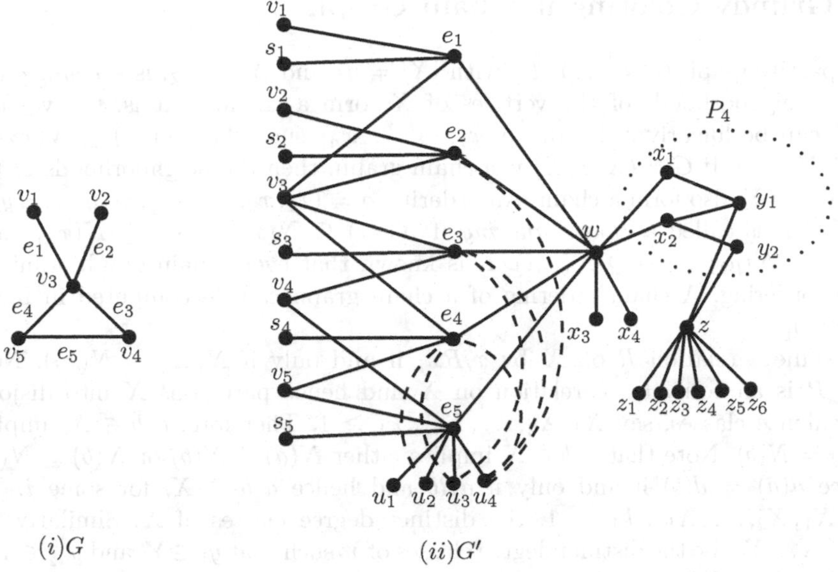

Fig. 1. An example of the Construction of G' from a given graph G

7. Add a vertex z and make z adjacent with y_1 and y_2 and add $(m+1)$ vertices $z_1, z_2, \ldots, z_{m+1}$ and $m+1$ edges $zz_i, 1 \leq i \leq m$.

The construction of G' from G is illustrated in Fig. 1. Note that the constructed graph is a perfect elimination bipartite graph, and $\sigma = (s_1e_1, s_2e_2, \ldots, s_me_m, wx_3, zz_1, x_1y_1, x_2y_2)$ is a perfect edge elimination scheme of G'.
Observe that,
For each i, $1 \leq i \leq m$, $d(e_i) = i + 3$.
For each i, $1 \leq i \leq m - 1$, $d(u_i) = i$.
$d(z) = m + 3$, $d(w) = m + 4$ and $d(v_j) \leq 4$ for all j, $1 \leq j \leq n$.
Thus, $\Delta(G') = m + 4$ and vertex w is the only vertex of degree $\Delta(G')$. □

Claim. G is 3-colorable graph if and only if $\delta\Gamma(G') = \Delta(G') + 1 = m + 5$.

Proof. The proof of the claim will appear in the journal version of the paper.

So, it is NP-complete to decide whether $\delta\Gamma(G) = \Delta(G) + 1$ for perfect elimination bipartite graphs NP-complete.

The following corollary follows immediately.

Corollary 1. PARTIAL GRUNDY NUMBER DECISION PROBLEM *is NP-complete even for perfect elimination bipartite graphs.*

4 Grundy Coloring in Chain Graph

A bipartite graph $G = (X, Y, E)$ with $|X| = n_1$ and $|Y| = n_2$, is a *chain graph* if the neighborhoods of the vertices of X form a chain, that is, the vertices of X can be linearly ordered, say $x_1, x_2, \ldots, x_{n_1}$ such that $N(x_1) \subseteq N(x_2) \subseteq \cdots \subseteq N(x_{n_1})$. If $G = (X, Y, E)$ is a chain graph, then the neighborhoods of the vertices of Y also form a chain. An ordering $\sigma = (x_1, x_2, \ldots, x_{n_1}, y_1, y_2, \ldots, y_{n_2})$ of $X \cup Y$ is called a *chain ordering* if $N(x_1) \subseteq N(x_2) \subseteq \cdots \subseteq N(x_{n_1})$ and $N(y_1) \supseteq N(y_2) \supseteq \cdots \supseteq N(y_{n_2})$. It is known that every chain graph admits a chain ordering. A chain ordering of a chain graph can be computed in linear time [10].

Define a relation R on X by $x_i R x_j$ if and only if $N(x_i) = N(x_j)$. Note that R is an equivalence relation on X and hence partitions X into disjoint equivalence classes, say $X_1, X_2, \ldots, X_{k_1}$, $k_1 \geq 1$. Therefore, $a, b \in X_i$ implies $N(a) = N(b)$. Note that $a, b \in X$ implies either $N(a) \subseteq N(b)$ or $N(b) \subseteq N(a)$. Hence $d(a) = d(b)$ if and only if aRb and hence $a, b \in X_i$ for some i. We call $X_1, X_2, \ldots, X_{k_1}$, $k_1 \geq 1$, the distinct degree classes of X. Similarly, let $Y_1, Y_2, \ldots, Y_{k_2}$ be the distinct degree classes of Y such that $y_1 \in Y_1$ and $y_{n_2} \in Y_{k_2}$. The number of distinct degree classes in a partite set X (or Y) is equal to the number of vertices in X (or in Y) of distinct degrees.

Proposition 1. *Let $G = (X, Y, E)$ be a chain graph. Then X and Y have the same number of distinct degree classes.*

Proof. We prove that X and Y have the same number of distinct degree classes by induction on the number of distinct degree classes of X. Let $G = (X, Y, E)$ be a chain graph such that X has exactly one degree class. So $d(a) = d(b)$ for every $a, b \in X$. So, G is a complete bipartite graph and hence the number of distinct degree classes of Y is also 1. Assume that the proposition is true for all chain graphs $G = (X, Y, E)$ with the number of distinct degree classes of X equals to $k, k > 1$. Let $G = (X, Y, E)$ be a chain graph such that X has $k + 1$ distinct degree classes. Let $X_i, 1 \leq i \leq k + 1$ and $Y_i, 1 \leq i \leq k_2$ be the distinct degree classes of X and Y, respectively such that $y_1 \in Y_1$ and $y_{n_2} \in Y_{k_2}$. Let $G' = G \backslash X_1$. Note that $G' = (X \backslash X_1, Y, E')$ is a chain graph and $X_i, 2 \leq i \leq k+1$ and $Y_i', 2 \leq i \leq k_2$ are the distinct degree classes of $X \setminus X_1$ and Y, respectively, where $Y_2' = Y_1 \cup Y_2$ and $Y_j' = Y_j, 3 \leq j \leq k_2$. By induction hypothesis, $k = k_2 - 1$. So, $k + 1 = k_2$. Hence, the number of distinct degree classes of X is equal to the number of distinct degree classes of Y. Hence, by induction principle, the proposition is true.

Proposition 2. *Let $G = (X, Y, E)$ be a chain graph and C be a partial Grundy k-coloring of G. If $x_i, x_j \in X$ with $i < j$ and $y_k, y_l \in Y$ with $k < l$, are the Grundy vertices then $C(x_i) \leq C(x_j)$ and $C(y_k) \geq C(y_l)$.*

Proof. Let us assume that $i < j$ and $C(x_i) > C(x_j)$. Since x_i is a Grundy vertex, x_i has a neighbor of color c' such that $c' = C(x_j) < C(x_i)$. Since $i < j$, by the chain property in X, $N(x_i) \subseteq N(x_j)$. So x_j also has a neighbor of color

c' ($c' = C(x_j)$), which is a contradiction to the fact that C is a proper coloring. Hence, $C(x_i) \leq C(x_j)$. Similarly, because of the chain property in Y, $k < l$ implies $N(y_k) \supseteq N(y_l)$ and hence $C(y_k) \geq C(y_l)$.

Lemma 1. *Let $G = (X, Y, E)$ be a chain graph and C be a partial Grundy k-coloring of G. Then there exists a partial Grundy k-coloring C' of G such that every Grundy vertex of color greater than 1 with respect to C' belongs to the partite set X.*

Proof. The proof of the lemma will appear in the journal version of the paper.□

Theorem 2. *Let $G = (X, Y, E)$ be a chain graph and k be the number of distinct degree classes in X (or Y). Then $\delta\Gamma(G) = k + 1$.*

Proof. Let us assume that $\delta\Gamma(G) \geq k + 2$. Then by Lemma 1, there exists a partial Grundy $\delta\Gamma(G)$-coloring C such that all the Grundy vertices of color greater than 1 belongs to same partite set, say X. Since there are at least $k+1$ Grundy vertices in X and X has k distinct degree classes, by Pigeon hole principle, there exists a distinct degree class X_i of X, $1 \leq i \leq k$ such that there are two Grundy vertices of different colors $x_j, x_l \in X_i$ where $1 \leq j < l \leq n_1$. Without loss of generality $C(x_j) < C(x_l)$. There exists a vertex $y \in N(x_l)$ such that $C(y) = C(x_j)$, since x_l is a Grundy vertex. But x_j and x_l have same neighborhood as they belong to same distinct degree class, that is, $yx_j \in E(G)$. But $C(y) = C(x_j)$, which is a contradiction to the fact that C is a proper coloring. Hence $\delta\Gamma(G) \leq k + 1$.

Next, we will show that $\delta\Gamma(G) \geq k + 1$. Let $X_i, 1 \leq i \leq k$ and $Y_i, 1 \leq i \leq k$ be the distinct degree classes of X and Y, respectively such that $y_1 \in Y_1$ and $y_{n_2} \in Y_k$. Define a coloring C' on $X \cup Y$ such that for every $x \in X_i$, $C'(x) = i+1$ and for every $y \in Y_i$, $C'(y) = i$ for each $1 \leq i \leq k$. Figure 2 illustrates the coloring C' of the chain graph G.

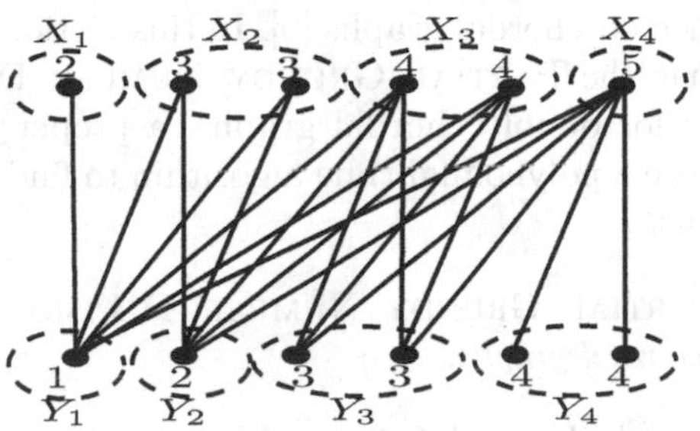

Fig. 2. An example of a chain graph G and the optimal partial Grundy coloring C'

We claim that the coloring C' is a partial Grundy $(k + 1)$-coloring. The coloring C' is a proper coloring, since $N(X_i) \subseteq Y_1 \cup Y_2 \cup \cdots, \cup Y_i$ for each $1 \leq i \leq k$ and clearly C' uses $(k+1)$ colors. We can easily verify that the vertices belongs to X_i are the Grundy vertex of color $(i+1)$ for all $1 \leq i \leq k$ and every vertex of color 1 is a Grundy vertex vacuously. Therefore, C' is a partial Grundy $(k+1)$-coloring. This implies that $\delta\Gamma(G) \geq k+1$. Hence $\delta\Gamma(G) = k+1$. □

All the distinct degree classes of X and Y of a chain graph $G = (X, Y, E)$ can be computed in $O(n+m)$ time. So, an optimal partial Grundy coloring of a chain graph can be computed in $O(n+m)$ time. In view of the above, we have the following result.

Corollary 2. *The optimal partial Grundy coloring of a chain graph can be computed in $O(n+m)$ time.*

5 Partial Grundy Coloring in Subclasses of Chordal Graphs

A graph G is said to be a *chordal graph* if every cycle in G of length at least four has a *chord*, that is, an edge joining two non-consecutive vertices of the cycle. A graph $G = (V, E)$ is a *split graph* if V can be partitioned into two sets I and C such that C is a clique and I is an independent set. Note that a split graph is a chordal graph. A vertex $v \in V(G)$ is a *simplicial* vertex of G if $N_G[v]$ is a clique of G. An ordering $\alpha = (v_1, v_2, \ldots, v_n)$ is a *perfect elimination ordering* (PEO) of G if v_i is a simplicial vertex of $G_i = G[\{v_i, v_{i+1}, \ldots, v_n\}]$ for all i, $1 \leq i \leq n$. It is characterized that a graph G is chordal if and only if it has a PEO [6]. A vertex $u \in N_G[v]$ is a *maximum neighbor* of v in G if $N_G[w] \subseteq N_G[u]$ for all $w \in N_G[v]$. A vertex v in G is called *doubly simplicial* if it is simplicial and has a maximum neighbor in G. An ordering $\alpha = \{v_1, v_2, \ldots, v_n\}$ of vertices of G is a *doubly perfect elimination ordering* (DPEO) if v_i is doubly simplicial vertex in the induced subgraph $G[\{v_i, v_{i+1}, \ldots, v_n\}]$ for each i, $1 \leq i \leq n$. A graph is *doubly chordal* if it admits a doubly perfect elimination ordering (DPEO) [2].

The Partial Grundy Number Decision Problem is known to be NP-complete for (disconnected)chordal graphs [9]. In this section, we strengthen this result by showing that the Partial Grundy Number Decision Problem remains NP-complete for doubly chordal graphs, a proper subclass of chordal graphs. We also propose a polynomial time algorithm to find an optimal Grundy coloring of split graphs.

Theorem 3. *The* Partial Grundy Number Decision Problem *is NP-complete for doubly chordal graphs.*

Proof. Given a function $f : V \rightarrow \{1, 2, 3, \ldots, k\}$, it can be checked in polynomial time whether f is a partial Grundy coloring of a given doubly chordal graph G. So, Partial Grundy Number Decision Problem for doubly chordal graph is in NP. Next we produce a polynomial reduction from the Partial Grundy Number Decision Problem for chordal graphs, which is known to be NP-complete [9], to the Partial Grundy Number Decision Problem for doubly chordal graphs.

Given a chordal graph $G = (V, E)$ and an integer k, we construct a doubly chordal graph G' as follows. Take a new vertex x and add the edges xv for all $v \in V(G)$. Formally, $G' = (V', E')$, where $V' = V \cup \{x\}, x \notin V$, and $E' = E \cup \{xv | v \in V\}$. Since G is chordal, there exists a PEO $\alpha = (v_1, v_2, \ldots, v_n)$ of G. Now $\beta = (v_1, v_2, \ldots, v_n, x)$ is a DPEO of G'. Also f is a partial Grundy k-coloring of G if and only f', defined by $f'(v) = f(v), v \in V(G)$ and $f'(x) = k+1$, is a Grundy $(k+1)$-coloring of G'.

Hence, Partial Grundy Number Decision Problem remains NP-complete for doubly chordal graphs. □

Next, we propose a $O(n+m)$ time algorithm for computing the partial Grundy number of a split graph.

Theorem 4. *For a split graph G with clique number $\omega(G)$, $\omega(G) \leq \delta\Gamma(G) \leq \omega(G) + 1$.*

Proof. Let (R, S) be the partition of $V(G)$ such that R is an independent set and S is a clique of size $\omega(G)$. If possible, let us suppose that $\delta\Gamma(G) \geq \omega(G) + 2$. Let f be an optimal partial Grundy coloring of G. Since $|S| = \omega(G)$ and $\delta\Gamma(G) \geq \omega(G) + 2$, there exists at least two different colored Grundy vertices, say v_c and $v_{c'}$ with color c and c' respectively, where $1 \leq c < c' \leq \delta\Gamma(G)$, in R such that these colors are not assigned to any vertex in S. We know that, every vertex in R has neighbors in S only and color c is not assigned to any vertex in S. Therefore, $v_{c'}$ does not have any neighbor with color c, contradicting the fact that $v_{c'}$ is a Grundy vertex. Hence, $\delta\Gamma(G) \leq \omega(G) + 1$. We know that, $\omega(G) \leq \chi(G)$ and $\chi(G) \leq \delta\Gamma(G)$. Hence, $\omega(G) \leq \delta\Gamma(G) \leq \omega(G) + 1$. □

Lemma 2. *Let (R, S) be a partition of the vertex set of the split graph G such that S is a clique of size $\omega(G)$ and R is an independent set. Then $\delta\Gamma(G) = \omega(G)$ if and only if there exists a vertex $x \in S$ such that x is not adjacent to any vertex in R.*

Proof. **Necessity:** Given that $\delta\Gamma(G) = \omega(G)$. We have to show that there exist a vertex $x \in S$ such that x is not adjacent to any vertex in R. If possible, every vertex x in S has a neighbor in R. Let $S = \{x_1, x_2, \ldots, x_{\omega(G)}\}$. Consider the coloring $f : V(G) \to \{1, 2, \ldots, \omega(G) + 1\}$ such that $f(x) = 1$ for $x \in R$ and $f(x_i) = i + 1, 1 \leq i \leq \omega(G)$. One can easily verify that this coloring f is a partial Grundy $(\omega(G)+1)$-coloring of G. This is a contradiction to the fact that $\delta\Gamma(G) = \omega(G)$. Hence, there exists a vertex $x \in S$ such that x is not adjacent to any vertex in R.

Sufficiency: Supposed that there exists a vertex $x \in S$ such that x is not adjacent to any vertex in R. We will show that $\delta\Gamma(G) = \omega(G)$. Let f be an optimal partial Grundy coloring of G. We have to show that the coloring f uses $\omega(G)$ colors. This will imply that $\delta\Gamma(G) = \omega(G)$. If possible, suppose that the coloring f uses $\omega(G) + 1$ colors. Let v be a Grundy vertex of color $\omega(G) + 1$. So v must have neighbors with colors 1 through $\omega(G)$. Now we have the following two cases to consider.

Case I: $v \in R$.

Since $v \in R$, v has its neighbors in S only. Since $|S| = \omega$, v must be adjacent to every vertex in S. This contradicts the fact that there exists a vertex $x \in S$ such that x is not adjacent to any vertex in R. So this case is not possible.

Case II: $v \in S$.

Since $|S| = \omega(G)$ and $f(v) = \omega(G)+1$, S does not contain a neighbor u of v of some color $c, 1 \leq c \leq \omega(G)$. Since v is adjacent to every vertex of S, S does not contain a vertex of color c. Hence S does not contain a Grundy vertex of color c. Let i be the largest color such that $1 \leq i \leq \omega(G)+1$ and S does not contain a Grundy vertex of color i. So S contains Grundy vertex of color j for each $j, i+1 \leq j \leq \omega(G)+1$. Let $w \in R$ be a Grundy vertex of color i. So, w has neighbors, say, $y_1, y_2, \ldots, y_{i-1}$ in S such that $f(y_j) = j, 1 \leq j \leq i-1$. Let z_j be a Grundy vertex in S of color $j, i+1 \leq j \leq \omega(G)$. Now $\{v, z_{i+1}, z_{i+2}, \ldots, z_{\omega(G)}, y_1, y_2, \ldots, y_{i-1}\} \subseteq S$. Since $|\{v, z_{i+1}, z_{i+2}, \ldots, z_{\omega(G)}, y_1, y_2, \ldots, y_{i-1}\}| = |S| = \omega(G)$, $\{v, z_{i+1}, z_{i+2}, \ldots, z_{\omega(G)}, y_1, y_2, \ldots, y_{i-1}\} = S$. Since, z_j is a Grundy vertex of color $j, i+1 \leq j \leq \omega(G)$, there exists a vertex $u_j \in R$ of color i which is adjacent to z_j for each $j, i+1 \leq j \leq \omega(G)$. Also v has a neighbor u in R and y_j has the neighbor w in R. So, each vertex in S has a neighbor in R. This is a contradiction to the fact that there is a vertex in S which is not adjacent to any vertex in R. So this case is also not possible. Hence, the sufficiency is also true. □

If each vertex in S has a neighbor in R, then the coloring f given in the proof of the necessity part of Lemma 2 is a partial Grundy $(\omega(G)+1)$-coloring of G and this is an optimal partial Grundy coloring of G by Theorem 4. If there is a vertex which is not adjacent to any vertex in R, then by Theorem 4, $\delta\Gamma(G) = \omega(G)$, and $f_1 : V \rightarrow \{1, 2, \ldots, \omega(G)\}$ defined by $f_1(x_i) = i, 1 \leq i \leq \omega(G)$ and $f_1(x) = f_1(x_j)$, where $S = \{x_1, x_2, \ldots, x_{\omega(G)}\}$, $x \in R$ and $x_j \in S$ has no neighbors in R is an optimal Grundy $\omega(G)$-coloring of G.

In view of this, we have the following result.

Theorem 5. *The optimal partial Grundy coloring of a split graph can be computed in $O(n+m)$ time.*

6 Conclusion

The Partial Grundy Number Decision problem is known to be NP-complete for bipartite graphs. In this paper, we strengthened this result by proving that this problem remains NP-complete even for perfect elimination bipartite graphs, a proper subclass of bipartite graphs. On the positive side, we proposed a linear time algorithm to determine the partial Grundy number of a chain graph, a proper subclass of perfect elimination bipartite graphs. It is also known that the Partial Grundy Number Decision problem is NP-complete for (disconnected) chordal graphs. We strengthened this result by proving that the Partial Grundy Number Decision problem remains NP-complete even for (connected) doubly chordal graphs, a proper subclass of chordal graphs. On the positive side, we proposed a linear time algorithm to determine the partial Grundy number of split graphs, a well known subclass of chordal graphs. It would be interesting to investigate the complexity of partial Grundy number problem on other important subclasses of bipartite graphs and chordal graphs.

References

1. Balakrishnan, R., Kavaskar, T.: Interpolation theorem for partial Grundy coloring. Discrete Math. **313**(8), 949–950 (2013)
2. Brandstädt, A., Dragan, F.F., Chepoi, V., Voloshin, V.I.: Dually chordal graphs. SIAM J. Discrete Math. **11**, 437–455 (1998)
3. Christen, C.A., Selkow, S.M.: Some perfect coloring properties of graphs. J. Combin. Theory Ser. B. **27**, 49–59 (1979)
4. Effantin, B., Gastineau, N., Togni, O.: A characterization of b-chromatic and partial Grundy numbers by induced subgraphs. Maths. arXiv:1505.07780v2 [cs.DM] (2016)
5. Erdös, P., Hedetniemi, S.T., Laskar, R.C., Prins, G.C.E.: On the equality of the partial Grundy and upper ochromatic numbers of graphs. Discrete Math. **272**(1), 53–64 (2003)
6. Fulkerson, D.R., Gross, O.A.: Incidence matrices and interval graphs. Pacific J. Math. **15**, 835–855 (1965)
7. Garey, M.R., Johnson, D.S., Stockmeyer, L.: Some simplified NP-complete graph problems. Theor. Comput. Sci. **1**, 237–267 (1976)
8. Golumbic, M.C., Goss, C.F.: Perfect elimination and chordal bipartite graphs. J. Graph Theory. **2**, 155–163 (1978)
9. Shi, Z., Goddard, W., Hedetniemi, S.T., Kennedy, K., Laskar, R., McRae, A.: An algorithm for partial Grundy number on trees. Discrete Math. **304**, 108–116 (2005)
10. Uehara, R., Uno, Y.: Efficient algorithms for the longest path problem. In: Fleischer, R., Trippen, G. (eds.) ISAAC 2004. LNCS, vol. 3341, pp. 871–883. Springer, Heidelberg (2004). doi:10.1007/978-3-540-30551-4_74

On Prime Distance Labeling of Graphs

A. Parthiban(✉) and N. Gnanamalar David

Department of Mathematics, Madras Christian College,
Chennai 600059, Tamil Nadu, India
mathparthi@gmail.com, ngdmcc@gmail.com

Abstract. A graph G is a prime distance graph if its vertices can be labeled with distinct integers in such a way that for any two adjacent vertices, the absolute difference of their labels is a prime number. It is known that cycles and bipartite graphs are prime distance graphs. In this paper we derive certain general results concerning prime distance labeling. We also investigate prime distance labeling of some cycle related graphs in the context of some graph operations, namely, power, fusion, duplication and vertex switching in cycle C_n.

1 Introduction

All graphs considered in this paper are simple, connected and undirected. The distance graph, first introduced by Eggleton et al. [3–5], is motivated by the well-known Hadwiger-Nelson plane coloring problem which asks for the minimum number of colors needed to color all points of the plane such that points at unit distance receive distinct colors. Motivated by the plane coloring problem, one can consider the analogue to the one-dimensional case by investigating the chromatic numbers of distance graphs on the real line R and the integer set Z.

If D is a subset of the set of positive integers, then the integer distance graph $G(Z, D)$ is defined to be the graph with vertex set Z, where two vertices u and v are adjacent if and only if $|u - v| \in D$. A particularly interesting problem is determining the chromatic number of $G(Z, D)$ for a given set D. The chromatic number of integer distance graphs, denoted $\chi(G(Z, D))$ has been studied extensively for different families of distance sets D. The prime distance graph $Z(P)$ is the distance graph with $D = P$, the set of all primes. Research in prime distance graphs has since focused on the chromatic number of $Z(D)$ where D is a non-empty proper subset of P. Note that these graphs are all infinite (non-induced) subgraphs of $Z(P)$.

In this paper we consider finite subgraphs of $Z(P)$. In [7] Laison et al. have defined a graph G to be a prime distance graph if there exists a one-to-one labeling of its vertices given by $L : V(G) \to Z$ such that for any two adjacent vertices u and v, the integer $|L(u) - L(v)|$ is a prime and L is called a prime distance labeling of G.

We need the following theorem.

Theorem 1 [7]. *Every cycle is a prime distance graph.*

S. Arumugam et al. (Eds.): ICTCSDM 2016, LNCS 10398, pp. 238–241, 2017.
DOI: 10.1007/978-3-319-64419-6_31

2 Main Results

If H is a subgraph of G and H is not a prime distance graph, then G is not a prime distance graph. In particular the square of any non-prime distance graph is again a non-prime distance graph.

Definition 1. *Let $G = (V, E)$ be a graph and let $v \in V$. Let H be the graph obtained from G by adding two vertices v', v'' and the edges $v'v'', v'v$ and $v''v$. Then H is called the graph obtained by duplication of v.*

Conjecture 1. (The Twin Prime Conjecture) [2] There are infinitely many pairs of primes that differ by 2.

Theorem 2. *The graph obtained by duplication of every vertex by an edge in any prime distance graph H is a prime distance graph if the Twin prime conjecture is true.*

Proof. Let $V(H) = \{v_1, v_2, ..., v_n\}$. Let f be a prime distance labeling and let $f(v_j) = \max_{1 \leq i \leq n} f(v_i)$. Let G be the graph obtained by duplication of every vertex v_k by an edge $v'_k v''_k$ for $k = 1, 2, ..., n$. Then G is a graph with $3n$ vertices and having n vertex disjoint cycles each of length three. Now we define a labeling $f' : V(G) \to Z$ as follows: Let $f'(v_i) = f(v_i)$ for all i. Let p_1 and p'_1 be any twin primes sufficiently larger than $f(v_j)$. Let $f'(v'_j) = f(v_j) + p_1$ and $f'(v''_j) = f(v_j) + p'_1$. Next let p_2, p'_2 be any twin primes sufficiently larger than $f'(v''_j)$. Now let $f(v_l) = \max_{1 \leq i \leq n, i \neq j} f(v_i)$. Let $f'(v'_l) = f(v_l) + p_2$ and $f'(v''_l) = f(v_l) + p'_2$. Continuing this process for all the vertices of G, we obtain prime distance labeling of G as there are infinitely many pairs of primes that differ by 2 by the Twin prime conjecture.

Definition 2. *Let $G = (V, E)$ be a graph and let $S \subseteq V$. A vertex-switching G_s of G is obtained by deleting from G all edges of G with exactly one end in S, and adding to G all edges of the complement of G with exactly one end in S.*

It has been proved in [8] that the fan graph F_n for $n \geq 12$ admits no prime distance labeling. Now for $n \geq 14$, the graph obtained by switching a vertex in C_n contains F_{12} as a subgraph and hence it is not a prime distance graph.

Fusing of two distinct vertices u and v in a graph G is the process of replacing u and v by a new vertex w such that $N(w) = N(u) \cup N(v)$.

Theorem 3. *The graph obtained by fusing any two vertices v_i and v_j, where $d(v_i, v_j) \geq 3$, of C_n is a prime distance graph.*

Proof. Let $C_n = \{v_1, v_2, ..., v_n, v_1\}$. Without loss of generality, let the vertex v_1 be fused with v_m where $m \leq \lceil \frac{n}{2} \rceil$. We denote the resultant graph as G. Then $|V(G)| = n-1$ and G is a connected graph which includes two edge disjoint cycles C_{m-1} and C_{n-m+1}. It follows from Theorem 1 that C_{m-1} and C_{n-m+1} are prime distance graphs. Let f' be a prime distance labeling of C_{m-1} with $f'(v_1) = 0$ and let f'' be a prime distance labeling of C_{n-m+1} with $f''(v_m) = 0$. Now the labeling $f : V(G) \to Z$ defined by $f(C_{m-1}) = f'(C_{m-1})$ and $f(C_{n-m+1}) = -f''(C_{n-m+1})$ is a prime distance labeling of G.

Definition 3. *Let $G = (V, E)$ be a graph and let $v \in V$. The duplication of v gives the graph G_1 obtained from G by adding a new vertex v_1 such that $N(v_1) = N(v)$.*

Conjecture 2. Goldbach's Conjecture [1,6] Every even number greater than 2 is the sum of two primes.

Theorem 4. *Let $C_n = \{u_1, u_2, ..., u_n, u_1\}$ be a cycle with $n \geq 6$. Let G be the graph obtained from C_n by duplicating u_n. Then G admits a prime distance labeling if Goldbach's conjecture is true.*

Proof. Let $V(G) = V(C_n) \cup \{u'_n\}$ where $N(u'_n) = N(u_n) = \{u_1, u_{n-1}\}$. Define $f : V(G) \to Z^+$ as follows: $f(u_i) = 2(i-1)$ for $1 \leq i \leq n-1$. If Goldbach's conjecture is true, then $f(u_{n-1})$ can be expressed as the sum of two primes. Let $f(u_{n-1}) = p_1 + p_2$. Now let $f(u_n) = p_1$ and $f(u'_n) = p_2$. Then f is a prime distance labeling of G.

Definition 4. *The graph G obtained from m copies of the cycle C_n by joining a vertex in the ith copy of C_n to a vertex in the $(i+1)^{th}$ copy of C_n where $1 \leq i \leq m-1$, is called the joint sum of m copies of C_n.*

Theorem 5 (Vinogradov's Theorem). *Every sufficiently large odd number is the sum of three primes.*

Theorem 6. *The joint sum G of any number of copies of C_n admits a prime distance labeling.*

Proof. Let $C_n = (v_1, v_2, \ldots, v_n, v_1)$. Let f be a prime distance labeling of C_n, which exists by Theorem 1. Let p be a prime number large enough such that $p+2n-8$ can be written as the sum of three primes, say $p+2n-8 = p_1+p_2+p_3$. By Theorem 5 such a prime exists. We also assume that $p > 4n$ and $p_1 \geq p_2 \geq p_3$, so that $p_1 > 2n-8$. Then C_n can be labelled with labels $0, 2, \ldots, 2n-8, p+2n-8, p_1+p_2$, and p_1 in cyclic order. Let the joint sum of m copies of C_n be G so that $V(G) = \{v_j^i : 1 \leq i \leq m, 1 \leq j \leq n\}$ and $E(G) = mE(C_n) \cup \{v_1^i v_1^{i+1} : 1 \leq i \leq m-1\}$. Define a one-to-one labeling $f^* : V(G) \to Z$ as follows. $f^*(v_j^1) = f(v_j)$ for $1 \leq j \leq n$. Next let $f^*(v_k^1) = \max_{1 \leq i \leq n} \{f^*(v_i^1)\}$ and let p_1^* be any prime larger than $f^*(v_k^1)$. Let $f^*(v_j^2) = f^*(v_j^1) + p_1^*$ for $1 \leq j \leq n$. Proceedings in this manner, let $f^*(v_k^{m-1}) = \max_{1 \leq i \leq n} \{f^*(v_i^{m-1})\}$ and let p_{m-1}^* be any prime larger than $f^*(v_k^{m-1})$. Let $f^*(v_j^m) = f^*(v_j^{m-1}) + p_{m-1}^*$ for $1 \leq j \leq n$. It is clear that f^* is a prime distance labeling of G.

Acknowledgement. The authors are very much grateful for the valuable and detailed comments of the reviewer. The comments have served to be very useful in correcting the errors and for improving the presentation of the paper. The first author A. Parthiban acknowledges with gratitude the award (No.: F1-17.1/2014-15/RGNF-2014-15-SC-TAM-65968/ (SA-III/Website)) of Rajiv Gandhi National Fellowship for SC students by UGC, India.

References

1. Burton, D.M.: Elementary Number Theory, 7th edn. McGraw-Hill, New York (2011)
2. Clement, P.A.: Congruences for sets of primes. Am. Math. Mon. **56**, 23–25 (1949)
3. Eggleton, R.B., Erdos, P., Skilton, K.: Colouring the real line. J. Comb. Theor. Ser. B. **39**(1), 86–100 (1985)
4. Eggleton, R.B., Erdos, P., Skilton, K.: Erratum: colouring the real line. J. Comb. Theor. Ser. B **41**(1), 139 (1986). MR0805458 (87b:05057)]. J. Combin. Theory Ser. B, 41(1), 139 (1986)
5. Eggleton, R.B., Erdos, P., Skilton, K.: Colouring prime distance graphs. Graphs Comb. **6**, 17–32 (1990)
6. Rosen, K.H.: Elementary Number Theory and its Applications, 6th edn. Addison Wesley, Reading (2010)
7. Laison, J.D., Starr, C., Walker, A.: Finite prime distance graphs and 2-odd graphs. Discret. Math. **313**, 2281–2291 (2013)
8. Parthiban, A., David, N.G.: On finite prime distance graphs. In: International Conference on Current Trends in Graph Theory and Computation organized. South Asian University, New Delhi (2016)

Dominator Colorings of Products of Graphs

P. Paulraja and K. Raja Chandrasekar(✉)

Department of Mathematics, Kalasalingam University,
Krishnankoil 626126, Tamilnadu, India
ppraja56@gmail.com, rajmath84@gmail.com

Abstract. A *dominator coloring* $\mathcal{C}$ of a graph G is a proper coloring of G such that closed neighborhood of each vertex of G contains a color class of $\mathcal{C}$. The minimum number of colors required for a dominator coloring of G is called the *dominator chromatic number* of G, denoted by $\chi_d(G)$. In this paper we obtain the exact value of χ_d for some classes of graphs, such as $K_m \times K_n$, $K_{m(n)} \times K_{r(s)}$, $(K_r \circ K_1) \times K_s$, $K_n \Box Q_{r+2}$, where $\times, \Box$ and $\circ$ denote the tensor product, Cartesian product and corona of graphs, respectively, and $K_{m(n)}$ denotes the complete m-partite graph in which each partite set has n vertices. Also we present an upper bound for $\chi_d(G \times K_m)$ in terms of $\chi_d(G)$ and $\gamma_t(G)$, where $\gamma_t(G)$ is the total domination number of G.

Keywords: Dominator coloring · Tensor product · Cartesian product · Corona of a graph

Mathematics Subject Classification (2000): 05C15, 05C69.

1 Introduction

All graphs $G = (V, E)$ considered here are finite, undirected simple graphs. The *open neighborhood* of v and the *closed neighborhood* of v are defined by $N(v) = \{u \in V : uv \in E\}$ and $N[v] = N(v) \cup \{v\}$, respectively. For $S \subseteq V(G)$, the open and closed neighborhoods of S are defined by $N(S) = \bigcup_{v \in S} N(v)$ and $N[S] = \bigcup_{v \in S} N[v]$.

A subset S of V is called a *dominating set* (resp. *total dominating set*) of G if every vertex in $V - S$ (resp. in V) is adjacent to a vertex in S. The *domination number*, $\gamma(G)$, (resp. *total domination number* $\gamma_t(G)$) is the minimum cardinality of a dominating set (resp. total dominating set) of G. The minimum cardinality of a dominating set (resp. total dominating set) is called a γ-set (resp. γ_t-set) of G. The corona $G \circ K_1$ is the graph obtained from the graph G by adding a new vertex v' to each vertex v of G and make v adjacent to v'.

A *proper vertex coloring* of a graph G is an assignment of colors to the vertices of G such that no two adjacent vertices receive the same color. The *chromatic number*, $\chi(G)$, is the minimum number of colors required for a proper coloring of G. Gera [6] introduced the concept of dominator coloring of a graph.

S. Arumugam et al. (Eds.): ICTCSDM 2016, LNCS 10398, pp. 242–250, 2017.
DOI: 10.1007/978-3-319-64419-6_32

A *dominator coloring* of a graph G is a proper coloring $\mathcal{C} = \{C_1, C_2, \dots, C_k\}$ of G in which each vertex of G dominates a color class C_i in $\mathcal{C}$, that is, for each $v \in V(G)$, there exists an i, such that $C_i \subset N[v]$. The minimum number of colors required for a dominator coloring of G is called the *dominator chromatic number* of G and its denoted by $\chi_d(G)$. If $C_i \subset N[v]$, then we say that the vertex v dominates the color class C_i or the color class C_i is dominated by the vertex v. A dominator coloring of G using $\chi_d(G)$ colors is called a χ_d-*coloring* of G.

If $\{C_1, C_2, \dots, C_{\chi_d}\}$ is a χ_d-coloring of G and if $v_i \in C_i$, then $S = \{v_1, v_2, \dots, v_{\chi_d}\}$ is a dominating set of G. Also if S is a γ-set of G, then $\mathcal{C}' \cup \{\{v\} : v \in S\}$, where $\mathcal{C}'$ is a proper coloring of $G - S$, gives a dominator coloring of G. This observation leads to the following bounds for $\chi_d(G)$, see [6]; $max\ \{\chi(G), \gamma(G)\} \leq \chi_d(G) \leq \chi(G) + \gamma(G)$. Consequently, $\chi_d(G) \in \{\gamma(G), \gamma(G)+1, \gamma(G)+2\}$ for a bipartite graph G. Further, the class of bipartite graphs for which $\chi_d(G) = \gamma(G)$ has been completely characterized in [7]. In general, the decision problem corresponding to the dominator coloring is proved to be NP-complete. Even if we consider the split graph, it is known to be NP-complete; see [1]. However, polynomial time algorithm exists for a graphs with $\chi_d(G) = 3$; see [5]. Also for a P_4-free graph G, polynomial time algorithm exists to find $\chi_d(G)$; see [5]. For related results see [2,3,8–10]. For graph theoretic terminology we refer to Chartrand and Lesniak [4].

Let G and H be two graphs. We define two fundamental products, namely, Cartesian product and tensor product of graphs G and H, denoted by $G \square H$ and $G \times H$, respectively. These graphs have their vertex sets as $V(G) \times V(H)$ and their edge sets are defined as follows:

$E(G \square H) = \{(g,h)(g',h') |\ g = g' \text{ and } hh' \in E(H) \text{ or, } gg' \in E(G) \text{ and } h = h'\}$, and $E(G \times H) = \{(g,h)(g',h') |\ gg' \in E(G) \text{ and } hh' \in E(H)\}$. Let $\{u_1, u_2, \dots, u_m\}$ and $\{v_1, v_2, \dots, v_n\}$ be the vertex sets of G and H respectively. We call $u_i \times V(H)$ (resp. $V(G) \times v_j$) as the ith-row (resp. j^{th}-column) of $G * H$, where $*$ represents the tensor product or the Cartesian product; see Fig. 1.

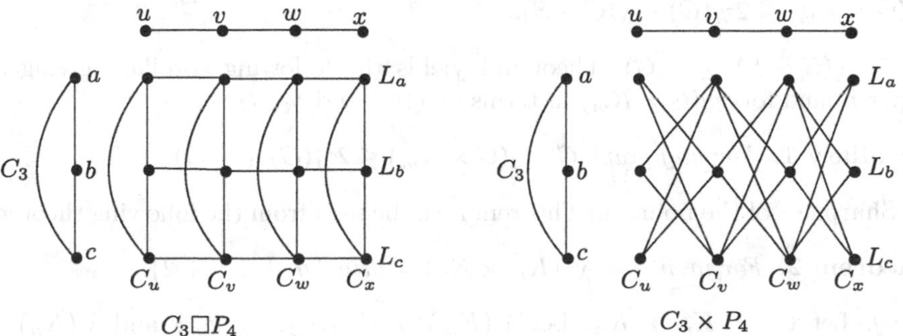

Fig. 1. Cartesian product and tensor product of the graphs C_3 and P_4 are shown in the figure. Here L_a denotes the vertex set $a \times V(P_4)$ and C_u denotes the vertex set $V(C_3) \times u$.

2 Dominator Chromatic Number of the Tensor Product of Graphs

In this section, we obtain the exact value of χ_d for tensor product graphs, such as $K_m \times K_n$, $K_{m(n)} \times K_{r(s)}$, $(K_r \circ K_1) \times K_s$. One can observe that the graph $K_m \times K_n$ is isomorphic to the complement of the graph $K_m \square K_n$. The following theorem gives an upper bound for dominator chromatic number of $G \times K_m$ in terms of the chromatic number and the total domination number of G.

Theorem 1. *For a graph G, $\chi_d(G \times K_m) \leq 2\gamma_t(G) + \chi(G - S)$, where S is a γ_t-set of G such that $\chi(G - S)$ is minimum among all γ_t-sets S of G.*

Proof. Let $V(G) = \{u_1, u_2, \ldots, u_n\}$ and $V(K_m) = \{v_1, v_2, \ldots, v_m\}$. Then $V(G \times K_m) = \{(u_i, v_j) : 1 \leq i \leq n,\ 1 \leq j \leq m\}$. Clearly, the subset $u_i \times V(K_m), 1 \leq i \leq n$, which we call a row of $G \times K_m$, is an independent set of $G \times K_m$. Let $S = \{x_1, x_2, \ldots, x_r\}$ be a γ_t-set of G as in the statement of the theorem. Let $\chi(G - S) = k$ and let $\{V_1, V_2, \ldots, V_k\}$ be a χ-coloring of $G - S$. As V_i is an independent set of $G - S$, each subset $V_i \times V(K_m), 1 \leq i \leq k$, is an independent set of $G \times K_m$. Now we present a dominator coloring $\mathcal{C}$ of $G \times K_m$ using $2\gamma_t(G) + \chi(G - S)$ colors as follows: let $\mathcal{C} = \{A_1, A_2, \ldots, A_k, B_1, B_2, \ldots, B_r, C_1, C_2, \ldots, C_r\}$, where $A_i = \{V_i \times V(K_m)\}, 1 \leq i \leq k$, $B_j = \{(x_j, v_1)\}, 1 \leq j \leq r$ and, $C_l = \{(x_l \times V(K_m)) \setminus (x_l, v_1)\}$, $1 \leq l \leq r$. Clearly $\mathcal{C}$ is a proper coloring of $G \times K_m$.

Next we claim that each vertex in $V(G \times K_m)$ dominates a color class in $\mathcal{C}$. Let $y \in V(G - S)$. As S is a total dominating of G, there exists an $x_i \in S$ such that $yx_i \in E(G)$. Clearly, (y, v_1) dominates the color class $\{(x_i \times V(K_m)) \setminus (x_i, v_1)\}$ and each of the vertices in $(y \times V(K_m)) \setminus (y, v_1)$ dominates the color class $\{(x_i, v_1)\}$. Hence each vertex in $y \times V(K_m)$ dominates a color class of $\mathcal{C}$. Let $x_j \in S$, for some j. As S is a total dominating set of G, there exists an x_k, $k \neq j$, in S such that $x_j x_k \in E(G)$. As above, each vertex in $x_j \times V(K_m)$ dominates either the color class $\{(x_k, v_1)\}$ or $\{(x_k \times V(K_m)) \setminus (x_k, v_1)\}$ in $\mathcal{C}$. Hence $\mathcal{C}$ is a dominator coloring of $G \times K_m$ using $2r + k = 2\gamma_t(G) + \chi(G - S)$ colors. Thus $\chi_d(G \times K_m) \leq 2\gamma_t(G) + \chi(G - S)$.

As $\chi(G - S) \leq \chi(G)$, Theorem 1 yields the following corollary giving an upper bound for $\chi_d(G \times K_m)$ in terms of $\chi(G)$ and $\gamma_t(G)$:

Corollary 1. *For any graph G, $\chi_d(G \times K_m) \leq 2\gamma_t(G) + \chi(G)$.*

Sharpness of the bound in Theorem 1 can be seen from the following theorem:

Theorem 2. *For $m, n \geq 3$, $\chi_d(K_m \times K_n) = min\ \{m + 2, n + 2\}$.*

Proof. Let $G = K_m \times K_n$. Let $V(K_m) = \{u_1, u_2, \ldots, u_m\}$ and $V(K_n) = \{v_1, v_2, \ldots, v_n\}$. Then $V(K_m \times K_n) = \{(u_i, v_j) : 1 \leq i \leq m, 1 \leq j \leq n\}$. As the tensor product is commutative, without loss of generality we assume that $m \leq n$. It is enough to show that $\chi_d(G) = m + 2$. The upper bound follows from Theorem 1, since $\gamma_t(K_m) = 2$ and, $\chi(K_m - S) = m - 2$, where S is any γ_t-set of K_m.

Next we shall prove that $\chi_d(G) \geq m+2$. Suppose that $\chi_d(G) = m+1$. Let $\mathcal{C}$ be a χ_d-coloring of G using $m+1$ colors. Observe that any color class of $\mathcal{C}$ is a subset of a row or a column, since any two vertices which do not lie on a common row or column are adjacent in G.

We claim that there exists a color class $C_i \in \mathcal{C}$ such that C_i is a proper subset of either a row or a column of G. Suppose not, then every color class C_i is either a row or a column of G, as two vertices which are not lie on a common row or a column are adjacent. Since each vertex of G can be adjacent to at most $m-1$ vertices of each row and at most $n-1$ vertices of each column of G, no vertex in G dominates a color class of $\mathcal{C}$, which is a contradiction. This completes the proof of the claim.

Without loss of generality, we assume that there exists a color class, say, C_{m+1}, which is a proper subset of the first column of G, if necessary relabel the vertices of K_n. Clearly, C_{m+1} (see Fig. 2) can be assumed to be $\{(u_1, v_1), (u_2, v_1), \ldots, (u_i, v_1)\}$ where $i < m$, if necessary relabel the vertices of K_m.

We observe that $S = \{(u_i, v_{m+1-i}) : 1 \leq i \leq m\}$, see Fig. 2, induces a clique of size m in G. Without loss of generality, we assume that the vertex $(u_i, v_{m+1-i}), 1 \leq i \leq m$, in S, receives the color i, as $\chi_d(G) = m+1$, C_{m+1} is a proper subset of the first column vertices of G and $(u_m, v_1) \notin C_{m+1}$.

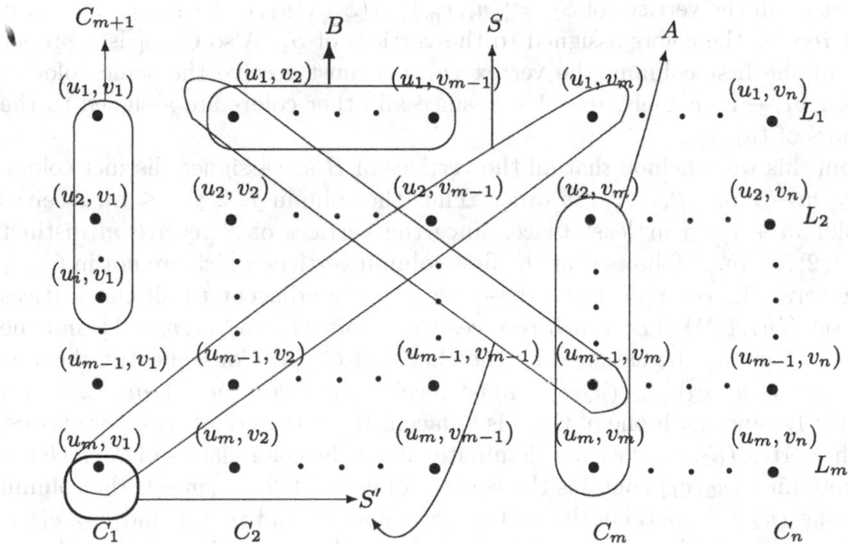

Fig. 2. In this figure, $L_i = u_i \times V(K_m)$ represents the ith row of G and $C_j = V(K_m) \times v_j$ represents the jth column of G. The first i consecutive vertices of column 1 represents the color class C_{m+1}, the $m-2$ consecutive vertices beginning from the second vertex of the first row denotes the set B and the $m-1$ consecutive vertices beginning from the second vertex of the mth column is A. The set S' is shown in two different portions.

Let $A = \{(u_i, v_m) : 2 \leq i \leq m\}$, that is, all the vertices of mth column except its first vertex. Let $B = \{(u_1, v_j) : 2 \leq j \leq m-1\}$, that is, the $m-2$ consecutive vertices beginning from the second vertex of the first row; see Fig. 2.

Now we claim that $\{(u_1, v_m)\}$ is not in $\mathcal{C}$, that is, the color class containing (u_1, v_m) is not a singleton set in $\mathcal{C}$. Suppose that $C_1 = \{(u_1, v_m)\} \in \mathcal{C}$. It is clear that $S' = \{(u_1, v_2), (u_2, v_3), \ldots, (u_{m-1}, v_m), (u_m, v_1)\}$ induces a clique of size m in G, as no two of these vertices lie on a common column or row; see Fig. 2 (Note that the set S' is shown in two portions). To color all the vertices of S', we need m colors besides the colors 1 and $m+1$, since $C_1 = \{(u_1, v_m)\}$ is a singleton color class having the color 1, and the first column contains the color class C_{m+1}. It follows that $|\mathcal{C}| \geq m+2$, which is a contradiction to the assumption $|\mathcal{C}| = m+1$. Hence $\{(u_1, v_m)\}$ cannot be a singleton color class of $\mathcal{C}$. If the color 1 assigned to (u_1, v_m) is neither in A nor in B, then the vertices of the clique $S_2 = \langle\{(u_1, v_2), (u_2, v_3), \ldots, (u_{m-1}, v_m), (u_m, v_1)\}\rangle$ must receive m colors besides 1 and $m+1$. Thus $\chi_d(G) \geq m+2$, which is again a contradiction to the assumption $|\mathcal{C}| = m+1$. Hence the color of (u_1, v_m), namely 1, appears more than once in A or B.

Case (i) A vertex, say, $(u_i, v_m) \in A$ receives the color of (u_1, v_m), namely, 1.

Choose an arbitrary vertex $(u_1, v_j) \in B$. We claim that (u_1, v_j), in B, and (u_{m+1-j}, v_j), in S, receive the same color, namely, $m+1-j$ (note that we have already assigned the color $m+1-j$ to the vertex (u_{m+1-j}, v_j)). As (u_1, v_j) is adjacent to all the vertices of $S_1 = \{(u_i, v_m)\} \cup (S \setminus \{(u_1, v_m), (u_{m+1-j}, v_j)\})$, it cannot receive the colors assigned to the vertices of S_1. Also C_{m+1} is a proper subset of the first column, the vertex (u_1, v_j) must receive the same color of $(u_{m+1-j}, v_j) \in S$, namely, $m+1-j$, since all other colors are assigned to the neighbors of (u_1, v_j).

From this we conclude that all the vertices of B are assigned distinct colors, namely, $m-1, m-2, \ldots, 2$, in order. Thus the column j, $2 \leq j \leq m$, receives the color $m+1-j$ at least twice, since the vertices of S receive m distinct colors $1, 2, \ldots, m$. It follows that the first column vertices which are not in C_{m+1} must receive the color m, since these vertices are adjacent to all the vertices of the set $\{(u_1, v_m)\} \cup B$, which receives the colors $\{1, 2, 3, \ldots, m-1\}$ and the color $m+1$ cannot be assigned by the choice of C_{m+1}. Consequently, the vertices $(u_2, v_2), (u_2, v_3), \ldots, (u_2, v_m)$ must receive the colors $m-1, m-2, \ldots, 1$, respectively, since each one of them is adjacent to vertices of other color classes. Now the vertex (u_2, v_1) does not dominate any of the color classes of $\mathcal{C}$, since the row containing (u_2, v_1) contains the vertices of colors $1, 2, \ldots, m-1$, the column containing (u_2, v_1) contains the vertices of colors m and $m+1$ and (u_2, v_1) is nonadjacent to all the vertices of the column and row in which it lies. Consequently the neighbors of (u_2, v_1) misses at least one vertex of each of the color classes of $\mathcal{C}$, which is a contradiction.

Case (ii) A vertex, say, (u_1, v_j), $2 \leq j \leq m-1$, in B, receives the color of (u_1, v_m), namely, 1.

Clearly, the vertex $(u_2, v_m) \in A$ is adjacent to all the vertices of $S_2 = S \setminus \{(u_1, v_m), (u_2, v_{m-1})\}$, which are assigned the colors $3, 4, \ldots, m$, in order.

Further, (u_2, v_m) is adjacent to the vertex (u_1, v_j) in B of color 1, by assumption, and the vertex (u_2, v_m) cannot be assigned the color $m+1$, by the choice of C_{m+1}. Hence the color of (u_2, v_m) must be 2. By similar arguments we can conclude that each vertex $(u_j, v_m) \in A$, $3 \leq j \leq m$, must receive the color j.

As the ith row, $1 \leq i \leq m$, of G contains at least two vertices, namely (u_i, v_{m+1-i}) and (u_i, v_m), having the same color i, all the vertices in the row i, except the vertices in the color class C_{m+1}, must receive the color i, since each one of them is adjacent to vertices of other color classes. It follows that (u_1, v_2) does not dominate any color class in $\mathcal{C}$, which is again a contradiction. Hence $\chi_d(G) \geq m+2$. Already we have shown that $\chi_d(G) \leq m+2$. Thus $\chi_d(G) = m+2$.

Let $K_{m(n)}$ denote the complete m-partite graph in which each partite set has n vertices. Let the partite sets of $K_{m(n)}$ (resp. $K_{r(s)}$) be $U_1, U_2, \ldots, U_m$ (resp. $V_1, V_2, \ldots, V_r$). Let $G = K_{m(n)} \times K_{r(s)}$. Clearly, $V(G) = V(K_{m(n)} \times K_{r(s)}) = \bigcup (U_i \times V_j), i = 1, 2, \ldots, m,\ j = 1, 2, \ldots, r$. Let $B_{i,j} = U_i \times V_j$; then $|B_{i,j}| = ns$.

Lemma 1. *Let $G = K_{m(n)} \times K_{r(s)}, n, s \geq 3$. Then any χ_d-coloring of G can be transformed to a χ_d-coloring of it in which all the vertices of $B_{i,j}$ receive the same color.*

Proof. Let $\mathcal{C}$ be a χ_d-coloring of G. Let $x, y, z \in B_{i,j}$ (it is possible as $n, s \geq 3$).

Case (i) $\{x\}, \{y\}$ and $\{z\}$ are singleton color classes in $\mathcal{C}$.

Let the colors of x, y and z be $1, 2$ and 3, respectively. Now we obtain a dominator coloring of G using $|\mathcal{C}| - 1$ colors which contradicts that $\mathcal{C}$ is a χ_d-coloring of G.

Recoloring the vertices of G as follows: assign the color 1 to all the vertices of $B_{i,j}$ and choose any vertex in $N(x)$, say a, and recolor the vertex a by color 2. Rest of the colors of the vertices of G are retained as in $\mathcal{C}$. The resulting coloring, say $\mathcal{C}'$, is also a dominator coloring of G; for each vertex in $B_{i,j}$ dominates the color class $\{a\}$ and if a vertex dominates a color class either $\{x\}$ or $\{y\}$ or $\{z\}$ in $\mathcal{C}$, then that vertex dominates the color class $\{B_{i,j}\}$ in $\mathcal{C}'$. Hence $\mathcal{C}'$ is a dominator coloring of G using $|\mathcal{C}| - 1$ colors, that is the color 3 is not used, which is a contradiction for the minimality of $|\mathcal{C}|$.

Case (ii) $\{x\}$ and $\{y\}$ are singleton color classes in $\mathcal{C}$.

By Case (i), $\{z\}$ cannot be a singleton color class in $\mathcal{C}$.

Let z dominate a color class C_k in $\mathcal{C}$. Then $C_k \subseteq N(z)(= N(x) = N(y))$. Now recolor all the vertices of $B_{i,j} - \{x\}$ by the color of x and the colors of the rest of the vertices of G are retained as in $\mathcal{C}$. Let the resulting coloring be $\mathcal{C}''$. Note that $\mathcal{C}''$ does not use the color of y. By the similar argument as in Case (i), $\mathcal{C}''$ is a dominator coloring using $|\mathcal{C}| - 1$ colors, which is again a contradiction.

Case (iii) $\{x\} \in \mathcal{C}$.

By Cases (i) and (ii), $\{y\}, \{z\} \notin \mathcal{C}$.

Clearly the vertices y and z dominate a color class $C_k \subseteq N(y)$. In this case simply recolor all the vertices of $B_{i,j}$ by the color of x and retain the colors of the other vertices of G. The resulting coloring $\mathcal{C}'''$ is a dominator coloring of G using $|\mathcal{C}|$ colors in which all the vertices of $B_{i,j}$ receive the same color.

Case (iv) $B_{i,j}$ does not contain a singleton color class of $\mathcal{C}$.

Fix a color, say k, of a vertex in $B_{i,j}$. Replace the color of every other vertex in $B_{i,j}$ with the color k and retain the colors of the other vertices of G. The resulting coloring is a χ_d-coloring of G.

By successively applying Cases (*iii*) and (*iv*) (since Cases (*i*) and (*ii*) are not possible) to each $B_{i,j}$, we get a χ_d-coloring of G such that all the vertices of $B_{i,j}$ have a common color.

Theorem 3. *For* $n, s \geq 3$, $\chi_d(K_{m(n)} \times K_{r(s)}) = min\ \{m+2, r+2\}$.

Proof. Let $G = K_{m(n)} \times K_{r(s)}$. We first claim that $\chi_d(G) \leq min\ \{m+2, r+2\}$. Without loss of generality, we assume that $m \leq r$.

We obtain a dominator coloring using $m+2$ colors as follows. Let $\mathcal{C} = \{\{B_{1,1}\}, \{\bigcup_{i=2}^{r} B_{1,i}\}, \{\bigcup_{i=2}^{m} B_{i,1}\}, \{\bigcup_{j=2}^{r} B_{2,j}\}, \{\bigcup_{j=2}^{r} B_{3,j}\}, \ldots, \{\bigcup_{j=2}^{r} B_{m,j}\}\}$, where the $B_{i,j}^{\,s}$ are as defined just above the Lemma 1. Clearly $\mathcal{C}$ is proper coloring of G. Also each vertex in $B_{1,1}$ dominates the color class $\{\bigcup_{j=2}^{r} B_{2,j}\}$, each vertex in $\bigcup_{i=2}^{r} B_{1,i}$ dominates the color class $\{\bigcup_{i=2}^{m} B_{i,1}\}$ and vice versa and, the remaining all vertices dominate the color class $\{B_{1,1}\}$. Hence $\mathcal{C}$ is a dominator coloring using $m+2$ colors. Thus $\chi_d(G) \leq m+2$.

Next we claim that $\chi_d(G) \geq m+2$. Let $\mathcal{C}$ be a χ_d-coloring of G such that all the vertices of $B_{i,j}$ receive the same color, by Lemma 1. Now we identify each $B_{i,j}$ into a vertex $v_{i,j}$ and join $v_{i,j}$ and $v_{k,l}$ by an edge if and only if $\langle B_{i,j} \cup B_{k,l} \rangle$ is a complete bipartite graph $K_{ns,ns}$. We can check that the resulting graph, say, G' is isomorphic to $K_m \times K_r$. Clearly, by Theorem 2, $min\ \{m+2, r+2\} = \chi_d(G') \leq \chi_d(G)$. If $\chi_d(G) \leq m+1$, then it would imply $\chi_d(G') \leq m+1$, that is, by assigning the color of the vertices in $B_{i,j}$ to the vertex $v_{i,j}$ of G', a contradiction to Theorem 2. Hence $\chi_d(G) \geq m+2$. Already we have shown that $\chi_d(G) \leq m+2$. Hence $\chi_d(G) = \chi_d(K_{m(n)} \times K_{r(s)}) = m+2$.

Theorem 4. *Let* $G = (K_r \circ K_1) \times K_s, r, s \geq 3, r \leq s$. *Then* $\chi_d(G) = 2r+1$.

Proof. Let $H = K_r \circ K_1$. Let $V(H) = \{x_1, x_2, \ldots, x_r, y_1, y_2, \ldots, y_r\}$, where $\langle \{x_1, x_2, \ldots, x_r\} \rangle = K_r$, $\langle \{y_1, y_2, \ldots, y_r\} \rangle = \overline{K}_r$, $x_i y_i \in E(H)$, *and* $x_i y_j \notin E(H)$, $i \neq j$. Then the vertex set of G can be partitioned into $2r$ vertex disjoint subsets, namely $x_i \times V(K_s)$, where $x_i \in V(H)$ and $deg_H(x_i) = r$ and, $y_i \times V(K_s)$, where $y_i \in V(H)$ and $deg_H(y_i) = 1$.

Let $A_i = x_i \times V(K_s) = \{u_1^i, u_2^i, \ldots, u_s^i\}$, $1 \leq i \leq r$, and let $B_i = y_i \times V(K_s) = \{v_1^i, v_2^i, \ldots, v_s^i\}$, $1 \leq i \leq r$. It is clear that each of the induced subgraphs $\langle \{A_i \cup A_j\} \rangle$, $1 \leq i \neq j \leq r$, and $\langle \{A_i \cup B_i\} \rangle$, $1 \leq i \leq r$, are isomorphic to $K_{s,s} - I$, where I is a perfect matching in $K_{s,s}$.

Since $N[v_1^1], N[v_2^2], \ldots, N[v_r^r]$ are r-vertex disjoint closed neighborhoods of G, every dominator coloring of G contains at least one of its color class in each of these closed neighborhoods. Clearly the subgraph $\langle V(G) \backslash (\bigcup_{i=1}^{r} N[v_i^i]) \rangle$ contains a subgraph which is isomorphic to $K_r = \langle \{u_1^1, u_2^2, \ldots, u_r^r\} \rangle$. As seen above, each $N[v_i^i], i = 1, 2, \ldots, r$, contains a color class and to color all the vertices of the complete subgraph $\langle \{u_1^1, u_2^2, \ldots, u_r^r\} \rangle$, we need another r new colors and hence $\chi_d(G) \geq 2r$.

Now we claim that $\chi_d(G) > 2r$. Suppose $\chi_d(G) = 2r$ and let $\mathcal{C} = \{V_1, V_2, \ldots, V_{2r}\}$ be a χ_d-coloring of G. Observe that the number of color classes in $\mathcal{C}$ which are dominated by vertices of each subset $B_i(= y_i \times K_s)$ is at least 2. Hence each closed neighborhood $N[B_i], 1 \leq i \leq r$, contains at least two color classes of $\mathcal{C}$. Also since $\chi_d(G) = 2r$ and each $\langle\{A_i \cup B_i\}\rangle$ induces a non-trivial bipartite graph, it follows that $\mathcal{C} = \{\{A_1\}, \{A_2\}, \ldots, \{A_r\}, \{B_1\}, \{B_2\}, \ldots, \{B_r\}\}$. Now each vertex in $B_j, 1 \leq j \leq r$, does not dominate any color class in $\mathcal{C}$, which is a contradiction. Hence $\chi_d(G) \geq 2r + 1$.

Now we claim that $\chi_d(G) \leq 2r+1$, by giving a dominator coloring using $2r+1$ colors as follows: $\mathcal{C}' = \{\{u_1^1\}, \{u_2^2\}, \ldots, \{u_r^r\}, \{A_1 \setminus \{u_1^1\}\}, \{A_2 \setminus \{u_2^2\}\}, \ldots, \{A_r \setminus \{u_r^r\}\}, \{\bigcup_{i=1}^{r} B_i\}\}$ is a dominator coloring of G, since each vertex v_i^i dominates the color class $\{A_i \setminus \{u_i^i\}\}$, $1 \leq i \leq r$ and remaining all vertices dominates any one of the color class from $\{\{u_1^1\}, \{u_2^2\}, \ldots, \{u_r^r\}\}$. Hence $\chi_d(G) = 2r + 1$.

3 Dominator Chromatic Number of the Cartesian Product of Graphs

In this section, we obtain the exact value of χ_d for the graph $K_n \Box Q_{r+2}$, where $Q_{r+2} = \underbrace{K_2 \Box K_2 \Box \cdots \Box K_2}_{r+2 \text{ times}}$ is the hypercube.

Theorem 5. *Let n and r be two integers such that $n \geq 2^{r+2}$. Then $\chi_d(K_n \Box Q_{r+2}) = 2^{r+2} + n - 1$, where $r \geq 0$.*

Proof. Let $G = K_n \Box Q_{r+2}$. First we claim that $\chi_d(G) \leq 2^{r+2} + n - 1$. One can easily see, using the definition of Cartesian product of graphs, G is obtained by replacing each vertex of Q_{r+2} by a copy of K_n and if there is an edge joining two vertices x and y of Q_{r+2}, then there is a perfect matching joining the corresponding vertices of the respective copies of K_n corresponding to x and y.

Let $V(Q_{r+2}) = \{x_1, x_2, \ldots, x_{2^{r+2}}\}$. Let $A_i = \{x_i^1, x_i^2, \ldots, x_i^n\}$ denote the vertices of G corresponding to the ith vertex x_i of Q_{r+2}. Let us consider the first vertex of each $A_i, 1 \leq i \leq 2^{r+2}$, namely, $S_1 = \{x_1^1, x_2^1, \ldots, x_{2^{r+2}}^1\}$.

Let $G' = G - S_1$. Clearly $G' = K_{n-1} \Box Q_{r+2}$. Then $\chi(G') = max\ \{\chi(K_{n-1}), \chi(Q_{r+2})\} = n - 1$; see Theorem 26.1 of [9]. Let this coloring of G' be $\mathcal{C}'$.

A χ_d-coloring $\mathcal{C}$ of G can be obtained by extending the coloring $\mathcal{C}'$ of G' to G by assigning 2^{r+2} new colors to the vertices of S_1. Clearly these 2^{r+2} colors of the vertices in S_1 give us 2^{r+2} singleton color classes in $\mathcal{C}$. As each copy of K_n, corresponding to a vertex of Q_{r+2}, contributes one vertex to S_1, each vertex of this K_n dominates a color class of $\mathcal{C}$. Thus $\mathcal{C} = \mathcal{C}' \cup \{\{v\} : v \in S_1\}$ is a dominator coloring of G and hence $\chi_d(G) \leq 2^{r+2} + n - 1$.

Now we claim that $\chi_d(G) \geq 2^{r+2} + n - 1$. Let $\mathcal{C}''$ be a χ_d-coloring of G.

Case (i) There exists an i, $1 \leq i \leq 2^{r+2}$, such that A_i contains no singleton color class in $\mathcal{C}''$.

Since any two vertices of A_i do not have a common neighbor in $V(G) \setminus A_i$, and there is no singleton color class in A_i, no two vertices of A_i can dominate a common color class in $\mathcal{C}''$. Hence $\mathcal{C}''$ contains at least n color classes which are dominated by the vertices of A_i. It is clear that the union of the above n color classes must be in $N[A_i]$, the closed neighborhood of A_i. It is clear that the induced subgraph $\langle V(G) \setminus N[A_i] \rangle$ contains a subgraph which is isomorphic to $A_j = K_n$, for some $j \neq i$, $1 \leq j \leq 2^{r+2}$. Hence $\mathcal{C}''$ contains another n new colors to color these vertices of K_n. It follows that $\chi_d(G) \geq 2n = n + n \geq n + 2^{r+2}$, since $n \geq 2^{r+2}$.

Case (ii) Each A_i, $1 \leq i \leq 2^{r+2}$, contains a singleton color class in $\mathcal{C}''$, say $\{x_i\}$.

Since $\langle A_i - \{x_i\} \rangle = K_{n-1}$, we need another $n - 1$ colors to color the vertices of $\langle A_i - \{x_i\} \rangle$. Hence $\chi_d(G) \geq 2^{r+2} + n - 1$, as there are 2^{r+2} singleton color classes, by assumption. Thus $\chi_d(G) = 2^{r+2} + n - 1$.

4 Conclusion

In this paper we discuss about the dominator coloring for tensor product and Cartesian product of complete graphs and complete multipartite graphs. Similar results may be obtained for wreath and strong products of graphs.

References

1. Arumugam, S., Chandrasekar, K.R., Misra, N., Philip, G., Saurabh, S.: Algorithmic aspects of dominator colorings in graphs. In: Iliopoulos, C.S., Smyth, W.F. (eds.) IWOCA 2011. LNCS, vol. 7056, pp. 19–30. Springer, Heidelberg (2011). doi:10.1007/978-3-642-25011-8_2
2. Arumugam, S., Chandrasekar, K.R.: Graph colorings. In: Thulasiraman, K.T., Arumugam, S., Brandstadt, A., Nishizeki, T. (eds.) Handbook of Graph Theory, Combinatorial Optimization, and Algorithms, pp. 449–474. CRC Press, New York (2016)
3. Arumugam, S., Bagga, J., Chandrasekar, K.R.: On dominator colorings in graphs. Proc. Indian Acad. Sci. (Math. Sci.) **122**(4), 561–571 (2012)
4. Chartrand, G., Lesniak, L.: Graphs and Digraphs, 4th edn. Chapman and Hall, CRC, Boca Raton (2005)
5. Chellai, M., Maffray, F.: Dominator coloring in some classes of graphs. Graphs Comb. **28**, 97–107 (2012)
6. Gera, R.M.: On dominator coloring in graphs. Graph Theor. Notes N.Y. **LII**, 25–30 (2007)
7. Gera, R.M.: On the dominator colorings in bipartite graphs. In: ITNG 2007, pp. 1–6. IEEE Computer Scoiety (2007)
8. Gera, R.M., Rasmussen, C., Horton, S.: Dominator colorings and safe clique partitions. Congr. Numer. **181**, 19–32 (2006)
9. Hammack, R., Imrich, W., Klavzar, S.: Handbook of Product Graphs, 2nd edn. Chapman and Hall, CRC, Boca Raton (2011)
10. Merouane, H.B., Chellali, M.: On the dominator colorings in trees. Discuss. Math. **32**, 677–683 (2012)

Efficient Hybrid Approach for Compression of Multi Modal Medical Images

B. Perumal(✉), M. Pallikonda Rajasekaran, and T. Arun Prasath

Department of Electroncis and Communication Engineering,
Kalasalingam University, Anand Nagar, Krishnankoil 626126, Tamil Nadu, India
{perumal,m.p.raja}@klu.ac.in, arun.aklu@gmail.com

Abstract. A Fractal based Neural Network Radial Basis Function (FNNRBF) for image compression is proposed through this work. Generally, a large amount of data are required to represent digital images where the transmission and storage of such images are time consuming and unrealizable. Hence, image compression technique can be used to reduce the storage and transmission costs. In order to overcome the difficulties a Hybrid Fractal with NNRBF image compression techniques FNNRBF is proposed. The implementation of this technique shows the effectiveness in terms of compression of medical images. Also, a comparative synthesis is performed to prove that the proposed system is capable of compressing the images effectively in terms of Compression Ratio (CR), Peak Signal to Noise Ratio (PSNR) and memory space.

Keywords: Image compression · Hybrid · Fractal NNRBF and NNRBF

1 Introduction

Compression refers to the process of reducing the file size by rearranging the information in the file. Compressing the images is different from zipping the files. Image compression changes the system and content of the information within a file. Image compressions may be used to rearrange the data to achieve the desired compression level, depending on the preferred compression ratio. The Loss of the data may or may not be noticeable. The quantity of the image compression can be influenced by the type of the images. Higher compression ratio can be achieved in the portions of the image that have similar tones, such as water area that has the same shade. The Images acquired need to be stored or transmitted over long distances. An untreated image occupies more memory and hence it needs to be compressed. Due to the demand for high quality video on mobile platforms, there is a need to compress the untreated images and reproduce the images without any degradation. Lossy compression is used to compress the images and the video files. Lossless compression permits the genuine data to be perfectly reconstructed from the compressed data. The superfluous information is removed in compression and added during decompression. Almost every lossless compression program does two things in sequence, the first step generates a statistical framework for the input information and the second step uses

S. Arumugam et al. (Eds.): ICTCSDM 2016, LNCS 10398, pp. 251–261, 2017.
DOI: 10.1007/978-3-319-64419-6_33

this framework to map input information to bit sequences in such a way that the "probable" (e.g. frequently encountered) information will make shorter output than the "improbable" information.

2 Related Works

In the case of hybrid wavelet transform, the image size of 256256 is produced. It can be generated by utilizing two component transforms of size 88 and 3232 respectively, Larger the size of first component transform, more the global features of an image. Larger the sizes of second component transform, local features are focused more [9]. FNNRBF using hybrid fuzzy clustering approach which concerns the structure of clusters in multidimensional feature space. To attain this task, these papers use a transition scheme from fuzzy mode. This transition is carried out through analytical conditions that are extracted by the minimization of a specialized objective function [11]. In the Image compression problem, the original 256256 gray image is divided into overlapping non blocks of 44 pixels. For training the Neural Network (NN), the connection weights and the thresholds are initially set in the range [−1, 1] and the maximal epochs as 250 [13]. The learning rate of the back propagation is set as 0.001. Compression Ratio (CR), Signal to Noise Ratio (SNR) and Peak Signal to Noise Ratio (PSNR) are used to assess the quality of the reconstructed image [7]. The original meaning of the Fractal is the similarity between the local and global, that is, the portion of the image is the result through affine transformation of a whole image, but as mentioned earlier, the natural image of this local and the whole image is its self-similarity [15] Number of neurons in any neural network are represented by M-N-P where, M, N, P indicate the number of neurons in the input layer, hidden layer and output layer respectively. In case of a multi layered perception type feed forward neural network, the number of connections between any two layers are obtained by summing the number of bias connections. [16]. Fuzzy Logic has been used to develop a transfer function that modifies the input image into a new form that provides higher Compression Ratio (CR). The only possible solution to achieve increased CR is to decrease the contrast of the input image while keeping the contrast information of the original image unaltered [5]. Hybrid coding refers to the techniques which combines transform coding and spatial coding techniques. This technique combines the advantage of hardware simplicity of spatial coding and the good performance of transform coding with respect to its low sensitivity to channel error [3]. The KLT is the optimal linear orthogonal transform and provides higher decorrelation and energy compaction than wavelet transforms [4]. Fractal image compression is used in the compression of medical and color images. Fractal compression, a lossy method is a technique used for decomposing the images color separation, edge detection, and spectrum and texture analysis [14]. The Fractal image compression is based on the fractals of various images. The merits of converting the images to fractal data are (1) Reduced memory space requirement of the compressed image. (2) Quantification of parameters like CR, PSNR, Bits Per Pixel (BPP) and others [10]. An Adaptive Fractal Image Compression (AFIC) algorithm accelerates the encoding action and attains a higher CR, with little decline in the quality of the reproduced image. So, AIFC performs better than other methods [2]. This technique uses the universal approximation characteristics of generalized Neural Networks Radial Basis Function (NNRBF) to approximate the empirical kernel map associated to the Kernel based Principal Components Analysis (KPCA) or SV machines Support vector machines (SVM) [17]. The number of RBFs used to encode a sub image is much lower than the number of data points that result in reduction of data size [1]. RBF networks configure a neural network architecture that is extensively used for modeling and controlling nonlinear systems [6]. The goal of image compression is to minimize the time of uploading to the website and downloading from there on the internet and saving large amount of data [18]. There are two key steps in the machine-learning-based image compression model: choosing the most informative color pixels and learning to prefigure the color values. They are sculptured and solved with two standard machine learning methods: active learning and semi supervised learning respectively [8]. The unique high spectral resolution has been used in a broad range of scientific research such as terrain classification, agricultural monitoring, and military surveillance. Therefore, compression of hyper spectral image is necessary to facilitate the storage and transmission [3].

In the existing Fractal Algorithm, Compression Ratio (CR) is good but Mean Square Error (MSE) is high and Peak Signal to Noise Ratio (PSNR) value is low. But the proposed RBFNN takes small convergence time during the training period. New hybrid approach combining the working principles of Fractal and RBFNN is implemented here. The comparisons of existing algorithms are also exhibited.

3 Methodologies for Medical Image Compression

3.1 Fractal Algorithm

A fractal is a model which represents using similar patterns that occur in many different sizes. The term fractal was first used by Benoit Mandelbrot to describe the repeated patterns that he noticed occurring in many different structures. Fractal encoding is a mathematical technique used to encode the bitmaps considering an original image as a set of mathematical data that explains the fractal properties of the image.

3.2 Radial Basis Function Neural Network (RBFNN) for Image Compression

In ANN every neuron in a MLP (Multilayer Perception) holds a weighted sum of its input values. That is, every input value is multiplied by a coefficient and all the outcomes are summed up. A single MLP neuron is a plain linear classifier, but the difficult non-linear classifiers can be constructed by introducing these neurons into a network. RBFN method is more spontaneous than the MLP. To classify a fresh input, every neuron estimates the Euclidean distance between the model and the input. Figure 1 shows the general structure of NNRBF Algorithm. An input vector x is employed as input to all radial basis functions with different properties. Each RBF neuron compares to the input model, and outputs a value in range [0, 1] which measures the similarity. If the input is identical to the model, then the RBF neuron's output will be 1. As the distance between the model and input increases, the output falls off exponentially towards 0. RBF neuron's output resembles a bell curve. The output of the network is composed of a set of nodes. Every output node calculates a score for the linked class. The score is calculated by taking a weighted total of the activation values from each RBF neuron. By weighted total we mean that an output node links a weight value with every RBF neuron, and multiplies the neuron's activation by this weight before adding it to the total output.

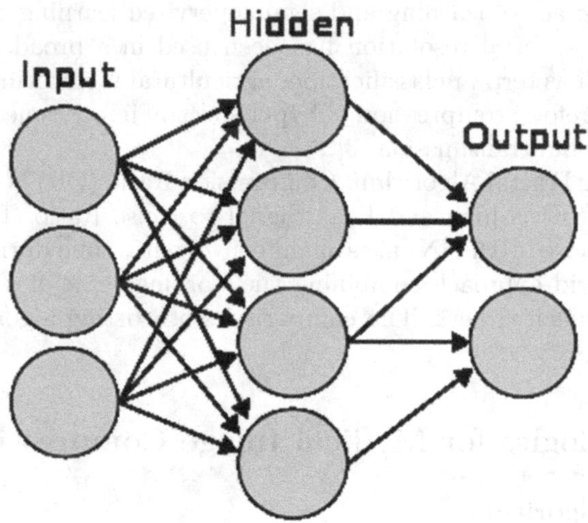

Fig. 1. General structure of Radial Basis Function Neural Network

3.3 Hybrid Image Compression

Figure 2 shows the proposed Hybrid image compression using FNNRBF. The NN-RBF algorithm is used to improve the transformation process, which increases the edge threshold. At the same time, the fractal coding and NN-RBF algorithm are combined to obtain hybrid FNNRBF coding, in order to get better quality in image compression.

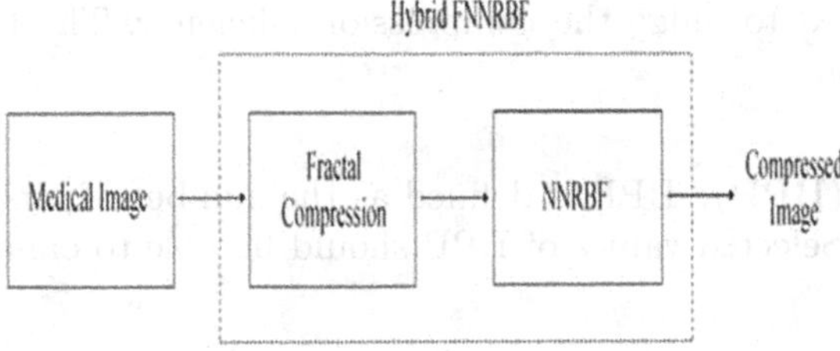

Fig. 2. Hybrid image compression using Fractal and Radial Basis Function Neural Network

4 Performance Parameters

Compression methods can be compared based on various parameters. Efficiency of Compression Algorithm (CA) is measured in terms of parameters such as Compression Ratio (CR), Peak Signal Noise Ratio (PSNR), Bits Per Pixel (BPP), Mean Square Error (MSE) and Testing and Training Time.

Mean Squared Error (MSE): Mean Squared Error is given by;

$$MSE = \frac{1}{m \times n} \sum_{i=0}^{m-1} \sum_{j=0}^{n-1} [I(i,j) - K(i,j)]^2 \tag{1}$$

Where I(i,j) is the original image, K(i, j) is the compressed image and the image is represented as an m n matrix. Low value of MSE implies low error values (PSNR).

Peak Signal to Noise Ratio (PSNR): Peak Signal to Noise Ratio is a measure of the peak error.

$$PSNR = 10 \log\{\frac{MAX_I^2}{MSE}\} \tag{2}$$

Here MAXI represents the maximum possible pixel value of the image. Pixels are represented as eight bits per sample. Logically, a greater value of PSNR is good because it indicates that the ratio of Signal to Noise is greater. Here, the 'signal' is the genuine image and the 'noise' is the error due to reconstruction.

Compression Ratio (CR): It is the ratio of the size of the genuine image to the size of the compressed image.

$$CR = \left(\frac{\text{Uncompressed file size}}{\text{Compressed file size}}\right) \tag{3}$$

CR can be used to judge the compression efficiency. The lower CR means better compression.

Bits Per Pixel (BPP): BPP is defined as the number of bits used to encode each pixel value. Selected values of BPP should be able to enhance the storage level [12].

5 Results

Images with the compression tables are also included. Figures 3, 4, 5 and 6 show the original and compressed images by using Fractal, and Hybrid Fractal with Neural Network Radial Basis Function. All the input images used in this work are of 512 × 512 image size.

5.1 Compression Ratio

Table 1 shows CR obtained using NNRBF, Fractal and Hybrid FNNRBF. CR is better with Hybrid FNNRBF.

Table 1. Compression results of multi modal medical images using three different algorithms

Images	Input memory size (KB)	Fractal	NNRBF	Hybrid Fractal & NNRBF
CT image 1	46.4	43.2	43.6	41.5
CT image 2	86.2	25	24.5	24.8
MR image 3	61.3	26.4	25.7	24.1
MR image 4	126	23	25.4	20.5
MR image 5	124	25.1	27.4	23.4
MR image 6	106	20.5	21.9	18.7
MR image 7	19.4	15.9	71.5	18.8
MR image 8	47.1	28.2	41.4	27.4
PET image 9	46.9	14.3	18.7	20.7

5.2 PSNR

Table 2 shows PSNR for NNRBF, Fractal and Hybrid FNNRBF. Here PSNR is higher with Hybrid FNNRBF.

Table 2. PSNR results of multi modal medical images using three different algorithms

Images	Fractal	NNRBF	Hybrid Fractal & NNRBF
CT image 1	35.8632	30.6855	34.7694
CT image 2	38.9335	19.5331	23.1454
MR image 3	41.796	22.9515	25.7061
MR image 4	39.049	22.3658	27.4178
MR image 5	42.3063	29.9071	36.3128
MR image 6	43.6769	24.6009	31.6895
MR image 7	41.3458	35.7803	32.3152
MR image 8	3.427	1.1379	22.5682
PET image 9	43.9157	24.4163	27.1644

5.3 Memory

Table 3 shows the Memory Usage of NNRBF, Fractal and Hybrid FNNRBF. Here the Compressed Image Size is much less in Hybrid FNNRBF.

Table 3. Memory usage of multi modal medical images results using three different algorithms

Images	Input memory size (KB)	Fractal	NNRBF	Hybrid Fractal & NNRBF
CT image 1	46.4	43.2	43.6	41.5
CT image 2	86.2	25	24.5	24.8
MR image 3	61.3	26.4	25.7	24.1
MR image 4	126	23	25.4	20.5
MR image 5	124	25.1	27.4	23.4
MR image 6	106	20.5	21.9	18.7
MR image 7	19.4	15.9	71.5	18.8
MR image 8	47.1	28.2	41.4	27.4
PET image 9	46.9	14.3	18.7	20.7

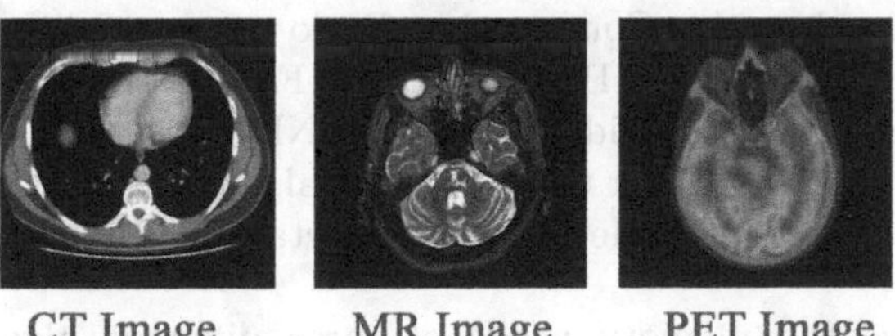

Fig. 3. Sample input images

Figure 3 show input (medical) images, Fig. 4 show compressed medical images obtained using Fractal, Fig. 5 show compressed medical images obtained using NNRBF and Fig. 6 show compressed medical images obtained using Hybrid Fractal and NNRBF algorithms.

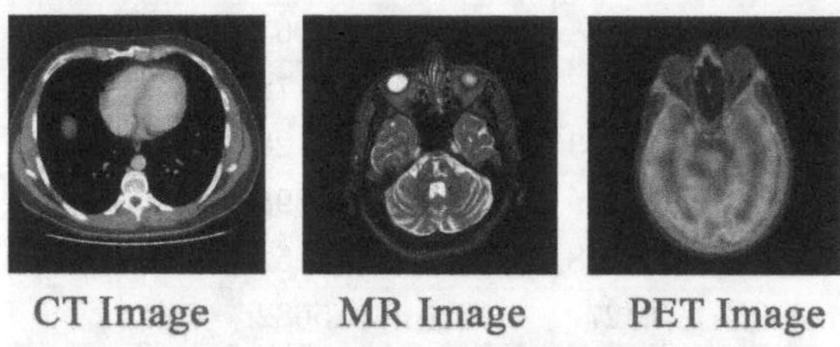

Fig. 4. Output obtained from Fractal

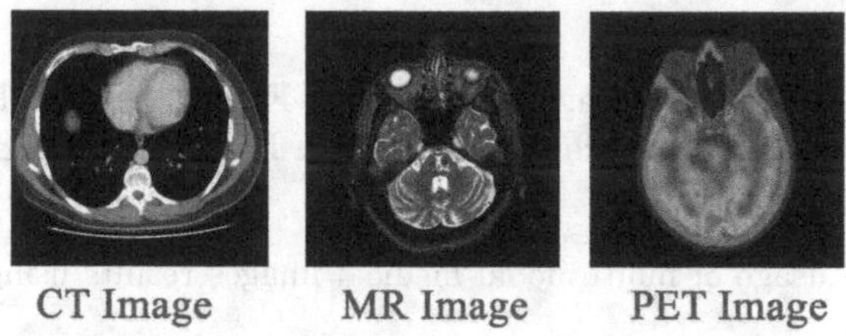

Fig. 5. Output obtained from NNRBF

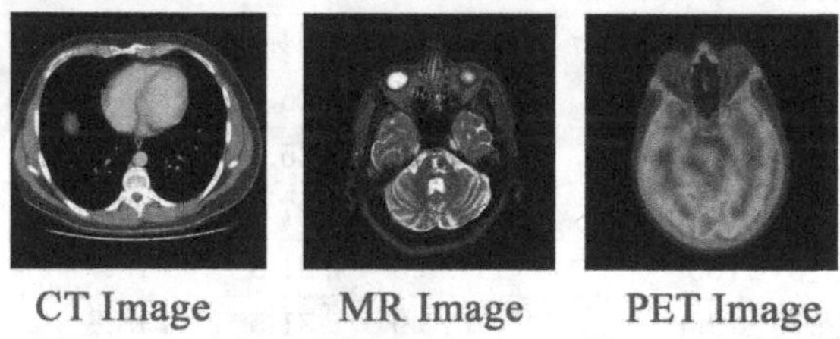

Fig. 6. Output obtained from Hybrid Fractal and NNRBF compression algorithms

Figure 7 shows that the Compression Ratio of three different algorithms namely Neural Network Radial Basis Function, Fractal and Hybrid Fractal and NNRBF. It is evident that Hybrid Fractal and NNRBF provide better CR values.

Figure 8 shows the PSNR for three different algorithms NNRBF, Fractal and Hybrid Fractal and NNRBF. Here Hybrid Fractal and NNRBF achieve better PSNR values.

Figure 9 shows the memory usage of three different algorithms namely, NNRBF, Fractal and Hybrid Fractal and NNRBF. It is noticed that Hybrid Fractal and NNRBF use less memory for image compression.

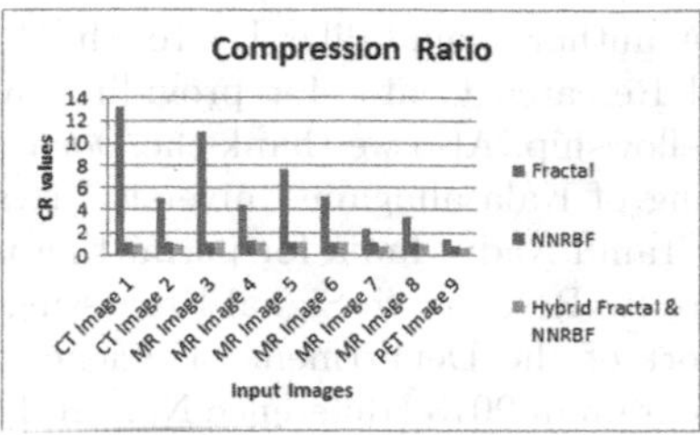

Fig. 7. Compression Ratio expressed in percentage

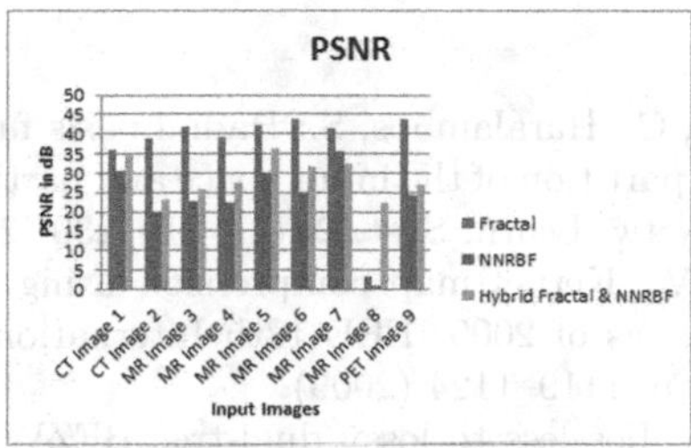

Fig. 8. PSNR expressed in decibel

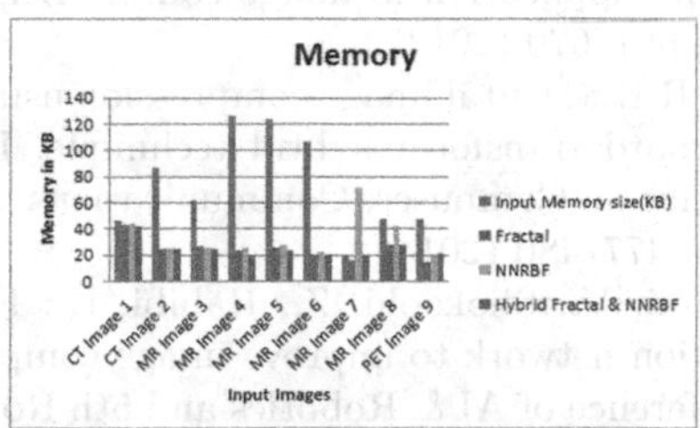

Fig. 9. Memory expressed in kilo byte

6 Conclusion and Future Work

In this paper, three different approaches such as Fractal, Radial Basis Function Neural Network and Hybrid Fractal & NNRBF are applied to medical image compression and the results compared. Here, MR and CT images are considered as Quality parameters such as Compression Ratio (CR), Peak Signal to Noise Ratio (PSNR), Execution time and Memory usage. It is observed that Hybrid approach has low CR and high PSNR values and is more efficient than Fractal and NNRBF methods.

Acknowledgement. The author would like [9] to thank the Sir.C.V. RAMAN KRISHNAN International Research Center for providing financial assistance under the University Research Fellowship. Also we thank the Department of Electronics and Communication Engineering of Kalasalingam University, (Kalasalingam Academy of Research and Education), Tamil Nadu, India for permitting to use the computational facilities available in Centre for Research in Signal Processing and VLSI Design which was setup with the support of the Department of Science and Technology (DST), New Delhi under FIST Program in 2013 (Reference No: SR/FST/ETI-336/2013 dated November 2013).

References

1. Alexandridis, A., Eva, C., Haralambos, S.: Radial basis function network training using a nonsymmetric partition of the input space and particle swarm optimization. IEEE Trans. Neural Netw. Learn. Syst. **24**(2), 219–230 (2013)
2. Arif, O., Patricio, A.V.: Kernel map compression using generalized radial basis functions. In: Proceedings of 2009 IEEE 12th International Conference on Computer Vision, Kyoto, pp. 1119–1124 (2009)
3. Cheng, K., Jeffrey, D.: Lossless to lossy dual-tree BEZW compression for hyperspectral images. IEEE Trans. Geosci. Remote Sens. **52**(9), 5765–5770 (2014)
4. Ding, J.J., Hsin-Hui, C., Wei-Yi, W.: Adaptive Golomb code for joint geometrically distributed data and its application in image coding. IEEE Trans. Circuits Syst. Video Technol. **23**(4), 661–670 (2013)
5. Ghrare, S.E., Ahmed, R.K.: Digital image compression using block truncation coding and Walsh Hadamard transform hybrid technique. In: Proceedings of 2014 International Conference on Computer, Communications, and Control Technology (I4CT), Langkawi, pp. 477–480 (2014)
6. Hormat, A.M., Rostami, V., Shokoohi, Z., Habibi, H.: Fuzzy modified forward-only counter propagation network to improve image compression. In: Proceedings of 2013 3rd Joint Conference of AI & Robotics and 5th RoboCup Iran Open International Symposium, Tehran, pp. 1–6 (2013)
7. Jiang, C., Shuxin, Y.: A hybrid image compression algorithm based on human visual system. In: IEEE International Conference on Computer Application and System Modeling (ICCASM), vol. 9, pp. 170–173 (2010)
8. Karayiannis, N.B., Mary, M.R.G.: On the construction and training of reformulated radial basis function neural networks. IEEE Trans. Neural Networks **14**(4), 835–846 (2003)
9. Kekre, H.B., TanujaSarode, P.: Image compression based on hybrid wavelet transform generated using orthogonal component transforms of different sizes. Int. J. Soft Comput. Eng. **3**(3), 53–57 (2013)
10. MohammedHasan, T., Xingqian, W.: An adaptive fractal image compression. IJCSI Int. J. Comput. Sci. **10**(2), 98–110 (2013)
11. Niros, A.D., Tsekouras, G.E.: A novel training algorithm for RBF neural network using a hybrid fuzzy clustering approach. Fuzzy Sets Syst. **193**, 62–84 (2012)
12. Perumal, B., Pallikonda Rajasekaran, M.: Efficient image compression techniques for compressing multimodal medical images using neural network radial basis function approach. Int. J. Imaging Syst. Technol. **25**(2), 115–122 (2015)
13. Rawat, C., Sukadev, M.: A hybrid image compression scheme using DCT and fractal image compression. Int. Arab J. Inf. Technol. **10**(6), 553–562 (2013)

14. Seeli, D.S., Jeyakumar, M.K.: A study on fractal image compression using soft computing techniques. IJCSI Int. J. Comput. Sci. **9**(6), 420–430 (2012)
15. Sridhar, S., Kumar, P.R., Ramanaiah, K.V., Nataraj, D.: Coiflets, artificial neural networks and predictive coding based hybrid image compression methodology. In: 2014 2nd International Conference on Devices, Circuits and Systems (ICDCS), Combiatore, pp. 1–6 (2014)
16. Thakur, S., Nilesh, K.D., Kavita, T.: A highly efficient gray image compression codec using neuro fuzzy based soft hybrid JPEG standard. In: Proceedings of Second International Conference, Emerging Research in Computing, Information, Communication and Applications, vol. 1, pp. 625–631 (2014)
17. Vascan, O., Ionel-Bujorel, P.: Image compression using radial basis function networks. Telecommun. Control **5**, 195–198 (2013)
18. Zhang, C., Xiaofei, H.: Image compression by learning to minimize the total error. IEEE Trans. Circuits Syst. Video Technol. **23**(4), 565–576 (2013)

The Median Problem on Symmetric Bipartite Graphs

K. Pravas[1,2] (✉) and A. Vijayakumar[1]

[1] Cochin University of Science and Technology, Cochin 682022, India
{pravask,vambat}@gmail.com
[2] K.K.T.M. Govt. College, Pullut, Thrissur 680 663, India
http://www.cusat.ac.in

Abstract. In a connected graph G, the status of a vertex is the sum of the distances of that vertex to each of the other vertices in G. The subgraph induced by the vertices of minimum (maximum) status in G is called the median (anti-median) of G. A bipartite graph G is symmetric if for a bi-partition (X, Y) of G, there is a map f from X onto Y such that if $(u, f(v)) \in E(G)$, then $(v, f(u)) \in E(G)$, where $u, v \in X$. In this paper we show, by construction, that any symmetric bipartite graph is a median (anti-median, center) of another symmetric bipartite graph. We also obtain results on median and anti-median problem on square graphs of bi-partite graphs with equal partitions.

Keywords: Distance · Median · Anti-median · Symmetric bipartite

1 Introduction

Let $G = (V, E)$ be a graph on n vertices with vertex set V and edge set E. A graph is bipartite if its vertex set can be partitioned into two nonempty subsets X and Y such that each edge of G has one end in X and the other in Y. A bipartite graph G is symmetric if for a bi-partition (X, Y) of G, there is a map f from X onto Y such that for every edge $(u, f(v))$ in G, there is an edge $(v, f(u))$ in G, where $u, v \in X$. Such a partition $(X, Y)_f$ is called a symmetric bi-partition of G. The ladder graph $L_n = \{x_i, y_i\}_{i=1}^{n}$ on $2n$ vertices is obtained by taking two paths $x_1, \ldots, x_n$ and $y_1, \ldots, y_n$ on n vertices and making x_i and y_i adjacent for each $i = 1, \ldots, n$. The square G^2 of a graph G has the same vertex set as G and two vertices $u, v \in V(G^2)$ are adjacent if $d_G(u, v) \leq 2$.

The degree of a vertex v, $d(v)$, is the number vertices adjacent to v and by $N(v)$ we denote the neighbor set of v. When H is a subgraph of G and $u, v \in V(H)$, the set of common neighbors of u and v in H is denoted by $N_H^*(u, v)$.

The distance between two vertices u and v is the number of edges on a shortest path between u and v, and it is denoted by $d(u, v)$. The eccentricity of u is $e(u) = \max_v d(u, v)$. The center $C(G)$ of a graph G is the subgraph of G induced by the vertices of minimum eccentricity. The status of a vertex $v \in V(G)$,

S. Arumugam et al. (Eds.): ICTCSDM 2016, LNCS 10398, pp. 262–270, 2017.
DOI: 10.1007/978-3-319-64419-6_34

denoted by $S(v)$, is the sum of the distances from v to all other vertices in G. The subgraph induced by the vertices of minimum (maximum) status in G is known as the median (anti-median) of G, denoted by $M(G)$ ($AM(G)$). The status difference [5] in a graph G is $\mathrm{SD}(G) = \max_{u,v \in V(G)} (S(u) - S(v))$.

Given a graph G the problem of finding a graph H such that $M(H) \cong G$ is referred to as the median problem. In [8], it is shown that any graph $G = (V, E)$ is the median of some connected graph. In [3] the notion of anti-median of a graph was introduced and proved that every graph is the anti-median graph of some graph. The problem of simultaneous embedding of median and anti-median is discussed in [1]. Another construction, which generalises all the previously mentioned constructions, can be seen in [6].

The median vertices represent facility locations with minimum average distance. In network theory the median problem is significant as it is related to the optimization problems involving the placement of network servers, the core of the entire networks, specially in very large interconnection networks. However, the median constructions for general graphs cannot be directly applied to many networks as their underlying graph belong to different classes of graphs. Most of the analysis in network communities are done using preference networks [4], which are modelled using bipartite graphs. The study of the median operators for some classes of graphs is in [5,7,9].

In this paper we show, by construction, that any symmetric bipartite graph is a median (or anti-median) of another symmetric bipartite graph. In addition, we provide constructions to embed another symmetric graph as center in both the constructions. Together with the additional properties of these constructions, we show that these constructions can be extended to any arbitrary graph and hence we prove some general results. As another application of these constructions, we provide the results on median and anti-median problem on square graphs of bi-partite graphs with equal partitions.

2 Median Problem on Symmetric Bipartite Graphs

Consider a ladder graph $L_n = \{x_i, y_i\}_{i=1}^n$. Let $X_L = \{x_i : i \text{ is odd}\} \cup \{y_i : i \text{ is even}\}$ and $Y_L = L_n \setminus X_L$, then $(X_L, Y_L)_f$ is a symmetric bi-partition of L_n, where $f(x_i) = y_i$, when i is odd, and $f(y_i) = x_i$, when i is even (Fig. 1).

Lemma 1. *Given a symmetric bipartite graph G, there exists a connected symmetric bipartite graph G' such that G is an induced subgraph of G' and all the vertices of G in G' have equal status in G'.*

Proof. Let $(X, Y)_f$ be a symmetric bi-partition of G. Let X', Y' be the copy of X, Y such that v' denote the copy of a vertex $v \in V(G)$. Consider two new vertices v_x and v_y. Let $A = X \cup X' \cup \{v_x\}$ and $B = Y \cup Y' \cup \{v_y\}$. Define a map g from A to B such that $g(v) = f(v)$, $g(v') = f(v)'$, $\forall v \in X$ and $g(v_x) = v_y$.

Then, make v_y adjacent to all the vertices in A and v_x adjacent to all vertices of B. Also, for each $v \in X$ (Y) make v' adjacent to $Y \backslash N(v) \cup \{g(v)\}$ ($X \backslash N(v) \cup$

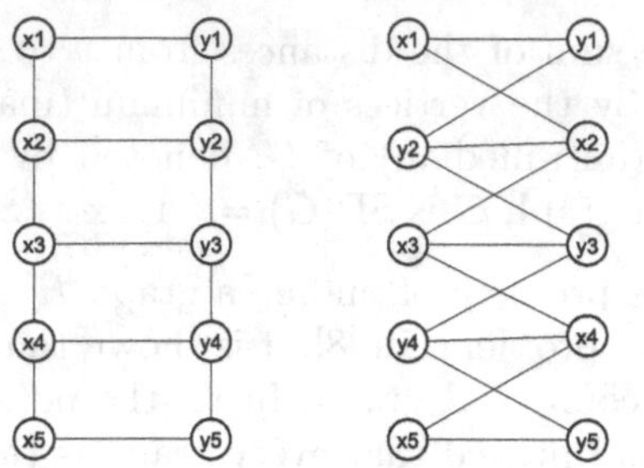

Fig. 1. An L_5 and a symmetric bi-partite representation of L_5.

$\{g^{-1}(v)\}$). Call this graph G'. It now follows that $(A, B)_g$ is a symmetric bi-partition of G' and $S_{G'}(v) = 4n + 1$, for all $v \in V(G)$.

The graph G' is called the symmetric bipartite gadget graph of G.

Theorem 1. *Given two symmetric bipartite graphs G and J there exists a symmetric bipartite graph H such that $M(H) \cong G$ and $C(H) \cong J$.*

Proof. The proof is by construction. Let G' be the symmetric bipartite gadget graph of G with symmetric bi-partition $(A, B)_f$ and $(R, S)_g$ be a symmetric bi-partition of J. For $k \geq 3$, introduce two ladder graphs $\{x_i, y_i\}_{i=1}^{k-1}$ and $\{u_i, v_i\}_{i=1}^{k+1}$ with symmetric bi-partitions $(X_1, Y_1)_{f_1}$ and $(X_2, Y_2)_{f_2}$ respectively.

Make x_1 adjacent to $X \cup \{v_x\}$, y_1 to $Y \cup \{v_y\}$, x_{k-1} to R, y_{k-1} to S, u_1 to R and v_1 to S. Denote this graph by H_0. Introduce s copies of K_2 and let a_ib_i, $i = 1, \ldots, s$ be the edges in sK_2. Make $\{a_i\}_{i=1}^{s}$ adjacent to all the vertices in X and $\{b_i\}_1^s$ adjacent to all the vertices in Y. Denote this new graph by H. Clearly $C(H) \cong J$ with $e(v) = k + 2$, for all $v \in V(J)$ and $S(x) = S(y) = 4n + 1 + (2k + 1)(2k + 2 + |R|) + 3s$, for all $x \in X$, $y \in Y$.

For a vertex $u \in V(H)$, let $S^*(u) = d(u, a_m) + d(u, b_m)$, where a_mb_m be an edge in the s copies of K_2 in H. Then, $S^*(u) = 3$, $u \in V(G)$ and $S^*(u) \geq 5$, $u \in V(H \backslash G) \backslash \{a_m, b_m\}$. Hence $M(H) = G$, when $s > \mathrm{SD}(H_0)/2$.

When k is even, let $A' = A \cup X_1 \cup X_2 \cup R \cup \{b_i\}$ and $B' = H \setminus A'$. Let h be the function defined on A' by $h(x) = f(x)$, when $x \in A$, $h(x) = g(x)$, when $x \in R$, $h(x) = f_i(x)$, when $x \in X_i$, $i = 1, 2$, and $h(b_i) = a_i$, $1 \leq i \leq s$. It is clear that $(A', B')_h$ is a symmetric bi-partition of H. When k is odd, the elements in R and S are interchanged in the bi-partition (A', B'). Re-defining $h(x) = g^{-1}(x)$, for the vertices $x \in S$, $(A', B')_h$ becomes a symmetric bi-partition of H (Fig. 2).

Lemma 2. *Given a symmetric bi-partite graph G, there exists a symmetric bipartite graph H such that $AM(H) = G$.*

Proof. Let G' be the symmetric bipartite gadget graph of G and let $(A, B)_f$ be a symmetric bi-partition of G'. Introduce a complete bipartite graph $K_{r,r}$ with symmetric bi-partition $(C, D)_g$. Make each vertex in C adjacent to $Y' \cup \{v_y\}$

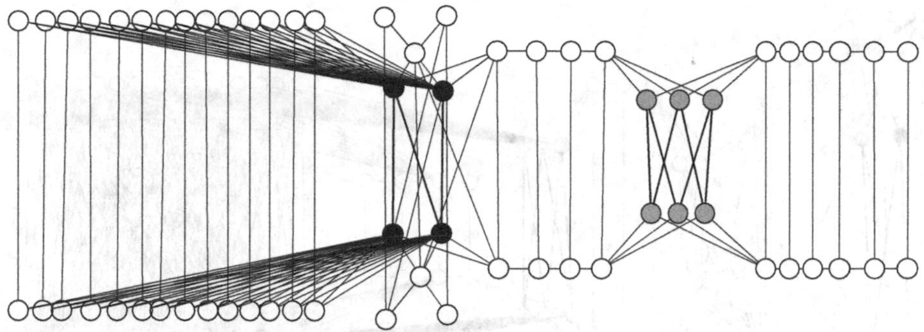

Fig. 2. A construction of H as in Theorem 1. Here the vertices of G are colored black and the vertices of J are colored grey.

and each vertex in D adjacent to $X' \cup \{v_x\}$. Call this graph H. We can see that $S_H(u) = S_{G'}(u) + 5r$, when $u \in V(G)$ and $S_H(u) = S_{G'}(u) + 3r$, when $u \in V(H \setminus G)$. Choosing $r > SD(G')/2$, we get $AM(H) = G$.

Let $P = A \cup C$ and $Q = H \setminus P$. Define h on P by $h(x) = g(x)$, when $x \in A$, and $h(x) = g(x)$, when $x \in C$. Then, $(P, Q)_h$ is a symmetric bi-partition of H.

Theorem 2. *Given two symmetric bipartite graphs G and J there exists a symmetric bipartite graph H such that $AM(H) \cong G$ and $C(H) \cong J$.*

Proof. The proof is by construction. Let H be the graph constructed in Lemma 2 with symmetric bi-partition $(P, Q)_h$ and let $(R, S)_g$ be a symmetric bi-partition of J. For $k \geq 3$, introduce two ladder graphs $\{x_i, y_i\}_{i=1}^{k-1}$ and $\{u_i, v_i\}_{i=1}^{k+1}$ with symmetric bi-partitions $(X_1, Y_1)_{f_1}$ and $(X_2, Y_2)_{f_2}$ respectively.

Make x_1 adjacent to P, y_1 to Q, x_2 to D and y_2 to C, x_{k-1} to R, y_{k-1} to S, u_1 to R, v_1 to S. Denote this graph by H_0.

Introduce a complete bipartite graph $K_{s,s}$ with symmetric bi-partition $(E, F)_f$. Make each vertex in E adjacent to $R \cup \{u_2\}$ and each vertex in F adjacent to $S \cup \{v_2\}$. Call this graph H. Clearly $C(H) \cong J$ with $e(v) = k + 2$, for all $v \in V(J)$ and $S(x) = S(y) = 4n + 5r + (2k + 1)(2k + R + s) + 2s$.

For a vertex $u \in V(H)$, let $S^*(u) = d(u, e) + d(u, f)$, where ef is an edge in $K_{s,s}$. Then, $S^*(u) = 13$, $u \in V(G')$ and $S^*(u) \leq 11$, $u \in V(H \setminus G') \setminus \{e, f\}$. Since ef is an arbitrary edge in $K_{s,s}$ and by Lemma 2, $S_{G'}(x) < S_{G'}(y)$, for every $x \in V(G), y \in V(G' \setminus G)$, for $r > SD(H_0)/2$, it follows that $AM(H) = G$.

When k is even, let $A' = P \cup X_1 \cup X_2 \cup R \cup E$ and $B' = H \setminus A'$. Let t be the function defined on A' by $t(x) = h(x)$, when $x \in P$, $t(x) = g(x)$, when $x \in R$, $t(x) = f_i(x)$, when $x \in X_i$, $i = 1, 2$, and $t(x) = f(x)$, when $x \in E$. It is clear that $(A', B')_h$ is a symmetric bi-partition of H. When k is odd, the elements in R and S are interchanged in the bi-partition (A', B'). Re-defining $h(x) = g^{-1}(x)$, for the vertices $x \in S$, $(A', B')_h$ becomes a symmetric bi-partition of H (Fig. 3).

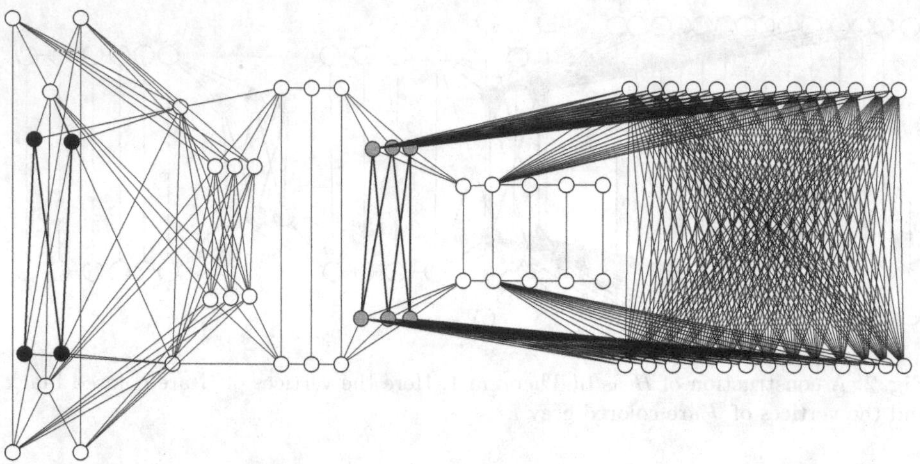

Fig. 3. A construction of H as in Theorem 2. Here the vertices of G are colored black and the vertices of J are colored grey.

3 Bipartite Graph of a Graph

The bipartite graph $B(G)$ of a graph G can be constructed as follows [2]. For each vertex $v \in V$, form $v' \in X$ and $v'' \in Y$ and let $N(v') = \{u'' \in Y : u \in N[v]\}$ and $N(v'') = \{u' \in X : u \in N[v]\}$.

Lemma 3. *Let G be a connected symmetric bipartite graph. Then $G \cong B(H)$ for some graph H if and only if there is a symmetric bi-partition $(X, Y)_f$ of G such that $uf(u)$ is an edge for all $u \in X$.*

Remark 1. Consider the graphs G and $B(G)$. Let $u, v \in V(G)$. If $d(u, v)$ is odd, then $d(u', v'') = d(u'', v') = d(u, v)$ and $d(u', v') = d(u'', v'') = d(u, v)+1$. Also, if $d(u, v)$ is even, then $d(u', v'') = d(u'', v') = d(u, v)+1$ and $d(u', v') = d(u'', v'') = d(u, v)$.

In the following theorem we prove that, for a connected graph, the operator $B(\cdot)$ commute with both $M(\cdot)$ and $AM(\cdot)$.

Theorem 3. *For any connected graph G, $B(M(G)) \cong M(B(G))$ and $B(AM(G)) \cong AM(B(G))$.*

Proof. Let $H = B(G)$. For a vertex $v \in V(G)$, let O_v and E_v be the set of vertices respectively at odd distance and even distance from v. By Remark 1,

$$\sum_{u \in O_v} d_H(v', u'') = \sum_{u \in O_v} d_G(v, u)$$
$$\sum_{u \in O_v} d_H(v', u') = \sum_{u \in O_v} d_G(v, u) + |O_v|$$
$$\sum_{u \in E_v} d_H(v', u') = \sum_{u \in E_v} d_G(v, u)$$
$$\sum_{u \in E_v} d_H(v', u'') = \sum_{u \in E_v} d_G(v, u) + |E_v|.$$

Thus $S_H(v') = 2S_G(v) + n$ and similarly $S_H(v'') = 2S_G(v) + n$.

Now, for each vertex v of a graph G, the status of the vertices v' and v'' in $B(G)$ depends only on $S_G(v)$ so that the analogous median properties are preserved. Hence $M(B(G)) \cong B(M(G))$ and $AM(B(G)) \cong B(AM(G))$.

Corollary 1. *For any connected graph G, $B(\cdot)$ commute with $C(\cdot)$.*

Corollary 2. *Let $G' \cong B(G)$ and $J' \cong B(J)$ be two connected graphs. Then the following results hold.*

1. *There exist graphs H_1 and H_1' such that $M(H_1') = G'$ and $C(H_1') = J'$ and $H_1' \cong B(H_1)$.*
2. *There exist graphs H_2 and H_2' such that $AM(H_2') = G'$ and $C(H_2') = J'$ and $H_2' \cong B(H_2)$.*

Proof. From Theorems 1 and 2, we can see that all the symmetric bipartite graphs introduced in these constructions satisfy the conditions of Lemma 3. Hence, starting with symmetric bipartite graphs G' and J' which are also bipartite graphs of some graphs, we obtain H_1' and H_2' satisfying the required conditions in the assertion.

We now show that a general solution of median and anti-median problems can be obtained from the results on symmetric bi-partite graphs.

Theorem 4. *Let G and J be two connected graphs. Then,*

1. *There exist a graph H_1 such that $M(H_1) \cong G$ and $C(H_1) \cong J$.*
2. *There exist a graph H_2 such that $AM(H_2) \cong G$ and $C(H_2) \cong J$.*

Proof. 1. Let G' and J' are the graphs such that $B(G) = G'$ and $B(J) = J'$. From Corollary 2, we can see that there exists a graph H_1 such that $H_1' \cong B(H_1)$ with $M(H_1') = G'$ and $C(H_1') = J'$. Using Theorem 3, $M(H_1) = B^{-1}BM(H_1) = B^{-1}MB(H_1) = B^{-1}M(H_1') = B^{-1}(G') = G$. See Fig. 4 for an illustration.

2. The proof can be obtained using the similar arguments as in (1).

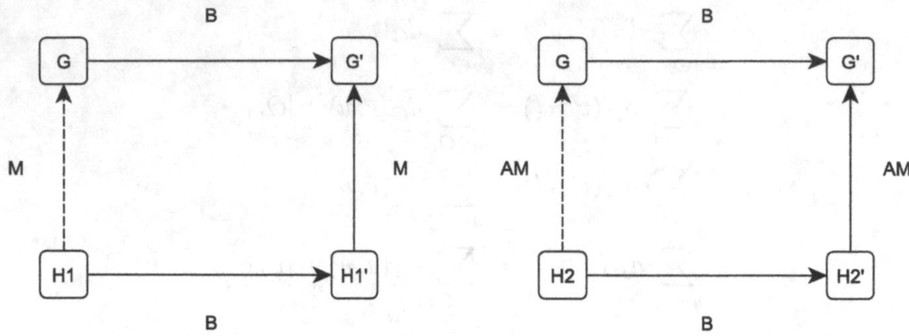

Fig. 4. Illustration of Theorem 4.

4 The Median Problem on Square of Bipartite Graphs

Lemma 4. *For a vertex $u \in V(G)$, $S_{G^2}(u) = \frac{1}{2}(S_G(u) + |O_u|)$, where $|O_u|$ is the number of vertices at odd distance from u in G.*

Proof. For a vertex $u \in V(G)$, let O_u be the set of all vertices at odd distance from u. Then

$$\begin{aligned} S_{G^2}(u) &= \sum_{v \in O_u} d_{G^2}(u,v) + \sum_{v \notin O_u} d_{G^2}(u,v) \\ &= \sum_{v \in O_u} \frac{1}{2} d_G(u,v) + \sum_{v \notin O_u} \frac{1}{2}(d_G(u,v)+1) \\ &= \frac{1}{2}(S_G(u) + |O_u|) \end{aligned}$$

Remark 2. If $|O_u|$ is a constant for all the vertices in $V(G)$, then it is immediate that the median set of G and G^2 are the same.

Definition 1. *A subgraph H of G is a square-subgraph of G if $H^2 \cong G^2[V(H)]$.*

Not all subgraphs of a graph are square-subgraphs. For, P_4 is not a square-subgraph of C_5 since $P_4^2 \cong K_4 - e$ is not induced in $C_5^2 \cong K_5$. The following result characterises square-subgraphs of graphs (Fig. 5).

Lemma 5. *H is square-subgraph of G if and only if for every non-adjacent vertices $u, v \in V(H)$ with $d_G(u,v) = 2$, $N_H^*(u,v) \neq \emptyset$.*

Proof. Let H be a subgraph of G. Then, it is clear that $E(H^2) \subseteq E(G^2[V(H)])$. Let u, v be two non-adjacent vertices of H such that $d_G(u,v) = 2$. That is, $uv \in E(G^2[V(H)])$. Then H is square-subgraph of G if and only if uv is an edge in H^2 if and only if $d_H(u,v) = 2$ if and only if $N_H^*(u,v) \neq \emptyset$.

Theorem 5. *Let G be a graph such that $|O_u|$ is a constant for all $u \in V(G)$. If $M(G)$ and $AM(G)$ are square-subgraphs of G, then $M(G^2) = (M(G))^2$ and $AM(G^2) = (AM(G))^2$.*

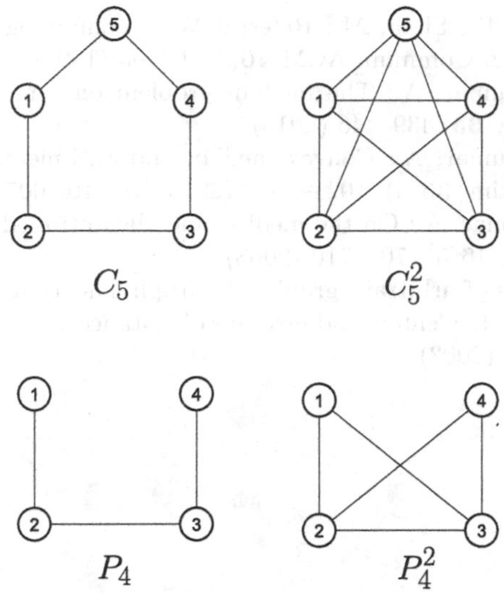

Fig. 5. Illustration of Definition 1.

Proof. Since $M(G)$ is a square-subgraph of G, $M(G)^2$ is induced in G^2. By Remark 2, the median sets of G and G^2 are the same.

Remark 3. We can see that the earlier constructions on symmetric bi-partite graphs are also valid for bipartite graphs with bi-partition (X, Y) and $|X| = |Y|$. Hence the following results holds.

Corollary 3. *Let G be a bipartite graph with bi-partition (X, Y) and $|X| = |Y|$, then there are bipartite graphs H_1 and H_2 such that Median set of H_1^2 is G^2 and Anti-median set of H_2^2 is G^2.*

Proof. The proof is by construction. By Remark 3, we apply the construction in Theorem 1 for symmetric bipartite graphs to obtain a graph H_1 such that $M(H_1) = G$. It is clear by the construction that H_1 is bi-partite with bi-partition (X', Y') and $|X'| = |Y'|$. Now by Remark 2, Median set of H_1^2 is G^2. Similarly using the construction in Theorem 2, the second part of the assertion also follows.

References

1. Balakrishnan, K., Brešar, B., Changat, M., Klavžar, S., Kovše, M., Subhamathi, A.R.: Simultaneous embeddings of graphs as median and antimedian subgraphs. Networks **56**, 90–94 (2010)
2. Balakrishnan, R., Ranganathan, K.: A Textbook of Graph Theory, 2nd edn. Springer, Heidelberg (2012). doi:10.1007/978-1-4614-4529-6
3. Bielak, H., Syslo, M.M.: Peripheral vertices in graphs. Studia Sci. Math. Hungar. **18**, 269–275 (1983)

4. Kautz, H., Selman, B., Shah, M.: Referral Web: combining social networks and collaborative filtering. Commun. ACM **40**(3), 63–65 (1997)
5. Pravas, K., Vijayakumar, A.: The median problem on k-partite graphs. Discuss. Math. Graph Theory **35**, 439–446 (2015)
6. Pravas, K., Vijayakumar, A.: Convex median and anti-median at prescribed distance. J. Comb. Optim. **33**(3), 1021–1029 (2016). doi:10.1007/s10878-016-0022-z
7. Rao, S.B., Vijayakumar, A.: On the median and the anti-median of a cograph. Int. J. Pure Appl. Math. **46**(5), 703–710 (2008)
8. Slater, P.J.: Medians of arbitrary graphs. J. Graph Theory **4**, 389–392 (1980)
9. Yeh, H.G., Chang, G.J.: Centers and medians of distance-hereditary graphs. Discrete Math. **265**, 297–310 (2003)

Intuitionistic Fuzzy Automaton with Unique Membership and Unique Nonmembership Transitions

K. Jency Priya(✉) and T. Rajaretnam

P.G. and Research Department of Mathematics,
St. Joseph's College (Autonomous), Tiruchirappalli 620002, Tamil Nadu, India
jencypriya9@gmail.com, T_rajaretnam@yahoo.com

Abstract. In this paper, we review an intuitionistic fuzzy finite state automaton which assigns a membership and nonmembership values in which there is a unique membership and unique nonmembership transition on an input symbol (IFAUMN) and also prove that there exists a complete IFAUMN for a given incomplete IFAUMN for the same fuzzy language.

Keywords: Intuitionistic · Complete

1 Introduction

The concept of a fuzzy set was first initiated by Zadeh [14,15]. Fuzzy set theory has been shown to be a useful tool to describe situations in which the data are imprecise or vague. Fuzzy sets handle such situations by attributing and degree to which a certain object belongs to a set. The fuzzy algebraic structures play a prominent role in mathematics with wide applications in many other branches such as theoretical Physics, Computer Science, Coding Theory, Topological Spaces, Logic, Set Theory, etc., In view of a fuzzy finite state automaton there may be more than one fuzzy state transition from a state on an input symbol with a given membership value given by Santos, Wee and Fu [11,13]. This development was followed by the postulation called deterministic fuzzy finite state automaton as in Malik and Mordeson [9], in which there can be atmost one fuzzy transition on a input symbol, which can be constructed equivalently from a fuzzy finite state automaton. However, it only acts as a deterministic fuzzy recognizer and a fuzzy regular languages accepted by the fuzzy finite state automaton and deterministic fuzzy finite automata need not necessarily be equal. (i.e., the degree of a string need not be same). Rajaretnam and Ayyaswamy [10] introduced fuzzy finite state automaton with unique membership transition on an input symbol, one kind of determinism of a given fuzzy finite automaton in which the membership value of any recognized string in both the system are the same.

Atanassov [1–5] initiated the concept of an intuitionistic fuzzy set (IFS). An Atanassov intuitionistic fuzzy set is considered as a generalization of fuzzy set

S. Arumugam et al. (Eds.): ICTCSDM 2016, LNCS 10398, pp. 271–280, 2017.
DOI: 10.1007/978-3-319-64419-6_35

and has been found to be useful to deal with vagueness. In the sense of Atanassov an IFS is characterized by a pair of functions valued in [0, 1], the membership function and the nonmembership function. Using the notions of intuitionistic fuzzy sets, Jun [6–8] introduced the concept of intuitionistic fuzzy finite state machines. In intuitionistic fuzzy automata with unique membership transition was introduced to reduce the length of uncertainty. We discussed Intuitionistic fuzzy finite state automata with unique membership transition on an input symbol (IFAUM) in [12]. In this paper, authors consider an intuitionistic fuzzy finite automaton with unique membership and unique nonmembership transition on an input symbol and also prove that there exists a complete IFAUMN for a given incomplete IFAUMN for the same fuzzy language.

1.1 Basic Definitions

Definition 1. *An Intuitionistic fuzzy sets (IFS) A^* in a nonempty set Σ is an object having the form $A^* = \{(x, \mu_A(x), \nu_A(x)) \mid x \in \Sigma\}$, where the functions $\mu_A : \Sigma \to [0,1]$ and $\nu_A : \Sigma \to [0,1]$ denote the degree of membership and nonmembership of each element $x \in \Sigma$ to the set A respectively, and $0 \leq \mu_A(x) + \nu_A(x) \leq 1$ for each $x \in \Sigma$. For the sake of simplicity, we use the notation $A^* = (\mu_A, \nu_A)$ instead of $A = \{(x, \mu_A(x), \nu_A(x)) \mid x \in \Sigma\}$.*

Definition 2. *An Intuitionistic fuzzy finite automaton (IFAUM) is an ordered 5-tuple $\mathscr{A} = (Q, \Sigma, A, i, f)$, where*

1. *Q is a finite nonempty set of states.*
2. *Σ is a finite nonempty set of input symbols.*
3. *$A = (\mu_A, \nu_A)$, each is an intuitionistic fuzzy subset of $Q \times \Sigma \times Q$.*
   (a) *the fuzzy subset $\mu_A : Q \times \Sigma \times Q \to [0,1]$ denotes the degree of membership such that $\mu_A(p, a, q) = \mu_A(p, a, q')$ for some $q, q' \in Q$ then $q = q'$.*
   (b) *$\nu_A : Q \times \Sigma \times Q \to [0,1]$ denotes the degree of nonmembership is a fuzzy subset of Q.*
4. *$i = (i_{\mu_A}, i_{\nu_A})$, each is an intuitionistic fuzzy subset of Q, i.e., $i_{\mu_A} : Q \to [0,1]$ and $i_{\nu_A} : Q \to [0,1]$ called the intuitionistic fuzzy subset of initial states.*
5. *$f = (f_{\mu_A}, f_{\nu_A})$, each is an intuitionistic fuzzy subset of Q, i.e., $f_{\mu_A} : Q \to [0,1]$ and $f_{\nu_A} : Q \to [0,1]$ called the intuitionistic fuzzy subset of final states.*

Definition 3. *Let $\mathscr{A} = (Q, \Sigma, A, i, f)$ be an IFAUM. Then the fuzzy behavior of IFAUM is $L_{\mathscr{A}} = (L_{\mu_{\mathscr{A}}}, L_{\nu_{\mathscr{A}}})$.*

Definition 4. *Let $\mathscr{A} = (Q, \Sigma, A, i, f)$ be an IFAUM. Define an IFS $A^* = (\mu_A^*, \nu_A^*)$ in $Q \times \Sigma^* \times Q$ as follows: $\forall p, q \in Q, x \in \Sigma^*, a \in \Sigma$.*

$$\mu_A^*(p, \lambda, q) = \begin{cases} 1, \text{ if } p = q \\ 0, \text{ if } p \neq q \end{cases}$$

$$\nu_A^*(p,\lambda,q) = \begin{cases} 0, \text{ if } p = q \\ 1, \text{ if } p \neq q \end{cases}$$

$$\mu_A^*(p,xa,q) = \vee\{\mu_A^*(p,x,r) \wedge \mu_A(r,a,q) | r \in Q\}$$

$$\nu_A^*(p,xa,q) = \wedge\{\nu_A^*(p,x,r) \vee \nu_A(r,a,q) | r \in Q\}$$

Definition 5. *Let* $\mathscr{A} = (Q, \Sigma, A, i, f)$ *be an IFAUM and* $x \in \Sigma^*$. *Then* x *is recognized by* $\mathscr{A}$ *if* $\vee\{i_{\mu_A}(p) \wedge \mu_A^*(p,x,q) \wedge f_{\mu_A}(q) \mid p,q \in Q\} > 0$ *and* $\wedge\{i_{\nu_A}(p) \vee \nu_A^*(p,x,q) \vee f_{\nu_A}(q) \mid p,q \in Q\} < 1$.

2 Unique Membership and Unique Non Membership Transitions

Definition 6. *An Intuitionistic fuzzy finite automaton with unique membership and unique nonmembership transition on an input symbol is an ordered 5-tuple (IFAUMN)* $\mathscr{A} = (Q, \Sigma, A, i, f)$, *where [(i)]*

1. Q *is a finite nonempty set of states.*
2. Σ *is a finite nonempty set of input symbols.*
3. $A = (\mu_A, \nu_A)$, *each is an intuitionistic fuzzy subset of* $Q \times \Sigma \times Q$.
4. *the fuzzy subset* $\mu_A : Q \times \Sigma \times Q \to [0,1]$ *and* $\nu_A : Q \times \Sigma \times Q \to [0,1]$ *denotes the degree of membership and degree of nonmembership such that* $\mu_A(p,a,q) = \mu_A(p,a,q')$ *and* $\nu_A(p,a,q) = \nu_A(p,a,q')$*for some* $q, q' \in Q$ *then* $q = q'$.
5. $i = (i_{\mu_A}, i_{\nu_A})$, *each is an intuitionistic fuzzy subset of* Q, *i.e.,* $i_{\mu_A} : Q \to [0,1]$ *and* $i_{\nu_A} : Q \to [0,1]$ *called the intuitionistic fuzzy subset of initial states.*
6. $f = (f_{\mu_A}, f_{\nu_A})$, *each is an intuitionistic fuzzy subset of* Q, *i.e.,* $f_{\mu_A} : Q \to [0,1]$ *and* $f_{\nu_A} : Q \to [0,1]$ *called the intuitionistic fuzzy subset of final states.*

Theorem 1. *Let* $\mathscr{A} = (Q, \Sigma, A, i, f)$ *be an intuitionistic fuzzy automata IFA and* $L_{\mathscr{A}}$ *be an intuitionistic fuzzy behaviour of* $\mathscr{A}$. *Then there exists an intuitionistic fuzzy finite automaton with unique membership and nonmembership transition on an input symbol IFAUMN* $\mathscr{A}_1 = (Q_1, \Sigma, A_1, i_1, f_1)$ *such that* $L_{\mathscr{A}_1} = L_{\mathscr{A}}$.

Proof. Let $\mathscr{A} = (Q, \Sigma, A, i, f)$ be an IFA with $A = (\mu_A, \nu_A), i = (i_{\mu_A}, i_{\nu_A})$, $f = (f_{\mu_A}, f_{\nu_A})$ and $L_{\mathscr{A}}$ be an intuitionistic fuzzy behaviour of $\mathscr{A}$. Let $Q_1 = P(Q)$, the set of all subsets of Q, every state in Q_1 is of the form $\{p_1, p_2, \ldots, p_r\}, r \geq 1, p_i \in Q, i = 1, 2, \ldots, r$.

Define $\mu_{A_1} : Q_1 \times \Sigma \times Q_1 \to [0,1]$ and $\nu_{A_1} : Q_1 \times \Sigma \times Q_1 \to [0,1]$ by

$$\mu_{A_1}(S,a,S') = \begin{cases} m, \text{if } S' = \{q \in Q \mid \mu_A(p,a,q) = m, p \in S\} \\ 0, \text{if } S' = \phi \end{cases}$$

and

$$\nu_{A_1}(S,a,S') = \begin{cases} n, \text{if } S' = \{q \in Q \mid \nu_A(p,a,q) = n, p \in S\} \\ 1, \text{if } S' = \phi \end{cases}$$

Define $i_{\mu_{A_1}} : Q_1 \to [0,1]$ and $i_{\nu_{A_1}} : Q_1 \to [0,1]$ by

$$i_{\mu_{A_1}}(S) = \begin{cases} \vee\{i_{\mu_A}(p) \mid p \in S\}, \text{if } S \neq \phi \\ 0, \text{if } S = \phi \end{cases}$$

$$i_{\nu_{A_1}}(S) = \begin{cases} \wedge\{i_{\nu_A}(p) \mid p \in S\}, \text{if } S \neq \phi \\ 1, \text{if } S = \phi \end{cases}$$

Define $f_{\mu_{A_1}} : Q_1 \to [0,1]$ and $f_{\nu_{A_1}} : Q_1 \to [0,1]$ by

$$f_{\mu_{A_1}}(S) = \begin{cases} \vee\{f_{\mu_A}(p) \mid p \in S\}, \text{if } S \neq \phi \\ 0, \text{if } S = \phi \end{cases}$$

$$f_{\nu_{A_1}}(S) = \begin{cases} \wedge\{f_{\nu_A}(p) \mid p \in S\}, \text{if } S \neq \phi \\ 1, \text{if } S = \phi \end{cases}$$

Define $\mathscr{A}_1 = (Q_1, \Sigma, A_1, i_1, f_1)$. Let $S, S_1, S_2 \in Q_1$, $\mu_{A_1}(S, a, S_1) = \mu_{A_1}(S, a, S_2)$ implies that $S_1 = \{q \in Q \mid \mu_A(p, a, q) = m, p \in S\}$ and $S_2 = \{q \in Q \mid \mu_A (p, a, q) = m, p \in S\}$.

Similarly, $\nu_{A_1}(S, a, S_1) = \nu_{A_1}(S, a, S_2)$ implies that $S_1 = \{q \in Q \mid \nu_A (p, a, q) = n, p \in S\}$ and $S_2 = \{q \in Q \mid \nu_A(p, a, q) = n, p \in S\}$.

Therefore, $S_1 = S_2$. Hence $\mathscr{A}_1$ is an IFAUMN.

Now, we prove for $\mu^*_A(p, x, q) = \mu^*_{A_1}(\{p\}, x, S), q \in S$. The result is proved by induction on $|x| = n$. Let $|x| = 1, x = \lambda$. If $p \neq q$, then the result is trivial. Let $p = q$. Therefore, $\mu_A(p, \lambda, q) = 1$. Let $S = \{q \in Q \mid \mu_A(p, \lambda, q) = 1\}$. Therefore $S = \{p\}, p = q \in S$, implies that $\mu_{A_1}(\{p\}, \lambda, S) = \mu_{A_1}(\{p\}, \lambda, \{p\}) = 1, p \in S$. Therefore, the result is true for $|x| = 0$. Assume that the result is true for any $x \in \Sigma^*, |x| \leq n$. Let $|x| = n, x = ya, |y| = n - 1, a \in \Sigma$. For $p, q \in Q$,

$$\begin{aligned} \mu^*_A(p, x, q) &= \mu^*_A(p, ya, q) \\ &= \vee\{\mu^*_A(p, y, q') \wedge \mu_A(q', a, q) \mid q' \in Q\} \end{aligned} \tag{1}$$

By induction $\mu^*_A(p, y, q') = \mu^*_{A_1}(\{p\}, y, S'), q' \in S'$. Let $\mu_A(q', a, q) = m$, $S = \{r \in Q \mid \mu_A(s, a, r) = m, s \in S'\}$. Therefore, $q' \in S', \mu_A(q', a, q) = m$ implies that $q \in S$. Therefore $\mu_{A_1}(S', a, S) = m, q' \in S'$. (1) implies that

$$\begin{aligned} \mu^*_A(p, x, q) &= \vee\{\mu^*_{A_1}(\{p\}, y, S') \wedge \mu_{A_1}(S', a, S) \mid S' \subset Q_1\} \\ &= \mu^*_{A_1}(\{p\}, ya, S) \\ &= \mu^*_{A_1}(\{p\}, x, S), q \in S. \end{aligned}$$

Therefore, $\mu^*_A(p, x, q) = \mu^*_{A_1}(\{p\}, x, S), q \in S$.

Similarly, we prove for $\nu^*_A(p, x, q) = \nu^*_{A_1}(\{p\}, x, S), q \in S$. The result is proved by induction on $|x| = n$. Let $|x| = 0, x = \lambda$. If $p \neq q$, then the result is trivial. Let $p = q$. Therefore, $\nu_A(p, \lambda, q) = 1$. Let $S = \{q \in Q \mid \nu_A(p, \lambda, q) = 1\}$. Therefore $S = \{p\}, p = q \in S$, implies that $\nu_{A_1}(\{p\}, \lambda, S) = \nu_{A_1}(\{p\}, \lambda, \{p\}) = 1, p \in S$.

Therefore, the result is true for $|x| = 0$. Assume that the result is true for any $x \in \Sigma^*, |x| \leq n$. Let $|x| = n, x = ya, |y| = n - 1, a \in \Sigma$. For $p, q \in Q$,

$$\begin{aligned} \nu_A^*(p, x, q) &= \nu_A^*(p, ya, q) \\ &= \wedge\{\nu_A^*(p, y, q') \vee \nu_A(q', a, q) \mid q' \in Q\} \qquad (2) \end{aligned}$$

By induction $\nu_A^*(p, y, q') = \nu_{A_1}^*(\{p\}, y, S'), q' \in S'$. Let $\nu_A(q', a, q) = n, S = \{r \in Q \mid \nu_A(s, a, r) = n, s \in S'\}$. Therefore, $q' \in S', \nu_A(q', a, q) = n$ implies that $q \in S$. Therefore $\nu_{A_1}(S', a, S) = n, q' \in S'$. (2) implies that

$$\begin{aligned} \nu_A^*(p, x, q) &= \wedge\{\nu_{A_1}^*(\{p\}, y, S') \vee \nu_{A_1}(S', a, S) \mid S' \subset Q_1\} \\ &= \nu_{A_1}^*(\{p\}, ya, S) \\ &= \nu_{A_1}^*(\{p\}, x, S), q \in S. \end{aligned}$$

Therefore, $\nu_A^*(p, x, q) = \nu_{A_1}^*(\{p\}, x, S'), q \in S$.
Let $L_{\mathscr{A}_1}(x)$ be an intuitionistic fuzzy behaviour of $\mathscr{A}_1$. Now for $x \in \Sigma^*$,

$$\begin{aligned} L_{\mu_{\mathscr{A}}}(x) &= \vee\{i_{\mu_A}(p) \wedge \mu_A^*(p, x, q) \wedge f_{\mu_A}(q) \mid q \in Q \mid p \in Q\} \\ &= \vee\{i_{\mu_{A_1}}(\{p\}) \wedge \mu_{A_1}^*(\{p\}, x, S) \wedge (\vee f_{\mu_{A_1}}(q) \mid q \in S) \mid p \in Q\} \\ &= \vee\{i_{\mu_{A_1}}(\{p\}) \wedge \mu_{A_1}^*(\{p\}, x, S) \wedge f_{\mu_{A_1}}(S) \mid S \in Q \mid \{p\} \in Q_1\} \\ &= L_{\mu_{A_1}}(x). \\ L_{\nu_{\mathscr{A}}}(x) &= \wedge\{i_{\nu_A}(p) \vee \nu_A^*(p, x, q) \vee f_{\nu_A}(q) \mid q \in Q \mid p \in Q\} \\ &= \wedge\{i_{\nu_{A_1}}(\{p\}) \vee \nu_{A_1}^*(\{p\}, x, S) \vee (\wedge f_{\nu_{A_1}}(q) \mid q \in S) \mid p \in Q\} \\ &= \wedge\{i_{\nu_{A_1}}(\{p\}) \vee \nu_{A_1}^*(\{p\}, x, S) \vee f_{\nu_{A_1}}(S) \mid S \in Q \mid \{p\} \in Q_1\} \\ &= L_{\nu_{A_1}}(x). \end{aligned}$$

Therefore $L_{\mathscr{A}_1} = L_{\mathscr{A}}$

Example 1. Consider an IFA $\mathscr{A} = (Q, \Sigma, A, i, f)$ where $Q = \{q_1, q_2, q_3\}$, $\Sigma = \{a, b\}, A = (\mu_A, \nu_A)$ where $\mu_A : Q \to [0, 1]$ and $\nu_A : Q \to [0, 1]$ are define as

$$\begin{array}{ll} \mu_A(q_1, a, q_1) = 0.8 & \nu_A(q_1, a, q_1) = 0.2 \\ \mu_A(q_1, b, q_1) = 0.7 & \nu_A(q_1, b, q_1) = 0.2 \\ \mu_A(q_1, a, q_2) = 0.8 & \nu_A(q_1, a, q_2) = 0.2 \\ \mu_A(q_1, b, q_3) = 0.6 & \nu_A(q_1, b, q_3) = 0.4 \\ \mu_A(q_1, b, q_2) = 0.6 & \nu_A(q_1, b, q_2) = 0.4 \\ \mu_A(q_2, a, q_3) = 0.5 & \nu_A(q_2, a, q_3) = 0.2 \\ \mu_A(q_3, b, q_3) = 0.4 & \nu_A(q_3, b, q_3) = 0.5 \end{array}$$

The initial and final values are given by $(i_{\mu_A}, i_{\nu_A})(q_1) = 0.3/0.6, (i_{\mu_A}, i_{\nu_A})(q_2) = 0.2/0.5, (f_{\mu_A}, f_{\nu_A})(q_3) = 0.5/0.4$. The intuitionistic fuzzy behaviour of $\mathscr{A}$ is $L_{\mu_{\mathscr{A}}} : \Sigma^* \to [0, 1]$ and $L_{\nu_{\mathscr{A}}} : \Sigma^* \to [0, 1]$ such that

$$L_{(\mu_{\mathscr{A}}, \nu_{\mathscr{A}})}(x) = \begin{cases} 0.3/0.6, \text{ if } x \in \{a,b\}^*aa\{a,b\}^* \\ 0.3/0.6, \text{ if } x \in \{a,b\}^*b\{a,b\}^* \\ 0.3/0.6, \text{ if } x \in \{a,b\}^*ba\{a,b\}^* \\ 0.2/0.5, \quad \text{if } x \in a\{a,b\}^* \\ 0, \quad \text{otherwise} \end{cases}$$

From the above theorem, we obtain the following IFAUMN $\mathscr{A}_1 = (Q_1, \Sigma, A, i, f)$ where $Q_1 = \{\{q_1\}, \{q_2\}, \{q_3\}, \{q_1, q_2\}, \{q_2, q_3\}\}$. The fuzzy transition diagram of $\mathscr{A}_1$ is shown in the following Fig. 1.

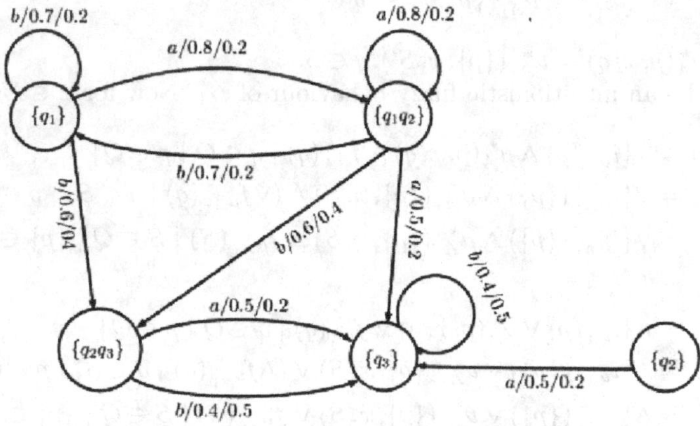

Fig. 1. Intuitionistic fuzzy finite automaton with unique membership and unique nonmembership transitions

For $(i_{\mu_A}, i_{\nu_A})(q_1) = 0.3/0.6, (i_{\mu_A}, i_{\nu_A})(q_2) = 0.2/0.5, (f_{\mu_A}, f_{\nu_A})(q_3) = 0.5/0.4$. The fuzzy behaviour of $\mathscr{A}_1$ is the same as $L_{\mathscr{A}}$.

3 Complete Intuitionistic Fuzzy Automaton with Unique Membership and Unique Nonmembership Transitions

Definition 7. *Let $\mathscr{A} = (Q, \Sigma, A, i, f)$ be an IFAUMN. $\mathscr{A}$ is called complete if for all $p \in Q, a \in \Sigma$, there exists $q \in Q$ such that $\mu_A(p, a, q) > 0$ and $\nu_A(p, a, q) < 1$.*

Theorem 2. *Let $\mathscr{A} = (Q, \Sigma, A, i, f)$ be an incomplete IFAUMN, then there exist an IFAUMN $\mathscr{A}^c$ which is the completion of $\mathscr{A}$ such that fuzzy regular behaviour accepted by $\mathscr{A}$ and $\mathscr{A}^c$ are equal.*

Proof. Let $\mathscr{A} = (Q, \Sigma, A, i, f)$ be an incomplete IFAUMN and the fuzzy behaviour accepted by it be $L_{\mathscr{A}}$. Let $Q^c = Q \cup \{t\}$, where t is a new state such that $t \notin Q$. Choose m,n such that $0 < m \leq 1, 0 < n \leq 1$ and $m \geq \vee\{\mu_A(p,a,q) \mid p,q \in Q\}, n \leq \wedge\{\nu_A(p,a,q) \mid p,q \in Q\}$.

Define $\mathscr{A}^c = (Q^c, \Sigma, A^c, i^c, f^c)$, where $A^c = (\mu_A^c, \nu_A^c)$ such that $\mu_A^c : Q^c \times \Sigma \times Q^c \rightarrow [0,1]$ and $\nu_A^c : Q^c \times \Sigma \times Q^c \rightarrow [0,1]$ are defined as follows: for all $p,q \in Q, a \in \Sigma$.

1. $\mu_A^c(p,a,q) = \mu_A(p,a,q)$ if $\mu_A(p,a,q) > 0$
   $\nu_A^c(p,a,q) = \nu_A(p,a,q)$ if $\nu_A(p,a,q) < 1$
2. $\mu_A^c(p,a,t) = m$ and $\nu_A^c(p,a,t) = n$
   if $\vee\{\mu_A(p,a,q) \mid q \in Q\} = 0$
   $\mu_A^c(p,a,t) = \nu_A^c(p,a,t) = 0$
   if $\vee\{\mu_A(p,a,q) \mid q \in Q\} > 0$
3.

$$\mu_A^c(t,a,p) = \begin{cases} m, \text{if } p = t \\ 0, \text{if } p \neq t \end{cases}$$

$$\nu_A^c(t,a,p) = \begin{cases} n, \text{if } p = t \\ 1, \text{if } p \neq t \end{cases}$$

$i_{\mu_A}^c : Q^c \rightarrow [0,1]$ and $i_{\nu_A}^c : Q^c \rightarrow [0,1]$ are defined as

$$i_{\mu_A^c}(p) = \begin{cases} i_{\mu_A}(p), \text{if } p \in Q \\ 0, otherwise \end{cases} .$$

$$i_{\nu_A^c}(p) = \begin{cases} i_{\nu_A}(p), \text{if } p \in Q \\ 1, otherwise \end{cases}$$

$f_{\mu_A}^c : Q^c \rightarrow [0,1]$ and $f_{\nu_A}^c : Q^c \rightarrow [0,1]$ are defined as

$$f_{\mu_A^c}(p) = \begin{cases} f_{\mu_A}(p), \text{if } p \in Q \\ 0, otherwise \end{cases}$$

$$f_{\nu_A^c}(p) = \begin{cases} f_{\nu_A}(p), \text{if } p \in Q \\ 1, otherwise \end{cases}$$

Clearly $\mathscr{A}^c$ is a complete IFAUMN. Let $L_{\mathscr{A}^c}$ be the fuzzy behaviour accepted by $\mathscr{A}^c$. Next we prove, $L_{\mathscr{A}} = L_{\mathscr{A}^c}$. (i.e.) to prove $L_{\mathscr{A}}(x) = L_{\mathscr{A}^c}(x) \forall x \in \Sigma^*$.

Case(i). $L_{\mu_{\mathscr{A}}}(x) = 0$.

$L_{\mu_{\mathscr{A}}}(x) = \vee\{i_{\mu_A}(p) \wedge \mu_A^*(p,x,q) \wedge f_{\mu_A}(q) \mid p,q \in Q\}$ $L_{\mu_{\mathscr{A}}}(x) = 0$ implies $i_{\mu_A}(p) = 0 \; \forall p \in Q$ or $\mu_A^*(p,x,q) = 0$ or $f_{\mu_A}(q) = 0$.

If $i_{\mu_A}(p) = 0 \; \forall p \in Q$ then $i_{\mu_A}^c(p) = 0 \; \forall p \in Q^c$, therefore $L_{\mu_{\mathscr{A}}}^c(x) = 0$.

Suppose $i_{\mu_A}(p) \neq 0, i_{\nu_A}(p) \neq 1$ and $\mu_A^*(p,x,q) \neq 0, \nu_A^*(p,x,q) \neq 1$. Let $x = a_1 a_2, \ldots, a_n, \mu_A^*(p,x,q) = \mu_A(p,a_1,p_1) \wedge \mu_A(p_2,a_2,p_3) \cdots \mu_A(p_n,a_n,q)$ implies for some $a_k, 1 \leq k \leq n$, there is no move in $\mathscr{A}$. Let j be the smallest integer such that $1 \leq j \leq n$ and there is no move in $\mathscr{A}$ on a_j from p_j. From the construction of $\mathscr{A}^c$, the automaton enters into the dead state t on a_j with membership and nonmembership value m_{p_j}, n_{p_j}, Thereafter $\mathscr{A}^c$ halts at t by reading the remaining input symbols. But $f^c_{\mu_A}(t) = 0$ and $f^c_{\nu_A}(t) = 1$. Similarly, the converse can be proved.

If $f_{\mu_A}(q) = 0\ \forall q \in Q$ then $f^c_{\mu_A}(q) = 0\ \forall q \in Q^c$, therefore $L^c_{\mu_{\mathscr{A}}}(x) = 0$. Similarly,$L_{\nu_{\mathscr{A}}}(x) = \wedge\{i_{\nu_A}(p) \vee \nu^*_A(p,x,q) \vee f_{\nu_A}(q) \mid q \in Q \mid p \in Q\}$ $L_{\nu_{\mathscr{A}}}(x) = 1$ implies $i_{\nu_A}(p) = 1 \forall p \in Q$ or $\nu^*_A(p,x,q) = 1$ or $f_{\nu_A}(q) = 1$.

If $i_{\nu_A}(p) = 1\ \forall p \in Q$ then $i^c_{\nu_A}(p) = 1\ \forall p \in Q^c$, therefore $L^c_{\nu_{\mathscr{A}}}(x) = 1$.

If $f_{\nu_A}(q) = 1\ \forall q \in Q$ then $f^c_{\nu_A}(q) = 1\ \forall q \in Q^c$, therefore $L^c_{\nu_{\mathscr{A}}}(x) = 1$.

Therefore $L_{\mathscr{A}}(x) = L^c_{\mathscr{A}}(x)$.

If $L_{\nu_{\mathscr{A}}}(x) < 1$ implies $i_{\nu_A}(p) < 1$ and $\nu^*_A(p,x,q) < 1$ and $f_{\nu_A}(q) < 1$ $i_{\nu_A}(p) < 1$ implies $i^c_{\nu_A}(p) = i_{\nu_A}(p) \forall p \in Q$ $f_{\nu_A}(q) < 1$ implies $f^c_{\nu_A}(q) = f_{\nu_A}(q) \forall q \in Q$ $\nu^*_A(p,x,q) < 1$ implies $\nu^*_A(p,x,q) \geq \nu^c_A{}^*(p,x,q)$

$$\begin{aligned}L_{\nu_{\mathscr{A}}}(x) &\geq i^c_{\nu_A}(p) \vee \nu^c_A{}^*(p,x,q) \vee f^c_{\nu_A}(q)\\ &\geq \wedge\{i^c_{\nu_A}(p) \vee \nu^c_A{}^*(p,x,q) \vee f^c_{\nu_A}(q) \mid q \in Q^c \mid p \in Q^c\}\\ &= L^c_{\nu_{\mathscr{A}}}(x).\end{aligned}$$

Therefore, $L_{\nu_{\mathscr{A}}}(x) \geq L^c_{\nu_{\mathscr{A}}}(x)$, Similarly we get $L^c_{\nu_{\mathscr{A}}}(x) \geq L_{\nu_{\mathscr{A}}}(x)$.

Hence $L_{\nu_{\mathscr{A}}}(x) = L^c_{\nu_{\mathscr{A}}}(x)$.

Case(ii). $L_{\mu_{\mathscr{A}}}(x) > 0$,

$L_{\mu_{\mathscr{A}}}(x) = \vee\{i_{\mu_A}(p) \wedge \mu^*_A(p,x,q) \wedge f_{\mu_A}(q) \mid q \in Q \mid p \in Q\}$. Since Q is finite, there exists $p, q \in Q$ such that
$L_{\mu_{\mathscr{A}}}(x) = i_{\mu_A}(p) \wedge \mu^*_A(p,x,q) \wedge f_{\mu_A}(q)$
$L_{\mu_{\mathscr{A}}}(x) > 0$ implies $i_{\mu_A}(p) > 0, \mu^*_A(p,x,q) > 0, f_{\mu_A}(q) > 0$.
$i_{\mu_A}(p) > 0$ implies $i^c_{\mu_A}(p) = i_{\mu_A}(p), p \in Q^c$
$f_{\mu_A}(q) > 0$ implies $f^c_{\mu_A}(q) = f_{\mu_A}(q), q \in Q^c$
$\mu^*_A(p,x,q) > 0$ from definition of $\mathscr{A}^c, \mu^c_A{}^*(p,x,q) = \mu^*_A(p,x,q)$. Thus

$$\begin{aligned}L_{\mu_{\mathscr{A}}}(x) &= i^c_{\mu_A}(p) \wedge \mu^c_A{}^*(p,x,q) \wedge f^c_{\mu_A}(q)\\ &\leq \vee\{i^c_{\mu_A}(p) \wedge \mu^c_A{}^*(p,x,q) \wedge f^c_{\mu_A}(q) \mid q \in Q^c \mid p \in Q^c\}\\ &= L^c_{\mu_{\mathscr{A}}}(x).\end{aligned}$$

$$\text{Hence } L_{\mu_{\mathscr{A}}}(x) \leq L_{\mu^c_{\mathscr{A}}}(x)\,. \tag{3}$$

Now $L^c_{\mu_{\mathscr{A}}}(x) = \vee\{i^c_{\mu_A}(p) \wedge \mu^c_A{}^*(p,x,q) \wedge f^c_{\mu_A}(q) \mid q \in Q^c \mid p \in Q^c\}$. If $L^c_{\mu_{\mathscr{A}}}(x) = 0$, then $L^c_{\mu_A}(x) \leq L_{\mu_{\mathscr{A}}}(x)$. Let $L^c_{\mu_A}(x) > 0$. Since Q^c is finite, there exists $p, q \in Q^c$ such that
$L^c_{\mu_{\mathscr{A}}}(x) = i^c_{\mu_A}(p) \wedge \mu^c_A{}^*(p,x,q) \wedge f^c_{\mu_A}(q)$
$L^c_{\mu_{\mathscr{A}}}(x) > 0$ implies $i^c_{\mu_A}(p) > 0, \mu^c_A{}^*(p,x,q) > 0, f^c_{\mu_A}(q) > 0$
$i^c_{\mu_A}(p) > 0$ implies $i_{\mu_A}(p) > 0$ and $p \in Q$
$f^c_{\mu_A}(q) > 0$ implies $f_{\mu_A}(q) > 0$ and $q \in Q$
$\mu^c_A{}^*(p,x,q) > 0$ and $f^c_{\mu_A}(q) > 0$ implies that $q \neq t$, therefore $\mathscr{A}^c$ never enters into

the dead state t in the sequence of moves. Therefore $\mu_A^{c\,*}(p,x,q) = \mu_A^*(p,x,q)$. We have

$$\begin{aligned} L^c_{\mu_{\mathscr{A}}}(x) &= i_{\mu_A}(p) \wedge \mu_A^*(p,x,q) \wedge f_{\mu_A}(q), p,q \in Q \\ &\leq \vee\{i_{\mu_A}(p) \wedge \mu_A^*(p,x,q) \wedge f_{\mu_A}(q) \mid q \in Q \mid p \in Q\} \\ &= L_{\mu_{\mathscr{A}}}(x). \end{aligned}$$

$$\text{Therefore } L^c_{\mu_{\mathscr{A}}}(x) \leq L_{\mu_{\mathscr{A}}}(x)\ . \tag{4}$$

Similarly, $L_{\nu_{\mathscr{A}}}(x) = \wedge\{i_{\nu_A}(p) \vee \nu_A^*(p,x,q) \vee f_{\nu_A}(q) \mid q \in Q \mid p \in Q\}$. Since Q is finite, there exists $p, q \in Q$ such that
$L_{\nu_{\mathscr{A}}}(x) = i_{\nu_A}(p) \vee \nu_A^*(p,x,q) \vee f_{\nu_A}(q)$
$L_{\nu_{\mathscr{A}}}(x) < 1$ implies $i_{\nu_A}(p) < 1, \nu_A^*(p,x,q) < 1, f_{\nu_A}(q) < 1$
$i_{\nu_A}(p) < 1$ implies $i^c_{\nu_A}(p) = i_{\nu_A}(p), p \in Q^c$,
$f_{\nu_A}(q) < 1$ implies $f^c_{\nu_A}(q) = f_{\nu_A}(q), q \in Q^c$
$\nu_A^*(p,x,q) < 1$ from definition of $\mathscr{A}^c, \nu_A^{c\,*}(p,x,q) = \nu_A^*(p,x,q)$

$$\begin{aligned} L_{\nu_{\mathscr{A}}}(x) &= i^c_{\nu_A}(p) \vee \nu_A^{c\,*}(p,x,q) \vee f^c_{\nu_A}(q) \\ &\geq \wedge\{i^c_{\nu_A}(p) \vee \nu_A^{c\,*}(p,x,q) \vee f^c_{\nu_A}(q) \mid q \in Q^c \mid p \in Q^c\} \\ &= L^c_{\nu_{\mathscr{A}}}(x). \end{aligned}$$

$$\text{Hence } L_{\nu_{\mathscr{A}}}(x) \geq L_{\nu^c_{\mathscr{A}}}(x)\ . \tag{5}$$

Now $L^c_{\nu_{\mathscr{A}}}(x) = \wedge\{i^c_{\nu_A}(p) \vee \nu_A^{c\,*}(p,x,q) \vee f^c_{\nu_A}(q) \mid q \in Q^c \mid p \in Q^c\}$.

If $L^c_{\nu_{\mathscr{A}}}(x) < 1$, Since Q^c is finite, there exists $p, q \in Q^c$ such that
$L^c_{\nu_{\mathscr{A}}}(x) = i^c_{\nu_A}(p) \vee \nu_A^{c\,*}(p,x,q) \vee f^c_{\nu_A}(q)$
$L^c_{\nu_{\mathscr{A}}}(x) < 1$ implies $i^c_{\nu_A}(p) < 1, \nu_A^{c\,*}(p,x,q) < 1, f^c_{\mu_A}(q) < 1$.
$i^c_{\nu_A}(p) < 1$ implies $i^c_{\nu_A}(p) = i_{\nu_A}(p)$
$f^c_{\nu_A}(q) < 1$ implies $f^c_{\nu_A}(q) = f_{\nu_A}(q)$ $\nu_A^{c\,*}(p,x,q) < 1$ and $f^c_{\mu_A}(q) < 1$ implies that $q \neq t$, therefore $\mathscr{A}^c$ never enters into the dead state t in the sequence of moves. Therefore $\nu_A^{c\,*}(p,x,q) = \nu_A^*(p,x,q)$. We have

$$\begin{aligned} L^c_{\nu_{\mathscr{A}}}(x) &= i_{\nu_A}(p) \vee \nu_A^*(p,x,q) \vee f_{\nu_A}(q), p,q \in Q \\ &\geq \wedge\{i_{\nu_A}(p) \vee \nu_A^*(p,x,q) \vee f_{\nu_A}(q) \mid q \in Q \mid p \in Q\} \\ &= L_{\nu_{\mathscr{A}}}(x). \end{aligned}$$

$$\text{Therefore } L^c_{\nu_{\mathscr{A}}}(x) \geq L_{\nu_{\mathscr{A}}}(x) \tag{6}$$

From Eqs. (3), (4), (5) and (6) $L_{\mathscr{A}}(x) = L^c_{\mathscr{A}}(x)$

4 Conclusion

In this paper, the authors have made an attempt to study an Intuitionistic fuzzy finite state automaton with unique membership and unique nonmembership transition on an input symbol and also proved that there exists a complete IFAUMN for a given incomplete IFAUMN for the same fuzzy language. We have made a humble beginning in this direction, however, many concepts are yet to be fuzzyfied in the context of IFAUMN.

References

1. Atanassov, K.T.: Intuitionistic fuzzy sets. Fuzzy Sets Syst. **20**, 87–96 (1986)
2. Atanassov, K.T.: More on intuitionistic fuzzy sets. J. Fuzzy Sets Syst. **33**, 37–46 (1989)
3. Atanassov, K.T.: Intuitionistic fuzzy relations. In: First Scientific Session of the Mathematical Foundation Artificial Intelligence, Sofia IM-MFAIS, pp. 1–3 (1989)
4. Atanassov, K.T.: New operations defined over the intuitionistic fuzzy sets. J. Fuzzy Sets Syst. **61**, 137–142 (1994)
5. Atanassov, K.T.: Intuitionistic Fuzzy Sets Theory and Applications. Physica-verlag, Heidelberg (1999)
6. Jun, Y.B.: Intuitionistic fuzzy finite state machines. J. Appl. Math. Comput. **17**, 109–120 (2005)
7. Jun, J.B.: Intuitionistic fuzzy finite switchboard state machines. J. Appl. Math. Comput. **20**, 315–325 (2006)
8. Jun, Y.B.: Quotient structures of intuitionistic fuzzy finite state machines. Inform. Sci. **177**, 4977–4986 (2007)
9. Malik, D.S., Mordeson, J.N.: Fuzzy Automata and Languages, Theory and Applications. CRC, Boca Raton (2002)
10. Rajaretnam, T., Ayyaswamy, S.K.: Fuzzy finite state automaton with unique membership transition on an input Symbol. J. Combin. Math. Combin. Comput. **69**, 151–164 (2009)
11. Santos, E.S.: Maximum automata. Inform. Control **12**, 367–377 (1968)
12. Telesphor, L., Jeny Jordon, A., Jency Priya, K., Rajaretnam, T.: Intuitionistic fuzzy finite automata with unique membership transition. In: IEEE Explore, pp. 103–107 (2014)
13. Wee, W.G., Fu, K.S.: A formulation of fuzzy automata and its application as a model of learning systems. IEEE Trans. Syst. Man Cybern. **5**, 215–223 (1969)
14. Zadeh, L.A.: Fuzzy sets. J. Inform. Control **8**, 338–353 (1965)
15. Zadeh, L.A.: Fuzzy languages and their relation to human and machine intelligence. Electrn. Research Laboratory University California, Berkeley, CA, Technical Report ERL - M302 (1971)

Independent 2-Point Set Domination in Graphs

Purnima Gupta and Deepti Jain(✉)

Department of Mathematics, Sri Venkateswara College, University of Delhi, New Delhi 110021, Delhi, India
purnimachandni1@gmail.com, djain@svc.ac.in

Abstract. A set D of vertices in a connected graph G is said to be an *independent 2-point set dominating set* (or in short i-2psd set) of G if D is an independent set and for every subset $S \subseteq V - D$ there exists a non-empty subset $T \subseteq D$ containing at most 2 vertices such that the induced subgraph $\langle S \cup T \rangle$ is connected. In this paper we explore graphs which possess an i-2psd set.

Keywords: Independent set · 2-point set domination

1 Introduction

By a graph G we mean a finite, undirected, connected and non-trivial graph with neither loops nor multiple edges. The order $|V(G)|$ and the size $|E(G)|$ of G are denoted by n and m respectively. For graph theoretic terminology we refer to West [9]. For the terms related to the concept of domination we refer to Haynes et al. [6].

In [8], Gupta and Jain define 2-point set domination in a graph as follows:

Definition 1. *A set* $D \subseteq V(G)$ *is a* 2-point set dominating set *(or, in short, 2-psd set) of* G *if for every subset* $S \subseteq V - D$ *there exists a non-empty subset* $T \subseteq D$ *containing at most* 2 *vertices such that* $\langle S \cup T \rangle$ *is connected. The* 2-point set domination number of G, *denoted by* $\gamma_{2ps}(G)$, *is the minimum cardinality of a 2-psd set of* G.

Definition 2. *A 2-psd set D of a graph G which is also independent is said to be an* ***independent 2-psd set*** *(abbreviated henceforth as i-2psd set) of G.*

The theory of independent domination was formalized by Berge [3] and Ore [7]. The independent domination number and the notation $i(G)$ were introduced by Cockayne and Hedetniemi in [4,5]. It is well know that *every finite graph has an independent dominating set.* However, such a statement is not true in the case of 2-psd sets in graphs. For example, the cycle C_7 has no independent 2-psd set.

We call a graph to be an **i-2psd graph** if it possesses an independent 2-psd set and a **non i-2psd graph** otherwise. Obviously, if G is a graph with a vertex of full degree, then G possesses an i-2psd set. Further, every complete bipartite graph $K_{m,n}$ is an i-2psd graph. The main aim of this paper is to explore graphs which possess an independent 2-psd set.

S. Arumugam et al. (Eds.): ICTCSDM 2016, LNCS 10398, pp. 281–288, 2017.
DOI: 10.1007/978-3-319-64419-6_36

In [1], Acharya and Gupta define a set D of vertices in a connected graph G to be an *independent point-set dominating set* (or in short i-psd set) of G if D is independent and for every subset $S \subseteq V - D$ there exists a vertex $v \in D$ such that the subgraph $\langle S \cup \{v\} \rangle$ is connected. A graph having an i-psd set is called an i-psd graph. A detailed study of graphs possessing i-psd sets has been done in [1,2]. Clearly by definition every i-psd graph is an i-2psd graph.

2 Basic Results

The following proposition is an immediate consequence of the definition of i-2psd set.

Proposition 1. *For any graph G, an independent subset $D \subseteq V$ is an i-2psd set of G if and only if for every independent subset $S \subseteq V - D$ there exists a subset $T \subseteq D$ such that $|T| \leq 2$ and $\langle S \cup T \rangle$ is connected.*

Theorem 1. *Let G be a separable graph. If an i-2psd set D is such that $V(B) \subseteq V - D$ for some block B of G, then $V - D = V(B)$, B is complete and each vertex of B is a cut-vertex.*

Proof. Let D be an i-2psd set of G such that $V(B) \subseteq V - D$ for some block B of G. Let $z \in (V - D) - V(B)$ and let $x \in V(B)$ be such that distance of z from x is least amongst all vertices of B. Then for any vertex $y \in V(B) - \{x\}$, the z-y path passes through x and $x \notin D$. Hence D is not a 2-psd set of G, a contradiction. Thus $V - D = V(B)$.

Now since each vertex of $V - D$ is adjacent to some vertex in D, each vertex of B is a cut-vertex of G. Next, if $x, y \in V(B)$ are such that $xy \notin E(G)$, then D being a i-2psd set there exists a vertex $d \in D (= V(G) - V(B))$ such that $x, y \in N(d)$ which implies that $d \in V(B)$. This is a contradiction to our hypothesis that $V(B) \subseteq V - D$. Thus, B is complete.

Theorem 2. *Let G be a separable graph. Let D be an i-2psd set such that $V - D \subsetneq V(B)$ for some block B of G and $V(B) \cap D$ is not an i-2psd set of B. Let*

$$P(B, D) := \{x \in V - D : N(x) \cap V(B) \cap D = \phi \ \text{ and } \ N(x) \cap (D - V(B)) \neq \phi\}.$$

Then $P(B, D)$ is non-empty, every vertex of $P(B, D)$ is adjacent to every other vertex of $V - D$ and $V(B) \cap D$ is an i-2psd set of $\langle V(B) - P(B, D) \rangle$.

Proof. If an i-2psd set D of G is such that $V - D \subsetneq V(B)$ for some block B of G and $V(B) \cap D$ is not an i-2psd set of B, then there exists a vertex $u \in V - D$ such that $N(u) \cap V(B) \cap D = \phi$ and $N(u) \cap (D - V(B)) \neq \phi$. Thus, $P(B, D)$ is non-empty.

Next we show that each vertex of $P(B, D)$ is adjacent to all other vertices of $V - D$. Suppose there exist vertices $u \in P(B, D)$ and $z \in (V - D) - \{u\}$ such that $uz \notin E(G)$. Since D is an i-2psd set of G, there exists a vertex $w \in D$

such that $u, z \in N(w)$ which implies $w \in D \cap V(B)$, a contradiction to the fact $u \in P(B, D)$. Hence the claim.

Lastly, we show that $V(B) \cap D$ is an i-2psd set of $\langle V(B) - P(B, D) \rangle$. By definition of $P(B, D)$, every vertex of $(V - D) - P(B, D)$ is adjacent to some vertex of $V(B) \cap D$. Also for every independent subset $S(|S| \geq 2)$ of $(V - D) - P(B, D)$, there exists a set $W \subseteq D$ such that $|W| \leq 2$ and $\langle S \cup W \rangle$ is connected. Since $S \subseteq V(B)$, W is a subset of $V(B)$. Therefore $V(B) \cap D$ is an i-2psd set of $\langle V(B) - P(B, D) \rangle$.

Observation 1. *If D is an i-2psd set of a graph G, then for any two vertices $x, y \in V - D$, $d(x, y) \leq 2$, since for any two non-adjacent vertices $x, y \in V - D$, there exists a vertex $w \in D$ such that $x, y \in N(w)$.*

We give an upper bound on the diameter of a graph for being an i-2psd graph.

Theorem 3. *For an i-2psd graph G, $diam(G) \leq 4$.*

Proof. Let D be an i-2psd set of G. Let $x, y \in V(G)$ be two non-adjacent vertices. If $x, y \in V - D$ then $d(x, y) \leq 2$. If $x \in D$ and $y \in V - D$, then there exists a vertex $z \in V - D$ such that $xz \in E(G)$ and therefore, $d(x, y) \leq d(x, z) + d(z, y) \leq 3$. If $x, y \in D$, then there exist $z_1, z_2 \in V - D$ such that $xz_1, yz_2 \in E(G)$ and therefore $d(x, y) \leq d(x, z_1) + d(z_1, z_2) + d(z_2, y) \leq 4$. Hence the theorem.

We observe that converse of Theorem 3 is not true. For example, C_7 has diameter 3 but fails to possess an i-2psd set.

In [1], Acharya and Gupta proved that a tree T is an i-psd graph if and only if $diam(T) \leq 4$. Since every i-psd graph is an i-2psd graph, we have the following theorem.

Theorem 4. *A tree T is an i-2psd graph if and only if $diam(T) \leq 4$.*

Theorem 5. *Let G be a graph which is not a tree and let $girth(G) \geq 7$. Then G is not a i-2psd graph.*

Proof. Let C_g be a cycle of length g in G. For any independent subset D of G, $|(V - D) \cap V(C_g)| \geq \lceil \frac{g}{2} \rceil$. Hence there exists a pair of vertices $x, y \in (V - D) \cap V(C_g)$ such that $d(x, y) \geq 3$. Thus, it follows from Observation 1 that G is not an i-2psd graph.

However, not every graph with girth less than 7 is i-2psd. For example, the corona $C_4 \circ K_1$ has girth 4 but is not an i-2psd graph.

Problem 1. Characterize i-2psd graphs with maximum possible girth.

The *independence number* of a graph G, denoted by $\beta_0(G)$, is the maximum cardinality of an independent subset of G. Since every i-2psd set is a 2-psd set, we have the following theorem.

Theorem 6. *A necessary condition for a graph to be an i-2psd graph is* $\beta_0(G) \geq \gamma_{2ps}(G)$.

For the corona $G = K_n \circ K_1$, we have $\beta_0(G) = n = \gamma_{2ps}(G)$. Also for each positive integer $n \geq 6$, $\bar{C}_n$ is i-2psd and $\beta_0(\bar{C}_n) = 2 = \gamma_{2ps}(\bar{C}_n)$.

Problem 2. Characterize i-2psd graphs G satisfying $\beta_0(G) = \gamma_{2ps}(G)$.

Based on Theorem 6, we have the following corollary giving some classes of non i-2psd graphs.

Corollary 1. *a. A cycle* C_n *is an i-2psd graph if and only if* $n \leq 6$.
b. $C_n \times K_2$ $(n \geq 5)$ *is not an i-2psd graph.*
c. $P_n \times P_m$ $(n \geq 4, m \geq 4)$ *is not an i-2psd graph.*

Note that $C_4 \circ K_1$ is a graph which satisfies all three necessary conditions given in Theorems 3, 5 and 6 but is not an i-2psd graph. Thus, all these three conditions together are not sufficient for a graph to be an i-2psd graph.

We now proceed to investigate separable graphs which are i-2psd graphs.

The set of all blocks of a separable graph G is denoted by $\mathcal{B}(G)$ and the set of all blocks at a cut-vertex w in G is denoted by $\mathcal{B}_w(G)$.

Theorem 7. *If a separable graph* G *is i-2psd, then each block of* G *is i-2psd.*

Proof. Let G be a separable i-2psd graph and $B \in \mathcal{B}(G)$ be such that B is not an i-2psd block. Let $D \in \mathcal{D}_{i2ps}(G)$. If $V(B) \cap D = \phi$, then by Theorem 1, $V - D = V(B)$ and B is complete which implies B is an i-2psd block, a contradiction to our choice of the block B. Hence $V(B) \cap D \neq \phi$. Since D is independent, $V(B) \cap (V - D) \neq \phi$.

Suppose $V - D \subsetneq V(B)$. Let $F = V(B) \cap D$. Since B is not an i-2psd block, F is not a 2-psd set of B and therefore by Theorem 2, $P(B, D) \neq \phi$. Let $V_1 = P(B, D)$ and $V_2 = (V - D) - P(B, D)$. Then again by Theorem 2, $\langle V_1 \rangle$ is complete, F is a 2-psd set of $\langle F \cup V_2 \rangle$ and for each $x \in V_1$, $N(x) \cap V_2 = V_2$ and $N(x) \cap F = \phi$. For any $x \in V_1$, the set $F \cup \{x\}$ is independent and $V(B) - [F \cup \{x\}] \subseteq N(x)$. This implies $F \cup \{x\}$ is an i-2psd set of B, which contradicts our choice of block B.

Next, suppose $V - D \nsubseteq V(B)$. Choose a vertex $y \in (V - D) \cap V(B')$ for some block B' $(B' \neq B)$. Let $x \in (V - D) \cap V(B)$ be such that $xy \notin E(G)$ (This is always possible, otherwise $B \cong K_2$ and hence an i-2psd block). Since D is an i-2psd set of G, there exists a vertex $w \in D$ such that $x, y \in N(w)$. We shall show that $(V - D) \cap V(B) \subseteq N(w)$. Suppose $z \in (V - D) \cap V(B)$ is such that $z \notin N(w)$. Then there exists $w' \in D$ $(w' \neq w)$ such that $z, y \in N(w')$. This yields a z-x path (z, w', y, w, x) containing vertices outside B, a contradiction. Thus, $(V - D) \cap V(B) \subseteq N(w)$. Now if $|(V - D) \cap V(B)| = 1$, then clearly $V(B) \cap D$ is an i-2psd set of B and if $|(V - D) \cap V(B)| \geq 2$, then $(V - D) \cap V(B) \subseteq N(w)$ implies that $w \in V(B)$ and therefore $V(B) \cap D$ is an i-2psd set of B. In both cases we get a contradiction to our assumption that B is not an i-2psd block. Hence every block of an i-2psd separable graph is an i-2psd block.

The converse of Theorem 7 is not true. For example, consider a separable graph G obtained by identifying a vertex of C_5 with a vertex of C_3. Both C_5 and C_3 are i-2psd graphs but G is not an i-2psd graph.

In next section we give some classes of i-2psd graphs by using the properties discussed in this section.

3 Some Classes of i-2psd Graphs

In this section we characterizes i-2psd cactus, i-2psd generalized theta graphs and i-2psd block graphs. Since we have discussed i-2psd cycles and trees, we consider separable cactus which contains at least one cycle.

Theorem 8. *Let G be a separable cactus with at least one cycle. Then G is an i-2psd graph if and only if one of the following holds:*

a. G has a unique cycle C_n, where $3 \leq n \leq 6$ and $V(G) - V(C_n)$ consists of pendant vertices with their support in $V(C_n) - F$ where F is a maximum independent subset of $V(C_n)$.
b. All cycles of G are of length at most 4 and are at a single cut-vertex, w and all vertices of $V(G) - \bigcup_{B \in \mathcal{B}_w(G)} V(B)$ are pendant vertices with their support in $N(w)$.

Proof. Necessity: Let G be a separable cactus with at least one cycle. Since each block of an i-2psd separable graph is i-2psd, each non-trivial block of G is isomorphic to a cycle of length 6 or less. Let D be an i-2psd set of G.

If $V - D = V(B)$ for some block B of G, then by Theorem 1, B is complete and each vertex of B is a cut-vertex. Therefore, $B \cong C_3$ or $B \cong K_2$. In this case $D = V(G) - V(B)$. Since D is independent, each vertex of $V(G) - V(B)$ is pendant with its support in $V(B)$. If $B \cong K_2$, then G is a double star and therefore non-acyclic. Therefore, $B \cong C_3$ and G has pendant vertices adjacent to each vertex of C_3. Thus G satisfies (b) where G has only C_3 as a non-trivial block, $w \in V(C_3)$ and all vertices of $V(G) - \cup_{B \in \mathcal{B}_w(G)} V(B)$ are pendant with their support in $V(C_3) - \{w\}$.

Next, if $V - D \subsetneq V(B)$, then $D \cap V(B) \neq \phi$. Let $F = D \cap V(B)$. Since $V(G) - V(B) \subset D$, it is independent and consists of pendant vertices with their support in $V(B) - F$. Therefore if $B \cong C_3$ or C_5 or C_6, then G satisfies (a). If $B \cong C_4$ and $|F| = 1$, then G satisfies (b) where G has only C_4 as a non-trivial block, C_4 has a cut-vertex w, $F \nsubseteq N(w)$ and $V(G) - \bigcup_{B \in \mathcal{B}_w(G)} V(B)$ are pendant with their support in $V(C_4) \cap N(w)$. If $B \cong C_4$ and $|F| = 2$, then G satisfies (a).

Lastly, suppose $V - D \nsubseteq V(B)$ for any block B of G. Let $x, y \in V - D$ be two non-adjacent vertices from different blocks of G. Since D is an i-2psd set of G, there exists $w \in D$ such that $x, y \in N(w)$. This implies that w is a cut-vertex of G and all x-y paths pass through w. We shall show that $V - D \subseteq N(w)$. Suppose there exists a vertex $z \in V - D$ such that $z \notin N(w)$. Since x and y belong to two different blocks of G, vertex z can be adjacent to at most one of x

and y. In either case, D being a 2-psd set, we get a x-y path passing through z and not containing w, a contradiction. Therefore, $V - D = N(w)$. This implies $V(G) - N(w) = D$ and is independent. This implies all non-trivial blocks are at a common cut-vertex w and are isomorphic to C_3 or C_4, and all vertices of $V(G) - (\bigcup_{B \in \mathcal{B}_w(G)} V(B))$ are pendant vertices with their support in $N(w)$. Hence in this case G satisfies (b).

Sufficiency: If G satisfies (a), then $F \cup \{V(G) - V(C_n)\}$ is an i-2psd set of G and if G satisfies (b), then $V(G) - N(w)$ is an i-2psd set of G. Hence the theorem.

A *generalized theta graph* $\mathbf{\Theta_{s_1,...,s_k}}$ consists of k internally disjoint u-v paths of lengths $s_1, ..., s_k$.

Theorem 9. *A generalized theta graph* $\Theta_{s_1,s_2,...,s_k} (k \geq 3)$ *is an i-2psd graph if and only if one of the following holds:*

a. $1 \leq s_i \leq 3$ *for each* i.
b. There exists exactly one $j \in \{1, 2, ..., k\}$ *such that* $s_j = 4$ *and* $s_i \leq 2$ *for all* $i \neq j$.

Proof. Let G be a generalized theta graph and let $P_1, P_2, ..., P_k$ be the internally disjoint u-v paths of lengths $s_1, s_2, ..., s_k$ respectively.

Necessity: Suppose G is an i-2psd graph and let D be an i-2psd set of G. It follows from Observation 1 that $s_i \leq 4$ for each i.

Now suppose P_j be a u-v path of length 4. We claim that $s_i \leq 2$ for all i, $i \neq j$.

Suppose $s_i > 2$ for some $i, i \neq j$. Since length of each u-v path is at most four, $s_i = 3$ or 4. In either case there exist vertices $x \in V(P_j) \cap (V - D)$ and $y \in V(P_i) \cap (V - D)$ such that $x, y \notin N(w)$ for any $w \in D$, which is a contradiction to the fact that D is an i-2psd set of G. Thus $s_i \leq 2$ for all $i \neq j$.

Thus, each u-v path P_i is of length at most four and if there exists a path of length four then all other u-v paths are of length two or less. This proves the necessity.

Sufficiency: Suppose G satisfies condition (a). If $s_i = 1$ for some i, then $V(G) - N(u)$ is an i-2psd set of G, if $s_i = 3$ for some i, then $V(G) - \{N(u) \cup \{v\}\}$ is an i-2psd set of G and if $s_i = 2$ for each i, then $\{u, v\}$ is an i-2psd set of G.

Now suppose G satisfies condition (b) and $s_j = 4$ for some j. Let $w \in V(P_j)$ be such that $d(u, w) = 2 = d(v, w)$. If $s_i = 1$ for some $i, i \neq j$, then $V(G) - \{u, v, w\}$ is an i-2psd set of G and if $s_i \neq 1$ for any $i, i \neq j$, then $\{u, v, w\}$ is an i-2psd set of G. Hence the theorem.

Theorem 10. *For a graph* G, *if there exists a vertex* $w \in V(G)$ *such that* $V(G) - N(w)$ *is independent then* $V(G) - N(w)$ *is an i-2psd set of* G *and therefore* G *is an i-2psd graph.*

Theorem 10 gives a sufficient condition for a graph to be an i-2psd graph and provides us with several classes of i-2psd graphs. Some classes are given by the following corollaries.

Corollary 2. *Every complete bipartite graph $K_{m,n} (m, n \geq 1)$ is an i-2psd graph.*

Corollary 3. *Every split graph is an i-2psd graph.*

The condition given by Theorem 10 is not necessary for an i-2psd graph in general but in case of block graphs it is necessary as well as sufficient.

A *block graph* is a graph in which every block is complete. We know that a complete graph is an i-2psd graph; here we characterize separable i-2psd block graphs.

Theorem 11. *A separable block graph G is an i-2psd graph if and only if there exists a vertex w such that $V(G) - N(w)$ is independent.*

Proof. Necessity: If G has no non-trivial block then G is a tree and therefore by Theorem 4, $diam(G) \leq 4$. In this case G clearly satisfies the condition of the theorem where w is a vertex in the center of G. Now we assume G has a non-trivial block, say B. Let D be an i-2psd set of G. If $D \cap V(B) = \phi$, then by Theorem 1, $V - D = V(B)$ and each vertex of B is a cut-vertex. Since $D = V(G) - V(B)$ and is independent, all vertices of $V(G) - V(B)$ are pendant vertices with their support in $V(B)$. Hence G satisfies the condition of the theorem with unique non-trivial block B and w be any vertex in $V(B)$.

Now if $D \cap V(B) \neq \phi$, then $|D \cap V(B)| = 1$. Let $D \cap V(B) = \{w\}$. We shall show that $V - D = N(w)$. Suppose there exists a vertex $z \in V - D$ such that $z \notin N(w)$. Then $z \in V(B')$ for some block $B', B' \neq B$. Now z can be adjacent to at most one vertex of B. Suppose z is adjacent to some vertex $y \in V(B)$. Choose a vertex $x \in V(B) - \{w\}$ such that $xz \notin E(G)$. Since D is an i-2psd set, for the set $\{x, z\}$ there exists $w' \in D$ such that $x, z \in N(w')$. Thus we get a cycle (x, y, z, w') containing vertices of different blocks, a contradiction. Therefore, z is not adjacent to any vertex of B. Then for any $x, y \in V(B)$, there exist $w', w'' \in D$ such that $x, z \in N(w')$ and $y, z \in N(w'')$. If $w'' = w'$, then $x, y \in N(w')$ implies that $w' \in V(B)$, a contradiction to the fact that $D \cap V(B) = \{w\}$ and if $w'' \neq w'$, then we get a cycle (x, w', z, w'', y, x) containing vertices of different blocks, again a contradiction. Thus $V - D = N(w)$ and therefore $D = V(G) - N(w)$ is independent. Since D is independent and each block is complete, every vertex in $V(G) - N[w]$ is a pendant vertex.

Sufficiency: If a block graph G satisfies the condition of the theorem, then $V(G) - N(w)$ is an i-2psd set of G and hence G is an i-2psd graph.

4 Conclusion

As seen in Theorem 11, for a separable i-2psd block graph G there exists a vertex w such that $D = V(G) - N(w)$ is independent and is an i-2psd set. Since D is independent and each block of G is complete, we may note that every vertex in $V(G) - N[w]$ is a pendant vertex with its support in $N(w)$. We will continue with our study of characterizing i-2psd graphs in our subsequent papers.

References

1. Acharya, B.D., Purnima, G.: On point set domination in graphs V: independent PSD-sets. J. Comb. Inf. Sys. Sci. **22**(2), 133–148 (1997)
2. Acharya, B.D., Purnima, G.: On point-set domination in graphs VI: quest to characterize blocks containing independent PSD-sets. Nat. Acad. Sci. Lett. **23**, 171–176 (2000)
3. Berge, C.: Graphs and Hypergraphs. North-Holland, Amsterdam (1985)
4. Cockayne, E.J., Hedetniemi, S.T.: Independence graphs. Congr. Numer. **X**, 471–491 (1974)
5. Cockayne, E.J., Hedetniemi, S.T.: Towards a theory of domination in graphs. Networks **7**, 247–261 (1977)
6. Haynes, T.W., Hedetniemi, S.T., Slater, P.J.: Domination in Graphs-Advanced Topics. Marcel Dekker, New York (1998)
7. Ore, O.: Theory of graphs. Am. Math. Soc. Transl. **38**, 206–212 (1962)
8. Purnima, G., Mukti, A., Deepti, J.: 2-point set domination number of a cactus. Gulf J. Math. **4**(3), 80–89 (2016)
9. West, D.B.: Introduction to Graph Theory. Prentice Hall Inc., Upper Saddle River (2001)

On Graphs Whose Graphoidal Length Is Half of Its Size

Purnima Gupta[1] and Rajesh Singh[2(✉)]

[1] Department of Mathematics, Sri Venkateswara College, University of Delhi, New Delhi 110021, Delhi, India
purnimachandni1@gmail.com

[2] Department of Mathematics, University of Delhi, New Delhi 110007, Delhi, India
singh_rajesh999@outlook.com

Abstract. Let $G = (V, E)$ be a finite graph. A graphoidal cover Ψ of G is a collection of paths (not necessary open) in G such that every vertex of G is an internal vertex of at most one path in Ψ and every edge of G is in exactly one path in Ψ. The graphoidal covering number η of G is the minimum cardinality of a graphoidal cover of G. The length $gl_{\Psi}(G)$ of a graphoidal cover Ψ of G is defined to be $\min\{l(P) : P \in \Psi\}$ where $l(P)$ is the length of the path P. The graphoidal length $gl(G)$ is defined to be $\max\{gl_{\Psi}(G) : \Psi$ is a graphoidal cover of $G\}$. For any graph G of size q, $gl(G) \le q$ and this bound is attained if and only if G is either a path or a cycle. Further if $gl(G) \ne q$, then $gl(G) \le \lfloor q/2 \rfloor$. In this paper we characterize graphs having graphoidal length $\lfloor q/2 \rfloor$. In the process we obtain that there are exactly 12 non homomorphic graphs having graphoidal covering number two.

Keywords: Graphoidal length · Graphoidal covering number · Graphoidal cover

1 Introduction

We consider finite, connected, undirected graphs without loops and multiple edges. The order and size of a graph G are denoted by p and q respectively. For terminology we refer to Chartrand and Lesniak [12], unless explicitly defined otherwise.

A *graphoidal cover* of a graph G is a collection Ψ of non-trivial paths in G are not necessarily open, such that every vertex of G is an internal vertex of at most one path in Ψ and every edge of G is in exactly one path in Ψ.

The set E of its edges is a graphoidal cover of G, called trivial graphoidal cover of G. A graphoidal cover Ψ of a graph containing at least one path of length

R. Singh is thankful to University Grants Commission (UGC) for providing research grant Schs/SRF/AA/139/F-212/2013-14/438.

S. Arumugam et al. (Eds.): ICTCSDM 2016, LNCS 10398, pp. 289–299, 2017.
DOI: 10.1007/978-3-319-64419-6_37

greater than one is called a non-trivial graphoidal cover. Clearly any graph of size at least 2 possesses a non-trivial graphoidal cover.

The concept of graphoidal covers [4] was introduced by Acharya and Sampathkumar as a close variant of another emerging discrete structure called *semigraphs* [18]. Many interesting notions based on the concept of graphoidal covers such as graphoidal covering number [4], graphoidal labeling [17], etc. were introduced and are being studied extensively. In particular, the notion of graphoidal covering number of a graph has attracted many researchers and numerous work is present in literature [7–10,14,15]. Acharya and Gupta in 1999 extended the concept of graphoidal covers to infinite graphs and introduced the notion of graphoidal domination in *graphoidally covered graphs* [1–3]. In 2016 the authors using the concept of graphoidal covers introduced a new graph invariant called graphoidal length of a graph [6]. A detailed study of graphoidal covers is given in [3,5].

Definition 1 [6]. *The length of a graphoidal cover Ψ of a non-trivial graph G, denoted by $gl_\Psi(G)$, is defined as $gl_\Psi(G) = \min\{l(P) \ : \ P \in \Psi\}$, where $l(P)$ is the length of the path P.*

Definition 2 [6]. *The graphoidal length of graph G, denoted by $gl(G)$, is defined by*

$$gl(G) = \max\left\{gl_\Psi(G) \ : \ \Psi \in \mathcal{G}_G\right\}.$$

For any graphoidal cover Ψ of G, $gl(G) \geq gl_\Psi(G)$. A graphoidal cover Ψ with $gl_\Psi(G) = gl(G)$ is called a gl-graphoidal cover of G.

By definition of the graphoidal length, for any graph G, $1 \leq gl(G) \leq q$ and $gl(G) = q$ if and only if G is a path or a cycle. Further for the star $K_{1,n} (n \geq 3), gl(K_{1,n}) = 1$. Hence the above bounds are sharp. If further $gl(G) \neq q$ for any graph G of size q, then every graphoidal cover Ψ of G contains at least two paths, whence length of at least one path in Ψ is less than or equal to $\lfloor q/2 \rfloor$. Thus if $gl(G) \neq q$ then length of every graphoidal cover of G is less than or equal to $\lfloor q/2 \rfloor$. It follows that $gl(G) \leq \lfloor q/2 \rfloor$ whenever $gl(G) \neq q$.

Proposition 1. *If $G \not\cong P_n, C_n$, then $1 \leq gl(G) \leq \lfloor q/2 \rfloor$.*

In this paper we characterize graphs G for which graphoidal length $gl(G) = \lfloor q/2 \rfloor$, where q is the size of the graph. For this purpose, we need some terminology.

1.1 Terminology and Notation

Definition 3. *A subdivision of an edge uv is obtained by removing edge uv, adding a new vertex w and adding edges uw and vw. Any graph derived from a graph G by a sequence of edge subdivisions is called a subdivision of G.*

Definition 4 [13]. *Two graphs G_1 and G_2 are homomorphic if there exists a graph G such that both G_1 and G_2 are subdivisions of G. We write $G_1 \sim G_2$ if the two graphs G_1 and G_2 are homomorphic.*

Example 1. The graphs G_1, G_2 and G_3 in Fig. 1 are the only mutually non homomorphic cycles with exactly two chords.

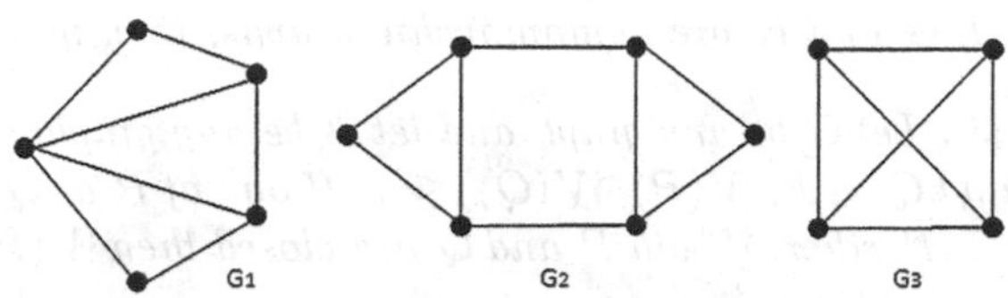

Fig. 1. Non homomorphic cycles with two chords

Definition 5 [11]. *A generalized theta graph or k-theta graph, denoted by $\Theta_{l_1,l_2,\ldots,l_k}$ consists of two vertices joined by k internally disjoint paths of lengths $l_1, l_2, \ldots, l_k$.*

Definition 6. *A guitar graph is a graph obtained by identifying an end vertex of path P_{n+1} with a vertex of k-theta graph $\Theta_{l_1,l_2,\ldots,l_k}$. If the k-degree vertex is identified then it is denoted by $\mathfrak{G}_{n,l_1,l_2,\ldots,l_k}$. If 2-degree vertex on path of length l_i at a distance r from the k-degree vertex is identified, then it is denoted by ${}_r\mathfrak{D}^{l_i}_{n,l_1,\ldots,l_k}$ or simply $\mathfrak{D}_{n,l_1,\ldots,l_k}$ if the position of the identified vertex is ineffectual.*

Definition 7. *A generalized tadpole or tadpole with k-tails, denoted by $\Gamma_{n,l_1,l_2,\ldots,l_k}$, is a graph obtained by identifying a vertex of the cycle C_n with a pendant vertex of each of the paths $P_{l_1+1}, P_{l_2+1}, \ldots, P_{l_k+1}$.*

Definition 8. *A spider with k-legs, denoted by $S_{l_1,l_2,,\ldots,l_k}$, is a tree obtained by identifying an end vertex of each of the paths $P_{l_1+1}, P_{l_2+1}, \ldots, P_{l_k+1}$. Each path is then called a leg of the spider $S_{l_1,l_2,,\ldots,l_k}$.*

Thus a spider with k-legs is a tree homomorphic to $K_{1,k}$ or subdivision of $K_{1,k}$.

Definition 9. *A double cycle, denoted by $C_{m,n}$ is a graph obtained by identifying a vertex of a cycle C_m and a vertex of a cycle C_n.*

Definition 10 [16]. *The bull is the connected graph consisting of a triangle and two nonadjacent pendant edges. A generalized bull is a graph homomorphic to the bull graph.*

A generalized bull can be partitioned into four internally disjoint paths of length l_1, l_2, l_3, l_4, where paths of lengths l_1, l_2 are between vertices of degree 3 and paths of length l_3, l_4 are between a 3-degree vertex and a pendant vertex. Hence based on this partition we denote a generalized bull as B_{l_1,l_2,l_3,l_4}.

Notation 1 [9]. *If $P = (v_0, v_1, ..., v_n)$ and $Q = (v_n = w_0, w_1, ..., w_m)$ are two paths in G, then the walk obtained by concatenating P and Q at v_n is denoted by $P \circ Q$ and the path $(v_n, v_{n-1}, ..., v_1, v_0)$ is denoted by P^{-1}.*

Definition 11 [4]. *The graphoidal covering number of a graph G, denoted by $\eta(G)$, is the minimum cardinality of a graphoidal cover of G. A graphoidal cover Ψ of a graph G is called an η-graphoidal cover if $|\Psi| = \eta(G)$.*

Proposition 2. *If G and H are homomorphic graphs, then $\eta(G) = \eta(H)$.*

Proposition 3 [3]. *Let G be any graph and let Ψ be any graphoidal cover of G. Then for any P and Q in Ψ, $|V(P) \cap V(Q)| \leq 4$. If one of P or Q is closed, then $|V(P) \cap V(Q)| \leq 3$. Further, if both P and Q are closed then $|V(P) \cap V(Q)| \leq 2$.*

Observation 1. *Let Ψ be any graphoidal cover of a (p, q)-graph G.*

1. *Every vertex of odd degree must be an end vertex of some open path in Ψ. If further G has k vertices of odd degree, then at least $\lceil k/2 \rceil$ paths in Ψ are open.*
2. *If $|\Psi| = 2$, then at most two vertices in G can have degree 4 and corresponding to each 4-degree vertex, there is a closed path in Ψ.*

2 Graphs with Graphoidal Length $\lfloor q/2 \rfloor$

In this section we characterize graphs with graphoidal length $\lfloor q/2 \rfloor$. For this purpose, we first give two lemmas.

Lemma 1. *If G is a graph with $gl(G) = \lfloor q/2 \rfloor$, then $\eta(G) = 2$.*

Proof. Since $gl(G) = \lfloor q/2 \rfloor$, $3 \leq \Delta(G) \leq 4$ and G has a graphoidal cover Ψ consisting of two paths (may be closed) of lengths $\lfloor q/2 \rfloor$ and $\lceil q/2 \rceil$. Since $\Delta(G) \geq 3$, $\eta(G) \geq 2$. Also, by definition, $\eta(G) \leq |\Psi| = 2$. Hence it follows that $\eta(G) = 2$.

Lemma 2. *Let G be any graph. Then $\eta(G) = 2$ if and only if G is homomorphic to one of the twelve graphs in Fig. 2.*

Proof. Let G be homomorphic to one of the graphs in Fig. 2. Then by Proposition 2, $\eta(G) = 2$. *Conversely,* let $\eta(G) = 2$ and let $\Psi = \{P, Q\}$ be an η-graphoidal cover of G. It follows from Proposition 3 that $1 \leq |V(P) \cap V(Q)| \leq 4$.

Case I. $|V(P) \cap V(Q)| = 1$.

Let $V(P) \cap V(Q) = \{v\}$. If both P and Q are open, then v must be an end vertex of one path and an internal vertex of the other and hence $G \sim K_{1,3}$ and we are through. If both P and Q are closed, then $G \sim C_{3,3}$. Now, suppose P is closed and Q is open. If v is an end vertex of Q, then $G \sim \Gamma_{3,1}$. If v is internal to Q, then it must be the coincident end vertex of P and hence $G \sim \Gamma_{3,1,1}$.

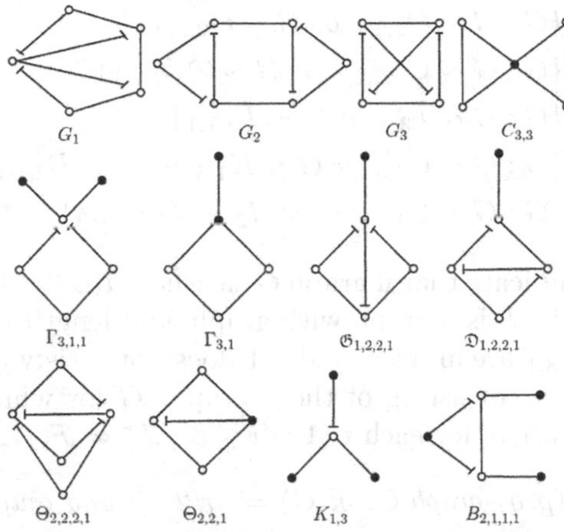

Fig. 2. η-graphoidal cover of every graph is shown.

Case II. $|V(P) \cap V(Q)| = 2$.

If both P and Q are open, then either $G \sim B_{2,1,1,1}$ or $\Gamma_{3,1}$. If both P and Q are closed, then $G \sim \Theta_{2,2,2,1}$. If exactly one of P or Q is open, then either $G \sim \Theta_{2,2,1}$ or $G \sim \mathfrak{G}_{1,2,2,1}$.

Case III. $|V(P) \cap V(Q)| = 3$.

Since $|V(P) \cap V(Q)| = 3$, both P and Q cannot be closed (Proposition 3). If exactly one of them is closed, then one common vertex must be coincident end vertex of the closed path and internal of the open and the other two common vertices must be end vertices of the open path and internal to closed path. Clearly, in this case $G \sim G_1$. If both P and Q are open, then without loss in generality we can assume that two common vertices are end vertices of P and the remaining common vertex an end vertex of Q. The other end vertex of Q, thus, must be a pendant vertex of G. Hence it follows that $G \sim \mathfrak{D}_{1,2,2,1}$.

Case IV. $|V(P) \cap V(Q)| = 4$.

Since $|V(P) \cap V(Q)| = 4$, both P and Q must be open. Further, two common vertices are end vertices of P and internal vertices of Q and the remaining two are end vertices of Q and internals to P. Then there are only two possible configuration for G i.e., $G \sim G_2$ or $G \sim G_3$.

Thus in all the possible cases, we arrive at the conclusion that G is homomorphic to one of the graphs listed in Fig. 2.

Notation 2. *Let $\mathcal{F}$ denote the family of all graphs G with $\eta(G) = 2$. Then in view of* Lemma *2, $\mathcal{F} = \cup_{i=1}^{5} \mathcal{F}_i$, where*

$$\mathcal{F}_1 = \{G : G \sim \Theta_{2,2,1} \ or \ G \sim \Theta_{2,2,2,1}\}$$
$$\mathcal{F}_2 = \{G : G \sim \mathfrak{G}_{1,2,2,1} \ or \ G \sim \mathfrak{D}_{1,2,2,1}\}$$
$$\mathcal{F}_3 = \{G : G \sim \Gamma_{3,1} \ or \ G \sim \Gamma_{3,1,1}\}$$
$$\mathcal{F}_4 = \{G : G \sim C_{3,3} \ or \ G \sim K_{1,3} \ or \ G \sim B_{2,1,1,1}\}$$
$$\mathcal{F}_5 = \{G : G \sim G_1 \ or \ G \sim G_2 \ or \ G \sim G_3\}.$$

Lemmas 1 and 2 indicate that if graph G satisfies $gl(G) = \lfloor q/2 \rfloor$, then $G \in \mathcal{F}$. But not every graph in $\mathcal{F}$ is a graph with graphoidal length $\lfloor q/2 \rfloor$. For example $\theta_{2,2,4,6}$, $\mathfrak{G}_{2,2,5,7}$ and $\Gamma_{3,1}$ are all in $\mathcal{F}$ and yet does not satisfy $gl(G) = \lfloor q/2 \rfloor$. Let $\mathcal{F}^*$ be the subset of $\mathcal{F}$ consisting of those graphs G for which $gl(G) = \lfloor q/2 \rfloor$. Then $\mathcal{F}^* = \cup_{i=1}^{5} \mathcal{F}_i^*$, where for each i $(1 \leq i \leq 5)$, $\mathcal{F}_i^* = \mathcal{F}_i \cap \mathcal{F}^*$.

Theorem 1. *For a (p, q)-graph G, $gl(G) = \lfloor q/2 \rfloor$ if and only if $G \in \mathcal{F}^*$.*

Thus to characterize graphs G with graphoidal length $gl(G) = \lfloor q/2 \rfloor$, we need to characterize graphs in $\mathcal{F}^*$. Hence we shall characterize graphs in $\mathcal{F}_i^*$ separately for each $i \in \{1, 2, 3, 4, 5\}$.

3 Characterization of Graphs in $\mathcal{F}^*$

Theorem 2. *A graph $G \in \mathcal{F}_1^*$ if and only if one of the following holds:*

(a). $G \cong \Theta_{l_1,l_2,l_3}$
(b). $G \cong \Theta_{l_1,l_2,l_3,l_4}$ *and the length of the paths is such that $|(l_i+l_j)-(l_r+l_s)| \leq 1$ for some distinct $i, j, r, s \in \{1, 2, 3, 4\}$.*

Proof. If $G \in \mathcal{F}_1^*$, then $gl(G) = \lfloor q/2 \rfloor$ and either $G \cong \Theta_{l_1,l_2,l_3}$ or Θ_{l_1,l_2,l_3,l_4}. If $G \cong \Theta_{l_1,l_2,l_3}$, then (a) holds. Thus let $G \cong \Theta_{l_1,l_2,l_3,l_4}$ and $\Psi = \{A_1, A_2\}$ be a gl-graphoidal cover of G. Since G has two vertices of degree 4, from Observation 1 both A_1 and A_2 are closed with coincident end vertices u and v, respectively. Consequently, $l(A_1) = l_i+l_j$ and $l(A_2) = l_r+l_s$ for distinct i, j, r, s. Consequently, $|(l_i + l_j) - (l_r + l_s)| = |l(A_1) - l(A_2)| \leq 1$. Thus (b) holds.

Conversely, suppose the hypothesis holds. Since G has a vertex of degree at least three, $gl(G) \leq \lfloor q/2 \rfloor$. Hence it s enough to show the existence of a graphoidal cover Ψ such that $gl_\Psi(G) = \lfloor q/2 \rfloor$.

If (a) holds then G has two vertices u and v of degree 3 and three internally disjoint paths P, Q, R of lengths l_1, l_2, l_3 respectively. Let $P = (u_0, u_1, ..., u_{l_1})$, $Q = (v_0, v_1, ..., v_{l_2})$ and $R = (w_0, w_1, ..., w_{l_3})$, where $u = u_0 = v_0 = w_0$ and $v = u_{l_1} = v_{l_2} = w_{l_3}$. Without loss of generality assume that $l_1 \geq l_2 \geq l_3$. Let

$$r = \left\lfloor \frac{l_1 + l_2 - l_3 + 2}{2} \right\rfloor.$$

By the choice of r, $\lfloor q/2 \rfloor = r + l_3 - 1$ and $r < l_1$. Then the graphoidal cover $\Psi = \{A_1, A_2\}$ of G, where $A_1 = Q \ o \ (u_0, u_1 ...., u_r)$ and $A_2 = R \ o \ (u_{l_1}, ..., u_r)$, is such that $l(A_1), l(A_2) \geq \lfloor q/2 \rfloor$ and consequently $gl_\Psi(G) = \lfloor q/2 \rfloor$.

Suppose (b) holds i.e., $G \cong \Theta_{l_1,l_2,l_3,l_4}$ and $|(l_i + l_j) - (l_r + l_s)| \leq 1$ for some distinct $i, j, r, s \in \{1, 2, 3, 4\}$. Let u and v be the two vertices of degree 4 in G and P, Q, R, S be internally disjoints paths joining u to v of respective lengths l_i, l_j, l_r, l_s. Consider the graphoidal cover $\Psi = \{A_1, A_2\}$ of G, where $A_1 = P\ o\ Q^{-1}$ is closed path with coincident end vertex u and $A_2 = R^{-1}\ o\ S$ is closed path with coincident end vertex v. By the hypothesis, $|l(A_1) - l(A_2)| \leq 1$ and consequently $l(A_1), l(A_2) \geq \lfloor q/2 \rfloor$. Thus $gl_\Psi(G) = \lfloor q/2 \rfloor$.

Theorem 3. *A graph $G \in \mathcal{F}_2^*$ if and only if one of the following holds:*

(a). $G \cong \mathfrak{G}_{n,l_1,l_2,l_3}$ *and* $|(n + l_i) - (l_j + l_k)| \leq 1$ *for some distinct* $i, j, k \in [3]$.
(b). $G \cong {}_r\mathfrak{D}^{l_i}_{n,l_1,l_2,l_3}$ $(1 \leq i \leq 3)$ *and one of the following holds:*
1. $|(n + 2r + l_j) - (l_i + l_k)| \leq 1$
2. $|(n + l_i + l_j) - (2r + l_k)| \leq 1$.
where $\{i, j, k\} = \{1, 2, 3\}$

Proof. If $G \in \mathcal{F}_2^*$, then $gl(G) = \lfloor q/2 \rfloor$ and $G \cong \mathfrak{G}_{n,l_1,l_2,l_3}$ or ${}_r\mathfrak{D}^{l_i}_{n,l_1,l_2,l_3}$. If $G \cong \mathfrak{G}_{n,l_1,l_2,l_3}$, then G has a pendant vertex u, a 3-degree vertex v, a 4-degree vertex w and all other vertices of degree 2. Let Ψ be a gl-graphoidal cover of G. Since $|\Psi| = 2$, from Observation 1, w must be coincident end vertex of a closed path P in Ψ and internal vertex of the other path Q in Ψ. Also, u and v are end vertices of Q and v is an internal vertex of P. Then for some i, j, k, $Q = P_{n+1}\ o\ P_{l_i+1}$ and $P = P_{l_j+1}\ o\ P_{l_k+1}$ and hence $l(Q) = n + l_i$ and $l(P) = l_j + l_k$. Since $|l(Q) - l(P)| \leq 1$, it follows that $|(n + l_i) - (l_j + l_k)| \leq 1$. Thus (a) holds.

If $G \cong {}_r\mathfrak{D}^{l_i}_{n,l_1,l_2,l_3}$, then G has 3 vertices of degree 3, one pendant vertex and all other vertices of degree 2. Also G has internally disjoint paths Q_1, Q_2, Q_3 and Q_4 of lengths l_1, l_2, l_3 and n, respectively. Let $Q_i = (u_{i_0}, u_{i_1}, ..., u_{i_{l_i}})$ for $i = 1, 2, 3$, where $u_{1_0} = u_{2_0} = u_{3_0}$ and $u_{1_{l_1}} = u_{2_{l_2}} = u_{3_{l_3}}$. Since G has four vertices of odd degree, both the paths in a gl-graphoidal cover $\Psi = \{A_1, A_2\}$ must be open with odd degree vertices as end vertices. Thus for some j, k either $A_1 = Q_4\ o\ (u_{1_r}, ..., u_{1_0})\ o\ Q_j$ and $A_2 = (u_{1_r}, ..., u_{1_{l_1}})\ o\ (Q_k)^{-1}$ or $A_1 = Q_4\ o\ (u_{1_r}, ..., u_{1_{l_1}})\ o\ (Q_j)^{-1}$ and $A_2 = (u_{1_r}, ..., u_{1_0})\ o\ Q_k$. Since $|l(A_1) - l(A_2)| \leq 1$, either $|(n+2r+l_j) - (l_1+l_k)| \leq 1$ or $|(n+l_1+l_j) - (2r+l_k)| \leq 1$ holds. Hence (b) holds.

Conversely, suppose the hypothesis holds. Then $gl(G) \leq \lfloor q/2 \rfloor$. If (a) holds, then the graphoidal cover $\Psi = \{A_1, A_2\}$ of G where A_1 is the closed Ψ-edge consisting of paths P_{l_j+1} and P_{l_k+1} with vertex of degree 4 as its coincident vertex and A_2 consists of paths P_{l_i+1}, P_{l_n+1}. Clearly, $gl_\Psi(G) = \lfloor q/2 \rfloor$.

If (b) holds, then again G has four internally disjoints paths Q_i, Q_j, Q_k and Q of lengths l_i, l_j, l_k and n respectively. If (1) is true, then for $A_1 = Q\ o\ (u_{i_r}, ..., u_{i_0})\ o\ Q_j$ and $A_2 = (u_{i_r}, ..., u_{i_{l_i}})\ o\ (Q_k)^{-1}$, $\Psi = \{A_1, A_2\}$ is a graphoidal cover of G such that $gl_\Psi = \lfloor q/2 \rfloor$. If (2) holds, then $\Psi = \{A_1, A_2\}$ is a graphoidal cover of length $\lfloor q/2 \rfloor$, where $A_1 = Q\ o\ (u_{i_r}, u_{i_{r+1}}, ..., u_{i_{l_i}})\ o\ Q_j^{-1}$ and $A_2 = (u_{i_r}, u_{i_{r-1}}, ..., u_{i_0})\ o\ Q_k$.

Hence in either case we get a graphoidal cover of length $\lfloor q/2 \rfloor$ and the theorem follows.

Theorem 4. *A graph $G \in \mathcal{F}_3^*$ if and only if one of the following holds:*

(a). $G \cong \Gamma_{n,l_1}$ *and* $l_1 \leq n+1$.
(b). $G \cong T_{n,l_1,l_2}$ *and* $|n-(l_1+l_2)| \leq 1$.

Proof. If $G \in \mathcal{F}_3^*$, then either $G \cong \Gamma_{n,l_1}$ or $G \cong T_{n,l_1,l_2}$. Let $\Psi = \{A_1, A_2\}$ be a *gl*-graphoidal cover of G. Without loss of generality assume that $V(A_1) \cap V(C_n) \neq \phi$ and $V(A_2) \cap V(P_{l_1+1}) \neq \phi$. Then $l(A_1) \leq n$ and $l(A_2) \geq l_1$. Consequently,

$$l_1 \leq l(A_2) \leq l(A_1) + 1 \leq n + 1.$$

If $G \cong \Gamma_{n,l_1}$, then we are through. Thus let $G \cong T_{n,l_1,l_2}$. Then $\Psi = \{A_1, A_2\}$, where $A_1 = C_n$ and $A_2 = (P_{l_1+1})^{-1}\ o\ P_{l_2+1}$, is the unique graphoidal cover of G with cardinality two. Since $|l(A_1) - l(A_2)| \leq 1$, we have $|n-(l_1+l_2)| \leq 1$.

Conversely, suppose the hypothesis holds. If (a) holds, then G has a 3-degree vertex u and a pendant vertex. Let $P_{l_1+1} = (u_0, u_1, ..., u_{l_1})$ and $C_n = (v_0, v_1, ..., v_{n-1}, v_0)$ be such that $u_0 = v_0 = u$. Since $l_1 \leq n+1$, it follows that $\lfloor q/2 \rfloor \leq n$. Let $k = \lfloor q/2 \rfloor$, $A_1 = (v_0, v_1, ..., v_k)$ and $A_2 = (v_k, ..., v_{n-1}, v_0 = u_0, u_1, ..., u_{l_1})$. Then $\Psi = \{A_1, A_2\}$ is a graphoidal cover of G having $gl_\Psi(G) = \lfloor q/2 \rfloor$. Hence we have $gl(G) = \lfloor q/2 \rfloor$. If (b) holds, then $\Psi = \{C_n, P_{l_1+1}^{-1}\ o\ P_{l_2+1}\}$ is the graphoidal cover of G with $gl_\Psi(G) = \lfloor q/2 \rfloor$, whence $gl(G) = \lfloor q/2 \rfloor$.

Theorem 5. *A graph $G \in \mathcal{F}_4^*$ if and only if one of the following holds:*

(a). $G \cong S_{l_1,l_2,l_3}$ *and* $|l_i - (l_j + l_k)| \leq 2$ *for some distinct* $i, j, k \in [3]$.
(b). $G \cong C_{m,n}$ *and* $|n-m| \leq 1$.
(c). $G \cong B_{l_1,l_2,l_3,l_4}$ *and one of the following holds:*
1. $|(l_i + l_r) - (l_j + l_s)| \leq 1$
2. $|(l_i) - (l_j + l_r + l_s)| \leq 1$

where $\{i, j\} = \{1, 2\}$ *and* $\{r, s\} = \{3, 4\}$.

Proof. Let $G \in \mathcal{F}_4^*$, then $G \cong S_{l_1,l_2,l_3}$ or $C_{m,n}$ or B_{l_1,l_2,l_3,l_4}. Let $\Psi = \{A_1, A_2\}$ be a *gl*-graphoidal cover of G. If $G \cong S_{l_1,l_2,l_3}$, then G has a 3-degree vertex which is internal to one path in Ψ (say) A_2. Then for some i, j, k, $A_1 = P_{l_i+1}$. But then $A_2 = (P_{l_j+1})^{-1}\ o\ P_{l_k+1}$. Thus

$$|l_i - (l_j + l_k)| = |l(A_1) - l(A_2)| \leq 1.$$

If $G \cong C_{m,n}$, then both A_1 and A_2 must be closed paths in Ψ and hence $A_1 = C_n$ and $A_2 = C_m$. Consequently, $|n-m| \leq 1$.

If $G \cong B_{l_1,l_2,l_3,l_4}$, then G has four vertices of odd degree and four internally disjoints paths Q_1, Q_2, Q_3 and Q_4 of lengths l_1, l_2, l_3 and l_4 respectively. Since G has four vertices of odd degree, both the paths A_1 and A_2 in Ψ must be open and with end vertices of odd degree. Thus either $A_1 = Q_r\ o\ Q_i$ and $A_2 = Q_s\ o\ (Q_j)^{-1}$ or $A_1 = Q_i$ and $A_2 = Q_r\ o\ Q_j\ (Q_s)^{-1}$, for some i, j, r, s such that $\{i, j\} = \{1, 2\}$ and $\{r, s\} = \{3, 4\}$. Since $|l(A_1) - l(A_2)| \leq 1$, either $|(l_i + l_r) - (l_j + l_s)| \leq 1$ or $|(l_i) - (l_j + l_r + l_s)| \leq 1$ holds.

Conversely, suppose the hypothesis hold. If (a) is true, then the graphoidal cover $\Psi = \{A_1, A_2\}$ of G, where $A_1 = P_{l_i+1}$ and $A_2 = (P_{l_j+1})^{-1} \ o \ P_{l_k+1}$, is such that $gl_\Psi(G) = \lfloor q/2 \rfloor$. Hence it follows that $gl(G) = \lfloor q/2 \rfloor$. If (b) is true, then the graphoidal cover $\Psi = \{C_n, C_m\}$ is such that $gl_\Psi(G) = \min\{m, n\}$. Since $|n - m| \leq 1$, it follows that $gl_\Psi(G) = \lfloor q/2 \rfloor$. Consequently, $gl(G) = \lfloor q/2 \rfloor$.

Let (c) holds and Q_i, Q_j, Q_r and Q_s be the four internally disjoints paths of lengths l_i, l_j, l_r and l_s respectively, where $\{i, j\} = \{1, 2\}$ and $\{r, s\} = \{3, 4\}$. If (1) holds, then for $A_1 = Q_r \ o \ Q_i$ and $A_2 = Q_s \ o \ (Q_j)^{-1}$, $\Psi = \{A_1, A_2\}$ is a graphoidal cover of G such that $gl_\Psi(G) = \lfloor q/2 \rfloor$. If (2) is true, then $\Psi = \{A_1, A_2\}$, where $A_1 = Q_i$ and $A_2 = Q_r \ o \ Q_j \ (Q_s)^{-1}$, is a graphoidal cover of G such that $l(A_1), l(A_2) \geq \lfloor q/2 \rfloor$. Thus in either case we get a graphoidal cover of length $gl_\Psi(G) = \lfloor q/2 \rfloor$. Hence $gl(G) = \lfloor q/2 \rfloor$.

We now proceed to characterize graphs in $\mathcal{F}_5^*$. If $G \in \mathcal{F}_5^*$, then $G \sim G_1$ or G_2 or G_3, given in Fig. 2.

Let G be G_1-subdivision graph with a vertex u of degree 4, two vertices v_1 and v_2 of degree 3 and all other vertices of degree 2. Let m and n be the lengths of the two u-v_1 paths which do not pass through v_2. Similarly, let r and s be the lengths of the two u-v_2 paths which do not pass through v_1. Let l be the length of the v_1-v_2 path which do not pass through u. Then with these notations in mind we denote the graph G by ${}_lA^{r,s}_{m,n}$.

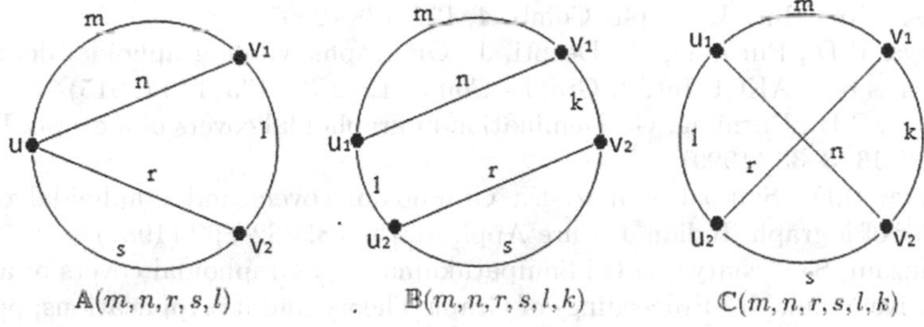

$\mathbb{A}(m, n, r, s, l)$. $\mathbb{B}(m, n, r, s, l, k)$ $\mathbb{C}(m, n, r, s, l, k)$

Let G be a G_2-subdivision graph with four vertices u_1, u_2, v_1 and v_2 of degree 3 and all other vertices of degree 2. We denote G by ${}^k_lB^{r,s}_{m,n}$, where m and n are lengths of the two u_1-v_1 paths which do not pass through u_2 and v_2, r and s are lengths of the two u_2-v_2 paths which do not pass through u_1 and v_1 and k and l are the respective lengths of u_1-u_2 path and v_1-v_2 path which do not pass through v_1 or v_2 and u_1 or u_2 respectively.

Let G be a G_3-subdivision graph with four vertices u_1, u_2, v_1 and v_2 of degree 3 and all other vertices of degree 2. We denote G by ${}^k_lC^{r,s}_{m,n}$, where m and n are lengths of u_1-v_1 and u_1-v_2 paths which do not pass through u_2 and v_1 respectively, r and s are lengths of u_2-v_1 and u_2-v_2 paths which do not pass through u_1 and v_1 respectively and k and l are the respective lengths of u_1-u_2 path and v_1-v_2 path which do not pass through v_1 and u_1 respectively.

The following three theorems give a characterization of graphs in $\mathcal{F}_5^*$ and we omit the proofs.

Theorem 6. *Let* $G \cong {}_lA^{n_1,n_2}_{m_1,m_2}$ *be a graph. Then* $gl(G) = \lfloor q/2 \rfloor$ *if and only if* $|(m_i + n_r) - (m_j + n_s + l)| \leq 1$ *for some* i, j, r, s *such that* $\{i, j\} = \{r, s\} = \{1, 2\}$.

Theorem 7. *Let* $G \cong {}_{l_1}^{l_2}B_{m_1,m_2}^{n_1,n_2}$ *be a graph. Then* $gl(G) = \lfloor q/2 \rfloor$ *if and only if* $|(m_i + n_j + l_k) - (m_r + n_s + l_t)| \leq 1$ *for some* i, j, k, r, s, t *such that* $\{i, r\} = \{j, s\} = \{k, t\} = \{1, 2\}$.

Theorem 8. *Let* $G \cong {}_{m_5}^{m_6}C_{m_1,m_2}^{m_3,m_4}$ *be a graph. Then* $gl(G) = \lfloor q/2 \rfloor$ *if and only if* $|(m_i + m_j + m_k) - (m_r + m_s + m_t)| \leq 1$ *for distinct* $i, j, k, r, s, t \in [6]$.

Theorems 1–8 completely characterize graphs G with $gl(G) = \lfloor q/2 \rfloor$. Following theorem summarizes all the above mentioned theorems.

4 Conclusion

In this paper graphs having graphoidal length $\lfloor q/2 \rfloor$ are characterized. Now if $\mathcal{H}$ denote the set of all graphs having graphoidal length q or $\lfloor q/2 \rfloor$, then for any $G \notin \mathcal{H}$, $gl(G) \leq \lfloor q/3 \rfloor$. One may consider characterizing extremal graphs for this bound. Also, as we noticed that $gl(G) = \lfloor q/2 \rfloor$ implied that $\eta(G) = 2$, similarly can we say that $\eta(G) = 3$ is necessary for a graph G to have $gl(G) = \lfloor q/3 \rfloor$.

References

1. Acharya, B.D., Purnima, G.: Further results on domination in graphoidally covered graphs. AKCE Int. J. Graphs Comb. **4**, 127–138 (2007)
2. Acharya, B.D., Purnima, G., Deepti, J.: On graphs whose graphoidal domination number is one. AKCE Int. J. Graphs Comb. **12**(2–3), 133–140 (2015)
3. Acharya, B.D., Purnima, G.: Domination in graphoidal covers of a graph. Discrete Math. **206**, 3–33 (1999)
4. Acharya, B.D., Sampathkumar, E.: Graphoidal covers and graphoidal covering number of a graph. Indian J. Pure Appl. Math. **18**, 882–890 (1987)
5. Arumugam, S., Acharya, B.D., Sampathkumar, E.: Graphoidal covers of a graph: a creative review. In: Proceedings of Graph Theory and its Applications, pp. 1–28. Tata McGraw-Hill, New Delhi (1997)
6. Arumugam, S., Purnima, G., Rajesh, S.: Bounds on graphoidal length of a graph. Electron. Notes Discrete Math. **53**, 113–122 (2016)
7. Arumugam, S., Pakkiam, C.: Graphoidal bipartite graphs. Graphs Combin. **10**, 305–310 (1994)
8. Arumugam, S., Pakkiam, C.: Graphs with unique minimum graphoidal cover. Indian J. Pure Appl. Math. **25**, 1147–1153 (1994)
9. Arumugam, S., Rajasingh, I., Pushpam, P.R.L.: Graphs whose acyclic graphoidal covering number is one less than its maximum degree. Discrete Math. **240**, 231–237 (2001)
10. Arumugam, S., Suseela, J.S.: Acyclic graphoidal covers and path partitions in a graph. Discrete Math. **190**, 67–77 (1998)
11. Brown, J.I., Hickman, C., Sokal, A.D., Wagner, D.G.: On the chromatic roots of generalized theta graphs. J. Combin. Theory Ser. B. **83**(2), 272–297 (2001)
12. Chartrand, G., Lesniak, L.: Graphs & Digraphs. Chapman & Hall/CRC, Boca Raton (2005)
13. Gross, J.L., Yellen, J., Zhang, P.: Handbook of Graph Theory. Discrete Mathematics and its Applications. CRC Press, Boca Raton (2014)

14. Pakkiam, C., Arumugam, S.: On the graphoidal covering number of a graph. Indian J. Pure Appl. Math. **20**, 330–333 (1989)
15. Purnima, G., Rajesh, S.: Domination in graphoidally covered graphs: least-kernel graphoidal covers. Electron. Notes Discrete Math. **53**, 433–444 (2016)
16. Reed, B., Sbihi, N.: Recognizing bull-free perfect graphs. Graphs Combin. **11**(2), 171–178 (1995)
17. Sahul Hamid, I., Anitha, A.: On label graphoidal covering number-I. Trans. Comb. **1**, 25–33 (2012)
18. Sampathkumar, E.: Semigraphs and their applications. Report on the DST Project (2000)

Point-Set Domination in Graphs. VIII: Perfect and Efficient PSD Sets

Purnima Gupta[1], Alka[2(✉)], and Rajesh Singh[2]

[1] Department of Mathematics, Sri Venkateswara College, University of Delhi, New Delhi 110021, Delhi, India
purnimachandni1@gmail.com

[2] Department of Mathematics, University of Delhi, New Delhi 110007, Delhi, India
09alka01@gmail.com, singh_rajesh999@outlook.com

(Dedicated to Dr. B.D. Acharya on his 69th birthday.)

Abstract. A perfect dominating set in a graph G is a dominating set D such that every vertex v in $V - D$ is adjacent to a unique vertex u in D. An efficient dominating set is a perfect dominating set D which is independent as well. A *point-set dominating set* (or, *psd*-set, in short) is a dominating set D for which every subset S of $V - D$ has a vertex $u \in D$ such that the induced subgraph $\langle S \cup \{u\} \rangle$ is connected. In this paper we determine necessary conditions for a graph to possess an efficient *psd*-set.

Keywords: Perfect dominating set · Efficient dominating set · Point-set dominating set

1 Introduction

For all terminology and notations in graph theory, we refer to West [12]. All graphs considered in this note are non-trivial simple graphs, in the sense that none of them contains a *self-loop* or a *multiple edge* and could be infinite unless mentioned otherwise.

Given any graph $G = (V, E)$, a subset D of V is called an *independent set* in G if no two vertices of D are adjacent in G, a *dominating set* if every vertex u of G is either in D or is adjacent to a vertex v in D, a *perfect dominating set* [6,7] (or, '*pd*-set' in short) if every vertex u of G is either in D or is adjacent to exactly one vertex v in D, an *efficient dominating set* [5] if it is a dominating set which is both independent and perfect; and a *point set dominating set* [1–4,10] (or '*psd*-set' in short) if for every subset S of $V - D$ there exists a vertex $u \in D$ such that the induced subgraph $\langle S \cup \{u\} \rangle$ is connected. For a detailed treatment

R. Singh is thankful to University Grants Commission (UGC) for providing the research grant with sanctioned letter number: Ref. No. Schs/SRF/AA/139/F-212/2013-14/438.

S. Arumugam et al. (Eds.): ICTCSDM 2016, LNCS 10398, pp. 300–304, 2017.
DOI: 10.1007/978-3-319-64419-6_38

of various shades of the notion of domination in graphs, the reader is referred to [8,9].

In this paper we characterize graphs possessing proper perfect psd-sets. We also investigate the structure of graphs G for which there exists an efficient psd-set.

2 Perfect and Efficient Psd-Sets

We shall now establish a necessary and sufficient condition for a dominating set to be both a perfect psd-set.

Theorem 1. *Let $G = (V, E)$ be a graph and let D be a proper nonempty subset of V. Then D is a perfect psd-set if and only if there exists a subset $D' = \{u_1, u_2, \ldots, u_t\}$ of D such that $\{X_1, X_2, \ldots, X_t\}$ is a partition of $V - D$ where $X_i \subset N(u_i)$ and for distinct $i, j \in \mathbf{t}$, each vertex of X_i is adjacent to each vertex of X_j.*

Proof. Sufficiency is obvious. Now, suppose there exists a proper perfect psd-set D in G. Let $D' = \{u \in D \ : \ (V - D) \cap N(u) \neq \emptyset\}$, where $N(u)$ is the open neighborhood of u. Since D is a perfect psd-set, the sets $X_v := (V - D) \cap N(v)$, $v \in D'$ form a partition of $V - D$. If $D = \{u\}$ then $N(u) = V - D$ and the result is trivial. So, let $|D| \geq 2$. Let $x \in X_u$ and $y \in X_v$ for distinct $u, v \in D$. Since $X_u \cap X_v = \emptyset$, we must have $x \neq y$. If x and y are nonadjacent, since D is a psd-set it follows that there must exist $z \in D - \{u, v\}$ such that $x, y \in N(z)$. Hence $x, y \in X_z$. But, this contradicts the fact that $X_z \cap X_u = \emptyset$ and $X_z \cap X_v = \emptyset$. Thus each vertex of X_u is adjacent to each vertex of X_v.

Corollary 1. *Let $G = (V, E)$ be a graph and let D be a proper nonempty subset of V. Then D is an efficient psd-set if and only if $\{N(u) : u \in D\}$ is a partition of $V - D$ and for any two distinct vertices $u, v \in D$ each vertex of $N(u)$ is adjacent to each vertex of $N(v)$.*

In 1979, Walkar et al. [11] proved the following interesting and fundamental theorem.

Theorem 2. [11] *For any finite graph $G = (V, E)$ and for any dominating set D of G, the following inequality holds:*

$$|V - D| \leq \sum_{u \in D} d(u).$$

Further, equality holds if and only if D is an efficient dominating set.

From the above theorem, the following characterization for efficient psd-sets follows immediately.

Theorem 3. *For any graph G, D is an efficient psd-set if and only if D is a psd-set of G satisfying*

$$|V - D| = \sum_{u \in D} d(u). \tag{3}$$

Our next result reveals another special nature of efficient psd-sets in a graph.

Theorem 4. *Every efficient psd-set of a graph G is a minimum psd-set of G.*

Proof. Let D be any efficient psd-set of G. Then D is an independent dominating set and hence D is a minimal dominating set of G. Now, suppose G possesses a psd-set D_1 such that $|D_1| < |D|$. Then, $(V - D_1) \cap D \neq \emptyset$. If there are distinct vertices $u, v \in D \cap (V - D_1)$, then there must exist $w_1 \in D_1$ such that $w_1 \in N(u) \cap N(v)$, a contradiction to the fact that, $\{N(u) : u \in D\}$ is a partition of $V - D$. Thus, $|D \cap (V - D_1)| = 1$ and hence $D_1 = D - \{u\}$ for some $u \in V$. But, this is a contradiction to the minimality of D as a dominating set of G. Hence the result follows.

We now proceed to investigate the structure of graphs possessing efficient psd-sets.

Lemma 1. *Let G be a connected graph having an efficient psd-set D of cardinality at least two. Then G is a nontrivial block if and only if the degree of each vertex in D is at least two.*

Proof. Suppose $d(u) \geq 2$ for every $u \in D$. Since D is an efficient psd-set, it follows from Corollary 1, that $\{N(u) : u \in D\}$ is a partition of $V - D$ and for any two distinct vertices u and v in D, every vertex of $N(u)$ is adjacent to every vertex of $N(v)$. Further, as $|N(w)| \geq 2$ for every $w \in D$, it is easy to see that $V - D$ is contained in a block B of G. But since $|V - D| \cap N(u)| = |N(u)| \geq 2$ it follows that $u \in V(B)$ for every $u \in D$. Hence, $V = D \cup (V - D) \subset V(B) \subset V$. Thus, G must be a block. The converse is obvious.

Theorem 5. *If a connected graph G has an efficient psd-set D, then exactly one of the following holds:*

1. *G has a vertex of full degree,*
2. *G is non-trivial block having no vertex of full degree,*
3. *Either $G \cong P_4$ or G is a separable graph with exactly one nontrivial block B, every vertex in $V(G) - V(B)$ is a pendant vertex and each vertex of B is a support to at most one vertex in $V(G) - V(B)$.*

Proof. Let D be an efficient psd-set. First suppose $|D| = 1$ and $D = \{u\}$, then trivially u is a vertex of full degree. In this case, thus, (i) holds. Next, assume that $|D| \geq 2$. If $d(u) \geq 2$ for every $u \in D$, then it follows Lemma 1 that G must be a nontrivial block. Further, since D is independent and has at least two vertices it follows that no vertex of D is of full degree. Also, since D is a perfect dominating set; $V - D$ does not contain a vertex of full degree too. Thus, (ii) holds.

Next, suppose that the set $D_1 = \{u \in D : d(u) = 1\}$ is nonempty. If $D = D_1$ then since D is an efficient psd-set of G it follows from Corollary 1 that $\langle V-D\rangle$ is a complete subgraph. Hence $G = K^+_{|V-D|}$, the complete of order $|V-D|$ together with one pendent edge attached at each of its vertices. Hence if $|D| = 2$, then $G = P_4$. Now, suppose $D \neq D_1$. Since $D_1 \subset D$ it then follows that there must exist $w \in D$ such that $d(w) \geq 2$. By Corollary 1, it follows that the induced subgraph $\langle (V-D) \cup (D-D_1)\rangle$ must be a nontrivial block, say B. Since D is an efficient psd-set of G the rest of the conclusions in condition (iii) follow.

Lastly, parallel to an ideal pursued by Acharya and Gupta [2], we examine below when an efficient psd-set is unique in an isolate-free graph.

Theorem 6. *Let G be a graph without isolated vertices. Then G has an unique efficient psd-set D if and only if either G has exactly one vertex of full degree or G is a nontrivial block in which $u \in D$ implies that $N(u) \not\subseteq N[x]$ for any $x \in N(u)$.*

Proof. Let D be an unique efficient psd-set of G. If $D = \{u\}$, then u is the only vertex of full degree in G. Now let $|D| = 2$. We claim that G is a nontrivial block. If G is not a nontrivial block, then by Lemma 1, there exists $w \in D$ such that $d(w) = 1$. Let v be the support of the pendent edge vw. Clearly, $v \in V - D$. Then, it follows from Theorem 4 that $D_1 = (D - \{w\}) \cup \{v\}$ is also an efficient psd-set of G, which is a contradiction. Thus, G must be a nontrivial block. Now suppose there exist $u \in D$ and $x \in N(u)$ such that $N(u) \subset N[x]$. Then, $(D - \{u\}) \cup \{x\}$ is an efficient psd-set, which is again a contradiction. Hence for any $u \in D, N(u) \not\subseteq N[x]$ for any $x \in N(u)$.

Conversely, let the conditions of the theroem be satisfied. If G has exactly one vertex of full degree, say u, then $D = \{u\}$ is the unique $\gamma_p(G)$-set. Hence, let G be a nontrivial block in which $u \in D \implies N(u) \not\subseteq N[x]$ for any $x \in N(u)$. Suppose there is another efficient psd-set D_1. Then, $|D \cap D_1| = |D| - 1$. Let $(V - D_1) \cap D = \{u\}$. If $D_1 = (D - \{u\}) \cup \{x\}$ for some $x \in V - D$, then D_1 is a dominating set of G and $x \in N(u)$. Hence $u \in N(x)$. Since D is independent, none of the vertices of $N(u)$ can be adjacent to any vertex of $D - \{u\}$. Also, since D is a pd-set we see that every vertex of $N(u)$ must be adjacent to x. Thus, $N[u] \subset N(x)$, contradicting our assumption and the proof is complete.

3 Conclusion and Scope

Theorem 5, gives some necessary conditions for a graph to possess an efficient psd-set. However, it is not difficult to see that the conditions are not sufficient and hence the problem of characterizing graphs possessing efficient psd-sets is open. Further, in view of condition 3 of Theorem 5, one may also try to characterize separable graphs possessing efficient psd-sets. Another interesting problem is to characterize graphs possessing perfect psd-sets.

References

1. Acharya, B.D., Gupta, P.: On point set domination in graphs V, independent psd-sets. J. Combin. Inform. Syst. Sci. **22**, 133–148 (1997)
2. Acharya, B.D., Gupta, P.: On point-set domination in graphs IV, separable graphs with unique minimum psd-sets. Discrete Math. **195**, 1–13 (1999)
3. Acharya, B.D., Gupta, P.: On point-set domination in graphs VI, quest to characterise blocks containing independent PSD-sets. Nat. Acad. Sci. Lett. **23**, 171–176 (2000)
4. Acharya, B.D., Gupta, P.: On point set domination in graphs: VII, some reflections. Electron. Notes Discrete Math. **15** (2003)
5. Bange, D.W., Barkauskas, A.E., Slater, P.J.: Efficient dominating sets in graphs. In: Ringeisen, R.D., Roberts, F.S. (eds.) Applications of Discrete Mathematics Clemson, pp. 189–199. SIAM, Philadelphia (1986)
6. Cockayne, E.J., Hartnell, B.L., Hedetniemi, S.T., Laskar, R.: Perfect domination in graphs. J. Combin. Inform. Syst. Sci. **18**, 136–148 (1993)
7. Fellows, M.R., Hoover, M.N.: Perfect domination. Australas. J. Combin. **3**, 141–150 (1991)
8. Haynes, T.W., Hedetniemi, S.T., Slater, P.J.: Fundamentals of Domination in Graphs. Marcel Dekker, New York (1998)
9. Haynes, T.W., Hedetniemi, S.T., Slater, P.J.: Domination in Graphs-Advanced Topics. Marcel Dekker, New York (1998)
10. Sampathkumar, E., Pushpa Latha, L.: Point-set domination number of a graph. Indian J. Pure Appl. Math. **24**, 225–229 (1993)
11. Walikar, H.B., Sampathkumar, E., Acharya, B.D.: Recent developments in the theory of domination in graphs and its applications. MRI Lecture Notes in Mathematics, pp. 189–199 (1979)
12. West, D.B.: Introduction to Graph Theory. Prentice Hall Inc., Upper Saddle River (1996)

Graphoidal Length and Graphoidal Covering Number of a Graph

Purnima Gupta[1(✉)], Rajesh Singh[2], and S. Arumugam[3]

[1] Department of Mathematics, Sri Venkateswara College, University of Delhi, New Delhi 110021, Delhi, India
purnimachandni1@gmail.com

[2] Department of Mathematics, University of Delhi, New Delhi 110007, Delhi, India
singh_rajesh999@outlook.com

[3] Kalasalingam University, Anand Nagar, Krishnankoil 626126, Tamil Nadu, India
s.arumugam.klu@gmail.com

Abstract. Let $G = (V, E)$ be a finite graph. A graphoidal cover Ψ of G is a collection of paths (not necessary open) in G such that every vertex of G is an internal vertex of at most one path in Ψ and every edge of G is in exactly one path in Ψ. The graphoidal covering number η of G is the minimum cardinality of a graphoidal cover of G. The length $gl_\Psi(G)$ of a graphoidal cover Ψ of G is defined to be $\min\{l(P) : P \in \Psi\}$ where $l(P)$ is the length of the path P. The graphoidal length $gl(G)$ is defined to be $\max\{gl_\Psi(G) : \Psi$ is a graphoidal cover of $G\}$. In this paper we investigate the existence of graphs which admit a graphoidal cover Ψ with $|\Psi| = \eta(G)$ and $gl_\Psi(G) = gl(G)$.

Keywords: Graphoidal cover · Graphoidal length · Graphoidal covering number

1 Introduction

Throughout this paper we consider only finite, undirected graphs with neither loops nor multiple edges. For graph theoretic terminology we refer to Chartrand and Lesniak [13].

A *graphoidal cover* of a graph G is a collection Ψ of non-trivial paths in G are not necessarily open, such that every vertex of G is an internal vertex of at most one path in Ψ and every edge of G is in exactly one path in Ψ.

The concept of graphoidal covers [4] was introduced by Acharya and Sampathkumar as a close variant of another emerging discrete structure called *semigraphs* [18]. Many interesting notions based on the concept of graphoidal covers such as graphoidal covering number [4], graphoidal labeling [17], graphoidal

R. Singh is thankful to University Grants Commission (UGC) for providing research grant Schs/SRF/AA/139/F-212/2013-14/438.

S. Arumugam et al. (Eds.): ICTCSDM 2016, LNCS 10398, pp. 305–311, 2017.
DOI: 10.1007/978-3-319-64419-6_39

signed graphs [16], etc. were introduced and are being studied extensively. In particular, the notion of graphoidal covering number of a graph has attracted many researchers and numerous work is present in literature [7–9,11,12,14]. Acharya and Gupta in 1999 extended the concept of graphoidal covers to infinite graphs and introduced the notion of graphoidal domination in *graphoidally covered graphs* [1–3]. In 2016 the authors using the concept of graphoidal covers introduced a new graph invariant called graphoidal length of a graph [6]. A detailed study of graphoidal covers is given in [3,5].

Definition 1 [6]. *The length of a graphoidal cover Ψ of a non-trivial graph G, denoted by $gl_\Psi(G)$, is defined as $gl_\Psi(G) = \min\{l(P) \ : \ P \in \Psi\}$, where $l(P)$ is the length of the path P.*

Definition 2 [6]. *The graphoidal length of graph G, denoted by $gl(G)$, is defined by*

$$gl(G) = \max\left\{gl_\Psi(G) \ : \ \Psi \in \mathcal{G}_G\right\}.$$

For any graphoidal cover Ψ of G, $gl(G) \geq gl_\Psi(G)$. A graphoidal cover Ψ with $gl_\Psi(G) = gl(G)$ is called a gl-graphoidal cover of G.

It follows from the definition that $gl(P_n) = n - 1 = l(P_n)$ and $gl(C_n) = n$ for all $n \geq 3$. Also for any (p, q)-graph G, $gl(G) \leq min\{p, q\}$ and equality holds if and only if G is either a path or a cycle.

For complete graph K_4 with $V(K_4) = \{a, b, c, d\}$, $\Psi = \{(a, b, c, d), (b, d, a, c)\}$ is a graphoidal cover of K_4. Hence $gl_\Psi(K_4) = 3$. Thus $gl(K_4) \geq gl_\Psi(K_4) = 3$. Also $gl(K_4) < \min\{p, q\} = 4$ and hence $gl(K_4) = 3$.

Definition 3 [4]. *The graphoidal covering number of G, denoted by $\eta(G)$, is the minimum cardinality of a graphoidal cover of G. A graphoidal cover of G with cardinality $\eta(G)$ is called an η-graphoidal cover.*

Since $gl(G) \geq gl_\Psi(G)$ for every graphoidal cover Ψ of a given graph G, in particular, $gl(G) \geq gl_\Phi(G)$ for every η-graphoidal cover Φ of G. In this paper we investigate the following problem "Does every graph possess an η-graphoidal cover Ψ such that $gl(G) = gl_\Psi(G)$?".

We start with some basic definitions and theorems which are used.

Definition 4 [5]. *Let Ψ be a graphoidal cover of G. A vertex v of G is an interior vertex of Ψ if v is an internal vertex of some path in Ψ. Any vertex which is not an interior vertex of Ψ is called an exterior vertex of Ψ. The set of all exterior vertices of Ψ is denoted by T_Ψ and its cardinality is denoted by t_Ψ.*

Notation 1. *Let $t = min\ t_\Psi$, where the minimum is taken over all graphoidal covers Ψ of G. Clearly, for any graph G, $t \geq e$, where e denotes the number of pendant vertices in G. Also, let $t_g = \min\{t_\Psi \ : \Psi$ is a gl-graphoidal cover$\}$.*

Theorem 1 [5]. *For any (p, q)-graph G, $\eta = q - p + t$.*

Pakkiam and Arumugum gave the following interesting and useful results on graphoidal covering number of graphs.

Theorem 2 [14]. *If G is a connected (p, q)-graph with $\delta(G) \geq 3$, then $\eta(G) = q - p$.*

Theorem 3 [14]. *For any tree T one has $\eta(G) = e(T) - 1$, where $e(T)$ denotes the number of pendant vertices in T.*

Theorem 4 [15]. *Let G be a unicyclic graph with unique cycle C. Let m be the number of vertices of degree at least 3 on C and let e denote the number of pendant vertices of G. Then*

$$\eta(G) = \begin{cases} 1 & \text{if } m = 0 \\ e + 1 & \text{if } m = 1 \text{ and } d(v) = 3 \text{ where } v \text{ is the unique vertex of} \\ & \text{degree at least 3 on } C, \\ e & \text{otherwise.} \end{cases}$$

Corollary 1. *If G is a unicyclic graph with e number of pendant vertices, then $e \leq t \leq e + 1$.*

Definition 5 [13]. *A theta graph is a graph having two vertices of degree 3 and three internally disjoint paths joining them, each of length at least 2.*

Arumugam et al. worked on the problem of characterizing the class of graphs with $\eta(G) = q - p$ in [10] and in the process obtained the following characterization for a 2-edge connected graph with $\delta = 2$ and $\eta(G) = q - p$.

Theorem 5 [10]. *Let G be a 2-edge connected graph with $\delta = 2$. Then $\eta(G) \neq q - p$ if and only if every block of G is a cycle or a cycle with exactly one chord or a theta graph and at most one block of G is not a cycle.*

Theorem 6 [10]. *Let G be a 2-edge connected graph with $\delta = 2$. Then either $\eta(G) = q - p$ or $\eta(G) = q - p + 1$.*

Arumugam et al. [10] determined the graphoidal length of a wheel.

Theorem 7 [6]. *For the wheel W_n $(n \geq 3)$,*

$$gl(W_n) = \left\lfloor \frac{q}{\eta(G)} \right\rfloor = \begin{cases} 3 \text{ if } n = 3, \\ 2 \text{ if } n \geq 4. \end{cases}$$

2 Main Results

In this section we are going to give basic results that will prove quite crucial in the development of the concept of graphoidal length of graphs.

Lemma 1. *Let G be a (p, q)-graph. Let Ψ_1 and Ψ_2 be two graphoidal covers of G. Then $|\Psi_1| > |\Psi_2|$ if and only if $t_{\Psi_1} > t_{\Psi_2}$*

Proof. Since $t_{\Psi_1} = p - q + |\Psi_1|$ and $t_{\Psi_2} = p - q + |\Psi_2|$ it follows that $|\Psi_1| > |\Psi_2|$ if and only if $t_{\Psi_1} > t_{\Psi_2}$. □

Corollary 2. *For any graphoidal cover Ψ of a (p, q)-graph G, $|\Psi| = \eta(G)$ if and only if $t_\Psi(G) = t(G)$.*

In the following lemma we show that if G is a graph having a cut edge x and a component H of $G - x$ such that every block of H is a cycle, then $t \geq 1$.

Lemma 2. *Let G be a graph having a cut edge x and let H be a component of $G - x$ such that every block of H is a cycle. Then for every graphoidal cover Ψ of G at least one vertex of H is an exterior vertex of Ψ.*

Proof. Let $k(G, H)$ denote the number of cycles in the component H of $G - x$. The proof is by induction on $k(G, H)$. If $k(G, H) = 1$ then trivially for any graphoidal cover Ψ of G at least one vertex of H is an exterior vertex of Ψ. Suppose the result holds whenever $k(G', H') \leq m$, where G' is any graph having a cut edge y and H' is a component of $G' - y$ such that every block of H' is a cycle.

Let $k(G, H) = m + 1$. Let C be a block of H having exactly one cut vertex (say) v. Consider any graphoidal cover Ψ of G. If one of the vertices of $V(C) - \{v\}$ is an exterior vertex of Ψ, then we are through. Let every vertex in $V(C) - \{v\}$ be an internal vertex of Ψ. Then C must be a Ψ-edge with v as its coincident end vertex. Let $G_1 = G - (V(C) - \{v\})$ and H_1 be the component of $G_1 - x$ such that H_1 is a subgraph of H. Then every block of H_1 is a cycle and $k(G_1, H_1) = m$, therefore by induction hypothesis the graphoidal cover $\Psi_1 = \Psi - \{C\}$ of G_1 must have an exterior vertex (say) w in H_1. Then $w \in V(H)$ is also an exterior vertex of graphoidal cover Ψ of G. Thus result is true for $k(G, H) = m + 1$. Hence lemma follows by induction.

Consider the wheel $W_n = C_n + \{u\}$, where $C_n = (v_1, v_2, \ldots, v_n, v_1)$. Then $\Psi = \{(u, v_1, v_2, v_3, u), (u, v_4, v_3), \ldots, (u, v_{n-1}, v_{n-2}), (v_{n-1}, v_n, v_1), (v_2, u, v_n)\}$ is graphoidal cover of W_n such that $gl_\Psi(W_n) = 2$ and $|\Psi| = n - 1$. From Theorems 2 and 7, it follows that Ψ is an η-graphoidal cover of W_n such that $gl_\Psi(G) = gl(G)$. Thus for every wheel there exists a graphoidal cover Ψ such that $gl_\Psi(G) = gl(G)$ and $|\Psi| = \eta$.

However there exists a graph G which has no graphoidal cover Ψ satisfying $gl_\Psi(G) = gl(G)$ and $|\Psi| = \eta$. For example, consider the graph $G = C_{10,4,10}$ given in Fig. 1. Then $\Psi = \{(u, w_1, v, v_1, \ldots, v_4), (v, w_1, u, u_1, \ldots, u_4), (u_4, \ldots, u_9, u), ((v_4, \ldots, v_9, v))\}$ is a gl-graphoidal cover of G with $gl_\Psi(G) = 6$. Further for any η-graphoidal cover Φ, $gl_\Phi(G) = 4$.

Thus not every graph possesses a graphoidal cover Ψ satisfying $gl_\Psi(G) = gl(G)$ and $|\Psi| = \eta$. If in a graph G there exists a graphoidal cover Ψ satisfying $gl_\Psi(G) = gl(G)$ and $|\Psi| = \eta$, then such a graphoidal cover (gl, η)-graphoidal cover.

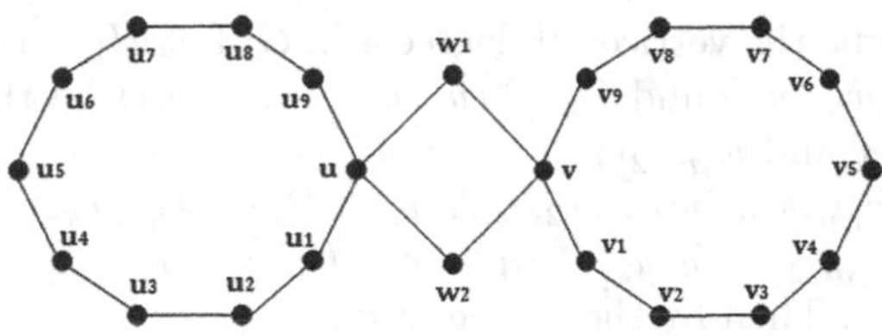

Fig. 1. $G = \mathcal{C}_{10,4,10}$.

If G is a graph with graphoidal length $gl(G) = 1$, then length of every graphoidal cover is 1. In particular, length of every η-graphoidal cover is 1. In [6], it has been proved that if G is a (p, q)-graph with $q > 2p$, then $gl(G) = 1$. Thus every (p, q)-graph G with $q > 2p$ possesses a (gl, η)-graphoidal cover. Consequently, every complete graph K_p $(p \geq 6)$ and every complete bipartite graph $K_{m,n}$ $(n, m \geq 5)$ possesses a (gl, η)-graphoidal cover.

We proceed to investigate this problem for graphs with $q \leq 2p$. We first consider trees.

Theorem 8. *Every tree possesses a (gl, η)-graphoidal cover.*

Proof. Let G be tree and let Ψ be a gl-graphoidal cover such that $t_\Psi = t_g = min\{t_\Phi : \Phi \text{ is a } gl\text{-graphoidal cover}\}$. We claim that $t_g = e$.

Suppose $t_g > e$. Then there exists a vertex x in V of degree at least 2 such that x is an exterior vertex of Ψ. Hence there exist paths P and Q in Ψ with end vertex x. Let $\Psi' = \Psi \cup \{P \cup Q\} - \{P, Q\}$, where $P \cup Q$ is the path in Ψ' with x as internal vertex. Clearly, $gl_{\Psi'}(G) = gl(G)$ and $t_{\Psi'} < t_\Psi = t_g$, contradicting the minimality of t_g. Hence follows that $t_g = e$ and Ψ is a (gl, η)-graphoidal cover of G.

We state without proof two theorems which give the existence of (gl, η)-graphoidal cover for unicyclic graph and cactus in which no two cycles intersect.

Theorem 9. *For any unicyclic graph G, there exists a graphoidal cover Ψ of G such that $gl_\Psi(G) = gl(G)$ and $|\Psi| = \eta(G)$.*

Theorem 10. *Let G be a cactus such that no two cycles intersect. Then there exists an η-graphoidal cover Ψ such that $gl_\Psi(G) = gl(G)$.*

Thus every tree, every unicyclic graph and every cactus having disjoint cycles possesses a (gl, η)-graphoidal cover. But not every cactus possesses a (gl, η)-graphoidal cover (for example the graph $C_{10,4,10}$). In what follows we provide a class of 2-edge connected cactus which does not possess any (gl, η)-graphoidal cover. Let $\mathcal{C}_{n,m,n}$ $(n, m \geq 3)$ be a 2-edge connected cactus having 2-copies of C_n and one copy of C_m with two non-adjacent vertices of degree 4 at distance $\lfloor m/2 \rfloor$ from each other and all other vertices are of degree 2.

Theorem 11. *Let $G = \mathcal{C}_{n,m,n}$ where $n \geq 2m \geq 8$. Then G does not possess a (gl, η)-graphoidal cover.*

Proof. Let x and y be the vertices of degree 4 in G. Let $H_1 = (v_1, v_2, \ldots, v_n, v_1)$, $H_2 = (w_1, w_2, \ldots, w_m, w_1)$ and $H_3 = (u_1, u_2, \ldots, u_n, u_1)$ be the three cycles in G with $v_1 = w_1 = x$ and $w_{\lfloor m/2 \rfloor + 1} = u_1 = y$.

Let $P_1 = (y, w_{\lfloor m/2 \rfloor} \ldots, w_2, x, v_2, v_3, \ldots v_k)$, $P_2 = (v_k, v_{k+1}, \ldots, v_n, v_1)$, $P_3 = (x, w_m, w_{m-1}, \ldots, w_{\lfloor m/2 \rfloor + 2}, y, u_2, \ldots, u_k)$ and $P_4 = (u_k, u_{k+1}, \ldots, u_n, u_1)$ where $2k = n + 2 - \lfloor m/2 \rfloor$. Then by the choice of k,

$$l(P_i) \geq 3 \lfloor m/2 \rfloor \qquad for\ i = 1, 2, 3, 4.$$

Thus $\Psi = \{P_1, P_2, P_3, P_4\}$ is a graphoidal cover of G with $gl_\Psi \geq 3 \lfloor m/2 \rfloor$. Hence $gl(G) \geq 3 \lfloor m/2 \rfloor$.

Since any η-graphoidal cover Φ of G, contains the three cycles $gl_\Phi(G) = m$ for any η-graphoidal cover Φ of G. Hence for any η-graphoidal cover Φ of G, $gl_\Phi(G) < gl(G)$, so that G does not possess a (gl, η)-graphoidal cover.

3 Conclusion and Scope

In view of Theorem 11, it would be interesting to characterize cactus having a (gl, η)-graphoidal cover. Also, the general problem of characterizing graphs G which admit a graphoidal cover Ψ with $|\Psi| = \eta(G)$ and $gl_\Psi(G) = gl(G)$ is open.

References

1. Acharya, B.D., Purnima, G.: Further results on domination in graphoidally covered graphs. AKCE Int. J. Graphs Comb. **4**, 127–138 (2007)
2. Acharya, B.D., Purnima, G., Deepti, J.: On graphs whose graphoidal domination number is one. AKCE Int. J. Graphs Comb. **12**(2–3), 133–140 (2015)
3. Acharya, B.D., Purnima, G.: Domination in graphoidal covers of a graph. Discrete Math. **206**, 3–33 (1999)
4. Acharya, B.D., Sampathkumar, E.: Graphoidal covers and graphoidal covering number of a graph. Indian J. Pure Appl. Math. **18**, 882–890 (1987)
5. Arumugam, S., Acharya, B.D., Sampathkumar, E.: Graphoidal covers of a graph: a creative review. In: Proceedings of Graph Theory and its Applications, pp. 1–28. Tata McGraw-Hill, New Delhi (1997)
6. Arumugam, S., Purnima, G., Rajesh, S.: Bounds on graphoidal length of a graph. Electron. Notes Discrete Math. **53**, 113–122 (2016)
7. Arumugam, S., Pakkiam, C.: Graphoidal bipartite graphs. Graphs Combin. **10**, 305–310 (1994)
8. Arumugam, S., Pakkiam, C.: Graphs with unique minimum graphoidal cover. Indian J. Pure Appl. Math. **25**, 1147–1153 (1994)
9. Arumugam, S., Rajasingh, I., Pushpam, P.R.L.: Graphs whose acyclic graphoidal covering number is one less than its maximum degree. Discrete Math. **240**, 231–237 (2001)
10. Arumugam, S., Rajasingh, I., Pushpam, P.R.L.: A note on the graphoidal covering number of a graph. J. Discrete Math. Sci. Cryptography **5**, 145–150 (2002)
11. Arumugam, S., Suseela, J.S.: Acyclic graphoidal covers and path partitions in a graph. Discrete Math. **190**, 67–77 (1998)

12. Purnima, G., Rajesh, S.: Domination in graphoidally covered graphs: least-kernel graphoidal covers. Electron. Notes Discrete Math. **53**, 433–444 (2016)
13. Chartrand, G., Lesniak, L.: Graphs & Digraphs. Chapman & Hall/CRC, Boca Raton (2005)
14. Pakkiam, C., Arumugam, S.: On the graphoidal covering number of a graph. Indian J. Pure Appl. Math. **20**, 330–333 (1989)
15. Pakkiam, C., Arumugam, S.: The graphoidal covering number of unicyclic graphs. Indian J. Pure Appl. Math. **23**, 141–143 (1992)
16. Reddy, P.S.K., Misra, U.K.: Graphoidal signed graphs. Proc. Jangjeon Math. Soc. **17**, 41–50 (2014)
17. Hamid, I.S., Anitha, A.: On label graphoidal covering number-I. Trans. Comb. **1**, 25–33 (2012)
18. Sampathkumar, E.: Semigraphs and their applications. Report on the DST Project (2000)

An Overview of the MapReduce Model

S. Rajeswari, K. Suthendran(✉), K. Rajakumar, and S. Arumugam

Kalasalingam University,
Anand Nagar, Krishnankoil 626126, Tamil Nadu, India
s.rajeswari30@gmail.com, k.suthendran@klu.ac.in, rajakumarkk@gmail.com,
s.arumugam.klu@gmail.com

Abstract. Data is getting accumulated fast in various domains all over the world and the data size varies from terabytes to yottabytes. Such huge size data are known as Big Data. Extraction of meaningful information from raw data using special patterns are called Data Mining and sophisticated algorithms have been designed for this purpose. In this paper, the time complexity for MapReduce-based data mining algorithm is presented.

Keywords: Big Data · MapReduce · Hadoop · Datamining · Time complexity

1 Introduction

The size of data, storage capacity, processing power and availability of data are rising day by day. The traditional data warehousing and management systems tools are not capable of dealing with huge data. The analytic methods used to deal with huge data is referred to as Big Data analytics. Doug Laney was the first one to discuss about 5 V's in Big Data management [1], namely, volume, variety, velocity, variability and veracity.

Some of the technologies to process big data are Hadoop HDFS, MapReduce, Hive, HBase, Pig, Flume, Hadoop Mahout, Windows Azure etc. [1]. MapReduce was first developed by Google but now it is incorporated by the Apache. In Parallel, the large clusters of hardware are available to process huge amount of data. It has two functions Map() and Reduce(). Map() is used for filtering and sorting the data and Reduce() is used for summarizing the data [2] (Fig. 1).

The MapReduce process is depicted in Fig. 1. Here the input is application specific while the output is a group of $<$ key, value $>$ pairs produced by the Map function. In the mapping process, each single key-value input pair (key, value) is mapped into several key-value pairs: $[(l_1, x_1), \ldots, (l_1, x_r)]$ with same key, but different values. These key-value pairs are used for reducing the functions.

Data mining techniques are applied on raw data for extracting useful information. The process of finding a model is to describe the data classes by using a classification algorithm [3].

Before using the classification algorithm, preprocessing steps such as categorization and feature selection are included. This process is helpful for getting an

S. Arumugam et al. (Eds.): ICTCSDM 2016, LNCS 10398, pp. 312–317, 2017.
DOI: 10.1007/978-3-319-64419-6_40

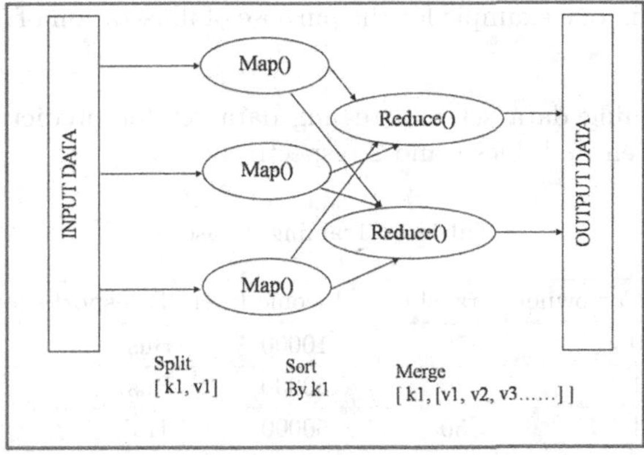

Fig. 1. MapReduce process

improved accuracy in the prediction. Categorization is the process of converting the data into a categorical format. Based on the condition, the data is categorized into a standard format. Feature selection is used to select the important features of the data and to remove the irrelevant attributes. MapReduce concept is used in the categorization and feature selection process.

Classification algorithm involves two steps, namely training set and testing set. The training set is used to build a model with the training data. Testing set is applied on the classification model and is used to check the accuracy [4].

A Decision tree is a classification model. It is mainly used to classify an object to a predetermined class. CHAID, CART, ID3, C4.5, C5.0 are decision tree algorithms and C5.0 is a widely used decision tree algorithm.

C5.0 algorithm handles continuous and categorical values. Feature selection is the basic step to construct a decision tree. The decision tree algorithm C5.0 is used to access the data and has higher speed when compared to ID3 and C4.5 [5]. MapReduce process is used to evaluate the data.

In this paper we discuss the time complexity for MapReduce - based C5.0 algorithm. MapReduce technique is a training model to be linked with performance for processing huge data sets. MapReduce concept runs on a large cluster of product machines and is highly scalable [6].

Matei Zaharia et al. proposed an improved scheduling algorithm that decreases the Hadoop reply time and the performance by using MapReduce [7].

2 Implementation Scrutiny

MapReduce with Datamining algorithm is used for tera bytes of data [8]. Using the categorization algorithm, attribute values can be grouped. Among categorized data, relevant attribute has to be picked up. For this purpose, feature selection algorithm can be used. These two algorithms are used for the prediction of class labels. We fix one example for the purpose of illustration of all the concepts presented here.

Example 1. Training data set and testing data set for prediction of mode of transport are given in Tables 1 and 2 respectively.

Table 1. Training dataset

Gender	Car owner	Travel cost	Income level	Transportation mode
Male	0	50	10000	Bus
Male	1	50	50000	Bus
Female	1	50	50000	Train
Female	0	50	10000	Bus
Male	1	50	50000	Bus
Male	0	500	50000	Train
Female	1	500	50000	Train
Female	1	1000	100000	Car
Male	2	1000	50000	Car
Female	2	1000	100000	Car

Table 2. Testing dataset

Gender	Car owner	Travel cost	Income level	Transportation mode
Male	1	Standard	High	?
Male	0	Cheap	Medium	?
Female	1	Cheap	High	?

2.1 Categorization

Categorization is the process of grouping objects into categories, usually for some specific purpose. Generally, categorization algorithm has two parts. First is Map part which is used to check the conditions of the attribute values and the second is reduce part which changes the numerical values to categorized values [9].

The attribute values are changed into categorized format based on the conditions. Finally the categorical values are stored in a new file.

For the case study given in Tables 1 and 2, for the categorization process, the travel cost and Income level attributes are used and the conditions are as follows.

Condition for travel cost	Condition for income level

if Travel cost == 500 then
Standard
else if Travel cost < 500 then
cheap
else
Expensive
end
	if Income == 100000 then
	high
	else if Income < 50000 then
	low
	else
	medium
	end

2.2 Feature Selection

Feature selections is also called attribute selection or variable selection. After categorization, the feature selection process is used to select a subset of relevant attributes [10]. Chi-square (χ^2) feature selection based on MapReduce concept is used for finding the relevant attributes. It is one of the popular feature selection methods. Statistical test is also used to decide whether observed frequencies are much dissimilar from expected frequencies [11]. The χ^2-filter is defined by

$$\chi^2 = \frac{\Sigma(O-E)^2}{E} \tag{1}$$

where for each attribute, O is the Observed Frequency and E is the Expected Frequency.

The weight for each attribute is calculated by using (1). Attributes whose weight is greater than a chosen threshold value are taken for further processing.

2.3 C5.0 Classifier

Classification is one of the major components in data mining and C5.0 is decision tree based classification algorithm. It can handle continuous values, categorical values and numerical attributes. Let C be a categorization of a set S of objects into categories $C_1, C_2, \ldots, C_r$. Let p_i be the probability that an object in S is in category C_i. The entropy of S is defined as

$$\text{Entropy}(S) = -\sum_{i=1}^{r} p_i \log_2 p_i.$$

The entropy measures the homogeneity of objects.

To determine the best attribute to be chosen for a node in a decision tree, we use the concept of information gain. For any given attribute A, consider the set $V(A)$ of possible values of A. For any $v \in V(A)$, let S_v be the set of all elements of S having value v for the attribute A. The information gain of A with respect to S is given by

$$\text{Inf.Gain }(S, A) = \text{ Entropy }(S) - \sum_{v \in V(A)} \frac{|S_v|}{|S|} \text{ Entropy }(S_v).$$

Thus the information gain measures the expected reduction in the entropy.

The attribute with highest information gain is taken as the root node in the decision tree and the values of the attribute are its children nodes. Then using the remaining attributes, the attribute with highest information gain is taken as a child node and the process is repeated. If $S(V_i) \neq \emptyset$, the tree constructed is added as a new branch at v_i. Each leaf node of the final decision tree gives a rule for predicting the class label.

For the example given in Table 1, the probability of an individual travelling by bus, car and train are respectively 0.4, 0.3 and 0.3. Hence the entropy value is $-(0.4\log_2(0.4) + 0.3\log_2(0.3) + 0.3\log_2(0.3)) = 1.571$.

For each of the attributes Gender, Car-owner ship, travel cost and income the information gain in computed and the results are given in Table 3.

Table 3. Information gain result

Attribute	Information gain
Gender	0.125
Car owner	0.21
Travel cost	1.21
Income level	0.695

The attribute with highest gain value is travel cost which is taken as the root node and its branches are its values, namely, standard, expensive and cheap.

The next step computation of informations gain for the removing three attributes is carried out. Table 4 gives the results for the attribute cheap.

Table 4. Information gain for the node "cheap"

Attribute	Information gain
Gender	0.322
Car owner	0.171
Income level	0.171

In this way the process can be continued, giving the decision tree. From the decision tree teh transportation mode can be predicted in terms of the attributes. For example for the object with attributes female, car owner, cheap the predicted transportation mode is bus.

3 Conclusion

The input for the MapReduce model is a set S of objects with $|S| = n$, attributes $A_1, A_2, \ldots, A_k$ where the attribute A_i takes α_i values and rules for categorization. The complexity of categorization process is $O(n)$. The complexity of feature selection is $O(k)$. The complexity of C5.0 classifier is $O(n^2)$. Thus the overall complexity of the algorithm is $O(n^2)$.

Acknowledgment. The first author is thankful to the management of Kalasalingam University for providing fellowship.

References

1. Miranda Lakshmi, T., Martin, A., Mumtaj Begum, R., Prasanna Venkatesh, V.: An analysis on performance of decision tree algorithms using students qualitative data. Int. J. Mod. Educ. Comput. Sci. **5**, 18–27 (2013)
2. Vaitheeswaran, G., Arockiam, L.: Big data for education in students perspective. Int. J. Comput. Appl. Adv. Comput. Commun. Tech. High Perform. Appl. **4**, 11–17 (2014)
3. Laney, D.: 3-D Data management: controlling data volume, velocity and variety. META Group Research Note, pp. 20–25 (2011)
4. Ladha, L., Deepa, T.: Feature selection methods and algorithms. Int. J. Comput. Sci. Eng. **3**, 1787–1790 (2011)
5. Dean, J., Ghemawat, S.: MapReduce: simplified data processing on large clusters. Commun. ACM **51**(1), 107–113 (2008)
6. Zaharia, M., Konwinski, A., Joseph, A.D., Katz, R., Stoica, I.: Improving MapReduce performance in heterogeneous environments. In: Proceedings of the 8th USENIX Conference on Operating Systems Design and Implementation, pp. 29–42 (2008)
7. Choura, R.: Use of data mining techniques for the evaluation of student performance: a case study. Int. J. Comput. Sci. Manag. Res. **1**(3), 425–433 (2012)
8. Naseriparsa, M., Bidgoli, A.M., Varaee, T.: A hybrid feature selection method to im-prove performance of a group of classification algorithms. Int. J. Comput. Appl. **69**(17), 28–35 (2014)
9. Azhagusundari, B., Thanamani, A.S.: Feature selection based on information gain. Int. J. Innov. Technol. Explor. Eng. **2**(2), 18–21 (2013)
10. Vanaja, S., Ramesh Kumar, K.: Analysis of feature selection algorithms on classification: a survey. Int. J. Comput. Appl. **96**(17), 29–35 (2014)
11. Yadav, S.K., Pal, S.: Data mining: a prediction for performance improvement of engineering students using classification. World Comput. Sci. Inf. Technol. J. **2**(2), 51–56 (2012)

Identification of Salinity Stress Tolerant Proteins in Sorghum Bicolor Computational Approach

S. Rajeswari[1](✉), M. Indhumathy[2], A. Somasundaram[1], Neeru Sood[1], and S. Arumugam[2]

[1] Department of Biotechnology, Birla Institute of Technology and Science, Dubai Campus, Dubai, UAE
{srajeshwari,asomasundaram,sood}@dubai.bits-pilani.ac.in
[2] Kalasalingam University, Krishnankoil 626126, Tamil Nadu, India
indhumathy.bio@gmail.com, s.arumugam.klu@gmail.com

Abstract. Sorghum bicolor is the fifth most important cereal crop in the world after rice, wheat, barley and maize and is grown worldwide in the semi-arid and arid regions. Functional identification of proteins and a detailed study of protein-protein interactions in Sorghum are very essential to understand the biological process underlying the various traits of crops such as salt stress, yield and drought response. The molecular mechanisms that exists among them are still unclear due to the limited studies available in the literature and databases. In this paper, salt stress tolerance responsive protein-protein interaction network has been constructed using STRING database. A comprehensive bioinformatics analysis of this network has been performed using Cytoscape, through the computation of centrality measures and functional enrichment study. This study has resulted in the identification of Sb02g023480.1, Sb01g040040.1, Sb09g027910.1, Sb03g031310.1, Sb10g002460.1 as salt stress tolerant proteins in Sorghum bicolor. However, experimental studies are required to confirm this observation.

Keywords: Sorghum bicolor · Computational approach · Salt stress tolerant proteins · Centrality measures · Gene Ontology

1 Introduction

Agricultural production has to be increased considerably to meet the food and feed demands of fast growing human population and livestock generation. Crop protection plays a key role in crop productivity, but it is mainly affected by environmental stresses such as drought, heat and salt [8]. Salt stress is a major abiotic stress that limits the productivity as well as and the geographical distribution of many crop species. Hence understanding the molecular mechanism of cereal crops plays a major role.

Sorghum bicolor is the fifth most important cereal crop in the world after rice, wheat, barley and maize and is grown worldwide in the semi-arid and arid regions [10]. The molecular mechanisms behind them are still unclear due to the

S. Arumugam et al. (Eds.): ICTCSDM 2016, LNCS 10398, pp. 318–325, 2017.
DOI: 10.1007/978-3-319-64419-6_41

limited studies available in the literature and databases. Functional identification of proteins and a detailed study of protein-protein interactions in Sorghum are very essential to understand the biological process underlying the various traits of crops such as salt stress, yield and drought resistance. Proteomics technology has developed considerably in recent years [3]. The combination of high-throughput technique with bioinformatics tools and databases can be used for functional identification of proteins. But the mechanism of salt-tolerance based on proteome has not improved as most of these data were not validated [6]. Interpretation and analysis of these data are time-consuming. Biological implications will facilitate the discovery and characterization of important physiological mechanisms and pathways. Therefore, the obstacle in data analysis should be completely overcome by evolving strategies through multi-disciplinary research. Currently, several computational approaches for functional identification of proteins such as sequence similarity, phylogenetic profiles, protein-protein interaction (PPI) and gene expression are available [4]. The study of protein interactions is fundamental to understand how proteins function within the cell. Particularly, graph theory has been used to identify important proteins in the PPI network and the function of an unknown protein can be identified on the basis of their interactions with known proteins [1]. For this purpose, centrality measures such as degree, stress, betweenness, radiality, closeness, subgraph and eigenvector are widely used.

A literature survey shows that SOD, CAT, GS and PEPC are Sorghum bicolor salt stress tolerant proteins [5]. In this paper, a PPI network using the proteins has been constructed from STRING database and a comprehensive bioinformatics analysis of this network has been performed using Cytoscape through the computation of centrality measures and functional enrichment study. This study has resulted in the identification of Sb02g023480.1, Sb01g040040.1, Sb09g027910.1, Sb03g031310.1, Sb10g002460.1 as salt stress tolerant proteins in Sorghum bicolor and however experimental studies are required to confirm this observation.

2 Materials and Methods

2.1 String Database

STRING is used for predicting protein-protein interactions. The interactions include direct (physical) and indirect (functional) associations based on computational prediction, from knowledge transfer between organisms and from interactions aggregated from DIP, BioGRID, IntAct, MINT and PDB databases. Interactions in STRING are derived from five main sources: Genomic Context Predictions, High-throughput Lab Experiments, (Conserved) Co-Expression, Automated Text mining, and Previous Knowledge in Databases. We have retrieved the interactions of four proteins with high confidence score (≥ 0.7) to reduce the impact of false positive and false negative results. The other parameters were set to default.

2.2 Cytoscape

Cytoscape is an open-source software package widely used to integrate and visualize diverse data-sets in biology [12]. Salt-response proteins network in *S.bicolor* was analyzed by Cytoscape Version 3.3.0. The PPI network contains 165 proteins as nodes and 1484 edges as interactions. This network is shown in Fig. 1. The target interactions and key proteins in this network are identified using the plugins in the software such as Centiscape [11] and CytoNAC [15].

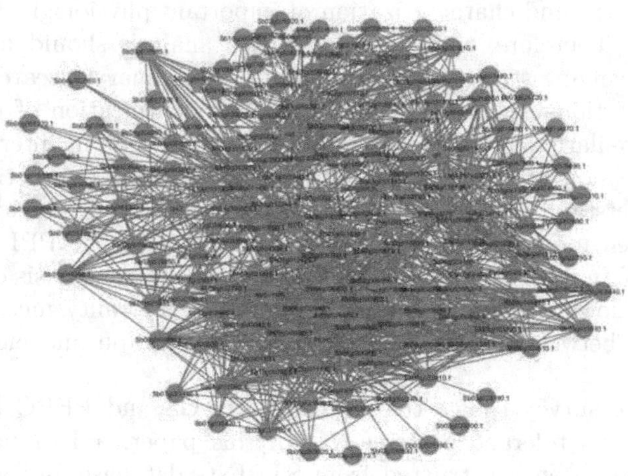

Fig. 1. Protein-protein interaction network constructed from string database

2.3 Topological Analysis

Graph theory is one of the most powerful tools for analyzing large networks. It uses network centrality and node centrality indices. A network centrality index informs about the overall nature of the network and node centrality index describes the property of the nodes. For network centrality, various parameters such as average distance, connectivity, diameter and clustering coefficient are calculated. Node centrality indices such as degree, stress, betweenness, eccentricity, radiality, closeness, eigenvector and subgraph are some of the centrality measures which are often used to identify the important proteins. The higher value for degree indicates that more number of edges (interactions) are incident with that node (protein). Protein which holds the communication between other proteins has higher value for radiality and stress. Betweenness explains the total number of shortest paths passing through a given protein. Hence any protein with highest score for these measures is important among all the proteins in the network. For more details and definitions of various centrality measures, we refer to [11].

2.4 Functional Enrichment Analysis

Functional enrichment analysis including domain analysis, biological processes, molecular functions, and pathways was performed using Gene Ontology Consortium for the *S.bicolor* proteins. Gene Ontology (GO) is a structured and controlled vocabulary, which identifies the functional annotation of proteins using standardized terms such as biological process, molecular function, and cellular components. This study was conducted only for the key nodes identified through the topological analysis.

3 Results

Topological characteristics of the PPI network given in Fig. 1 are given below. This reveals the organization of the network.

Number of nodes	165
Number of edges	1484
Diameter	5
Density	0.110
Clustering coefficient	0.509
Network heterogeneity	0.721
Characteristic path length	2.604

The network was found to have scale-free organization as their degree distribution followed power law. Clustering coefficient distribution is also used to identify scale-free nature of the network, which decreases when node degree increases. The average clustering coefficient of the network is 0.509. It suggests that low degree nodes belong to dense subgraphs and these subgraphs are connected to the hubs in the network which tend to possess scale-free nature and small world property. Diameter of the network is found to be 5 which is the maximum length of a shortest path between any pair of nodes. It reveals that network is highly connected which also indicates the small world property of the network. Other parameters such as network heterogeneity, density and characteristic path length unveil the hierarchical modularity of the analyzed network. Modular analysis of the network suggests that highly connected proteins may be functionally related. Hence, the proteins which interact with the four salt stress proteins may also contain the characteristics similar to them [7].

Seven node centrality indices namely degree, betweenness, stress, radiality, closeness, eigenvector and subgraph, were calculated to identify the important proteins of the network. For each of these seven parameters, the maximum value, mean value and minimum value are computed and are given in Table 1.

For all measures, the mean value is taken as the threshold value. For a given parameter, any protein whose value exceeds the threshold value is considered to be an important protein with respect to that parameter. For each

Table 1. Seven centrality measures

Parameters	Maximum value	Mean value	Minimum value
Degree	61	17.98	1
Betweenness	2981.604175	263.05	0
Stress	135684	18759	0
Radiality	4.164634146	3.396	2.719512
Subgraph	3.77187E+11	4.80E+10	1.32E+07
Closeness	0.5448505	0.39	0.304833
Eigenvector	0.218172981	0.051	0.001286

protein, its combined centrality score (CCS) is the average of the seven centrality measures [2]. The proteins are arranged on the basis of CCS from high score to low score. The top six proteins in this sorted list are taken as the most influential proteins which are given in Table 2.

Among six proteins, it is already known that PEPC is salt stress tolerant. Hence five proteins are identified as important proteins in the network. All of them are interacting with PEPC. Two proteins are interacting with GS and both PEPC and GS are known to be salt stress tolerant proteins. This implies that the five identified proteins are functionally similar. To validate this property, functional enrichment studies have been performed.

Table 2. The top six proteins with high scores

Proteins	Betweenness	Closeness	Degree	Eigen vector	Radiality	Stress	Subgragh	Average
PEPC	2809.390397	0.475362	57	0.218173	3.896341	96166	377187237888	53883905275
Sb02g023480.1	699.8880746	0.421594	37	0.152202	3.628049	55382	183566450688	26223786687
Sb01g040040.1	699.8880746	0.421594	37	0.152202	3.628049	55382	183566450688	26223786687
Sb09g027910.1	2981.604175	0.544851	61	0.149362	4.164634	93080	176794206208	25256328905
Sb03g031310.1	2981.604175	0.544851	61	0.149362	4.164634	93080	176794140672	25256319543
Sb10g002460.1	672.9164465	0.420513	36	0.148932	3.621951	55000	175764635648	25109241623

Further, gene ontology studies were conducted to identify molecular function, biological process and cellular components. KEGG pathway, InterPro and Pfam domain studies were also done. The investigation of functional enrichment of proteins uncovers that proteins involved in the same cellular processes frequently interact with each other. Proteins commonly have one or more functional regions, termed domains. The identification of domains that occur within proteins can therefore provide insights into their function. Pfam indicates eight statistically significant domains with pfam id's PF00113 (Enolase, C-terminal TIM barrel domain), PF03952 (Enolase, N-terminal domain), PF14691 (Dihydroprymidine dehydrogenase domain II, 4Fe-4S cluster), PF01493 (GXGXG motif), PF01645 (Conserved region in glutamate synthase), PF04898 (Glutamate synthase central domain), PF00310 (Glutamine amido transferases class-II) and PF07992 (Pyridine nucleotide-disulphide oxidoreductase).

KEGG analysis indicated that nine statistically significant pathways id's such as 1230 (biosynthesis of amino acids), 1120 (microbial metabolism in diverse environments), 1110 (biosynthesis of secondary metabolites), 910 (nitrogen metabolism), 250 (alanine, aspartate and glutamate metabolism), 1100 (metabolic pathways), 3018 (RNA degradation), 10 (glycolysis/ gluconeogenesis) and 1200 (carbon metabolism) were associated with *S.bicolor* salt-response proteins. Molecular function analysis revealed that most of the proteins are involved in ion binding such as magnesium, iron and iron-sulfur cluster. Most of the *S.bicolor* proteins have subcellular localization on cytoplasm and their biological process are related to glycolytic process, glutamate biosynthetic process, photosynthesis and carbon fixation source. These results are given in Table 3.

Table 3. Biosynthetic path way Enzyme analysis

Proteins	Molecular function	Biological process	Cellular component
Sb02g023480.1	Magnesium ion binding, phosphopyruvate, hydratase activity	Glycolytic process, response to cytokinin, trichome morphogenesis	Chloroplast stroma, phosphopyruvate hydratase complex
Sb01g040040.1	Magnesium ion binding, phosphopyruvate, hydratase activity	Glycolytic process,	Nucleus phosphopyruvate hydratase complex
Sb09g027910.1	FMN binding, flavin adenine binding, glutamate synthase (NADH) activity, iron ion binding, iron-sulfur cluster binding	Ammonia assimilation cycle, glutamate biosynthetic process	Cytoplasm
Sb03g031310.1	FMN binding, flavin adenine binding, glutamate synthase (NADH) activity, iron ion binding, iron-sulfur cluster binding	Ammonia assimilation cycle, glutamate biosynthetic process	Cytoplasm
Sb10g002460.1	Not yet identified	Not yet identified	Not yet identified
PEPC	Phosphoenolpyruvate carboxylase activity	Carbon fixation, photosynthesis, tricarboxylic acid cycle	Cytoplasm

4 Discussion

Although salt-tolerant proteins in *S.bicolor* have been studied in the past years, there has been no systematic attempt to organize them in protein pathways. Proteomics studies can significantly contribute to unravel the possible relationships and interactions between protein abundance and plant stress. The network indicates that there are 165 nodes, 1484 edges with average node degree of 17.8 with p-value 0. From the string network analysis, the expected numbers of edges are found to be 43, but the number of actual edges is 1484 which is significantly very high. This means that the proteins have more interactions among themselves than what would be expected for a random set of proteins of similar size, drawn from the genome. Such enrichment indicates that the proteins are at least partially biologically connected, as a group. The results of the enriched protein domains, KEGG pathways, molecular functions, and cell localizations were comprehensive. Bioinformatics analysis indicated that most of the proteins responsible to salt stress could be clustered into different function groups and may be related to plant physiology.

Pfam (protein families) database is a large collection of protein families, each protein being represented by multiple sequence alignments. Proteins are generally composed of one or more functional regions, termed domains. Different combinations of domains give rise to the diverse range of proteins found in nature. Pfam also generates higher-level groupings of related entries, known as clans related by similarity of sequence, structure or profile (pfam.xfam.org). Domain analysis showed some significant enriched domains in the S. bicolor salt-response proteins.

InterPro is a powerful diagnostic tool and integrated resource as it combines signatures from multiple, diverse databases to provide functional analysis of protein sequences into families. Plants use signaling pathways to acclimate to changing environmental conditions.

Salt stress reduces water availability and leads to the inhibition of plant growth by increasing the threshold pressure for wall yielding in expanding cells or inducing hydraulic limitations to water uptake [14]. Rigid cell wall protects plant roots from dehydration [9]. The mechanism of structural molecules remodeling in response to salt stress need to be further studied. In general, most salinity induced genes are also induced by drought stress and many drought inducible genes are also induced by abscisic acid [13].

Interestingly, among 41 salt responsive proteins, many of them also responded to other stresses, such as bacterium stress, light stress, metal ion stress, radiation stress, temperature stress, water stress, and wounding stress, which have wider cross-talks, and revealed more potential biomarkers participating in signaling and metabolism pathways. We assumed that the proteins responding to other stresses may also associate with salt stress, and this would provide us a wider regulatory network to discover the salt response mechanisms. Salt-response proteins have been located at more than one cell component, which possibly indicated that they played different functions in various places and indicate the versatile nature of the proteins. It is possible that the proteins responded to multiple stresses and localized at multiple places.

5 Conclusion

This work has provided new insight into *Sorghum bicolor* salt-response mechanisms and could urge scientists in this field to integrate all the existing data to explore significant functions and pathways involved in salt stress response. Two new understandings should be noted: one was that salt-response proteins might become potential biomarkers for other stresses and other stress-responsive proteins might also participate in salt stress; the other one was that multiple cellular localizations indicate that the protein functions under stress are versatile. Cross-talk effects informed us that *S.bicolor* deals with salt stress and other stresses in a reciprocal economic way. Further studies are necessary to substantiate the enriched functions and pathways. Protein functional characterization helps us to understand the processes of plant stress acclimatization and stress tolerance acquisition in a better way.

References

1. Barabasi, A.L., Zoltan, N.O.: Network biology: understanding the cell's functional organization. Nat. Rev. Genet. **5**(2), 101–113 (2004)
2. Bhattacharyya, M., Saikat, C.: Identification of important interacting proteins (IIPs) in Plasmodium falciparum using large-scale interaction network analysis and in-silico knock-out studies. Malaria J. **14**, 1–17 (2015)
3. Brewis, I.A., Brennan, P.: Proteomics technologies for the global identification and quantification of proteins. Adv. Protein Chem. Struct. Biol. **80**(15), 1–44 (2010)
4. Deng, M., Zhang, K., Mehta, S., Chen, T., Sun, F.: Prediction of protein function using protein-protein interaction data. J. Comput. Biol. **10**(6), 947–960 (2003)
5. El-Omari, R., Nhiri, M.: Adaptive response to salt stress in sorghum (Sorghum bicolor). Am.-Eur. J. Agric. Environ. Sci. **15**(7), 1351–1360 (2015)
6. Jiang, Y., Yang, B., Harris, N.S., Deyholos, M.K.: Comparative proteomic analysis of NaCl stress-responsive proteins in Arabidopsis roots. J. Exp. Bot. **58**(13), 359–3607 (2007)
7. Karabekmez, M.E., Betul, K.: A novel topological centrality measure capturing biologically important proteins. Mol. BioSyst. **12**(2), 666–673 (2016)
8. Kasuga, M., Qiang, L., Setsuko, M., Kazuko, Y.S., Kazuo, S.: Improving plant drought, salt, and freezing tolerance by gene transfer of a single stress-inducible transcription factor. Nat. Biotechnol. **17**(3), 287–291 (1999)
9. Neumann, P.M., Azaizeh, H., Leon, D.: Hardening of root cell walls: a growth inhibitory response to salinity stress. Plant Cell Environ. **17**(3), 303–309 (1994)
10. Rai, K.N., Murty, D.S., Andrews, D.J., Bramel-Cox, P.J.: Genetic enhancement of pearl millet and sorghum for the semi-arid tropics of Asia and Africa. Genome. **42**(4), 617–628 (1999)
11. Scardoni, G., Michele, P., Carlo, L.: Analyzing biological network parameters with CentiScaPe. Bioinformatics **25**(21), 2857–2859 (2009)
12. Shannon, P., Markiel, A., Ozier, O., Baliga, N.S., Wang, J.T., Ramage, D., Amin, N., Schwikowski, B., Ideker, T.: Cytoscape: a software environment for integrated models of biomolecular interaction networks. Genome Res. **13**(11), 2498–2504 (2003)
13. Shinozaki, K., Kazuko, Y.S., Motoaki, S.: Regulatory network of gene expression in the drought and cold stress responses. Curr. Opin. Plant Biol. **6**(5), 410–417 (2003)
14. Steudle, E.: Water uptake by roots: effects of water deficit. J. Exp. Bot. **51**, 1531–1542 (2000)
15. Tanga, Y., Lia, M., Wanga, J., Pana, Y., Wu, F.X.: CytoNCA: a cytoscape plugin for centrality analysis and evaluation of protein interaction networks. Biosystems. **127**, 67–72 (2015)

On Total Roman Domination in Graphs

P. Roushini Leely Pushpam[1] and S. Padmapriea[2(✉)]

[1] Department of Mathematics, D.B. Jain College, Chennai 600 097, India
roushinip@yahoo.com

[2] Department of Mathematics, Sri Sairam Engineering College, Chennai 600 044, India
prieas@gmail.com

Abstract. A *Roman dominating function* (RDF) on a graph $G = (V, E)$ is a function $f : V \rightarrow \{0, 1, 2\}$ satisfying the condition that every vertex u for which $f(u) = 0$ is adjacent to at least one vertex v for which $f(v) = 2$. A *total Roman dominating function* on a graph $G = (V, E)$ is a Roman dominating function $f : V \rightarrow \{0, 1, 2\}$ satisfying the condition that every vertex u for which $f(u) > 0$ is adjacent to at least one vertex v for which $f(v) > 0$. The *weight* of a total Roman dominating function is the value $f(V) = \sum_{u \in V} f(u)$. The minimum weight of a total Roman dominating function on a graph G is called the *total Roman domination number* of G and denoted by $\gamma_{tR}(G)$. In this paper, we establish some bounds on the Total Roman domination number in terms of its order and girth.

Keywords: Roman domination · Total domination

1 Introduction

By a graph $G = (V, E)$, we mean a simple, finite, undirected, connected graph with $|V| = n$. For graph theoretic terminology we refer to Charatrand and Lesniak [3].

A set of vertices S is a *dominating set* if $N[S] = V$, or equivalently, every vertex in $V \setminus S$ is adjacent to at least one vertex in S. The *domination number* $\gamma(G)$ is the minimum cardinality of a dominating set in G and a dominating set S of minimum cardinality is called a $\gamma(G)$-set. For fundamentals of domination in graphs we refer to [6].

Cockayne et al. [4] defined a *Roman dominating function* (RDF) on a graph G to be a function $f : V \rightarrow \{0, 1, 2\}$ satisfying the condition that every vertex u for which $f(u) = 0$ is adjacent to at least one vertex v for which $f(v) = 2$. The weight of f is $f(V) = \sum_{v \in V} f(v)$. The *Roman domination number*, denoted by

The original version of this chapter was revised: The name of the second author was corrected. An erratum to this chapter can be found at https://doi.org/10.1007/978-3-319-64419-6_58

S. Arumugam et al. (Eds.): ICTCSDM 2016, LNCS 10398, pp. 326–331, 2017.
DOI: 10.1007/978-3-319-64419-6_42

$\gamma_R(G)$ is the minimum weight of an RDF in G. An RDF of weight $\gamma_R(G)$ is called a $\gamma_R(G)$-function. This definition of a Roman dominating function was motivated by an article in Scientific American by Ian Stewart [16]. Roman domination has also been studied in [4,8–16].

The notion of total domination in graphs was introduced by Cockayne et al. [5]. A set $S \subseteq V$ is a *total dominating set* (TDS) of G if for any vertex $v \in V$ there exists a vertex $u \in S$ such that $uv \in E$. The *total domination number* of G, denoted by $\gamma_t(G)$ is the minimum cardinality of a TDS in G; a TDS of G of minimum cardinality is called a $\gamma_t(G)$-set.

Liu and Chang [7] defined a *total Roman dominating function* (TRDF) on a graph G to be a Roman dominating function $f : V \to \{0, 1, 2\}$ satisfying the condition that every vertex u for which $f(u) > 0$ is adjacent to at least one vertex v for which $f(v) > 0$. The *weight* of a total Roman dominating function is the value $f(V) = \sum_{u \in V} f(u)$. The minimum weight of a total Roman dominating function on a graph G is called the *total Roman domination number* of G and denoted by $\gamma_{tR}(G)$. This parameter has been further studied by Abdollahzadeh Ahangar et al. [1]. Let (V_0, V_1, V_2) be the ordered partition of V induced by f, where $V_i = \{v \in V : f(v) = i\}$ for $i = 0, 1, 2$. Note that there exists a 1-1 correspondence between the function $f : V \to \{0, 1, 2\}$ and the ordered partition (V_0, V_1, V_2) of V. Thus, we write $f = (V_0, V_1, V_2)$. In this paper, we establish some bounds on the Total Roman domination number in terms of its order and girth.

2 Bounds on Total Roman Domination Number

Theorem 1 [2]. *For a connected graph G,* $\gamma_t(G) \geq \left\lceil \frac{n}{\Delta(G)} \right\rceil$.

Theorem 2. *If G is a graph of order n,* $n \geq 3$ *with no isolated vertex, then* $\gamma_{tR}(G) \geq \lceil \frac{n}{\triangle(G)} \rceil + 1$ *where* $\triangle(G)$ *is the maximum degree of the graph G. Also equality holds if* $\triangle(G) = n - 1$.

Proof. Let $f = (V_0, V_1, V_2)$ be a total Roman dominating function of G. By Theorem 1, we have $|V_1| + |V_2| \geq \gamma_t(G) \geq \left\lceil \frac{n}{\triangle(G)} \right\rceil$.

Now

$$\begin{aligned}\gamma_{tR}(G) &= |V_1| + 2|V_2| \\ &\geq \left\lceil \frac{n}{\triangle(G)} \right\rceil + |V_2| \\ &\geq \left\lceil \frac{n}{\triangle(G)} \right\rceil + 1\end{aligned}$$

When $\triangle(G) = n - 1$, $\left\lceil \frac{n}{\triangle(G)} \right\rceil = 2$ and $\gamma_{tR}(G) = 3$. Hence equality holds.

Theorem 3. *Let G be a graph with $\delta(G) = 1$ and $G \neq K_{1,n-1}$. Then $\gamma_{tR}(G) \leq n - l + s$ where l is the number of leaves and s is the number of support vertices in G.*

Proof. Let L denote the set of leaves of G and W denote the set of supports of G. Define $f = (V_0, V_1, V_2)$ by $V_0 = L, V_2 = W$ and $V_1 = V \setminus (V_0 \cup V_2)$. Clearly f is a TRDF of G of weight $2s + (n - l - s) = n - l + s$. Hence $\gamma_{tR}(G) \leq n - l + s$.

Abdollahzadeh Ahangar et al. [1] have proved that for paths P_n, $\gamma_{tR}(P_n) = n$. Hence paths attain the upper bound $n - l + s$. Now consider a tree T^* obtained by joining the heads of two star by a path [Refer Fig. 1]. Then $\gamma_{tR}(T^*) = n$.

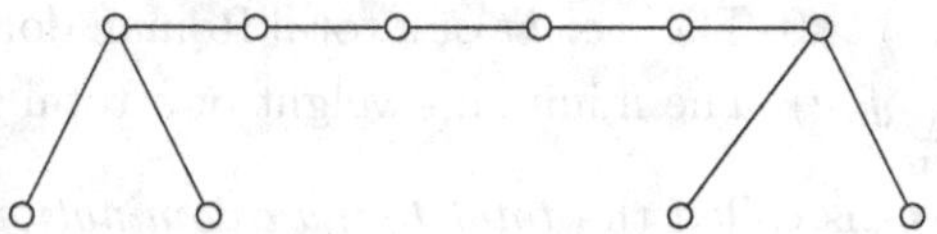

Fig. 1. A tree T^* with $\gamma_{tR}(T^*) = n - l + s$

In the following theorem, we characterize trees which are neither star, paths nor isomorphic to T^* for which $\gamma_{tR}(T) = n - l + s$. A vertex v is called an *isolated strong support vertex* if v is adjacent to at least two leaf vertices and has exactly one non-leaf neighbor which is not a support vertex.

Theorem 4. *For any tree T, which is neither star, path nor isomorphic to T^*, $\gamma_{tR}(T) = n - l + s$ if and only if the following conditions hold:*

1. *Every non-leaf neighbour of a non-isolated strong support vertex is either a support vertex or adjacent to an isolated strong support vertex.*
2. *For every non-support vertex u of T, at least $deg(u) - 1$ neighbours of u are either a non isolated strong support vertices or adjacent to an isolated strong support vertex.*

Proof. Let $f = (V_0, V_1, V_2)$ be a $\gamma_{tR}(T)$-function. Let W and L denote the set of supports and leaves of T respectively. Suppose $\gamma_{tR}(T) = n - l + s$.

Suppose (i) is not true. Then there exists a non-isolated strong support $u \in V$ such that u has at least one non-leaf neighbour say w which is neither a support nor adjacent to an isolated strong support. Let $v \in N(w) \setminus \{u\}$.

If v is a weak support, then there exists a vertex $x \in N(v) \setminus \{w\}$. We define a function $f : V \to \{0, 1, 2\}$ by

$$f(z) = \begin{cases} 2 & \text{if } z \in W \setminus \{v\} \\ 0 & \text{if } z \in (L \cup \{w\}) \setminus \{x\} \\ 1 & \text{otherwise.} \end{cases}$$

Then f is a TRDF with $f(V) < n - l + s$, which is a contradiction.

If v is not a support, then there exist two vertices say x, y such that v, x, y form a path in that order. We define a function $f : V \to \{0, 1, 2\}$ by

$$f(z) = \begin{cases} 2 & \text{if } z \in W \\ 0 & \text{if } z \in (L \cup \{w\}) \\ 1 & \text{otherwise.} \end{cases}$$

Then f is a TRDF with $f(V) < n - l + s$, which is a contradiction. Hence (i) holds.

Suppose (ii) is not true. Then there exists a non-support vertex say $u \in V$ such that at most $\deg(u) - 2$ neighbours of u are either a nonisolated strong support vertices or adjacent to an isolated strong support vertex. Let u_1 and u_2 be the two neighbours of u such that u_1 and u_2 are neither non isolated strong support vertices nor adjacent to an isolated strong support vertex. Therefore there exist vertices x_1, y_1, x_2 and y_2 such that u_1, x_1, y_1 and u_2, x_2, y_2 form paths in that order.

Now if y_1, y_2 are leaves, we define a function $f : V \to \{0, 1, 2\}$ by

$$f(z) = \begin{cases} 2 & \text{if } z \in (W \setminus \{x_1, x_2\}) \cup \{u\} \\ 0 & \text{if } z \in (L \setminus \{y_1, y_2\}) \cup \{u_1, u_2\} \\ 1 & \text{otherwise.} \end{cases}$$

Then f is a TRDF with $f(V) < n - l + s$, which is a contradiction.

Now if y_1, y_2 are not leaves, we define a function $f : V \to \{0, 1, 2\}$ by

$$f(z) = \begin{cases} 2 & \text{if } z \in (W \cup \{u\} \\ 0 & \text{if } z \in L \cup \{u_1, u_2\} \\ 1 & \text{otherwise.} \end{cases}$$

Then f is a TRDF with $f(V) < n - l + s$, which is a contradiction. Hence (ii) holds.

Conversely, suppose the given conditions are satisfied. We define a function $f : V \to \{0, 1, 2\}$ by

$$f(z) = \begin{cases} 2 & \text{if } z \in W \\ 0 & \text{if } z \in L \\ 1 & \text{otherwise.} \end{cases}$$

Then f is a $\gamma_{tR}(T)$-function and

$$\begin{aligned} \gamma_{tR}(T) &= 2|V_2| + |V_1| \\ &= 2s + n - l + s = n - l + s. \end{aligned}$$

3 Bounds in Terms of Girth

In this section, we present bounds for total Roman domination number of a graph G in terms of its girth. The following observation is immediate.

Observation 1. For a graph G of order n with $g(G) \geq 3$, we have $\gamma_{tR}(G) \geq g(G)$.

Theorem 5. *Let G be a graph with $g(G) = 3$. Then $\gamma_{tR}(G) = 3$ if and only if $\Delta(G) = n - 1$.*

Proof. Let $C = (v_1 v_2 v_3 v_1)$ be a cycle in G. Suppose G has a vertex v_1 of degree $n - 1$. We define $f : V \to \{0, 1, 2\}$ by $f(v_1) = 2$, $f(v_2) = 1$ and $f(x) = 0$ for all $x \in V \setminus \{v_1, v_2\}$. Then f is a TRDF and hence $\gamma_{tR}(G) = 3$.

Conversely, let $\gamma_{tR}(G) = 3$ and let $f = (V_0, V_1, V_2)$ be a $\gamma_{tR}(G)$-function. Since $\gamma_{tR}(G) = 3$, one of the following holds.

(i) $|V_2| = 0$ and $|V_1| = 3$.
(ii) $|V_2| = 1$ and $|V_1| = 1$.

If $|V_2| = 0, |V_1| = 3$, clearly $G = C_3$ and hence G has a vertex of degree $n - 1$.

Now, let $|V_2| = 1, |V_1| = 1$. Assume that $V_2 = \{u\}$ and $V_1 = \{v\}$. Clearly u and v are adjacent and every vertex $x \in V \setminus \{u, v\}$ is adjacent to u. Hence $\deg(u) = n - 1$.

Theorem 6. *For any graph G with $g(G) = 4$, $\gamma_{tR}(G) = 4$ if and only if G is a bipartite graph with partite sets X and Y such that both X and Y have at least one vertex of degree $|Y|$ and degree $|X|$ respectively.*

Proof. Suppose that G satisfies the given conditions. Let $\deg(u) = |Y|$ and $\deg(v) = |X|$ where $u \in X$ and $v \in Y$. Now define $f : V \to \{0, 1, 2\}$ by $f(u) = f(v) = 2$ and $f(x) = 0$ for all $x \in V(G) \setminus \{u, v\}$. Then f is $\gamma_{tR}(G)$ - function and hence $\gamma_{tR}(G) = 4$.

Conversely, let $\gamma_{tR}(G) = 4$ and $f = (V_0, V_1, V_2)$ be a $\gamma_{tR}(G)$-function. Since $\gamma_{tR}(G) = 4$, one of the following holds.

(i) $|V_2| = 0$ and $|V_1| = 4$.
(ii) $|V_2| = 1$ and $|V_1| = 2$.
(iii) $|V_2| = 2$ and $|V_1| = 0$.

If $|V_2| = 0$ and $|V_1| = 4$, then $G = C_4$ and hence G satisfies the given condition.

Suppose $|V_2| = 1$ and $|V_1| = 2$. Let $V_2 = \{u\}$ and $V_1 = \{v, w\}$. Since $\gamma_{tR}(G) = 4$, u is adjacent to either v or w but not to both. Without loss of generality, let u and w be non-adjacent vertices in G. Hence u and v are contained in exactly one four cycle of G. Also every vertex in $V \setminus \{u, v, w\}$ is adjacent to u. Therefore G is a bipartite graph with partite sets $X = \{u, w\}$ and $Y = V \setminus \{u, w\}$ and hence G is of the required type.

Suppose $|V_2| = 2$ ande $|V_1| = 0$. Let $V_2 = \{u, v\}$. Since $\gamma_{tR}(G) = 4$, u and v are adjacent and every vertex in $V \setminus \{u, v\}$ is adjacent to either u or v but not to both. Also no two members of $N(u)$ or $N(v)$ are adjacent. Hence G is a bipartite graph with partite sets $X = N(v)$ and $Y = N(u)$ and hence G is of the required type.

References

1. Aahangar, H.A., Henning, M.A., Samodivkin, V., Yero, I.G.: Total roman domination in graphs. Appl. Anal. Discrete Math. (to appear)
2. Atapour, M., Soltaankhah, N.: On total dominating set in graphs. Int. J. Contemp. Math. Sci. **4**, 253–257 (2009)
3. Chartand, G., Lesniak, L.: Graphs and Digraphs. CRC Press, Boca Raton (2005)
4. Cockayne, E.J., Dreyer, A., Hedetniemi, S.M., Hedetniemi, S.T.: Roman domination in graphs. Discrete Math. **78**, 11–12 (2004)
5. Cockayne, E.J., Dawes, R.M., Hedetniemi, S.T.: Total domination in graphs. Networks **10**, 211–219 (1980)
6. Haynes, T.W., Hedetniemi, S.T., Slater, P.J.: Fundamentals of Domination in Graphs. Marcel Dekker, Inc., New York (1998)
7. Liu, C.H., Chang, G.J.: Roman domination on strongly chordal graphs. J. Comb. Optim. **26**, 608–619 (2013)
8. Mobaraaky, B.P., Sheikholeslami, S.M.: Bounds on roman domination numbers of graphs. Matematnykn Bechnk **60**, 247–253 (2008)
9. Pushpam, P.R.L., Mai, T.N.M.M.: On efficient roman dominatable graphs. J. Comb. Math. Comb. Comput. **67**, 49–58 (2008)
10. Pushpam, P.R.L., Mai, T.N.M.M.: Edge roman domination in graphs. J. Comb. Math. Comb. Comput. **69**, 175–182 (2009)
11. Pushpam, P.R.L., Mai, T.N.M.M.: Weak roman domination in graphs. Discuss. Math. Graph Theory **31**, 115–128 (2011)
12. Pushpam, P.R.L., Mai, T.N.M.M.: Weak edge roman domination in graphs. Australas. J. Comb. **51**, 125–138 (2011)
13. Pushpam, P.R.L., Mai, T.N.M.M.: Roman domination in unicyclic graphs. J. Discrete Math. Sci. Crypt. **15**, 237–257 (2012)
14. Pushpam, P.R.L., Padmapriea, S.: Restrained roman domination in graphs. Trans. Comb. **4**, 1–17 (2015)
15. Pushpam, P.R.L., Padmapriea, S.: Global roman domination in graphs. Discrete Appl. Math. **200**, 176–185 (2016)
16. Stewart, I.: Defend the Roman Empire !. Sci. Am. **281**(6), 136–138 (1999)

Restrained Differential of a Graph

P. Roushini Leely Pushpam[1] and D. Yokesh[2(✉)]

[1] Department of Mathematics, D.B. Jain College, Chennai 600 097, Tamil Nadu, India
roushinip@yahoo.com

[2] Department of Mathematics, Sairam Institute of Technology, Chennai 600 044, Tamil Nadu, India
yokeshzabin@yahoo.com

Abstract. Let $G = (V, E)$ be an arbitrary graph. For any subset X of V, let $B(X)$ be the set of all vertices in $V - X$ that have a neighbour in X. Mashburn et al. defined the *differential of a set* X to be $\partial(X) = |B(X)| - |X|$, and the *differential of a graph* is $\max\{\partial(X)\}$, where the maximum is taken over all subsets X of V. Motivated by this parameter we define the *restrained differential of graph* as follows. For any subset X of V, let $\overline{B}(X)$ be the set of all vertices in $V - X$ that have a neighbor in X and a neighbour in $V - X$. We define the *restrained differential of a set* X to be $\overline{\partial}(X) = |\overline{B}(X)| - |X|$ and the restrained differential of a graph is $\max\{\overline{\partial}(X)\}$, where the maximum is taken over all subsets X of V. In this paper, we initiate a study of this parameter.

Keywords: Differential · Restrained domination number · Restrained differentials

1 Introduction

For any subset S of V, the boundary $B(S)$, of a set S is defined to be the set of vertices in $V - S$ dominated by vertices in S, that is $B(S) = (V - S) \cap N(S)$. The differential $\partial(S)$ of S equals the value $\partial(S) = |B(S)| - |S|$. The *differential of a graph* of G is defined as $\partial(G) = max\{\partial(S)|S \subseteq V\}$. The *differential of a set* was defined by Hedetniemi and later studied by Mashburn et al. [6] and Goddard and Henning [2]. The differential in certain classes of graphs is studied in [7] and some bound on the differential in graphs are shown in [1]. The parameter $\{\partial(S)\}$ is also considered in [8] and the minimum differential of an independent set was also motivated by Zhang [9]. This parameter was studied in [4,5].

We define the restrained differential of a graph as follows. For any subset X of V, let $\overline{B}(X)$ be the set of all vertices in $V - X$ that have neighbour in X and a neighbour in $V - X$. We define the *restrained differential of a set* X to be $\overline{\partial}(X) = |\overline{B}(X)| - |X|$ and the *restrained differential of a graph* equal the $max\{\overline{\partial}(X)\}$, for any subset X and V. A subset X of V is said to be $\overline{\partial}-$ set if $\overline{\partial}(G) = \overline{\partial}(X)$. One can easily observe that $\overline{\partial}(G) \leq \partial(G)$. In this paper, we initiate a study of this parameter.

S. Arumugam et al. (Eds.): ICTCSDM 2016, LNCS 10398, pp. 332–340, 2017.
DOI: 10.1007/978-3-319-64419-6_43

2 Notations

Let $G = (V, E)$ be a graph. For graph theoretical terminology not given here refer to Harary [3]. For a vertex $v \in V$, the *open neighbourhood* of v is the set $N(v) = \{u \in V | uv \in E\}$ and the *closed neighbourhood* is the set $N[v] = N(v) \cup \{v\}$. For a set $S \subseteq V$, its open neighbourhood is $N(S) = \bigcup_{v \in S} N(v)$ and the closed neighbourhood is $N[S] = N(S) \cup S$. The *degree* of a vertex v in a graph G is the number of edges of G incident with v and is denoted by $deg(v)$. A vertex of degree zero in G is called an *isolated vertex*, while a vertex of degree one is called a *leaf vertex* or a *pendant vertex* of G. We denote $\delta(G)$ and $\Delta(G)$, the minimum and maximum degree of the graph respectively. The sub graph induced by a set $S \subseteq V$ will be denoted by $< S >$ and any vertex $v \subseteq V(G)$, we denote by $N_S(v)$, the set of neighbours that v has in S. The *complement* $\overline{G}$ of G is defined to be the graph with the same vertex set as G and where two vertices u and v adjacent precisely when they are not adjacent in G.

A graph G is *bipartite* if the vertex set can be partitioned into two disjoint subsets A and B such that the vertices in A are only adjacent to vertices in B and vice versa. $K_{m,n}$ denotes the *complete bipartite graph* where $V = A \cup B$, $|A| = m$, $|B| = n$, A and B are independent sets and every vertex in A is adjacent to every vertex in B. This can be extended to a *complete k-partite graph* $K_{n_1,n_2,\ldots,n_k}$ where the vertex set V is the disjoint union of k independent sets V_i of size n_i; (called the partite classes of $K_{n_1,n_2,\ldots,n_k}$) and every vertex in V_i is adjacent to every vertex in $V - V_i$. A complete bipartite graph is said to be a *star* if $|X|$=1 and $|Y| = n - 1$ and is denoted by $K_{1,n-1}$.

A *split graph* is a graph $G = (V, E)$ whose vertices can be partitioned into two sets X and Y where the vertices in X are independent and vertices in Y form a complete graph.

For arbitrary graphs G and H, the *cartesian product* of G and H is defined to be the graph $G \Box H$ with vertices $\{(u, v) | u \in G, v \in H\}$. Two vertices (u_1, v_1) and (u_2, v_2) are adjacent in $G \Box H$ if and only if one of the following is true: $u_1 = u_2$ and v_1 is adjacent to v_2 in H; or $v_1 = v_2$ and u_1 is adjacent to u_2 in G. If $G = P_m$ and $H = P_n$, then the cartesian product of $G \Box H$ is called the $m \times n$ *grid graph*.

A *Complete binary tree* is a rooted tree in which all leaves have the same depth and all internal vertices have degree three except the root vertex which is of degree two. If T is a complete binary tree with root vertex v, the set of all vertices with depth k are called vertices at level k.

A *support* is a vertex which is adjacent to at least one leaf vertex. A *weak support* is a vertex which is adjacent to exactly one leaf vertex. A *strong support* is a vertex which is adjacent to at least two leaves.

For a positive integer t, a *wounded spider* is a star $K_{1,t}$ with at most $t - 1$ of its edges subdivided. Similarly for an integer $t \geq 2$, a healthy spider is a star $K_{1,t}$ with all of its edges subdivided. In a wounded spider, a vertex of degree t will be called the *head vertex* and the vertices at a distance two from the head vertex will be called the *foot vertices.*

A *caterpillar* is a tree with the property that the removal of the end vertices leaves a path called the *spine* of the caterpillar.

A *restrained dominating set* is a set $S \subseteq V$ in which every vertex in $V - S$ is adjacent to a vertex in S as well as in $V - S$. The *restrained domination number* of the graph is defined as the smallest cardinality of a restrained dominating set of G. It is denoted as $\gamma_r(G)$.

3 Restrained Differential Values of Some Standard Graphs

In this section we determine the value of $\overline{\partial}(G)$ where G is a path, cycle, complete graph, k-partite graph, complete binary tree and caterpillar.

Observation 1. For paths P_n, $n \geq 3$,

$$\overline{\partial}(P_n) = \begin{cases} \lfloor \frac{n}{3} \rfloor & n = 3k+2 \\ \lfloor \frac{n}{3} \rfloor - 1 & \text{otherwise} \end{cases}$$

Observation 2. For cycles C_n, $\overline{\partial}(C_n) = \lfloor \frac{n}{3} \rfloor$.

Observation 3. For complete graph K_n, $\overline{\partial}(K_n) = n - 2$.

Theorem 1. *For any k-partite graph with partition $(X_1, X_2, \ldots, X_k)$, $|X_i| = m_i$, where $i = 1, 2, \ldots, k$, $\overline{\partial}(G) \leq max\{N_1 - 1, N_2 - 1, N_k - 1, N_i + m_i - 4\}$, where $N_j = \sum_{\substack{i=1 \\ i \neq j}}^{k} M_i$, the bound is sharp when G is a complete k-partite graph.*

Proof. If we take $v \in X$, and $S = \{v\}$, we have that $\partial(S) \leq N_1 - 1$ and it is clear that even if we add one more vertex of X_1 to S, then $\partial(S) < N_1 - 1$. A similar argument holds for every set X_i, $2 \leq i \leq k$. Suppose $u \in X_1$ and $v \in X_2$ and $S = \{u, v\}$, we have $\partial(S) \leq N_1 - 1 + m_1 - 1 - 2 = N_1 + m_1 - 4$. It is clear that, if the graph is a complete bipartite graph, all the above inequalities are equalities.

Theorem 2. *For any caterpillar T of order n*

$$\overline{\partial}(T) = \begin{cases} \lfloor \frac{m}{3} \rfloor - 1 & m = 3k \text{ and both ends are week supports} \\ \lfloor \frac{m}{3} \rfloor & \text{otherwise} \end{cases}$$

where m denotes the number of vertices of the spine.

Proof. Clearly $\overline{\partial}(T) \leq \lfloor \frac{m}{3} \rfloor$. If both ends of the spine are supports and m = 3k, then $S_1 = \{v_3, v_6, v_9, ..., v_{m-3}\}$ is a $\overline{\partial}$-set of T. Hence $|S_1| = \frac{m}{3} - 1$. If at least end of the spine is a week support and m = 3k+1 or m = 3k+2, then $S_2 = \{v_2, v_5, v_8, \ldots, v_{m-2}\}$ is a $\overline{\partial}$-set of T. Hence $\overline{\partial}(T) = \lfloor \frac{m}{3} \rfloor$. If both ends of the spine are strong supports, then $S = \{v_2, v_5, v_8, \ldots, v_k\}$ is a $\overline{\partial}$-set of T where $k = m - 1$ when $m \equiv 0 \ (mod\ 3)$, $k = m - 2$ when $m \equiv 1 \ (mod\ 3)$ and $k = m - 4$ when $m \equiv 2 \ (mod\ 3)$. In all cases $\overline{\partial}(T) = \lfloor \frac{m}{3} \rfloor$.

Theorem 3. *For any complete binary tree G,*

$$\overline{\partial}(G) = \begin{cases} \frac{3(8)^n-10}{7} & k = 3n \\ \frac{3(8)^n-10}{14} & k = 3n-1 \\ \frac{3}{7}(2(8)^n - 1) & k = 3n+1 \end{cases}$$

Proof. Let G be a complete binary tree with k levels. Let S_i be the set of all vertices in level i and $|S_i| = n_i$, then $n_i = 2^i$. Now $n_k > n_{k-1} > n_{k-2} > \cdots > n_0$. If $k = 3n$, then clearly $S_{k-2} \cup S_{k-4} \cup \cdots \cup S_1$ is $\overline{\partial}$-set of G. If $k = 3n-1$, then clearly $S_{k-2} \cup S_{k-4} \cup \cdots \cup S_2$ is $\overline{\partial}$-set of G. If $k = 3n+1$, then clearly $S_{k-2} \cup S_{k-4} \cup S_1$ is a $\overline{\partial}$-set of G.
If $k = 3n$, then

$$\begin{aligned}
\overline{\partial}(G) &= (2^{k-1} + 2^{k-3} + 2^{k-4} + 2^{k-6} + 2^{k-7} + \cdots 2^3 + 2^2) \\
&\quad - (2^{k-2} + 2^{k-5} + 2^{k-8} + \cdots + 2) \\
&= (2^{k-1} + 2^{k-2} + 2^{k-3} + 2^{k-4} + \cdots 2^2 + 2) \\
&\quad - 2(2^{k-2} + 2^{k-5} + 2^{k-8} + \cdots + 2) \\
&= \frac{2^k}{2}\left[1 + \frac{1}{2} + \frac{1}{2^2} + \cdots + \frac{1}{2^{k-1}}\right] - \frac{2(2^k)}{2}\left[1 + \frac{1}{2^3} + \frac{1}{2^5} + \cdots + \frac{1}{2^{k/3-1}}\right] \\
&= \frac{2^k}{2}\left(\left[\frac{1-(1/2)^{k-1}}{1-1/2}\right] - \left[\frac{1-(1/8)^n}{1-1/8}\right]\right) \\
&= \frac{2^k}{2}\left[(2^k - 2) - \frac{4}{7}\left(\frac{2^k-1}{2^k}\right)\right] = \frac{3(2)^k - 10}{7} \\
&= \frac{3(8)^n - 10}{7}
\end{aligned}$$

If $k = 3n-1$, then

$$\begin{aligned}
\overline{\partial}(G) &= (2^{k-1} + 2^{k-3} + 2^{k-4} + 2^{k-6} + 2^{k-7} + \cdots 2^2 + 2) \\
&\quad - (2^{k-2} + 2^{k-5} + 2^{k-8} + \cdots + 2^3 + 2^1) \\
&= \frac{2^k}{2}\left[\frac{1-(1/2)^k}{1-1/2}\right] - \frac{2(2^k)}{2^2}\left[\frac{1-(1/2^3)^n}{1-1/2^3}\right] \\
&= \frac{3(2)^k - 5}{7} = \frac{3(8)^n - 10}{14}
\end{aligned}$$

If $k = 3n+1$, then

$$\begin{aligned}
\overline{\partial}(G) &= (2^{k-1} + 2^{k-3} + 2^{k-4} + 2^{k-6} + 2^{k-7} + \cdots 2^3 + 2) \\
&\quad - (2^{k-2} + 2^{k-5} + 2^{k-8} + \cdots + 2^5 + 2) \\
&= \frac{2^k}{2}\left[\frac{1-(1/2)^{k-1}}{1-1/2}\right] - \frac{2(2^k)}{2}\left[\frac{1-(1/2^3)}{1-1/2^3}\right]
\end{aligned}$$

$$= \frac{2^k}{2}\left[\frac{2^{k-1}-1}{2^{k-1}} \times 2 - \frac{8}{7}\left(\frac{2^{3n}-1}{2^{3n}}\right)\right]$$
$$= \frac{2^k}{2}\left[\frac{2^{k-1}-1}{2^{k-1}}\right]\left(\frac{6}{7}\right)$$
$$= \frac{6}{7}\left(\frac{2(8)^n-1}{2}\right)$$
$$= \frac{3}{7}(2(8)^n-1)$$

4 Bounds on $\overline{\partial}(G)$

In this section we obtain some bounds for $\overline{\partial}(G)$.

Theorem 4. *For any graph G, $0 \le \overline{\partial}(G) \le n-2$.*

Proof. Let S be any $\overline{\partial}$-set of G, then by definition

$$\overline{\partial}(G) = |\overline{B}(S)| - |S| \le n-1-1 \le n-2.$$

Theorem 5. *For any graph G, $\overline{\partial}(G) + 2\gamma_r(G) \ge n$.*

Proof. Let X be a any γ_r-set. Notice that,

$$\begin{aligned}\overline{\partial}(G) \ge \overline{\partial}(X) &= |\overline{B}(X)| - |X| \\ &= (n-\gamma_r(G)) - \gamma_r(G) \\ &= n - 2\gamma_r(G).\end{aligned}$$

Theorem 6. *For any two positive integers a and b with $a < b$, there exists a graph G such that $\overline{\partial}(G) = a$ and $\partial(G) = b$.*

Proof. Let a and b be any two positive integers with $a < b$. Consider the star $K_{1,b+2}$ and subdivide $a+1$ edges, clearly $\overline{\partial}(G) = a$ and $\partial(G) = b$.

Theorem 7. *Given any positive integer k and n and $0 \le k < n-2$, there exists a graph G on n vertices with $\overline{\partial}(G) = k$.*

Proof. We now construct a graph with $\overline{\partial}(G) = k$. Consider a star on $k+2$ vertices and subdivided all its edges. Clearly $\overline{\partial}(G) = k$.

A set $S \subset V$ is a minimum $\overline{\partial}$-set if

$$|S| = min\{|X| : X \subset V \text{ and } \overline{\partial}(X) = \overline{\partial}(G)\}.$$

Theorem 8. *If S is a minimum $\overline{\partial}$-set, then*

(i) $|\overline{B}(S)| \ge 2|S|$
(ii) $|S| \le \frac{n}{3}$
(iii) $|S| \le \overline{\partial}(G)$

Proof. We know that every vertex in S is of degree at least two. Let $v \in S$, then clearly, $|epn(v, S)| \geq 2$ and hence $|\overline{B}(S)| \geq 2|S|$. Hence (i) holds.
Now to prove (ii), $3|S| = |S| + 2|S| \leq |S| + |\overline{B}(S)| = n$. Hence (ii) holds.
Now to prove (iii), $|S| \leq 2|S| - |S| \leq |\overline{B}(S)| - |S| = \overline{\partial}(G)$.

Theorem 9. *Let G be any graph and S be a minimum $\overline{\partial}$-set of G. Then*

$$\overline{\partial}(G) + 1 \leq |\overline{B}(S)| \leq 2\overline{\partial}(G).$$

Proof

$$|\overline{B}(S)| = 2|\overline{B}(S)| - |\overline{B}(S)| \leq 2|\overline{B}(S)| - 2|S| \leq 2\overline{\partial}(G)$$
$$\overline{\partial}(G) + 1 = |\overline{B}(S)| - |S| + 1 = |\overline{B}(S)| - (|S| - 1).$$

Hence $\overline{\partial}(G) + 1 \leq |\overline{B}(S)| \leq 2\overline{\partial}(G)$.

Theorem 10. *For any graph G, $\overline{\partial}(G) = 0$ if and only if G is isomorphic to a star or a bistar.*

Proof. Suppose $\overline{\partial}(G) = 0$. We claim that G is isomorphic to a star or a bistar. Let v be a vertex in G such that $deg(v) = \Delta(G)$. First we claim that at most one vertex in $N(v)$ is of degree more than one. Suppose no vertex in $N(v)$ is of degree more than one, then G reduces to a star. Otherwise there exists two vertices x, y in $N(v)$ are of degree more than one. Then clearly $\overline{\partial}(\{v\}) = 2 - 1 = 1$, which is a contradiction. Next we claim that each neighbour of x except v is of degree one. Suppose there exists a vertex, say z in $N(x)$ of degree more than one. Then clearly $\overline{\partial}(\{x, v\}) = 2 - 1 = 1$, which is a contradiction. Hence G is isomorphic to a star or a bistar. Converse is obvious.

Theorem 11. *For any graph G, $\overline{\partial}(G) = n - 2$, if and only if $\delta(G) > 1$ and there exists a vertex $v \in V(G)$ with $deg(v) = n - 1$.*

Proof. Let $v \in V(G)$ be such that $deg(v) = \Delta(G)$. First we assume that $\overline{\partial}(G) = n - 2$. We claim that $deg(v) = n - 1$. Suppose there exists a vertex u such that $u \notin N(v)$ and $\overline{\partial}(\{v\}) = |\overline{B}(S)| - |S| \leq (n - 1 - 1) - 1 = n - 3$, which is a contradiction. Next we claim $\delta(G) > 1$. Suppose not. then v is a support. Let w be the leaf which is adjacent to v. Then $\overline{\partial}(G) \leq (n - 1 - 1) - 1 = n - 3$, which is a contradiction.

Conversely, Suppose there exist a vertex $v \in V(G)$ with $deg(v) = n - 1$ and $\delta(G) > 1$, then $\overline{\partial}(\{v\}) = n - 1 - 1 = n - 2$.

Theorem 12. *For any graph G, $\overline{\partial}(G) = n - 3$ if and only if there exists a vertex v such that $deg(v) = n - 1$ or $n - 2$. Further if $deg(v) = n - 1$, then v is a weak support, otherwise v is not a support.*

Proof. First we assume that $\overline{\partial}(G) = n - 3$. Let $v \in V(G)$ with $deg(v) = \Delta(G)$. First we claim that $deg(v) = n - 1$ or $n - 2$. Suppose not. Then there exists two vertices $x, y \in V(G)$ such that $x, y \notin N(v)$. Then, clearly $\overline{\partial}(\{v\}) < (\Delta - 1 - 2) - 1 < n - 3$, which is a contradiction. Next we claim that if $deg(v) = n - 1$, then v is a weak support. Suppose not. Then v is adjacent to more than one pendant vertices say m vertices. Then $\overline{\partial}(\{v\}) = n - m - 1 - 1 = n - m - 2 < n - 3$, which is a contradiction. If $deg(v) = n - 2$, then v is not a support. For otherwise $N(v)$ contains m pendant vertices and $\overline{\partial}(\{v\}) = (n - m - 2) - 1 < n - 3$, which is a contradiction.

Conversely if $deg(v) = n - 1$ and v is a week support, then $\overline{\partial}(\{v\}) = (n - 1) - 1 - 1 = n - 3$. If $deg(v) = n - 2$ and v is not a support, then $\overline{\partial}(\{v\}) = n - 2 - 1 = n - 3$.

Theorem 13. *For any graph G, with $diam(G) = 2$, $0 \leq \overline{\partial}(G) \leq \Delta(\Delta - 2)$, where Δ denotes the maximum degree of G.*

Proof. If G is a wounded spider with exactly one foot vertex, then $\overline{\partial}(G) = 0$, hence $\overline{\partial}(G) \geq 0$. To prove the upper bound, we consider a vertex $v \in V(G)$ with $deg(v) = \Delta$. Define $A_1 = \{x|x \in N(v)\}$, $A_2 = \{y|y \in N(A_1)\}$. Since $diam(G) = 2$, $V(G) = A_1 \cup A_2 \cup \{v\}$ and the graph will have a maximum restrained differential value if it is a Δ-regular graph. In this case A_1 is a $\overline{\partial}$-set. Hence $\overline{\partial}(G) \leq \Delta(\Delta - 1) = \Delta(\Delta - 1 - 1) = \Delta(\Delta - 2)$.

Theorem 14. *For any graph G, with $diam(G) = 3$, $1 \leq \overline{\partial}(G) \leq (\Delta - 1)^3 + 1$. where Δ denotes the maximum degree of G.*

Proof. Since $diam(G) = 3$, clearly $\overline{\partial}(G) \geq 1$. For the upper bound, consider a vertex $v \in V(G)$, with $deg(v) = \Delta$. Let A_1, A_2 be two sets as defined in Theorem 13. Now define $A_3 = \{z|z \in N(A_3)\}$. Since $diam(G) = 3$, $V(G) = A_1 \cup A_2 \cup A_3 \cup \{v\}$ and the graph will have the maximum restrained differential, when it is a Δ-regular graph. We take all the vertices adjacent to $N(v)$ as a restrained differential set. Tthen $\partial(G) \leq \Delta + \Delta(\Delta - 1)^2 - \Delta(\Delta - 1) = (\Delta - 1)^3 + 1$.

Theorem 15. *Let G be any graph and $v \in V(G)$ with $deg(v) = \Delta(G)$ and v is not a support. Then $\overline{\partial}(G) \geq \Delta(G) - 1$ and equality holds if and only if there exists a $\overline{\partial}$-set S satisfying the following conditions:*

(i) For each $w \in \overline{B}(S)$ at most two vertices in $N_{C(S)}(w)$ has at least one neighbour in $C(S)$, where $C(S) = V \backslash (S \cup \overline{B}(S))$.
(ii) Each component of $C(S)$ is either a K_1 or a star or a wounded spider with at most two foot vertices. If there is a wounded spider with exactly two foot vertices, then the corresponding w in $\overline{B}(S)$ has no neighbours in $C(S)$.
(iii) $< C(S) >$ has at most $2\Delta(G) - 1$ wounded spiders.

Proof. Let $v \in V(G)$ be a non-support, such that $deg(v) = \Delta(G)$. It is obvious that $\overline{\partial}(G) \geq \Delta(G) - 1$. To prove equality, we assume that $\overline{\partial}(G) = \Delta(G) - 1$. Let S be any $\overline{\partial}$-set. First we claim that for each $w \in \overline{B}(S)$ at most two vertices in $N_{C(S)}(w)$ has at least one neighbour in $C(S)$. Suppose not. Then at least three vertices in $N_{C(S)}(w)$ have more than one neighbour in $C(S)$, which imply that $\overline{\partial}(\{v, w\}) = \Delta(G) - 1 + 3 - 2 = \Delta(G) > \Delta(G) - 1$, which is a contradiction.

Suppose x and y are two vertices in $N_{C(S)}(w)$ which have at least one neighbour in $C(S)$, then no vertex in $C(S)$ is a common neighbour of both x and y. For, otherwise $\overline{\partial}(\{v,z\}) = \Delta(G)-1+2-1 = \Delta(G) > \Delta(G)-1$, which is a contradiction, where $z \in N(x) \cap N(y) \cap C(S)$.

Further at most two neighbours of x have neighbours in $C(S)$. For, otherwise $\overline{\partial}(\{v,x\}) \geq \Delta(G) - 1 + 3 - 2 = \Delta(G) > \Delta(G) - 1$, which is a contradiction. Similarly, at most two neighbours of y have neighbour in $C(S)$. Suppose y_1, y_2 be two neighbours of y having neighbours in $C(S)$. Then we claim that no vertex in $C(S)$ is a common neighbour of both y_1 and y_2. Suppose not. Let u be the vertex in $C(S)$ which is a common neighbour of both y_1 and y_2. Then $\overline{\partial}(\{v,u\}) = \Delta(G) - 1 + 2 - 1 = \Delta(G) > \Delta(G) - 1$, which is a contradiction. Hence each component of $C(S)$ is either a K_1 or a star or a wounded spider with at most two foot vertices. If a component of $C(S)$ is a wounded spider with two foot vertices, then we claim that corresponding $w \in \overline{B}(S)$ has no neighbour in $C(S)$. Suppose not. Let $z \neq x$ be a neighbour of w in $C(S)$, then $\overline{\partial}(\{x\}) = \Delta(G)+3-2 = \Delta(G)+1 > \Delta(G)-1$, which is a contradiction. Finally we claim that $< C(S) >$ have at most $2\Delta(G)-1$ wounded spiders. Suppose not. Then $\overline{\partial}(\{N_2(v)\}) \geq 2\Delta(G)-\Delta(G) = \Delta(G) > \Delta(G)-1$, which is a contradiction.

Conversely, let us assume that the given conditions hold, then clearly $S = \{v\}$ and $\overline{\partial}(G) = \Delta(G) - 1$. Hence proved.

Theorem 16. *Let G be any bipartite graph with bipartition (X,Y) and $v \in V(G)$ with $deg(v) = \Delta(G)$ and v is not a support. Then $\overline{\partial}(G) = \Delta(G) - 1$, if and only if the following conditions hold,*

(i) Corresponding to each $w \in N(v)$ at most two vertices in $N(w)$ has neighbours in $Y - N(v)$.

(ii) At most two supports in $Y - N(v)$ have a common neighbour. If two supports have a common neighbour say two, then the vertex $x \in N(v) \cap N(z)$ is of degree two.

(iii) At most $2\Delta(G)$ vertices in $Y - N(v)$ are supports.

Proof. Without loss of generality, let $v \in X$. Suppose $\overline{\partial}(G) = \Delta(G) - 1$. Condition (i) follows from Theorem 15. Next we claim that at most two supports in $Y - N(v)$ have a common neighbour. Suppose there exists three vertices which are supports in $Y - N(v)$ having a common neighbour say w, then $\overline{\partial}(\{v,w\}) = \Delta(G) - 1 + 3 - 2 = \Delta(G) > \Delta(G) - 1$, which is a contradiction. Further, if two support have a common neighbour say z, then the vertex x in $N(v) \cap N(z)$ is of degree two. Otherwise, $\overline{\partial}(\{v,z\}) = \Delta(G) - 1 + 3 - 2 = \Delta(G) > \Delta(G) - 1$, which is a contradiction. Finally we claim that at most $2\Delta(G)$ vertices in $Y - N(v)$ are supports. Suppose not. Let S_1 be the set of all vertices in $Y - N(v)$ such that each $x \in S_1$ in degree at least two. Then $\overline{\partial}(S_1) = 2\Delta(G) - \Delta(G) = \Delta(G) > \Delta(G) - 1$, which is a contradiction. Converse is obvious.

Theorem 17. *Let G be a split graph with bipartition (X,Y), where X is an independent and Y is complete and $|X| = m_1$, $|Y| = m_2$. Let $v_1, v_2, v_3, \ldots, v_m$ be the vertices in Y such that $deg(v_1) \geq deg(v_2) \geq deg(v_3) \geq \cdots \geq deg(v_m)$.*

Then $\overline{\partial}(G) = \partial(G)$ *if and only* v_1 *is not a support and at most two vertices of* $N_X(v_i) - \left[\bigcup_{j=1}^{i-1} N_X(v_j)\right]$ *are pendant vertices, where if* $2 \le i \le m_2$.

Proof. We assume that $\overline{\partial}(G) = \partial(G)$. First we claim that v_1 is not a support. Suppose v_1 is a support. Then clearly v_1 belongs to every ∂ set. If $N_X(v_1)$ contains m pendant vertices then $\overline{\partial}(\{v\}) = N_X(v_1) + (m_2 - 1) - m - 1$. But $\partial(\{v\}) = N_X(v_1) + (m_2 - 1) - 1$, which is a contradiction. Hence v_1 is not a support. Next we claim that at most two vertices of $N_X(v_i) - \left[\bigcup_{j=1}^{i-1} N_X(v_j)\right]$ are two pendent vertices. Suppose not. Let $v_i \neq v_1 \in Y$ such that $N_X(v_i) - \bigcup_{j=1}^{i-1} N_X(v_j)$ contains more than two pendant vertices, then

$$\begin{aligned}\partial(\{v_1, v_i\}) &\ge |N_X(v_1)| - 1 + 3 - 2\\ &= |N_X(v_1)|\\ \overline{\partial}(\{v_1, v_i\}) &= |N_X(v_1)| - 1 \le \partial(\{v_1, v_i\})\end{aligned}$$

which imply that $\overline{\partial}(G) \neq \partial(G)$ which is a contradiction.

Conversely, assume that the given conditions holds. To prove $\overline{\partial}(G) = \partial(G)$. Let S be any $\overline{\partial}$ set of G. By the given condition $v_1 \in S$, and let S_1 be the set of all vertices in Y except v_1 such that $N_X(v_i) - \bigcup_{j=1}^{i-1} N_X(v_j)$ has at least one neighbour in X. Then $S = \{v_1\} \cup S_1$ is also a $\overline{\partial}$-set of G. Hence $\partial(G) = \overline{\partial}(G)$.

References

1. Bermudo, S., Rodriguez, J.M., Sigarreta, J.M.: On the differential in graphs. Util. Math. (To appear)
2. Goddard, W., Henning, M.A.: Generalized domination and independence in graphs. Congr. Numer. **123**, 161–171 (1997)
3. Harary, F.: Graph Theory. Addison Wesley, Reading (1972)
4. Haynes, T.W., Hedetniemi, S.T., Slater, P.J.: Fundamentals of Domination of Graphs. Marcel Dekker, New York (1998)
5. Haynes, T.W., Hedetniemi, S.T., Slater, P.J.: Domination of Graphs: Advanced Topics. Marcel Dekker, New York (1998)
6. Mashburn, J.L., Haynes, T.W., Hedetniemi, S.M., Hedetniemi, S.T., Slater, P.J.: Differentials in graphs. Util. Math. **69**, 43–54 (2006)
7. Pushpam, P.R.L., Yokesh, D.: Differential in certain classes of graphs. J. Math. **41**(2), 129–138 (2010). Tamkang
8. Slater, P.J.: Locating dominating sets and locating-dominating sets. In: Graph Theory, Combinatorics and Algorithms, vol. 1, pp. 1073–1079. Wiley, New York (1995). Wiley-Interscience Publications, Kalmazoo (1992)
9. Zhang, C.Q.: Finding critical independent sets and critical vertex subsets are polynomial problems. SIAM J. Discrete Math. **3**(3), 431–438 (1990)

The Distinguishing Number of Kronecker Product of Two Graphs

Saeid Alikhani(✉) and Samaneh Soltani

Department of Mathematics, Yazd University, 89195-741 Yazd, Iran
alikhani@yazd.ac.ir

Abstract. The distinguishing number $D(G)$ of a graph G is the least integer d such that G has a vertex labeling with d labels that is preserved only by a trivial automorphism. The Kronecker product $G \times H$ of two graphs G and H is the graph with vertex set $V(G) \times V(H)$ and edge set $\{\{(u,x),(v,y)\} | \{u,v\} \in E(G)$ *and* $\{x,y\} \in E(H)\}$. In this paper we study the distinguishing number of Kronecker product of two graphs.

Keywords: Distinguishing number · Kronecker product

1 Introduction

Let $G = (V, E)$ be a simple graph of order $n \geqslant 2$. We use the following notations: The set of vertices adjacent in G to a vertex of a vertex subset $W \subseteq V$ is the open neighborhood $N_G(W)$ of W. The $\mathrm{Aut}(G)$ denotes the automorphism group of G. A labeling of G, $\phi : V \to \{1, 2, \ldots, r\}$, is said to be r-distinguishing, if no non-trivial automorphism of G preserves all of the vertex labels. The point of the labels on the vertices is to destroy the symmetries of the graph, that is, to make the automorphism group of the labeled graph trivial. Formally, ϕ is r-distinguishing if for every non-trivial $\sigma \in \mathrm{Aut}(G)$, there exists x in V such that $\phi(x) \neq \phi(\sigma(x))$. The distinguishing number of a graph G is defined by

$$D(G) = \min\{r : \ G \text{ has a labeling that is } r\text{-distinguishing}\}.$$

This number has defined by Albertson and Collins [1]. If a graph has no non-trivial automorphisms, its distinguishing number is 1. In other words, $D(G) = 1$ for the asymmetric graphs. The other extreme, $D(G) = |V(G)|$, occurs if and only if $G = K_n$. Also $D(P_n) = 2$ for every $n \geqslant 3$, and $D(C_n) = 3$ for $n = 3, 4, 5$, $D(C_n) = 2$ for $n \geqslant 6$. A graph and its complement, always have the same automorphism group while their graph structure usually differs, hence $D(G) = D(\overline{G})$ for every simple graph G. The distinguishing number of some graph products has been studied in literature (see [2–4,9,10]). The Cartesian product of graphs G and H is a graph, denoted $G\square H$, whose vertex set is $V(G) \times V(H)$. Two vertices (g, h) and (g', h') are adjacent if either $g = g'$ and $hh' \in E(H)$, or $gg' \in E(G)$ and $h = h'$. Denote $G\square G$ by G^2, and recursively define the k-th Cartesian power of G as $G^k = G\square G^{k-1}$. A non-trivial graph G is called prime

S. Arumugam et al. (Eds.): ICTCSDM 2016, LNCS 10398, pp. 341–346, 2017.
DOI: 10.1007/978-3-319-64419-6_44

if $G = G_1 \square G_2$ implies that either G_1 or G_2 is K_1. Two graphs G and H are called relatively prime if K_1 is the only common factor of G and H. We need knowledge of the structure of the automorphism group of the Cartesian product, which was determined by Imrich [8], and independently by Miller [13].

Theorem 1 [8,13]. *Suppose ψ is an automorphism of a connected graph G with prime factor decomposition $G = G_1 \square G_2 \square \ldots \square G_r$. Then there is a permutation π of the set $\{1, 2, \ldots, r\}$ and there are isomorphisms $\psi_i : G_{\pi(i)} \to G_i$, $i = 1, \ldots, r$, such that*

$$\psi(x_1, x_2, \ldots, x_r) = (\psi_1(x_{\pi(1)}), \psi_2(x_{\pi(2)}), \ldots, \psi_r(x_{\pi(r)})).$$

The Kronecker product is one of the (four) most important graph products and seems to have been first introduced by K. Čulik, who called it the cardinal product [6]. Weichsel [14] proved that the Kronecker product of two nontrivial graphs is connected if and only if both factors are connected and at least one of them possesses an odd cycle. If both factors are connected and bipartite, then their Kronecker product consists of two connected components. The Kronecker product $G \times H$ of graphs G and H is the graph with vertex set $V(G) \times V(H)$ and edge set $\{\{(u, x), (v, y)\} | \{u, v\} \in E(G) \text{ and } \{x, y\} \in E(H)\}$. The terminology is justified by the fact that the adjacency matrix of a Kronecker graph product is given by the Kronecker matrix product of the adjacency matrices of the factor graphs; see [14] for details. However, this product is also known under several different names including categorical product, tensor product, direct product, weak direct product, cardinal product and graph conjunction. The Kronecker product is commutative and associative in an obvious way. It is computed that $|V(G \times H)| = |V(G)|.|V(H)|$ and $|E(G \times H)| = 2|E(G)|.|E(H)|$. We recall that graphs with no pairs of vertices with the same open neighborhoods are called R-thin. In continue, we need the following theorem:

Theorem 2 [7]. *Suppose φ is an automorphism of a connected non-bipartite R-thin graph G that has a prime factorization $G = G_1 \times G_2 \times \ldots \times G_k$. Then there exists a permutation π of $\{1, 2, \ldots, k\}$, together with isomorphisms $\varphi_i : G_{\pi(i)} \to G_i$, such that*

$$\varphi(x_1, x_2, \ldots, x_k) = (\varphi_1(x_{\pi(1)}), \varphi_2(x_{\pi(2)}), \ldots, \varphi_k(x_{\pi(k)})).$$

2 Main Results

We begin with the distinguishing number of Kronecker product of complete graphs.

Theorem 3. *Let k, n, and d be integers so that $d \geqslant 2$ and $(d-1)^k < n \leqslant d^k$. Then*

$$D(K_k \times K_n) = \begin{cases} d & \text{if } n \leqslant d^k - \lceil \log_d k \rceil - 1 \\ d+1 & \text{if } n \geqslant d^k - \lceil \log_d k \rceil + 1 \end{cases}$$

If $n = d^k - \lceil \log_d k \rceil$, then $D(K_k \times K_n)$ is either d or $d + 1$.

Proof. It is easy to see that $K_k \times K_n$ is the complement of Cartesian product $K_k \square K_n$. By Theorem 1.1 in [9], $D(K_k \square K_n) = \begin{cases} d & \text{if } n \leqslant d^k - \lceil \log_d k \rceil - 1 \\ d+1 & \text{if } n \geqslant d^k - \lceil \log_d k \rceil + 1 \end{cases}$, so we have the result.

It is known that connected non-bipartite graphs have unique prime factor decomposition with respect to the Kronecker product [12]. If such a graph G has no pairs u and v of vertices with the same open neighborhoods, then the structure of automorphism group of G depends on that of its prime factors exactly as in the case of the Cartesian product. As said before graphs with no pairs of vertices with the same open neighborhoods are called R-thin and it can be shown that a Kronecker product is R-thin if and only if each factor is R-thin.

Theorem 4. *Let G and H be two simple connected, relatively prime graphs, non-bipartite R-thin graphs, then $D(G \times H) = D(G \square H)$.*

Proof. By hypotheses and Theorems 1 and 2, it can be concluded that $\text{Aut}(G \times H) = \text{Aut}(G \square H)$. Therefore $D(G \times H) = D(G \square H)$.

Imrich and Klavžar in [10] proved that the distinguishing number of k-th power with respect to the Kronecker product of a non-bipartite, connected, R-thin graph different from K_3 is two.

Theorem 5 [10]**.** *Let G be a nonbipartite, connected, R-thin graph different from K_3 and $\times G^k$ the k-th power of G with respect to the Kronecker product. Then $D(\times G^k) = 2$ for $k \geqslant 2$. For the case $G = K_3$ we have $D(K_3 \times K_3) = 3$ and $D(\times K_3^k) = 2$ for $k \geqslant 3$.*

Now we want to obtain the distinguishing number of Kronecker product of two complete bipartite graphs. We need the following lemma:

Lemma 1 [11]**.** *If $G = (V_0 \cup V_1, E)$ and $H = (W_0 \cup W_1, F)$ are bipartite graphs, then $(V_0 \times W_0) \cup (V_1 \times W_1)$ and $(V_0 \times W_1) \cup (V_1 \times W_0)$ are vertex sets of the two components of $G \times H$.*

Proposition 1. *If $K_{m,n}$ and $K_{p,q}$ are complete bipartite graphs such that $q \geqslant p$ and $m \geqslant n$ then the distinguishing number of $K_{m,n} \times K_{p,q}$ is*

$$D(K_{m,n} \times K_{p,q}) = \begin{cases} mq+1 & m = n, p = q \\ mq & \text{otherwise.} \end{cases}$$

Proof. The Kronecker product $K_{m,n} \times K_{p,q}$ is disjoint union of two complete bipartite graphs $K_{mp,nq}$ and $K_{mq,np}$ by Lemma 1. Hence if $m \neq n$ and $p \neq q$, then $K_{mp,nq}$ and $K_{mq,np}$ are the two non-isomorphic graphs, and so $D(K_{m,n} \times K_{p,q}) = \max\{D(K_{mp,nq}), D(K_{mq,np})\} = mq$. If $m = n$ or $p = q$, then $K_{mp,nq}$ and $K_{mq,np}$ are isomorphic to $K_{mp,mq}$ or $K_{mp,np}$, respectively. In fact

$$K_{m,n} \times K_{p,q} = \begin{cases} K_{mp,mq} \cup K_{mp,mq} & m = n, p \neq q \\ K_{mq,nq} \cup K_{mq,nq} & m \neq n, p = q \\ K_{mq,mq} \cup K_{mq,mq} & m = n, p = q. \end{cases}$$

Thus using the value of the distinguishing number of complete bipartite graphs we have

$$D(K_{m,n} \times K_{p,q}) = \begin{cases} mq & m = n, p \neq q \text{ or } m \neq n, p = q \\ mq + 1 & m = n, p = q. \end{cases}$$

Therefore the result follows.

Corollary 1. *Let $m, n \geqslant 3$ be two integers. The distinguishing number of Kronecker product of star graphs $K_{1,n}$ and $K_{1,m}$ is $D(K_{1,n} \times K_{1,m}) = mn$.*

The following result shows that the distinguishing number of Kronecker product of complete bipartite graphs is an upper bounds for the distinguishing number of Kronecker product of bipartite graphs.

Corollary 2. *If $G = (V_0 \cup V_1, E)$ and $H = (W_0 \cup W_1, F)$ are bipartite graphs such that $|V_0| = m$, $|V_1| = n$, $|W_0| = p$, and $|W_1| = q$, then $D(G \times H) \leqslant D(K_{m,n} \times K_{p,q})$.*

Proof. It is sufficient to note that $\mathrm{Aut}(G \times H) \subseteq \mathrm{Aut}(K_{m,n} \times K_{p,q})$, and $G \times H$ and $K_{m,n} \times K_{p,q}$ have the same size. Now we have the result by Proposition 1.

Before we prove the next result we need some additional information about the distinguishing number of complete multipartite graphs. Let $K_{a_1{}^{j_1}, a_2{}^{j_2}, \ldots, a_r{}^{j_r}}$ denotes the complete multipartite graph that has j_i partite sets of size a_i for $i = 1, 2, \ldots, r$ and $a_1 > a_2 > \ldots > a_r$.

Theorem 6 [5]. *Let $K_{a_1{}^{j_1}, a_2{}^{j_2}, \ldots, a_r{}^{j_r}}$ denote the complete multipartite graph that has j_i partite sets of size a_i for $i = 1, 2, \ldots, r$, and $a_1 > a_2 > \ldots > a_r$. Then*

$$D(K_{a_1{}^{j_1}, a_2{}^{j_2}, \ldots, a_r{}^{j_r}}) = \min\{p : \binom{p}{a_i} \geqslant j_i \text{ for all i}\}.$$

Theorem 7. *If G and H are two simple connected, relatively prime graphs such that $G \times H$ has j_i R-equivalence classes of sie a_i for $i = 1, \ldots, r$, and $a_1 > a_2 > \ldots > a_r$ then*

$$D(G \Box H) \leqslant D(G \times H) \leqslant \min\{p : \binom{p}{a_i} \geqslant j_i \text{ for all i}\}.$$

Proof. Since $\mathrm{Aut}(G \Box H) \subseteq \mathrm{Aut}(G \times H)$, so $D(G \Box H) \subseteq D(G \times H)$. To prove the second inequality, it is sufficient to consider each R-equivalence classes of $G \times H$ as a partite set. Thus graph $G \times H$ can be considered as multipartite graph that has j_i partite sets of size a_i such that every two partite sets of this multipartite graph is complete bipartite or there exists no edge between the two partite sets. So the automorphism group of this multipartite graph is subset of the automorphism group of complete multipartite graph with the same partite sets. Therefore $D(G \times H) \leqslant D(K_{a_1^{j_1}, \ldots, a_r^{j_r}})$, and the result follows from Theorem 6.

By using the concept of the Cartesian skeleton we can obtain an upper bound for Kronecker product of R-thin graphs. For this purpose we need the following preliminaries from [7]: The Boolean square of a graph G is the graph G^s with $V(G^s) = V(G)$ and $E(G^s) = \{xy \mid N_G(x) \cap N_G(y) \neq \emptyset\}$. An edge xy of the Boolean square G^s is dispensable if it is a loop, or if there exists some $z \in V(G)$ for which both of the following statements hold:

(1) $N_G(x) \cap N_G(y) \subset N_G(x) \cap N_G(z)$ or $N_G(x) \subset N_G(z) \subset N_G(y)$.
(2) $N_G(y) \cap N_G(x) \subset N_G(y) \cap N_G(z)$ or $N_G(y) \subset N_G(z) \subset N_G(x)$.

The Cartesian skeleton $S(G)$ of a graph G is the spanning subgraph of the Boolean square G^s obtained by removing all dispensable edges from G^s.

Proposition 2 [7]. *If H and K are R-thin graphs without isolated vertices, then $S(H \times K) = S(H) \square S(K)$.*

Proposition 3 [7]. *Any isomorphism $\varphi : G \to H$, as a map $V(G) \to V(H)$, is also an isomorphism $\varphi : S(G) \to S(H)$.*

Now we are ready to give an upper bound for Kronecker product of R-thin graphs.

Theorem 8. *If G and H are R-thin graphs without isolated vertices, then $D(G \times H) \leqslant D(S(G) \square S(H))$.*

Proof. By Proposition 3 we have $\mathrm{Aut}(G \times H) \subseteq \mathrm{Aut}(S(G \times H))$, and so $D(G \times H) \subseteq D(S(G \times H))$. The result follows immediately from Proposition 2. $\square$

References

1. Albertson, M.O., Collins, K.L.: Symmetry breaking in graphs. Electron. J. Combin. **3**, R18 (1996)
2. Alikhani, S., Soltani, S.: Distinguishing number and distinguishing index of certain graphs. Filomat (to appear). http://arxiv.org/abs/1603.04005
3. Alikhani, S., Soltani, S.: Distinguishing number and distinguishing index of lexicographic product of two graphs. https://arxiv.org/abs/1606.08184
4. Alikhani, S., Soltani, S.: Distinguishing number and distinguishing index of neighbourhood corona of two graphs. https://arxiv.org/abs/1606.03751
5. Collins, K.L., Trenk, A.N.: The distinguishing chromatic number. Electron. J. Combin. **13**, R16 (2006)
6. Čulik, K.: Zur theorie der graphen. Časopis Pešt. Mat. **83**, 135–155 (1958)
7. Hammack, R., Imrich, W., Klavžar, S.: Handbook of Product Graphs, 2nd edn. Taylor & Francis group, New York (2011)
8. Imrich, W.: Automorphismen und das kartesische Produkt von Graphen, O sterreich. Akad. Wiss. Math.-Natur. Kl. S.-B. II, vol. 177, pp. 203–214 (1969)
9. Imrich, W., Jerebic, J., Klavžar, S.: The distinguishing number of cartesian products of complete graphs. Eur. J. Combin. **29**(4), 922–929 (2008)
10. Imrich, W., Klavžar, S.: Distinguishing cartesian powers of graphs. J. Graph Theory. **53**(3), 250–260 (2006)

11. Jha, P.K.: Klavžar, S., Zmazek, B.: Isomorphic components of Kronecker product of bipartite graphs. Discuss. Math. Graph Theory. **17**(2), 301–309 (1997)
12. McKenzie, R.: Cardinal multiplication of structures with a reflexive relation. Fund Math. **70**, 59–101 (1971)
13. Miller, D.J.: The automorphism group of a product of graphs. Proc. Am. Math. Soc. **25**, 24–28 (1970)
14. Weichsel, P.M.: The Kronecker product of graphs. Proc. Am. Math. Soc. **13**, 47–52 (1962)

Grammar Systems Based on Equal Matrix Rules and Alphabetic Flat Splicing

G. Samdanielthompson, N. Gnanamalar David, and K.G. Subramanian(✉)

Department of Mathematics, Madras Christian College,
Tambaram, Chennai 600059, India
samdanielthompson@gmail.com, ngdmcc@gmail.com, kgsmani1948@gmail.com

Abstract. Studies on the concept of splicing on words, established important theoretical results on computational universality. A specific kind of splicing, called flat splicing on strings and in particular, alphabetic flat splicing, were recently considered and studied for their properties. On the other hand, in the study of language-oriented modelling of distributed complex systems, grammar systems were proposed. Here we introduce a grammar system, called alphabetic flat splicing equal matrix grammar system ($AFSEMGS$), as a new model of language generation, based on the operation of alphabetic flat splicing on words and equal matrix grammar (EMG) type of rules. The components of a $AFSEMGS$ generate in parallel words using the EMG rules while two different components of the $AFSEMGS$ "communicate" using the alphabetic flat splicing operation on the words. We derive some comparison results that bring out the generative power of $AFSEMGS$ and as an application construct a $AFSEMGS$ to generate certain "chain code pictures".

Keywords: Flat splicing · Formal Languages · Grammar systems · Equal matrix grammar

1 Introduction

When Adleman solved a 7-node instance of the directed Hamilton path problem [2] using DNA sequence and simple bio-operations, it signalled a significant development in the field of DNA computing [11,19,20]. Formal language theory based modelling of the DNA recombination process under the action of restriction enzymes and a ligase was proposed by Head in his seminal work [12–14], by introducing a new operation on words, called splicing. This operation was utilized in developing theoretical models of computation in the framework of formal language theory [9,17,21], which is considered to be the backbone of theoretical computer science. A special kind of splicing, called flat splicing on words, was recently introduced in [3] inspired by the splicing operation on circular words. The idea of flat splicing on two words $u = u_1\alpha\beta u_2$ and $v = \gamma w\delta$ for some words $u_1, u_2, w, \alpha, \beta, \gamma, \delta$, is to "insert" v with a specified "prefix" γ and a specified "suffix" δ, into u between α and β. When $\alpha, \beta, \gamma, \delta$ are letters of an alphabet,

S. Arumugam et al. (Eds.): ICTCSDM 2016, LNCS 10398, pp. 347–356, 2017.
DOI: 10.1007/978-3-319-64419-6_45

the flat splicing is referred to as alphabetic. Several properties associated with flat splicing in the context of grammars and languages, have been studied in [3].

On the other hand, in the formal language based study of modelling of distributed complex systems, grammar systems [5] were proposed. The general idea of a grammar system is to have components, each with a certain type of grammar rules, and there are different types of communication between components, allowing the generation of different classes of languages. In [8], splicing grammar system is considered, with context-free or regular rules in the components and communication between components achieved by the use of splicing on words. Recently, using the flat splicing operation on words for the communication between components and context-free or regular rules in the components, a variant called flat splicing grammar system has been introduced [4] and studied.

Different classes of grammars specified by control mechanisms have been introduced in the theory of formal grammars and languages with a view to regulate rewriting and thereby increase the language generative capability of grammars [6,7]. One such control feature is known as matrix grammars [1]. A special kind of matrix grammar, known as equal matrix grammar, was introduced in [22] and an equivalent class called right-linear simple matrix grammar was introduced in [15]. Here we consider equal matrix grammar in the components of a flat splicing grammar system and alphabetic flat splicing for communication between components. We thus introduce alphabetic flat splicing equal matrix grammar system ($AFSEMGS$) as a new model of language generation. The language class of $AFSEMGS$ is compared with certain other language classes and as an application we construct an $AFSEMGS$ for describing "chain code pictures" [18], which play an important role in different problems related to images (see, for example, [10,16,23]).

2 Preliminaries

We refer to [21], for concepts and results related to formal grammars and languages. In this section, we recall some basic notions and results.

A word w is a finite sequence of symbols of a finite set Σ, referred to as an alphabet in formal language theory. We denote by Σ^*, the set of all words over Σ, including the empty word λ and $V^+ = V^* - \{\lambda\}$. The length $|w|$ of a word w is the number of symbols in w counting repetitions of symbols in w. Clearly, $|\lambda| = 0$.

We now recall the notion of flat splicing on words [3]. The idea is that a word with a specified "prefix" as well as a "suffix" is inserted into another word in a pre-specified position. In formal terms, a flat splicing rule r is of the form $(\alpha|\gamma - \delta|\beta)$, where $\alpha, \beta, \gamma, \delta$ are words over an alphabet Σ. For two words $u = x\alpha\beta y; v = \gamma z\delta$, an application of the flat splicing rule $r = (\alpha|\gamma - \delta|\beta)$ to the pair (u, v) yields the word $w = x\alpha\gamma z\delta\beta y$ and we write $(u, v) \vdash_r w$. A flat splicing rule $r = (\alpha|\gamma - \delta|\beta)$, where $\alpha, \beta, \gamma, \delta$ are letters in Σ or the empty word, is called alphabetic.

A flat splicing system (FSS) [3] is a triple $S = (\Sigma, I, R)$, where Σ is an alphabet; I, called initial set, is a set of words over Σ, and R is a finite set of flat splicing rules [3]. The FSS S is respectively called finite, regular or context-free according as I is a finite set, regular set or a context-free language. The language L generated by S is the smallest language containing I and such that for any two words $u, v \in L$, the word w is also in L, if $(u, v) \vdash_r w$. When all the flat splicing rules are alphabetic, the FSS is called an alphabetic flat splicing system ($AFSS$). The families of languages generated by FSS and $AFSS$ are respectively denoted by $L(FSS, X)$ and $L(AFSS, X)$ for $X = FIN$, REG or CF according as the initial set is finite, regular or context-free.

We illustrate an alphabetic flat splicing system and its work with an example.

Example 1. Consider the alphabetic flat splicing system

$$\Gamma = (\{a, c, x, y\}, \{a, c, xay\}, \{r_1, r_2\})$$

where

$$r_1 = (x|c - \lambda|a), r_2 = (x|a - \lambda|c)$$

Initially, the rule r_1 is applicable to the pair of words (xay, c). Application of the rule r_1 inserts c (the only axiom which begins with c) between x and a in the first word xay yielding $xcay$. Applying the rule r_2 to the pair $(xcay, a)$ yields $xacay$. Thus, proceeding in this way, the words generated will be of the form $xa(ca)^n y$, $n \geq 0$ or of the form $x(ca)^n y$, $n \geq 1$. The language generated by Γ is

$$L(\Gamma) = \{a, c\} \cup \{xa(ca)^n y | n \geq 0\} \cup \{x(ca)^n y | n \geq 1\}.$$

Note that all the axiom words are in the language $L(\Gamma)$.

In formal language theory, in order to increase the generative power of context-free grammars, additional mechanisms have been introduced under the category of regulated rewriting. One such mechanism is the matrix grammar initially introduced in [1] in which a sequence $[r_1, r_2, \cdots, r_n]$ of context-free rules is specified and is referred to as a matrix rule. The application of such a matrix rule to a word is done by applying the rules $r_1, r_2, \cdots, r_n$ one after another in the same order to constitute a single derivation step. A restricted class of matrix grammars, known as simple matrix grammars of degree $n \geq 1$ ($n - SMG$) was introduced in [15] and have been investigated by many researchers. It is known [15] that there is a hierarchy of classes of languages generated by simple matrix grammars. More precisely, the family of languages generated by simple matrix grammars of degree $n \geq 1$ is denoted by $n - SML$. The $n - EMG$ [22](or equivalently $n-$right-linear simple matrix grammars [15]) constitute a subclass of $n - SMG$. We now recall the definition of equal matrix grammar (EMG) [22].

Definition 1 [22]. *An equal matrix grammar of degree* $n \geq 1$ ($n - EMG$) *is a construct of the form* $G = (N_1, \cdots, N_n, T, S, M)$ *where* N_i, $1 \leq i \leq n$ *and* T *pairwise disjoint alphabets; the elements of* $N = N_1 \cup N_2 \cup \cdots \cup N_n$ *are nonterminals and those of* T *are terminal symbols;* $S \notin N \cup T$ *is the initial symbol or start symbol.* M *consists of the following types of matrix rules:*

(i) a set of initial matrix rules of the form $[S \rightarrow w], w \in T^*$ *or* $[S \rightarrow A_1A_2 \cdots A_n]$ *where* $A_i \in N_i, 1 \leq i \leq n$;;
(ii) a set nonterminal equal matrix rules of the form $[A_1 \rightarrow w_1B_1, \cdots, A_k \rightarrow w_nB_n]$ *where* $w_i, 1 \leq i \leq n$ *are elements of* T^* *and* $A_i, B_i \in N_i, 1 \leq i \leq n$;
(iii) a set of terminal matrix rules of the form $[A_1 \rightarrow w_1, \cdots, A_n \rightarrow w_n]$ *where* $w_i, 1 \leq i \leq n$ *are elements of* T^*.

A derivation starts with a matrix rule of type (i) of the form $[S \rightarrow A_1A_2 \cdots A_n]$ *and can be continued by matrix rules of type (ii). The derivation can be terminated by a matrix rule of type (iii). Note that an application of a matrix rule* m *to a word* ζ *means that all the rules of* m *are applied one by one in the same sequence in which they are given in* m *to constitute a single step of derivation yielding a word* η *from* ζ.

The language $L(G)$ *generated by* G *consists of all words* w *such that* w *is derived from* S *in a finite number of steps. The family of languages generated by* $n-EMGs$ *is denoted by* $n-EML$. *It is known that* $n-EML \subset (n+1)-EML$.

Note that $1-EML$ coincides with the class REG of regular sets of the Chomsky hierarchy. Also the family $n-EML$ is known to coincide with the family of right-linear simple matrix languages considered in [15] but it is also known [15] that there are context-free languages which cannot be generated by any equal matrix grammar. We denote the class of context-free languages by CFL.

Example 2. Consider the $2-EMG$ $G_1 = (N_1, N_2, T, S, M)$ where $N_i = \{A_i\}$, $i \in \{1, 2\}$, $T = \{a, b, c\}$ and

$$M = \{m_1 : [S \rightarrow A_1A_2], m_2 : [A_1 \rightarrow aA_1, A_2 \rightarrow aA_2],$$

$$m_3 : [A_1 \rightarrow bA_1, A_2 \rightarrow bA_2], m_4 : [A_1 \rightarrow cA_1, A_2 \rightarrow cA_2],$$

$$m_5 : [A_1 \rightarrow a, A_2 \rightarrow a], m_6 : [A_1 \rightarrow b, A_2 \rightarrow b], m_7 : [A_1 \rightarrow c, A_2 \rightarrow c]\}.$$

Then $L(G_1) = \{ww | w \in T^+\}$. A sample derivation generating the word $bcabca$ is as follows:

$$S \Rightarrow A_1A_2 \Rightarrow bA_1bA_2 \Rightarrow bcA_1bcA_2 \Rightarrow bcabca$$

where $\alpha \Rightarrow \beta$ means that the word β is derived from the word α. The sequence of matrix rules used is m_1, m_3, m_4, m_5.

3 Alphabetic Flat Splicing Equal Matrix Grammar Systems

Flat splicing grammar systems with context-free or regular rules in the components have been considered in [4]. In fact essentially, alphabetic flat splicing rules are considered in [4], although this is not explicitly mentioned. We refer to these as alphabetic flat splicing context-free or regular grammar systems. When the number of components is n, $n \geq 1$, we denote the corresponding families of languages respectively by $L_n(AFSCFGS)$ and $L_n(AFSRGS)$. We now introduce a variant called alphabetic flat splicing equal matrix grammar system which has equal matrix kind of rules in the components. Rewriting is done in parallel in the components but two different components "communicate" by flat splicing rules. We now formally define this grammar system.

Definition 2. *An alphabetic flat splicing $k-$equal matrix grammar system (AFSk-EMGS) of degree n is a construct*

$$G = (N_1, \cdots, N_n, T, (S_1, M_1), \cdots, (S_n, M_n), F)$$

where N_i, $1 \leq i \leq n$ and T pairwise disjoint alphabets; the elements of N_i, $1 \leq i \leq n$, are nonterminals and those of T are terminal symbols; the $k-$equal matrix grammars $G_i = (N_i, T, S_i, M_i)$ are called the component grammars of G; $S_i \notin N \cup T, 1 \leq i \leq n$ is the initial symbol or start symbol in the ith component, where $N = N_1 \cup \cdots \cup N_n$; For $1 \leq i \leq n$, M_i consists of the $k-$equal matrix types of initial, nonterminal and/or terminal rules with the matrix rules in the component involving nonterminals from G_i; F is a finite set of alphabetic flat splicing rules.

A configuration in G is an $n-$tuple of words over $N \cup T$, with the initial configuration given by $(S_1, \cdots, S_n)$. For two configurations $u = (u_1, \cdots, u_n)$ and $v = (v_1, \cdots, v_n)$, we define $u \Rightarrow_G v$ (or simply, $u \Rightarrow v$) if and only if one of the following conditions holds:

(i) for each i, $(1 \leq i \leq n)$, $u_i \Rightarrow_{G_i} v_i$;
(ii) there exist some $j, k\, (1 \leq j \leq n; 1 \leq k \leq n)$, and $(\alpha|\gamma - \delta|\beta) \in F$ such that $u_j = x_j\alpha\beta y_j$; $u_k = \gamma z\delta$ and $v_j = x_j\alpha\gamma z\delta\beta y_j$; $v_i = u_i$ for all $i \neq j$.

$\Rightarrow^*$ is the reflexive transitive closure of $\Rightarrow$. There is no priority in the application of the rewriting equal matrix rules and alphabetic flat splicing rules. The language generated by the ith component is given by

$$L_i(G) = \{v_i | (S_1, \cdots, S_n) \Rightarrow^* (v_1, \cdots, v_n), v_j \in T^*, 1 \leq j \leq n\}.$$

Without loss of generality, we take the language generated by the first component as the language of G. We denote by L_n(AFSk-EMGS), the family of languages generated by alphabetic flat splicing equal matrix grammar systems with at most n components and $k-$equal matrix type of rules in the components. Also an alphabetic flat splicing equal matrix grammar system of degree n is an AFSk-EMGS of degree n, for some $k \geq 1$.

We give an example of $AFS2 - EMGS$ of degree 2.

Example 3. Consider the alphabetic flat splicing 2−equal matrix grammar system *AFS2-EMGS* of degree 2, given by

$$G_2 = (\{A, B\}, \{C, D\}, \{a, b, c\}, (S_1, M_1), (S_2, M_2)\}, F)$$

where $M_1 = \{[S_1 \rightarrow AB], [A \rightarrow aA, B \rightarrow bB], [A \rightarrow a, B \rightarrow b]\}$,
$M_2 = \{[S_2 \rightarrow CD], [C \rightarrow cC, D \rightarrow D], [C \rightarrow c, D \rightarrow \lambda]\}$ and $F = \{(a|\lambda - c|b)\}$

Starting from S_1 the first component generates AB using the initial rule. Then by the application of the rule $[A \to aA, B \to bB]$, $(n-1)$ times $(n \geq 1)$, the word $a^{n-1}Ab^{n-1}B$ can be generated. On applying the rule $[A \to a, B \to b]$, the generation terminates yielding the word $a^n b^n$. Similarly, at the same time, in the second component the word c^n is generated. The alphabetic flat splicing rule $(a|\lambda - c|b)$ becomes applicable, generating the word $a^n c^n b^n$. The language generated is $L(G_2) = \{a^n b^n | n \geq 1\} \cup \{a^n c^n b^n | n \geq 1\}$. Note that the alphabetic flat splicing rule is applicable only when in both the components, derivations terminate together and so no other sequence of rule applications is successful.

Theorem 1. *(i)* $REG = L_1(AFS1\text{-}EMGS) \subset L_1(AFS2\text{-}EMGS)$
(ii) $L_2(AFS2\text{-}EMGS) \setminus L_2(AFSRGS) \neq \emptyset$
(iii) $L_2(AFS3\text{-}EMG) \setminus L_2(AFSCFGS) \neq \emptyset$

Proof. Since it is known that [15] the language family of $1 - EMGs$ is exactly the family REG, the equality $REG = L_1(AFS1\text{-}EMGS)$ in statement (i) of the theorem holds. Also, the inclusion $L_1(AFS1\text{-}EMGS) \subseteq L_1(AFS2\text{-}EMGS)$ in statement (i) follows as it is clear that a $1 - EMG$ can be modified into a $2 - EMG$ without altering the language generated. In order to prove the proper inclusion in $L_1(AFS1\text{-}EMGS) \subset L_1(AFS2\text{-}EMGS)$, consider the one component $AFS2 - EMGS$ $G_3 = (N_1, T, (S_1, M_1), F)$ where $N_1 = \{A, B\}$, $T = \{a, b\}$, $M_1 = \{[S_1 \to AB], [A \to aA, B \to bB], [A \to a, B \to b]\}$, $F = \emptyset$. The language generated by G_3 is the non-regular language $L(G_3) = \{a^n b^n | n \geq 1\}$. This proves the proper inclusion as $REG = L_1(AFS1\text{-}EMGS)$. Note that there is only one component in G_3 and there are no alphabetic flat splicing rules and the derivations start from S_1 and yield the words in $L(G_3)$ with the application of equal matrix rules in the first component, suitable number of times.

In order to prove statement (ii), consider the language $L(G_2)$ in Example 3 which is generated by G_2, a $AFS2 - EMGS$ of degree 2. This language cannot belong to $L_2(AFSRGS)$. In fact in a $AFSRGS$, the components can have only right-linear rules so that in a component only words of the form a^n or b^n or c^n can be generated and so two components are not enough to generate the words of the form $a^n c^n b^n$, $n \geq 1$, as the alphabetic flat splicing rules can only insert words generated in one component into words generated in another component.

In order to prove statement (iii), consider the language generated by the $AFS3\text{-}EMGS$ $G_4 = (N_1, N_2, T, (S_1, M_1), (S_2, M_2), F)$ where $N_1 = \{A, D, E\}$, $N_2 = \{B, C, F\}$, $T = \{a, b, c, d, e\}$ $M_1 = \{[S_1 \to ADE], [A \to aA, D \to dD, E \to eE], [A \to a, D \to d, E \to e]\}$, $M_2 = \{[S_2 \to BCF], [B \to bB, C \to cC, F \to F], [B \to b, C \to c, F \to \lambda]\}$, $F = \{(a|b - c|d)\}$. Then the language generated is $L(G_4) = \{a^n d^n e^n | n \geq 1\} \cup \{a^n b^n c^n d^n e^n | n \geq 1\}$. In fact in the first component the $3 - EMG$ rules generate a word $a^n d^n e^n$ while at the same time, in the second component the $3 - EMG$ rules generate a word $b^n c^n$ for the same n. At this stage the flat splicing rule $(a|b - c|d)$ is applicable and its application yields the word $a^n b^n c^n d^n e^n$. This language $L(G_4)$ can not be in $L_2(AFSCFGS)$ as in a component the context-free rules can generate at the most only words of the form $a^n b^n$ and so two components are not enough to generate words of the form

$a^nb^nc^nd^ne^n$ as the alphabetic flat splicing rules can insert a word generated in a component into another word generated in another component.

Theorem 2. $L_2(AFSn - EMGS) \setminus n - EML \neq \emptyset$

Proof. Consider the $AFSn - EMGS$ $G_5 = (N_1, N_2, T, (S_1, M_1), (S_2, M_2), F)$ where $N_1 = \{A_i \mid 1 \leq i \leq n\}$, $N_2 = \{A_{n+1}, F_i \mid 1 \leq i \leq n-1\}$, $T = \{a_i \mid 1 \leq i \leq n+1\} \cup \{b\}$,

$$M_1 = \{[S_1 \to A_1A_2 \cdots A_n], [A_1 \to a_1A_1, \cdots, A_n \to a_nA_n],$$
$$[A_1 \to a_1, \cdots, A_n \to a_nb], \}$$

$$M_2 = \{[S_2 \to A_{n+1}F_1 \cdots F_{n-1}], [A_{n+1} \to a_{n+1}A_{n+1}, F_1 \to F_1, \cdots, F_{n-1} \to F_{n-1}],$$
$$[A_{n+1} \to a_{n+1}, F_1 \to \lambda, \cdots, F_{n-1} \to \lambda]\}, F = (a_n|a_{n+1} - \lambda|b).$$

The language generated by G_5 is

$$L(G_5) = \{a_1^ka_2^k \cdots a_n^kb \mid k \geq 1\} \cup \{a_1^ka_2^k \cdots a_n^ka_{n+1}^kb \mid k \geq 1\}$$

since it can be shown that $L(G_5)$ cannot be generated by any $n-EMG$ by closely following the argument in [15] in showing that the language $\{a_1^ka_2^k \cdots a_n^ka_{n+1}^k \mid k \geq 1\}$ cannot be generated by any $n - EMG$.

Theorem 3. $L_3(AFSn - EMGS) \setminus n - SML \neq \emptyset$

Proof. Consider the $AFSn - EMGS$

$$G_6 = (N_1, N_2, N_3, T, (S_1, M_1), (S_2, M_2), (S_3, M_3), F)$$

where $N_1 = \{A_i \mid 1 \leq i \leq n\}$, $N_2 = \{C_i \mid 1 \leq i \leq n\}$, $N_3 = \{D_i \mid 1 \leq i \leq n\}$, $T = \{a_i, c_i, b, d \mid 1 \leq i \leq n\}$,

$$M_1 = \{[S_1 \to A_1A_2 \cdots A_n], [A_1 \to a_1A_1, \cdots A_n \to a_nA_n],$$
$$[A_1 \to a_1, \cdots A_n \to a_nd]\}$$
$$M_2 = \{[S_2 \to C_nC_{n-1}C_{n-2} \cdots C_1], [C_n \to c_nC_n, \cdots C_1 \to c_1C_1],$$
$$[C_n \to c_n, \cdots C_1 \to c_1]\}$$
$$M_3 = \{[S_3 \to D_1 \cdots D_n], [D_1 \to bD_1, D_2 \to D_2, \cdots, D_n \to D_n],$$
$$[D_1 \to b, D_2 \to \lambda, \cdots, D_n \to \lambda]$$

$F = \{(a_n|c_n - \lambda|d), (a_n|b - \lambda|c_n)\}$. The language generated by G_6 is

$$L(G_6) = \{a_1^k \cdots a_n^kd \mid k \geq 1\} \cup \{a_1^k \cdots a_n^kc_n^k \cdots c_1^kd|k \geq 1\}$$
$$\cup\{a_1^k \cdots a_n^kb^kc_n^k \cdots c_1^k \mid k \geq 1\}$$

since it can be shown that $L(G_6)$ cannot be generated by any $n - SMG$ by closely following the argument in [15] in showing that the language

$$\{a_1^ka_2^k \cdots a_n^kb^kc_n^kc_{n-1}^k \cdots c_1^k \mid k \geq 1\}$$

cannot be generated by any $n - SMG$.

4 Application

Pictures in the two-dimensional plane given by chain codes have been of interest and investigation [16,18], since these chain-code pictures are described by the well-developed Chomsky and other kinds of string grammars [21] and have applications in different problems [10,23]. A chain code picture [18] p is made of unit horizontal and vertical lines in the two-dimensional plane and can be encoded by words over the alphabet $\{l, r, u, d\}$ with the symbols l, r, u, d respectively interpreted as instructions to draw a horizontal or vertical unit line to the left, right, up or down directions from the present position in the chain code picture. A chain code picture language is a set of chain code pictures. Here we provide an example illustrating the application of flat splicing equal matrix grammar system in generating chain code picture languages.

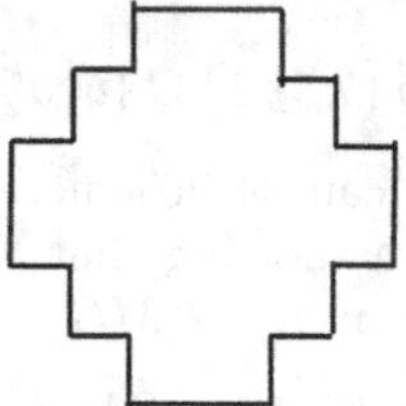

Fig. 1. A diamond shaped chain-code picture with four equal sized stairs

Consider the $AFS2 - EMGS$ of degree 2, $G_7 = (N_1, N_2, T, (S_1, M_1), (S_2, M_2), F)$ where $N_1 = \{A, B\}$, $N_2 = \{C, D\}$, $T = \{r, l, u, d\}$,

$$M_1 = \{[S_1 \to AB], [A \to ruA, B \to ulB],$$

$$[A \to rrll, B \to uu]\}$$

$$M_2 = \{[S_2 \to CD], [C \to drC, D \to ldD],$$

$$[C \to dd, D \to \lambda]\}$$

$F = \{(r|d - d|l)\}$. It can be seen that the language generated is

$$L(G_7) = \{(ru)^n rrll(ul)^n uu | n \geq 1\} \cup \{(ru)^n rr(dr)^n dd(ld)^n ll(ul)^n uu | n \geq 1\}.$$

The words of this language correspond to "diamond shaped chain-code pictures with four equal sized stairs", one member of which is shown in Fig. 1 which corresponds to $n = 2$.

Acknowledgements. The first author acknowledges with gratitude the award (No.: F1-17.1/2016-17/MANF-2015-17-TAM-51257/(SA-III/Website)) of Maulana Azad National Fellowship for minority students by UGC, India. The third author is grateful to UGC, India, for the award of Emeritus Fellowship (No. F.6-6/2016-17/EMERITUS-2015-17-GEN-5933/(SA-II)).

References

1. Abraham, S.: Some questions of phrase structure grammars I. Comput. Linguist. **4**, 61–70 (1965)
2. Adleman, L.M.: Molecular computation of solutions to combinatorial problems. Sci. New Series **266**(5187), 1021–1024 (1994)
3. Berstel, J., Boasson, L., Fagnot, I.: Splicing systems and the Chomsky hierarchy. Theoret. Comput. Sci. **436**, 2–22 (2012)
4. Ceterchi, R., Subramanian, K.G.: Grammar systems with context-free rewriting and flat splicing. In: Gheorghe, M., et al. (ed.) Multidisciplinary Creativity, pp. 221–227. Spandugino, Bucuresti (2015)
5. Csuhaj-Varju, E., Dassow, J., Kelemen, J., Păun, G.: Grammar systems: a grammatical approach to distribution and cooperation. Gordon and Breach Science Publishers (1994)
6. Dassow, J.: Grammars with regulated rewriting. In: Martin-Vide, C., et al. (eds.) Formal Languages and Applications. Studies in Fuzziness and Soft Computing, vol. 148, pp. 249–273. Springer, Heidelberg (2004). doi:10.1007/978-3-540-39886-8_13
7. Dassow, J., Păun, G.: Regulated Rewriting in Formal Language Theory. Springer, Berlin (1989)
8. Dassow, J., Mitrana, V.: Splicing grammar systems. Comput. Artif. Intell. **15**, 109–122 (1996)
9. Esik, Z., Martin-Vide, C., Mitrana, V.: Recent Advances in Formal Languages and Applications. Studies in Computatioinal Intelligence. Springer, Heidelberg (2006)
10. Fating, K., Ghotkar, A.: Performance analysis of chaincode descriptor for handshape classification. Int. J. Comp. Graphics Animation. **4**(2), 9–19 (2014)
11. Freund, R., Kari, L., Păun, G.: DNA computing based on splicing: the existence of universal computers. Theory Comput. Syst. **32**(1), 69–112 (1999)
12. Head, T.: Formal language theory and DNA: an analysis of the generative capacity of specific recombinant behaviours. Bull. Math. Biol. **49**, 735–759 (1987)
13. Head, T.: Circular suggestions for DNA Computing. In: Carbone, A., et al. (ed.) Pattern Formation in Biology, Vision and Dynamics, pp. 325–335. World Scientific (2000)
14. Head, T., Păun, G., Pixton, D.: Language theory and molecular genetics: generative mechanisms suggested by DNA recombination. In: Rozenberg, G., Salomaa, A. (eds.) Handbook of Formal Languages, pp. 295–358. Springer, Heidelberg (1997). doi:10.1007/978-3-662-07675-0_7
15. Ibarra, O.H.: Simple matrix languages. Inform. Control. **17**, 359–394 (1970)
16. Kim, C., Sudborough, I.H.: The membership and the equivalence problem for picture languages. Theor. Comput. Sci. **52**, 177–191 (1987)
17. Martin-Vide, C., Mitrana, V., Păun, G.: Formal Languages and Applications. Springer, Heidelberg (2004)
18. Maurer, H.A., Rozenberg, G., Welzl, E.: Using string languages to describe picture languages. Inf. Contro. **54**, 155–185 (1982)
19. Păun, G.: DNA computing based on splicing: universality results. Theor. Comput. Sci. **231**(2), 275–296 (2000)
20. Păun, G., Rozenberg, G., Salomaa, A.: DNA Computing - New Computing Paradigms. Texts in Theoretical Computer Science. An EATCS Series. Springer, Heidelberg (1998)

21. Rozenberg, G., Salomaa, A.: Handbook of Formal Languages. Springer, Heidelberg (1997)
22. Siromoney, R.: On equal matrix languages. Inf. Control **14**, 135–151 (1969)
23. Vaddi, R.S., Boggavarapu, L.N.P., Vankayalapati, H.D., Anne, K.R.: Contour detection using Freeman chaincode and approximation methods for the real time object detection. Asian J. Comp. Sci. Inform. Tech. **1**(1), 15–17 (2011)

3-Simple 2-Fold 5-Cycle Systems

R. Sangeetha[1(✉)] and A. Muthusamy[2]

[1] Department of Mathematics, A.V.V.M Sri Pushpam College, Poondi, Thanjavur, Tamil Nadu, India
jaisangmaths@yahoo.com

[2] Department of Mathematics, Periyar University, Salem, Tamil Nadu, India
ambdu@yahoo.com

Abstract. A decomposition of λK_n into cycles of length k is called a *λ-fold k-cycle system* of λK_n. A λ-fold k-cycle system of λK_n is *t-simple*, $t < k$, if any two cycles in the decomposition have at most t vertices in common. We denote a t-simple λ-fold k-cycle system of λK_n by (n, k, λ, t)-cycle system. In this paper, it is shown that an $(n, 5, 2, 3)$-cycle system exists, for $n = 5r,\ 5r + 1$ when (i) $r \equiv 2$ or $6 \pmod{12}$ or (ii) $r \equiv 4$ or $12 \pmod{24}$.

Keywords: k-cycle system · t-simple

1 Introduction and Preliminaries

We denote the complete p-partite graph with m vertices in each partite set by $K(m, p)$. Define $|i-j|_n = \min\{|i-j|, n-|i-j|\}$. If $D = \{\{1, 2, \ldots, \lfloor \frac{mp}{2} \rfloor\}\} \backslash \{ip : 1 \leq i \leq \lfloor \frac{m}{2} \rfloor\}\}$, then $K(m, p) \cong \langle D \rangle_{mp}$, where $\langle D \rangle_{mp}$ is a graph with vertex set Z_{mp} and edge set $\{\{i, j\} : |i - j|_{mp} \in D$, for $i, j \in Z_{mp}\}$. The *wreath product* of two graphs G and H is a graph $G \otimes H$ with vertex set $V(G) \times V(H)$, and (u_1, v_1) is adjacent to (u_2, v_2) whenever (i) $\{u_1, u_2\} \in E(G)$, or (ii) $u_1 = u_2$ and $\{v_1, v_2\} \in E(H)$. A graph G with edge-multiplicity λ is called λ-fold graph and is denoted as $G(\lambda)$. The notation rG denotes r copies of the graph G. A cycle of length k is denoted by C_k. A *decomposition* of a graph G is a family $H_1, \ldots, H_k$ of subgraphs of G such that each edge of G is contained in exactly one member of $H_1, \ldots, H_k$. We denote it by $G = H_1 \oplus \cdots \oplus H_k$ and we say that $H_1, \ldots, H_k$ *decompose* G. For $1 \leq i \leq k$, if $H_i \cong H$, we say that G has a *H-decomposition.* A *balanced incomplete block design* of block-size k and index λ (in short, (v, k, λ)-BIBD) is a pair $(X, \mathfrak{B})$, where X is a set of v points, $\mathfrak{B}$ is a collection of k-subsets (called blocks) of X with the property that any pair of points of X is contained in exactly λ blocks. In graph theoretical terminology, existence of (v, k, λ)-BIBD is equivalent to the existence of K_k-decomposition of $K_v(\lambda)$, $k < v$ and is denoted as (v, k, λ) K_k-design [15]. If G is a graph on k vertices, then the existence of G-decomposition in $K_v(\lambda)$ is denoted as (v, k, λ) G-design. If $G \cong C_k$, then it is called a (v, k, λ) C_k-design. A (v, k, λ)-BIBD (respectively,

S. Arumugam et al. (Eds.): ICTCSDM 2016, LNCS 10398, pp. 357–361, 2017.
DOI: 10.1007/978-3-319-64419-6_46

(v, k, λ) C_k-design) is said to be *simple* if any pair of blocks (respectively, C_k's) contains at most $k-1$ elements (respectively, vertices) in common. A (v, k, λ)-BIBD (respectively, (v, k, λ) C_k-design) is said to be *super-simple* if any pair of blocks (respectively, C_k's) contains at most two elements (respectively, vertices) in common. Now we generalize this idea as follows: A (v, k, λ) C_k-design is said to be t-simple if any pair of C_k's in the decomposition has at most t vertices in common. A t-simple (v, k, λ) C_k-design is also called (v, k, λ, t)-cycle system. A (v, k, λ, t)-cycle system is simple or super-simple according as $t = k-1$ or $t = 2$. Further, a t-simple λ-fold k-cycle system of $K(m, p)(\lambda)$ is denoted as $(m, p; k, \lambda, t)$-multipartite cycle system. Super-simple designs were introduced by Gronau and Mullin [8] in 1992. The existence of super-simple designs is an interesting extremal problem by itself and also have useful applications. For example, such designs are used in the construction of perfect hash families [14], coverings [2] and superimposed codes [10,11], etc. The existence of super-simple designs have been investigated by many authors [4–6,9]. But there are only few results on the existence of super-simple cycle systems [1,7]. In 2007, Chen and Wei [7] have proved the existence of an $(n, 4, \lambda, 2)$-cycle system for $7 \le n \le 41$ and all admissible λ with few exceptions. Recently Billington et al. [1] have proved the existence of an $(n, 4, 2, 2)$-cycle system. For the given n, k, λ, t, the existence of an (n, k, λ, t)-cycle system is not guaranteed, even though it satisfies the obvious edge divisibility condition, $k|\lambda(\frac{n(n-1)}{2})$. For example, an $(n, 4, 2, 2)$-cycle system does not exist, when $n = 5, 6$ or 9, see [1]. Also an $(n, 4, \lambda, 2)$-cycle system does not exist, when $(n, \lambda) = (9, 3), (13, 5)$, see [7]. In this paper, it is shown that an $(n, 5, 2, 3)$-cycle system exists for $n = 5r,\ 5r+1$ when (i) $r \equiv 2$ or $6 \pmod{12}$ or (ii) $r \equiv 4$ or $12 \pmod{24}$. For $n \equiv 1 \pmod 5$, let $V(K_n) = Z_n$ and $\sigma = (0, 1, 2, 3, \ldots, n-2, n-1)$ be the permutation on Z_n. For small values of n, we construct a set of cycles C, say starter, so that $\{C, \sigma(C), \sigma^2(C), \ldots, \sigma^{n-1}(C)\}$ gives the required $(n, 5, 2, 3)$-cycle system. We have constructed the starter C, with the help of computer using C-Program. If $n \equiv 0 \pmod 5$, then take $V(K_n) = Z_{n-1} \cup \infty$, and let $\tau = (0, 1, 2, \ldots, n-3, n-2)(\infty)$ be the permutation on $Z_{n-1} \cup \infty$. Then $\{C, \tau(C), \tau^2(C), \ldots, \tau^{n-2}(C)\}$ gives the required $(n, 5, 2, 3)$-cycle system. The following is the necessary and sufficient condition for the existence of a C_3-decomposition in K_n.

Lemma 1 [12]. *K_n has a C_3-decomposition if and only if $n \equiv 1$ or $3 (mod\ 6)$.*

2 Main Results

The following lemma shows the existence of an $(n, 5, 2, 3)$-cycle system, for some small values of n.

Lemma 2. *There exists an $(n, 5, 2, 3)$-cycle system when $n \in \{10, 11, 15, 16, 20, 21\}$.*

Proof

(a) If $n = 10$, then $V(K_{10}) = Z_9 \cup \infty$ and the starter set is $C = \{(0,1,3,7,\infty),(0,3,6,1,2)\}$.
(b) If $n = 11$, then $V(K_{11}) = Z_{11}$ and the starter set is $C = \{(0,6,1,10,3),(0,2,3,4,8)\}$.
(c) If $n = 15$, then $V(K_{15}) = Z_{14} \cup \infty$ and the starter set is $C = \{(0,4,9,1,\infty),(0,7,5,6,8),(0,1,12,2,5)\}$.
(d) If $n = 16$, then $V(K_{16}) = Z_{16}$ and the starter set is $C = \{(0,1,9,4,11),(0,2,5,1,3),(0,6,7,3,10)\}$.
(e) If $n = 20$, then $V(K_{20}) = Z_{19} \cup \infty$ and the starter set is $C = \{(0,6,13,14,\infty),(1,7,2,4,11),(0,4,1,3,14),(0,1,11,15,12)\}$.
(f) If $n = 21$, then $V(K_{21}) = Z_{21}$ and the starter set is $C = \{(0,4,1,14,3),(0,1,6,14,7),(0,2,11,17,12),(0,2,8,7,17)\}$.

Thus $\{C,\sigma(C),\sigma^2(C),\ldots,\sigma^{n-1}(C)\}$ (respectively, $\{C,\tau(C),\tau^2(C),\ldots,\tau^{n-2}(C)\}$) gives a required $(n,5,2,3)$-cycle system when $n \equiv 1 \pmod 5$ (respectively, when $n \equiv 0 \pmod 5$).

Decomposition of Complete tripartite graphs into 5-cycles has been discussed in [3,13]. The following lemma shows the existence of 3-simple 2-fold 5-cycle decomposition of $K(10,3)$ and $K(20,3)$.

Lemma 3. *There exists an $(m,p;5,2,3)$-multipartite cycle system, when $(m,p) = (10,3),(20,3)$.*

Proof

(a) If $(m,p) = (10,3)$, then $V(K(10,3)) = Z_{30}$ and the starter set is $C = \{(0,1,6,2,13),(0,2,12,5,10),(0,7,5,6,14),(0,13,2,10,14)\}$.
(b) If $(m,p) = (20,3)$, then $V(K(20,3)) = Z_{60}$ and the starter set is $C = \{(0,29,3,28,47),(0,28,2,24,1),(0,17,1,15,2),(0,29,1,20,4),(0,23,1,21,4),(0,25,5,12,1),(0,5,15,1,8),(0,11,1,6,8)\}$.

Thus $\{C,\sigma(C),\sigma^2(C),\ldots,\sigma^{mp-1}(C)\}$ gives a required $(m,p;5,2,3)$-multipartite cycle system.

One can check that the condition $n = 5r$ or $5r+1$ is necessary for the existence of an $(n,5,2,3)$-cycle system. Further, it is obvious that an $(n,5,2,3)$-cycle system does not exist, when $n = 5$ or 6, since the 3-simple property is not satisfied. The following theorem shows the existence of $(n,5,2,3)$-cycle system in certain cases.

Theorem 1. *An $(n,5,2,3)$-cycle system exists, for $n = 5r,\ 5r+1$ when (i) $r \equiv 2$ or 6 (mod 12) or (ii) $r \equiv 4$ or 12 (mod 24).*

Proof. We split the proof in two cases.

Case 1. $r \equiv 2$ or 6 (mod 12).
Case 1(a). $n = 5r$.

When $r = 2$, there exists an $(10, 5, 2, 3)$-cycle system by Lemma 2. When $r = 6$, we write $K_{30}(2) = (K_3 \otimes \overline{K}_{10})(2) \oplus 3K_{10}(2) = K(10, 3)(2) \oplus 3K_{10}(2)$. By Lemma 3, a 3-simple 2-fold 5-cycle system exists in $K(10, 3)(2)$. By Lemma 2, a 3-simple 2-fold 5-cycle system exists in $K_{10}(2)$. Therefore, a 3-simple 2-fold 5-cycle system exists in $K_{30}(2)$.

When $r \geq 14$, we write $K_n(2) = (K_{\frac{r}{2}} \otimes \overline{K}_{10})(2) \oplus \frac{r}{2}K_{10}(2) = ((C_3 \oplus C_3 \oplus \cdots \oplus C_3) \otimes \overline{K}_{10})(2) \oplus \frac{r}{2}K_{10}(2)$(by Lemma 1) $= K(10, 3)(2) \oplus \cdots \oplus K(10, 3)(2) \oplus \frac{r}{2}K_{10}(2)$. By Lemmas 2 and 3, we get the required cycle system.

Case 1(b). $n = 5r + 1$.

When $r = 2$, there exists an $(11, 5, 2, 3)$-cycle system by Lemma 2. When $r \geq 6$, we write $K_n(2) = (K_{\frac{r}{2}} \otimes \overline{K}_{10})(2) \oplus \frac{r}{2}K_{11}(2)$, see Fig. 1. By Lemmas 1, 2 and 3, we get the required cycle system.

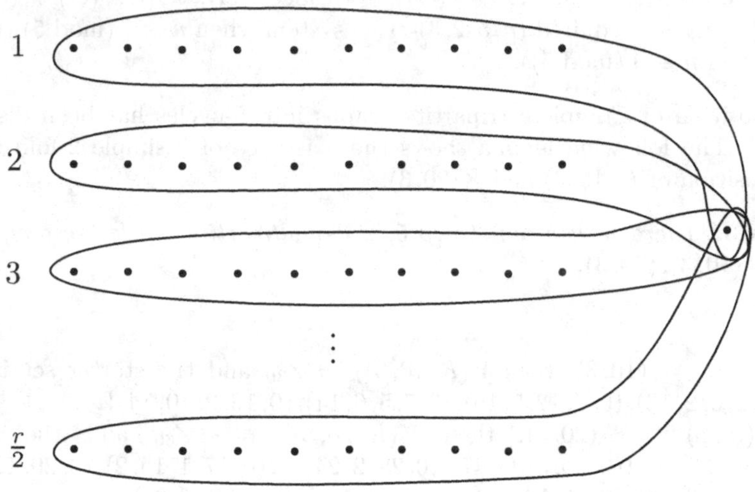

Fig. 1. $K_n(2) = (K_{\frac{r}{2}} \otimes \overline{K}_{10})(2) \oplus \frac{r}{2}K_{11}(2)$

Case 2. $r \equiv 4$ or $12 \pmod{24}$.

Case 2(a). $n = 5r$.

When $r = 4$, there exists an $(20, 5, 2, 3)$-cycle system by Lemma 2. When $r \geq 12$, we write $K_n(2) = (K_{\frac{r}{4}} \otimes \overline{K}_{20})(2) \oplus \frac{r}{4}K_{20}(2) = ((C_3 \oplus C_3 \oplus \cdots \oplus C_3) \otimes \overline{K}_{20})(2) \oplus \frac{r}{2}K_{20}(2)$ (by Lemma 1) $= K(20, 3)(2) \oplus \cdots \oplus K(20, 3)(2) \oplus \frac{r}{2}K_{20}(2)$. By Lemmas 2 and 3, we get the required cycle system.

Case 2(b). $n = 5r + 1$.

When $r = 4$, there exists an $(21, 5, 2, 3)$-cycle system by Lemma 2. When $r \geq 12$, we write $K_n(2) = (K_{\frac{r}{4}} \otimes \overline{K}_{20})(2) \oplus \frac{r}{4}K_{21}(2)$. By Lemmas 1, 2 and 3, we get the required cycle system.

Acknowledgments. The first author thank the University Grants Commission, New Delhi for its financial support (No: 4-4/2014-15 (MRP-SEM/UGC-SERO)) and the second author thank the DST-SERB, New Delhi for its financial support (No. SR/S4/MS: 828/13).

References

1. Billington, E.J., Cavenagh, N.J., Khodkar, A.: Super-simple twofold 4-cycle systems. Bull. ICA **63**, 48–50 (2011)
2. Bluskov, I., Hämäläinen, H.: New upper bounds on the minimum size of covering designs. J. Combin. Des. **6**, 21–41 (1998)
3. Cavenagh, N.J., Billington, E.J.: On decomposing complete tripartite graphs into 5-cycles. Australas. J. Combin. **22**, 41–62 (2000)
4. Chen, K.: On the existence of super-simple $(v, 4, 3)$-BIBDs. J. Combin. Math. Combin. Comput. **17**, 149–159 (1995)
5. Chen, K.: On the existence of super-simple $(v, 4, 4)$-BIBDs. J. Stat. Plann. Inference **51**, 339–350 (1996)
6. Chen, K., Cao, Z., Wei, R.: Super-simple balanced incomplete block designs with block size 4 and index 6. J. Stat. Plann. Inference **133**, 537–554 (2005)
7. Chen, K., Wei, R.: Super-simple cyclic designs with small values. J. Stat. Plann. Inference **137**, 2034–2044 (2007)
8. Gronau, H.D.O.F., Mullin, R.S.: On super-simple $2-(v, 4, \lambda)$ designs. J. Combin. Math. Combin. Comput. **11**, 113–121 (1992)
9. Hartmann, S.: Superpure digraph designs. J. Combin. Des. **10**, 239–255 (2000)
10. Kim, H.K., Lebedev, V.: On optimal superimposed codes. J. Combin. Des. **12**, 79–91 (2004)
11. Kim, H.K., Lebedev, V., Oh, D.Y.: Some new results on superimposed codes. J. Combin. Des. **13**, 276–285 (2005)
12. Kirkman, T.P.: On a problem in combinatorics. Camb. Dublin Math. J. **2**, 191–204 (1847)
13. Mahmoodian, E.S., Mirzakhani, M.: Decomposition of complete tripartite graphs into 5-cycles. In: 1994 Combinatorics Advances. Mathematics and Its Applications, Tehran, vol. 329, pp. 235–241. Kluwer Academic Publishers, Dordrecht (1995)
14. Stinson, D.R., Wei, R., Zhu, L.: New constructions for perfect hash families and related structures using combinatorial designs and codes. J. Combin. Des. **8**, 189 200 (2000)
15. Ushio, K.: G-designs and related designs. Discrete Math. **116**, 299–311 (1993)

ANN-Based Bias Correction Algorithm for Precipitation in the Yarra River Basin, Australia

P. Saravanan[1], C. Sivapragasam[1(✉)], M. Nitin[2], S. Balamurali[3], R.K. Ragul[1], S. Sundar Prakash[1], G. Selva Ganesan[1], and V. Vel Murugan[1]

[1] Centre for Water Technology, Kalasalingam University, Srivilliputtur, India
p.saravanan.me.1986@gmail.com, sivapragasam25@gmail.com

[2] College of Engineering and Science, Victoria University, Melbourne, Australia
Nitin.Muttil@vu.edu.au

[3] Department of Computer Applications, Kalasalingam University, Srivilliputtur, India
sbmurali@rediffmail.com

Abstract. Regional Climate Models (RCM) applied to simulate future climate parameters such as precipitation and temperature are reported to suffer from bias. Bias correction is necessary for using such data for climate change impact studies. In this study, a new ANN based bias correction algorithm is suggested and is compared with other three conventional methods, namely linear scaling, local intensity and power transformation. The proposed method outperforms conventional methods with mean, standard deviation and the RMSE of bias corrected time series more closely matches with that of the observed precipitation.

Keywords: Artificial Neural Network · Bias correction · Precipitation · Simulation · Climate change projection

1 Introduction

Precipitation is an important meteorological parameter which influences the flood and drought situations of any country. Climate change is believed to have perceptible effect on the occurrence of precipitation and its distribution and its quantification is a direct concern of water resources managers. Conventionally, future projected precipitation is simulated by Regional Climate Models (RCM) for climate change scenarios suggested by Intergovernmental Panel on Climate Change (IPCC). Precipitations simulated by RCMs are found to contain systematic error (also called as bias) due to inaccurate parameterization of the climatic process during the model development [5]. Bias means, deviation of statistics like the mean, variance, covariance of the model from the corresponding observed value [9,13]. Such simulated information need to be bias corrected before it can be used for any hydrological studies [1,4,7,14,18].

S. Arumugam et al. (Eds.): ICTCSDM 2016, LNCS 10398, pp. 362–370, 2017.
DOI: 10.1007/978-3-319-64419-6_47

Considerable research has already been done for bias correction. Some authors compared existing methods to find the best method for their own model [20], whereas other authors have proposed new bias correction methods [16,17,19,21,28]. Teutschbein and Seibert in [27] reviewed different bias correction methods such as quantile mapping, power transformation, local intensity and linear scale by comparing their performance both in terms of deviation from the observed precipitation as well as end use application of stream flow simulation. The performance of quantile mapping and power function are found to be the most robust. Some studies such as Tschoke et al., in [26] focused on developing new methodologies for error reduction during dry periods. Methods adopted in bias correction are seen to vary from very simple methods to advanced methods. While simple methods are found to perform poorly in summer season, the advanced methods offer difficulties in terms of long data length required to calibrate [3].

In a recently reported work, Um et al., in [28] proposed a hybrid bias correction method and compared that with other two conventional methods viz., linear scaling and quantile mapping. While linear scaling yielded the best result for estimating annual average precipitation, the hybrid method was reported to be optimal for predicting the variation in annual precipitation.

The bias corrections also find many applications other than for precipitations. For instance, Ahmed et al. [1] bias corrected a data set of daily maximum and minimum temperature for direct use of climate change impact studies for the future period of 2046–2065. Macias et al. [22] simulated the sea surface temperature using different ocean model and compared the result with satellite observed data and identified the bias of different models. They also applied simple bias correction to atmospheric variables of the model, to know the importance of each variable and found that wind velocity is the most important variable to bias correct.

Although the conventional bias correction methods are most popularly adopted, of late, researchers started applying black box methods such as Artificial Neural Network (ANN). Sanaz Moghim, in [24] have applied ANN for bias correction of Precipitation and Temperature. Also Chitra and Santhosh in [6] used ANN to downscale the simulated data. They applied Delta Method for bias correction.

To the author's knowledge, the application of ANN for bias correction is still in its inception with very few reported works. In this study we proposed a new robust bias correction algorithm which while reducing the Root Mean Squared Error(RMSE) between observed and simulated data, also tries to map the mean and standard deviation of the observed precipitation. The results are compared to the conventional methods of local intensity, linear scaling, and power transformation.

2 Study Area and Data Set

The middle segment of Yarra River Catchment is used for this study (Fig. 1). The water resources management is complex in this catchment due to the need of downstream supply for Melbourne as well as environmental flow provisions [2]. This catchment is the fourth highest productive in Victoria, even though its area is comparatively less than other catchments [23].

Observed daily precipitation data from 1980 to 2012 are available for the Yarra River Catchment, and are used in this study. Simulated data are obtained from the Climate change Data for SWAT (CMIP3) [8] database for the year 1961 to 2000 using CCCMA CGCM3.1 model. For the bias correction analysis, only the overlapping time period (1980 to 2000) is used for analysis.

For the analysis we considered two stations located within the middle segment of Yarra River Catchment, which are Toolangi (S -37.57, E 145.5) and Black Spur (S-37.59, E 145.62). The locations of these stations can be seen in Fig. 1. The observed daily rainfall data of Toolangi and Black Spur and the nearest available simulated data of the catchment is taken for this analysis.

Fig. 1. Location of study area

3 Methods

In this section, the methodologies of four bias correction methods used in this study are explained. Three of them are conventional methods. The methodology of the three conventional methods is explained based on [10,27].

3.1 Linear Scaling Method

This method is a mean based method which considerably reduces the deviation in the mean of observed and simulated data. The observed precipitation is corrected by a factor which is the ratio of long term monthly mean of observed and raw simulated precipitations

$$P_{correct(m)(d)} = P_{raw(m)(d)} \frac{\mu_{(P_{obser})(m)}}{\mu_{(P_{raw})(m)}} \tag{1}$$

Here, $P_{correct(m)(d)}$ and $P_{raw(m)(d)}$, are the corrected and raw simulated precipitation for d^{th} day of mth month respectively.

$\mu_{()}$, represents mean operator. For example $\mu_{(P_{raw})(m)}$ is the mean value of raw precipitation for the given month.

3.2 Local Intensity Method

This method is an improved version of linear scaling which not only corrects the monthly mean but also the wet day frequency and wet day intensity. The bias correction is done as follows:

Initially a threshold $P_{threshold(m)}$ of simulated data is calculated such that the number of days in a specified month which is more than the threshold precipitation equals the number of wet days (day of non zero precipitation) in the observed data.

Then the corrected simulated data is calculated as follows.

$$P_{correct(m)(d)} = \begin{cases} 0 & \text{if } P_{raw(m)(d)} < P_{thresold(m)} \\ (P_{raw(m)(d)}).(S) & \text{Otherwise} \end{cases}$$

Here 'S' is a scaling factor considering only wet day of observed and wet day of corrected simulated data and is calculated as follows,

$$S = \frac{\mu_{((P_{obser})(m)(d)\,|\,(P_{obser})(m)(d)>0))}}{\mu_{((P_{raw})(m)(d)\,|\,(P_{raw})(m)(d)>(P_{threshold})(m))}} \tag{2}$$

3.3 Power Transformation

The power transformation method can correct the standard deviation which is difficult to be ensured through linear scaling and local intensity methods. However, this method uses the bias corrected data by local intensity as input.

Initially a parameter b is calculated for each month m by using coefficient of variation (CV) of data corrected by local intensity and CV of observed data:

$$f(b_m) = CV_{P_{obser(m)}} - (CV_{P_{correct(LOI)(m)}})^{b_m} \tag{3}$$

Here, b_m is an exponent of mth month.

After the determination of 'b'_m. 'S' is calculated as follows:

$$S = \frac{\mu_{(P_{obser})(m)}}{\mu_{(P_{correct(LOI)(m)})^{b_m}}} \tag{4}$$

The bias corrected value obtained by power transformation is calculated as follows:

$$P_{correct(m)(d)} = (S).(P_{correct(LOI)(m)(d)})^{b_m} \tag{5}$$

where $P_{correct(m)(d)} and P_{correct(LOI)}$ are the corrected data by Power Transformation and Local Intensity respectively

3.4 ANN-based Bias Correction Algorithm

Fundamental Concepts of ANN: Artificial Neural Network (ANN) is a network of many simple elements called neurons, each having a small amount of local memory. The neurons are connected through communication channels or connections which carry numeric data encoded by various means. Each neuron operates only when it receives data through the communication channels. The architecture of ANN is motivated by the structure of the human brain and nerve cells. Historically, much of the inspiration to build ANNs came from the desire to produce artificial systems capable of sophisticated computations similar to those of the human brain.

Neural network modeling is based on the use of architecture and learning paradigms which allow the extraction of statistical structure present in the data set. This 'connectionist' philosophy is based on the idea that with a general architecture, often biologically motivated and with no prior information about the phenomenon of interest, it is possible to 'learn' the underlying structure of the data in an implicit form. The acquired information about the data is 'stored' at the connections between the elements of the neural architecture. The architecture is initially not structured, and the learning algorithm is responsible for the extraction of the regularities present in the data by finding a suitable set of synapses during the process of observation of the examples. Thus, ANNs solve problems by self-learning and self-organization. They derive their 'intelligence' from the collective behavior of simple computational mechanisms at individual neurons.

Multilayer feed-forward network with back-propagation learning algorithm is one of the most popular neural network architectures, which has been deeply studied and widely used in many fields. The feed-forward architecture allows connections only in one direction, that is, there is no back-coupling between neurons, and the neurons are arranged in layers, starting from an input layer and ending at the final output layer with one or more hidden layers. The information passes from the input to the output side. Each layer is made-up several neurons, and the layers are interconnected by a set of weights. The neurons in the input layers receive input directly from the input variables. The neurons in the hidden and output layers receive input from interconnections. Neurons operate on the input and transform it to produce an analog output. More details on ANN can be found in [11,12,15,25].

Model Development: Two different ANN models are employed in this study, which are discussed below.

Model 1(ANN-M1): Since the aim is to determine the bias corrected precipitation from the simulated precipitation, in the first model, a direct mapping is done with simulated precipitation as the input and observed precipitation as the output. It is desired to minimize the Root Mean Square Error (RMSE) of the corrected precipitation besides ensuring a closer match with standard deviation and mean of the observed precipitation. A three layer feed forward network with 11 hidden nodes gives the best output. Out of a total of 253 data of monthly precipitation, 108 is used for training, 24 for testing and remaining for verification.

Model 2(ANN-M2): Instead of directly mapping simulated and observed precipitation, it is desired to map the simulated precipitation to the absolute value of difference between observed and simulated (i.e. error in the simulated and observed precipitation, $|\Delta e|$). The number of training, testing and verification data is kept same as that used in Model 1. The output from the training has to be re-corrected to obtain the actual precipitation for which the following algorithm is used:

$$P_{corrected} = \begin{cases} P_{sim} + \Delta e, & \text{if } P_{sim} \epsilon (P_{sim,i}) - (P_{sim,j}) \text{ and } \theta(\Delta e) > 50 \\ P_{sim} + \Delta e, & \text{Otherwise} \end{cases} \quad (6)$$

where the $(P_{sim,i})$ and $(P_{sim,j})$ are the ranges adopted in this study is listed in Table 1;$\Delta e = (P_{obs,i}) - (P_{sim,i})$;$\theta(\Delta e)$ is the percentage of positive error in the specified range i.e.$(P_{sim,i}) - (P_{sim,j})$

This algorithm might induce significant error for those ranges of simulated precipitation for which $\theta(\Delta e)$ is in the neighborhood of 50.

4 Results and Discussions

In this section, the bias corrected simulated precipitation by the methods of linear scaling, local intensity, power transformation, ANN-M1, ANN-M2 are compared with the corresponding observed precipitation and is tabulated in Table 2.

Model 1 (ANN-M1): From Table 2, it is observed that the ANN-M1 has not only the lowest RMSE when compared to all the other methods, but also the SD for both the stations is also very low. Low SD indicates that the prediction has failed in capturing the variations in the precipitation as seen from Figs. 2 and 3. The conventional methods perform almost equally well for both the stations in terms of all the performance measure considered. Figures 2,3 indicates that

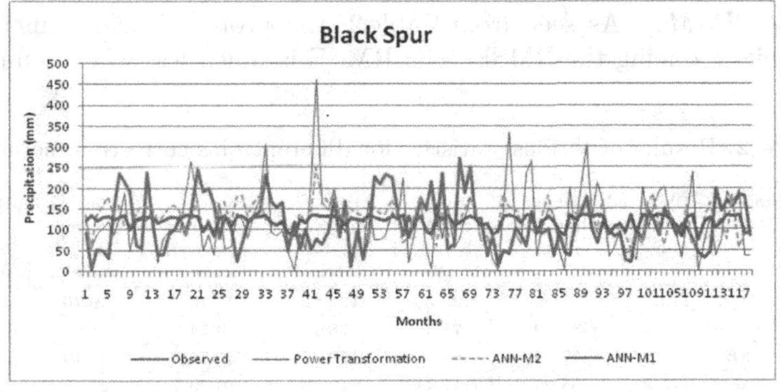

Fig. 2. Comparison of various methods for Black Spur Station

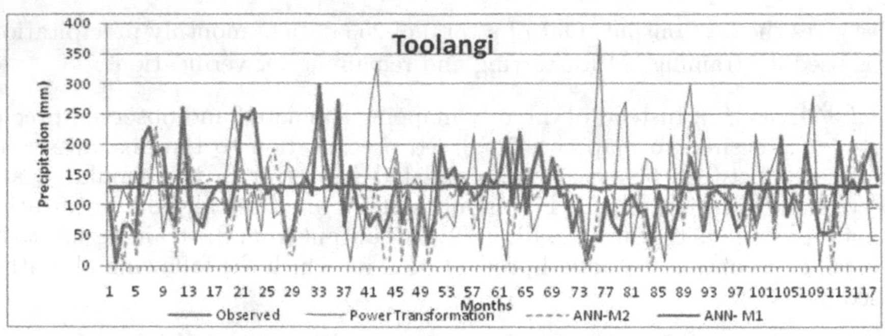

Fig. 3. Comparison of various methods for Toolangi Station

Table 1. Applied error sign for various range

Station name	$(P_{sim,i}) - (P_{simj})$	$\theta(\Delta e)$	Error Sign (+/−)
Toolangi	0–30	5.55	+
	30–60	11.42	+
	60–90	12.50	+
	90–120	28.57	+
	> 120	76.92	−
Black Spur	0–30	10.52	+
	30–60	18.42	+
	60–90	19.23	+
	90–120	33.33	+
	> 120	88.00	−

power transformation method overestimates the peaks to a much higher degree. This behavior can be reduced if RMSE can be reduced while maintaining the mean and SD.

Model 2 (ANN-M2): As seen from Table 2, the proposed algorithm increases the SD while reducing the RMSE. The RMSE is much less when compared to

Table 2. Results of various statistics for different bias correction methods

Station	Statistics	Observed precipitation	Simulated precipitation	Bias corrected by linear scaling	Bias corrected by local intensity	Bias corrected by power transformation	Bias corrected by ANN model 1	Bias corrected by ANN model 2
Toolangi	*Mean*	119.14	75.99	125.36	122.82	125.76	129.70	127.41
	SD	58.48	44.00	78.44	78.13	70.54	0.82	34.69
	RMSE		87.84	99.67	99.46	95.50	59.10	67.76
Black Spur	*Mean*	115.27	85.96	117.41	114.36	110.70	121.63	127.20
	SD	62.80	55.97	79.95	76.23	76.14	15.65	37.43
	RMSE		89.53	95.94	94.90	95.31	64.28	70.22

the conventional methods, and the SD has been considerably increased when compared to ANN-M1. The effect of this can be clearly seen in Figs. 2 and 3 in terms of mapping the peaks and other values more closely when compared to the conventional methods.

5 Conclusions

Based on this study the following conclusions can be drawn:

(a) ANN seems to be a potential tool for bias correction
(b) The proposed algorithm is able to correct the simulated precipitation to map more accurately with the observed value when compared to the conventional methods of bias correction.

References

1. Ahmed, K.F., Wang, G., Silander, J., Wilson, A.M., Allen, J.M.: Statistical downscaling and bias correction of climate model outputs for climate change impact assessment in the US northeast. Global Planet. Change **100**, 320–332 (2013)
2. Barua, S., Muttil, N., Ng, A.W.M., Perera, B.J.C.: Rainfall trend and its implications for water resource management within the Yarra River catchment. Aust. Hydrol. Process. **27**(12), 1727–1738 (2013)
3. Berg, P., Feldmann, H., Panitz, H.J.: Bias correction of high resolution regional climate model data. J. Hydrol. **448**, 80–92 (2012)
4. Ceglar, A., Kajfež-Bogataj, L.: Simulation of maize yield in current and changed climatic conditions: addressing modelling uncertainties and the importance of bias correction in climate model simulations. Eur. J. Agron. **37**(1), 83–95 (2012)
5. Chen, J., Brissette, F.P., Chaumont, D., Braun, M.: Finding appropriate bias correction methods in downscaling precipitation for hydrologic impact studies over North America. Water Resour. Res. **49**(7), 4187–4205 (2013)
6. Chithra, N.R., Santosh, G.T.: Bias correction of ANN based statistically downscaled precipitation data for the Chaliyar river basin. Int. J. Innov. Res. Sci. Eng. Technol. **2**, 6–11 (2013)
7. Christensen, J.H., Boberg, F., Christensen, O.B., Lucas-Picher, P.: On the need for bias correction of regional climate change projections of temperature and precipitation. Geophys. Res. Lett. **35**(20), 1–6 (2008)
8. Climate change Data for SWAT (CMIP3) Database. http://globalweather.tamu.edu/cmip (Viewed October 2016)
9. Ehret, U., Zehe, E., Wulfmeyer, V., Warrach-Sagi, K., Liebert, J.: HESS opinions "Should we apply bias correction to global and regional climate model data?". Hydrol. Earth Syst. Sci. **16**(9), 3391–3404 (2012)
10. Fang, G., Yang, J., Chen, Y.N., Zammit, C.: Comparing bias correction methods in downscaling meteorological variables for a hydrologic impact study in an arid area in China. Hydrol. Earth Syst. Sci. **19**(6), 2547–2559 (2015)
11. Govindaraju, R.S.: Artificial neural networks in hydrology. I: preliminary concepts. J. Hydrol. Eng. **5**(2), 115–123 (2000)
12. Govindaraju, R.: Artificial neural networks in hydrology: II, hydrologic applications (2000)

13. Haerter, J.O., Hagemann, S., Moseley, C., Piani, C.: Climate model bias correction and the role of timescales. Hydrol. Earth Syst. Sci. **15**(3), 1065–1079 (2011)
14. Hawkins, E., Osborne, T.M., Ho, C.K., Challinor, A.J.: Calibration and bias correction of climate projections for crop modelling: an idealised case study over Europe. Agric. For. Meteorol. **170**, 19–31 (2013)
15. Haykin, S.S.: A Comprehensive Foundation. Tsinghua University Press, Bejing (2001)
16. Hoffmann, H., Rath, T.: Meteorologically consistent bias correction of climate time series for agricultural models. Theor. Appl. Climatol. **110**(1–2), 129–141 (2012)
17. Ines, A.V., Hansen, J.W.: Bias correction of daily GCM rainfall for crop simulation studies. Agric. For. Meteorol. **138**(1), 44–53 (2006)
18. Johnson, F., Sharma, A.: What are the impacts of bias correction on future drought projections? J. Hydrol. **525**, 472–485 (2015)
19. Kim, K.B., Kwon, H.H., Han, D.: Bias correction methods for regional climate model simulations considering the distributional parametric uncertainty underlying the observations. J. Hydrol. **530**, 568–579 (2015)
20. Lafon, T., Dadson, S., Buys, G., Prudhomme, C.: Bias correction of daily precipitation simulated by a regional climate model: a comparison of methods. Int. J. Climatol. **33**(6), 1367–1381 (2013)
21. Li, J., Sharma, A., Evans, J., Johnson, F.: Addressing the mischaracterization of extreme rainfall in regional climate model simulations-a synoptic pattern based bias correction approach. J. Hydrol. doi:10.1016/j.jhydrol.2016.04.070
22. Macias, D., Garcia-Gorriz, E., Dosio, A., Stips, A., Keuler, K.: Obtaining the correct sea surface temperature: bias correction of regional climate model data for the Mediterranean Sea. Clim. Dyn. 1–23 (2016). doi:10.1007/s00382-016-3049-z
23. Water, M.: Port Phillip and Westernport Regional River Health Strategy. Yarra catchment, Richmond (2013)
24. Moghim, S.: Bias Correction of Global Circulation Model Outputs Using Artificial Neural Networks (Doctoral dissertation, Georgia Institute of Technology) (2015)
25. Sivapragasam, C., Vanitha, S., Muttil, N., Suganya, K., Suji, S., Selvi, M.T., Sudha, S.J.: Monthly flow forecast for Mississippi River basin using artificial neural networks. Neural Comput. Appl. **24**(7–8), 1785–1793 (2014)
26. Tschöke, G.V., Kruk, N.S., de Queiroz, P.I.B., Chou, S.C., de Sousa Junior, W.C.: Comparison of two bias correction methods for precipitation simulated with a regional climate model. Theore. Appl. Climatol. **127**, 1–12 (2015)
27. Teutschbein, C., Seibert, J.: Bias correction of regional climate model simulations for hydrological climate-change impact studies: review and evaluation of different methods. J. Hydrol. **456**, 12–29 (2012)
28. Um, M.J., Kim, H., Heo, J.H.: Hybrid approach in statistical bias correction of projected precipitation for the frequency analysis of extreme events. Adv. Water Resour. **94**, 278–290 (2016)

Some Diameter Notions of the Generalized Mycielskian of a Graph

K.S. Savitha[1(✉)], M.R. Chithra[2], and A. Vijayakumar[3]

[1] Department of Mathematics, St. Paul's College, Kalamassery, Cochin 683503, India
savithaks2009@gmail.com

[2] Department of Mathematics, School of Arts and Sciences, Amrita University, Cochin 682024, Amrita Vishwa Vidyapeetham, India
chithramohanr@gmail.com

[3] Department of Mathematics, Cochin University of Science and Technology, Cochin 682022, India
vambat@gmail.com

Abstract. Generalized Mycielskian of a graph is the natural generalization of the Mycielskian of a graph, which preserves some nice properties of a good interconnection network. Diameter is an important parameter for communication in an interconnection networks as it determines maximum communication delay between any pair of components in the network. In this paper, we study the diameter and its variability by the addition and deletion of edges in the generalized Mycielskian of a graph.

Keywords: Mycielskian · Generalized Mycielskian · Diameter · Diameter variability

1 Introduction

The topological structure of an interconnection network can be modeled by a connected graph, where the vertices represent components of the network and the edges represent the communication links between them. The diameter of a graph determines the maximum communication delay between any pair of processors in a network. The fact that the diameter of a graph can be affected by the addition or deletion of edges, give rise to the concept of diameter variability in graphs. The study of diameter variability in a network becomes important as it determines the communication efficiency when an addition or deletion of a link occurs.

In a search for triangle-free graphs with arbitrarily large chromatic number, Mycielski developed a graph transformation $\mu(G)$, called the Mycielskian of G. A natural generalization of this transformation is the generalized Mycielskian $\mu_m(G)$. In [12], it is observed that the Mycielskian produce large networks and preserve some nice properties of networks such as fast multi-path communication, high fault tolerance and reliable resource sharing.

S. Arumugam et al. (Eds.): ICTCSDM 2016, LNCS 10398, pp. 371–382, 2017.
DOI: 10.1007/978-3-319-64419-6_48

2 Background

In recent times, there has been an increasing interest in the study of the Mycielskian and generalised Mycielskian of a graph. In [6], Fisher et al. proved that if G is Hamiltonian, then so is $\mu(G)$ and diameter of $\mu(G) = \min(\max(2, diam(G)), 4)$. Balakrishnan and Francis Raj determined the vertex connectivity and edge connectivity of Mycielskian in [1]. In [10], L. Guo et al. showed that for a connected graph G with $|V(G)| \geq 2$, $\mu(G)$ is super connected if and only if $\delta(G) < 2\kappa(G)$ and $\mu(G)$ is super edge connected if and only if $G \ncong K_2$. Recently, Chithra M.R. studied the diameter variability of Mycielskian in [5].

Several parameters of generalized Mycielskian such as circular clique number, total domination number, open packing number and spectrum are determined in [11]. Francis Raj [7] investigated the vertex connectivity and edge connectivity of the generalised mycielskian of digraphs, which turned out to be a generalisation of the results due to Guo and Guo [9].

Graham and Harary [8] studied how the diameter of hypercubes $(Q_n, n \geq 1)$ can be affected by adding or deleting edges. They considered changing the diameter with out considering the extent of the change and showed that $D^{-1}(Q_n) = 2, D^{+1}(Q_n) = n - 1$ and $D^{+0}(Q_n) \geq (n-3)2^{n-1} + 2$. Bouabdallah et al. [2] improved the lower bound of $D^{+0}(Q_n)$ and furthermore gave an upper bound. Diameter variability of cycles and tori was determined by Wang et al. [13]. In [14], authors studied the changing of the diameter of a diagonal mesh network. Diameter variability of various graph products was studied in [4].

3 Preliminaries

Let $G = (V, E)$ be a graph with vertex set $V(G)$ and edge set $E(G)$. A vertex $u \in V(G)$ is called a *neighbor* of a vertex v in G, if uv is an edge of G, and $u \neq v$. The set of all neighbors of v is called the *neighbor set* of v, and is denoted by $N(v)$. The *degree* of a vertex v, denoted by $d(v)$, is the number of edges incident at v. The minimum degree of G, denoted by $\delta(G)$, is $\min\{d(v) : v \in V\}$ and the maximum degree of G, denoted by $\Delta(G)$, is $\max\{d(v) : v \in V\}$. A vertex of degree one is called a *pendant vertex* of G and the unique edge incident to such a vertex of G is a *pendant edge* of G.

A *path* from u to v, given by $u = x_0 - x_1 - x_2 - \cdots - x_k = v$, is a sequence of distinct vertices such that $x_i x_{i+1}$ is an edge for $0 \leq i \leq k-1$. The *length of a path* is the number of edges in it. The *distance* between two vertices u and v in G, denoted as $d_G(u, v)$, is the length of a shortest path joining u and v. The maximum distance between any two vertices is called the *diameter* of G and is denoted by $\text{diam}(G)$. Two vertices in G which are at a distance equal to the diameter of G are called *diametral vertices.* A graph in which every pair of vertices are joined by a path is called a *connected graph.*

Let k be an arbitrary integer. The *diameter variability* arising from change of edges of a graph G is defined as follows [13]:

- $D^{-k}(G)$: the least number of edges whose addition to G decreases the diameter by (at least) k;
- $D^{+0}(G)$: the maximum number of edges whose deletion from G does not change the diameter;
- $D^{+k}(G)$: the least number of edges whose deletion from G increases the diameter by (at least) k;

For example, consider the $m-$cycle C_m with vertex set $\{0, 1, 2, \cdots, m-1\}$ and edge set $\{(i, i+1)|0 \leq i \leq m-1\}$, where addition is in integer modulo m. Then $\text{diam}(C_m) = \lfloor \frac{m}{2} \rfloor$. If P_m is the path on m vertices with vertex set $\{0, 1, 2, \cdots, m-1\}$ and edge set $\{(i, i+1)|0 \leq i \leq m-2\}$, then $\text{diam}(P_m) = m-1$. It is easy to see that $D^{-1}(P_m) = D^{-2}(P_m) = \cdots = D^{-(m-1-\lfloor \frac{m}{2} \rfloor)}(P_m) = 1$ and $D^{+1}(C_m) = D^{+2}(C_m) = \cdots = D^{+(m-1-\lfloor \frac{m}{2} \rfloor)}(C_m) = 1$.

For a graph $G = (V, E)$, the Mycielskian of G [11] is the graph $\mu(G)$ with the vertex set $V(\mu(G)) = V \cup V' \cup \{z\}$, where $V' = \{u' : u \in V\}$ and the edge set $E(\mu(G)) = E \cup \{uv' : uv \in E\} \cup \{v'z : v' \in V'\}$. The vertex v' is called the twin of the vertex v and vice versa. The vertex z is called the root of $\mu(G)$. For $n \geq 2$, $\mu^n(G)$ is defined iteratively by setting $\mu^n(G) = \mu(\mu^{n-1}(G))$.

The generalized Mycielskian of a graph is defined as follows [11]:

Let G be a graph with vertex set $V^0 = \{v_1^0, v_2^0, \cdots v_n^0\}$ and edge set E^0. Given an integer $m \geq 1$, the m- Mycielskian of G, denoted by $\mu_m(G)$ is the graph with vertex set $V^0 \cup V^1 \cup V^2 \cdots V^m \cup \{z\}$, where $V^i = \{v_j^i : v_j^0 \in V^0\}$ is the ith distinct copy of V^0 for $i = 1, 2, \cdots, m$ and edge set $E^0 \cup \left(\bigcup_{i=0}^{m-1} \{v_j^i v_{j'}^{i+1} : v_j^0 v_{j'}^0 \in E^0\}\right) \cup \{v_j^m z : v_j^m \in V^m\}$. $\mu_0(G)$ is defined to be the graph obtained from G by adding a universal vertex z and the Mycielskian of G is simply $\mu_1(G)$. We call the vertices of V^i as vertices of level i (Fig. 1).

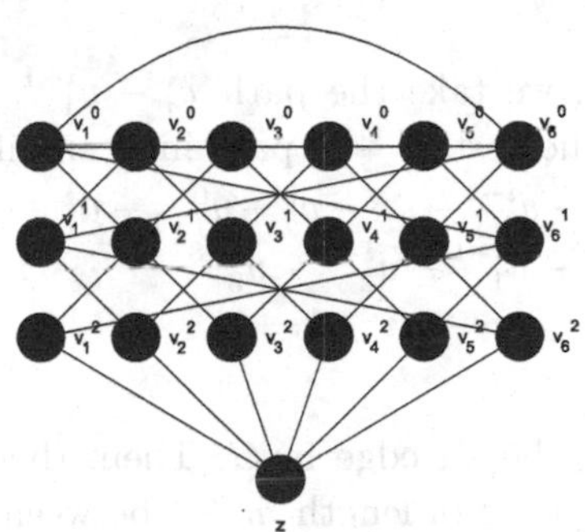

Fig. 1. $\mu_2(C_6)$

In this paper, we first determine the diameter of the generalized Mycielskian of a graph. In the next sections, we investigate the effect of addition and deletion of edges in the diameter of the generalized Mycielskian. All graphs considered in this paper are simple, finite and undirected.

4 Diameter of the Generalized Mycielskian

In this section, we determine the diameter of the generalized Mycielskian.

Theorem 1. *Let G be a connected graph with diameter $diam(G)$. Then diameter of $\mu_m(G)$ is given by*

$$diam\,(\mu_m(G)) = \min\{\max\{m+1, diam(G)\}, 2(m+1)\}.$$

Proof. We prove this result by considering the following cases.

Case 1. $\text{diam}(G) \leq m+1$.

In this case, we claim that $\text{diam}(\mu_m(G)) = m+1$. For this, consider the vertices v_i^0 and z and let $v_i^0 v_{i+1}^0 \in E^0$. Then $d(v_i^0, z) = m+1$ by taking the path $v_i^0 - v_{i+1}^1 - v_i^2 - \cdots - v_i^m(\text{or } v_{i+1}^m) - z$ according as m is even (or odd). Therefore $\text{diam}(\mu_m(G)) \geq m+1$.

Next, we show that for any pair of vertices u and v, $d_{\mu_m(G)}(u,v) \leq m+1$.

Case 1a. $u, v \in V^i$.

Let $u = v_j^i$ and $v = v_k^i$. If $u, v \in V^0$, then the distance between u and v in $\mu_m(G)$ is same as that in G. Hence $d_{\mu_m(G)}(u,v) \leq d(G) \leq m+1$. For $u, v \in V^1$, ifv_j^0 and v_k^0 are adjacent, then $d(v_j^1, v_k^1)$ is 3 by taking the path $v_j^1 - v_k^0 - v_j^0 - v_k^1$. If they are non adjacent, we consider the path $v_j^1 - u_1^0 - \cdots - u_{n-1}^0 - v_k^1$, where $v_j^0 - u_1^0 - \cdots - u_{n-1}^0 - v_k^0$ is a shortest $v_j^0 - v_k^0$ path in G. So $d_{\mu_m(G)}(v_j^1, v_k^1) = d_G(v_j^0, v_k^0)$ and hence $d_{\mu_m(G)}(u,v) \leq \text{diam}(G) \leq m+1$.

Now Let $u = v_j^i$ and $v = v_k^i$, $i > 1$, then

$$d_{\mu_m(G)}(u,v) = \begin{cases} d_G(v_j^0, v_k^0) & \text{if } d_G(v_j^0, v_k^0) \text{ is even,} \\ \min\{2(m-i)+2, 2i+1\} & \text{if } d_G(v_j^0, v_k^0) \text{ is odd.} \end{cases}$$

for, if $d_G(v_j^0, v_k^0)$ is even, we take the path $v_j^i - u_1^{i+1} - u_2^i - \cdots - v_k^i$, where $v_j^0 - u_1^0 - u_2^0 \cdots - v_k^0$ is a shortest $v_j^0 - v_k^0$ path in G and if $d(u,v)$ is odd, we have to take either the path $v_j^i - u_1^{i-1} - \cdots - u_j^0 - u_{j+1}^0 - u_j^1 - \cdots - v_k^i$ or the one which pass through z namely $v_j^i - u_1^{i+1} - u_2^{i+2} - u_3^{i+3} - \cdots - z - u_i^m - u_{i\pm 1}^{m-1} - \cdots - v_k^i$. This shows that $d(v_j^i, v_k^i) \leq m+1, i \geq 1$.

Case 1b. $u \in V^i, v = z$.

Let $u = v_j^i$ and $v_j^0 v_{j+1}^0$be an edge in G. Then, there exists the path $v_j^0 - v_{j+1}^1 - v_j^2 - \cdots - v_j^m(v_{j+1}^m) - z$ of length $m+1$ between v_j^0 and z. For all other $v_j^i, i > 1$ there exists the path $v_j^i - v_{j+1}^{i+1} - v_j^{i+2} - \cdots - v_j^m(v_{j+1}^m) - z$ of length less than $m+1$ between v_j^i and z. Thus $d(v_j^i, z) \leq m+1$.

Case 1c. $u \in V^i, v \in V^j, i \geq 0, j \geq 1, i < j$.

Case 1c(i). $u = v_k^i$ and $v = v_k^j$.

It is easy to see that $d_{\mu_m(G)}(v_k^0, v_k^1) = 2$. Now, consider $d(v_k^i, v_k^j), i \geq 0, j > 1$. Let v_k^0 be adjacent to v_l^0 in G. If $j - i$ is even, then we have the path $v_k^i - v_l^{i+1} -$

$v_k^{i+2} - \cdots - v_k^j$ between v_k^i and v_k^j of length $j-i$. If $j-i$ is odd, either we can take the path $v_k^i - v_l^{i-1} - v_k^{i-2} \cdots - v_k^0(v_l^0) - v_l^0(v_k^0) - v_k^1(v_l^1) - v_l^2(v_k^2) - \cdots - v_k^j$ or we can take $v_k^j - v_l^{j+1} - \cdots v_k^m(v_l^m) - z - v_l^m(v_k^m) - v_l^{m-1}(v_k^{m-1}) - \cdots - v_k^i$. Hence for $i \geq 0, j > 1, d(v_k^i, v_k^j) \leq \min\{i+j+1, 2(m+1)-(i+j)\}$. Thus, we get $d(v_k^i, v_k^j) \leq m+1$.

Case 1c(ii). $u = v_k^i$ and $v = v_l^j, k \neq l$.

If $v_k^0 - u_1^0 - u_2^0 \cdots - u_{n-1}^0 - v_l^0$ is a path in G, then $v_k^0 - u_1^0 - u_2^0 - \cdots u_{n-1}^0 - v_l^1$ is a path in $\mu_m(G)$ and hence $d(v_k^0, v_l^1) \leq \operatorname{diam}(G)$. Now, consider the pair $(v_k^i, v_l^j), i \geq 0, j \geq 2$. First suppose that v_k^0 and v_l^0 are adjacent. If $j-i$ is odd, then we have the path $v_k^i - v_l^{i+1} - v_k^{i+2} - v_l^{i+3} \cdots - v_l^j$ and hence $d(v_k^i, v_l^j) \leq j-i \leq m+1$. If $j-i$ is even, either we take $v_k^i - v_l^{i-1} - v_k^{i-2} - \cdots - v_k^0(v_l^0) - v_l^1(v_k^1) - \cdots - v_l^j$ or we take $v_k^i - v_l^{i+1} - v_k^{i+2} - \cdots - v_k^m(v_l^m) - z - v_l^m(v_k^m) - \cdots - v_k^{n-1} - v_l^j$. Hence $d(v_k^i, v_l^j) \leq min\{j+i+1, 2(m+1)-(j+i)\}$. If v_k^0 and v_k^l are not adjacent in G, then take a shortest $v_k^0 - v_l^0$ path say $v_k^0 - u_1^0 - u_2^0 - \cdots - u_{n-1}^0 - v_l^0$ in G. Corresponding to this path, we have, the path say $P = v_k^i - u_1^{i+1} - u_2^{i+2} - \cdots - v_l^j$ of length $d(u,v)$ in $\mu_m(G)$ if $d(v_k^0, v_l^0) \leq j-i$. If $j-i < d(v_k^0, v_l^0)$, instead of P we have the path $P' = v_l^j - u_{n-1}^{j-1} - \cdots - v_k^r - u_1^{r-1} - \cdots - u^0 - u_1^1 - u_2^2 - \cdots - v_k^i$, where $r = j-i-d(u,v)$. Therefore $d(v_k^i, v_l^j) \leq m+1$ and hence $\operatorname{diam}(\mu_m(G)) \leq m+1$.

Case 2. $m+1 < \operatorname{diam}(G) < 2(m+1)$.

In this case, proceeding on similar lines as in case 1, we get $d(u,v) \leq \operatorname{diam}(G)$, $u, v \in V(\mu_m(G))$. If v_i^0 and v_j^0 are the diametral vertices in G, then in $\mu_m(G)$ also, we have v_i^0 and v_j^0 at distance $\operatorname{diam}(G)$ and hence it follows that $\operatorname{diam}(\mu_m(G)) = d(G)$.

Case 3. $\operatorname{diam}(G) \geq 2(m+1)$.

The diametral vertices in G in this case, are at a distance $2(m+1)$ as the shortest path between them is through z in $\mu_m(G)$. For every pair of vertices, we can show that there exists a path of length less than or equal to $2(m+1)$ as in case 1. Hence $\operatorname{diam}(\mu_m(G)) = 2(m+1)$ in this case.

5 Diameter Variability

Here, we determine $D^{+0}(\mu_m(G)), D^{+1}(\mu_m(G)), D^{-1}(\mu_m(G))$.

Theorem 2. *Let G be a connected graph such that $G \not\cong K_{1,n-1}$ and $m \geq 1$ be an integer. Then*

$$D^{+0}(\mu_m(G)) \geq \begin{cases} 2e+k-(n+1+\sum_{i=1}^k e_i), & if\, diam(G) \leq m+1, \\ t(2e+k-(n+1+\sum_{i=1}^k e_i)), & if\, diam\,(G) > m+t, 1 \leq t \leq m, \\ \max\{m(2e+k-(n+1+\sum_{i=1}^k e_i)), e\}, & if\, diam\,(G) \geq 2(m+1). \end{cases}$$

where n is the number of vertices in G, e the number of edges in G and e_i's, $i = 1, 2, 3 \cdots, k$ are the number of pendant edges attached to the vertex v_i of G.

Proof. If $\text{diam}(G) \leq m+1$, then remove all the edges of the form $u_i - u_{i+1}$ from the $\lceil \frac{m-1}{2} \rceil^{th}$ level except the pendant edges, one edge from all but one vertex with $d(v) > 1$ and two from one vertex with $d(v) > 1$. More specifically, let $v_1 - v_2 - \cdots - v_d$ be a diametral path in G where, $v_{e_1}, v_{e_2}, \cdots, v_{e_k}$ are the k vertices v_i with e_i pendant vertices. Then remove all the edges except one from $\{v_1, v_2, \cdots, v_d\} \setminus \{v_{e_1}, v_{e_2}, \cdots, v_{e_k}, v_{d-1}\}$ and remove all the edges except two from v_{d-1} (See Fig. 2). This removal of edges will not affect the shortest path between the vertices in $\mu_m(G)$ is clear from the discussion of shortest paths in the Sect. 4.

If $\text{diam}(G) > m+t, 1 \leq t \leq m$, these set of edges can be removed from t levels $m-1, m-2, \cdots, m-t$. If $\text{diam}(G) \geq 2(m+1)$, either the removal of the edges of the above type from m levels or the removal of the edges from the copy of G in $\mu_m(G)$ will not change the diameter of $\mu_m(G)$.

Illustration

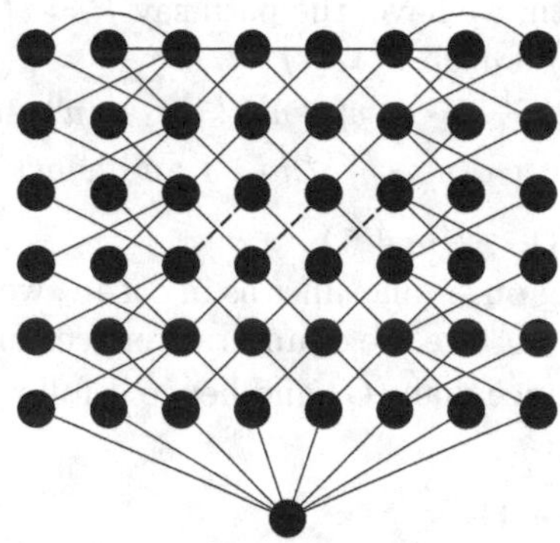

Fig. 2. Dotted lines are the deleted edges

Theorem 3. *Let G be any connected graph and $m \geq 2$ be an integer. Then $D^{+1}(\mu_m(G)) = 1$ if and only if diam $(G) \leq m+1$ and G has at least one pendant edge.*

Proof. First, suppose that G has at least one pendant edge and $\text{diam}(G) \leq m+1$. Then $\text{diam}(\mu_m(G)) = m+1$. Let $v_i^0 - v_j^0$ be a pendant edge in G. Consider the pair of vertices (v_j^1, z) in $\mu_m(G)$, which are at distance m. If the edge $v_i^2 - v_j^1$ is deleted, then, $d(v_j^1, z) = m+2$ by the path $v_j^1 - v_i^0 - v_k^1 - v_i^2 - v_j^3 - \cdots - v_i^m - z$ or $v_j^1 - v_j^0 - v_k^1 - v_i^2 - v_j^3 - \cdots - v_j^m - z$ according as m is odd or even respectively (Fig. 3). For all other vertices x in $\mu_m(G)$, $d(v_j^1, x) \leq m+2$, since the distance between any other pair is not affected by the removal of this edge. Therefore $D^{+1}(\mu_m(G)) = 1$.

Conversely, assume that $D^{+1}(\mu_m(G)) = 1$. If possible, let $\text{diam}(G) \leq m+1$ and G has no pendant edges. Then, $\text{diam}(\mu_m(G)) = m+1$. Let an edge $v_i^0 - v_{i+1}^0$,

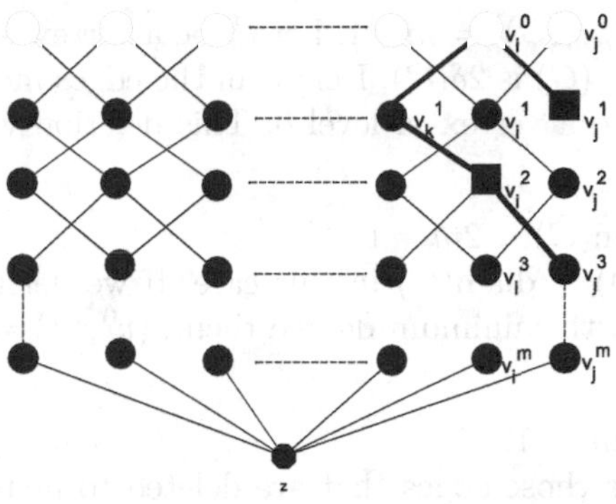

Fig. 3.

be deleted. Then, $d(v_x^0, v_y^0) \leq diam(G)$ by a path, $v_x^0 - v_{(x+1)}^0 - v_{(x+2)}^0 \cdots - v_i^0 - v_{i+1}^1 - v_{i+2}^0 - \cdots - v_y^0$. Clearly, the distance between any other pair is not affected by the removal of this edge. If an edge $v_k^i - v_l^{i+1}$ is deleted, then $d(v_k^i, v_l^{i+1})$ is affected. Since, $\delta(G) \geq 2$, v_k^i is adjacent to some other vertex v_j^{i+1}, in the $i + 1^{th}$ level. Thus, $d(v_k^i, v_l^{i+1}) = 3$, by a path, $v_k^i - v_j^{i+1} - v_k^{i+2} - v_l^{i+1}$. For any other pair (v_k^i, v_x^j), the edge $v_k^i - v_l^{i+1}$ in any $v_k^i - v_x^j$ path can be replaced by the edge $v_k^i - v_j^{i+1}$ for some $v_j^{i+1} \in N(v_k^i)$ and hence $d(v_k^i, v_x^j) \leq m + 1$. The removal of an edge $v_i^m - z$ also will not affect the diameter as it changes only the distance between v_i^m and z to 3. Thus we get a contradiction to the fact that $D^{+1}(\mu_m(G)) = 1$ and therefore G has at least one pendant edge.

Next, suppose that $\text{diam}(G) > m + 1$ and G has pendant edges. Since $\text{diam}(G) > m + 1$, $\text{diam}(\mu_m(G)) = \text{diam}(G)$. If any edge $v_x^0 - v_y^0$ is removed, the distance is unaffected in $\mu_m(G)$ as alternate paths exist through the duals. Let $v_k^0 - v_l^0$ be a pendant edge in G. If an edge of the form $v_k^{i+1} - v_l^i$ is removed from $\mu_m(G)$, then $d(v_k^{i+1}, v_l^i) = 3$ by the path $v_l^i - v_k^{i-1} - v_x^i - v_k^{i+1}$, where $v_x^0 \in N(v_k^0)$ for $i \neq 0$ and for $i = 0$, $d(v_k^1, v_l^0) = 3$ by the path $v_l^0 - v_k^0 - v_x^0 - v_k^1$. Thus the distance between v_l^0 and z becomes $m+2$ by the path $v_l^0 - v_k^0 - v_l^1 - v_k^2 - \cdots - z$. For $i > 0$, $d(v_l^i, z) < m+2$, by the path $v_l^i - v_k^{i-1} - v_x^i - v_k^{i+1} - v_l^{i+2} - \cdots - v_k^m - z$ or $v_l^i - v_k^{i-1} - v_x^i - v_k^{i+1} - v_l^{i+2} - \cdots - v_l^m - z$ according as m is odd or even respectively, where $d(v_k^{i+1}, v_l^i) = 3$ and $d(v_k^{i+1}, z) < m - 1$. The other edge removals will not affect the distance as there are alternate paths. Thus the removal of any single edge does not change the diameter and hence a contradiction.

Theorem 4. $D^{+1}(\mu_m(G)) \leq \begin{cases} 2\delta(G) - 1 & if\, diam\,(G) \leq m + 1, \\ \delta(G) & if\, m + 1 < \ diam(G) < 2m + 1, \\ D^1(G) & if\, diam\,(G) = 2m + 1, \\ \Delta(G) & if\, diam\,(G) \geq 2(m + 1). \end{cases}$

Proof. To obtain this upper bound, we consider the following cases.

Case 1. $\text{diam}(G) \leq m + 1$.

In this case, $\text{diam}(\mu_m(G)) = m+1$. Let v_i^0 be a vertex with minimum degree in G. Then $d(v_i^0)$ in $\mu_m(G)$ is $2\delta(G)$. Delete all the edges incident with v_i^0 except one that is adjacent to a vertex in level 0. This deletion will result in a graph with $d(v_i^0, z) = m+2$.

Case 2. $m+1 < \text{diam}(G) < 2m+1$.

Since $\text{diam}(\mu_m(G)) = \text{diam}(G)$ in this case, If we delete all the edges incident on a vertex v_i^0 with minimum degree then $d(v_i^0, v_i^1) = 2m+1$. Therefore, $D^{+1}(\mu_m(G)) \leq \delta(G)$.

Case 3. $\text{diam}(G) = 2m+1$.

In this case, delete those edges that are deleted to increase the diameter of G by at least 1 from level 0 of $\mu_m(G)$. This will clearly increase the diameter of $\mu_m(G)$ by at least 1 and hence $D^{+1}(\mu_m(G)) \leq D^{+1}(G)$.

Case 4. $\text{diam}(G) \geq 2(m+1)$.

Here, the shortest paths are through z. Let u^0, v^0 be a pair of diametral vertices in G and let $d(u^0) \leq d(v^0)$. Delete all the edges $u_i^m - z$, $u_i \in N(u^0)$. Then $d_{\mu_m(G)}(u^0, v^0) > 2(m+1)$. Hence $D^{+1}(\mu_m(G)) \leq d(u^0) \leq \Delta(G)$.

Theorem 5. *Let G be a connected graph with $D^{-1}(G) = 1, e = v_i^0 - v_j^0$ be an edge in G such that diam$(G+e) =$ diam$(G) - 1$ and $k = \min\{d_G(v_i^0, v_x^0), d_G(v_j^0, v_x^0)\}$, where v_x^0 is an end point of any diametral path in G.*

Then $D^{-1}(\mu_m(G)) = 1$ if and only if $m \leq \begin{cases} k + \frac{diam\,(G)-1}{2} \; if\; diam\,(G)\, is\; odd, \\ k + \frac{diam\,(G)-2}{2} \; if\; diam\,(G)\, is\; even. \end{cases}$

Proof. Let $\text{diam}(G)$ be odd and consider the edge $e = v_i^0 - v_j^0$ in G such that $\text{diam}(G+e) = \text{diam}(G) - 1$. Let $k = \min\{d_G(v_i^0, v_x^0), d_G(v_j^0, v_x^0)\}$, where v_x^0 is an end point of any diametral path in G. Let $m \leq k + \frac{\text{diam}(G)-1}{2}$. Then by adding e to $\mu_m(G)$, $d_{\mu_m(G)}(v_i^p, v_j^q) \leq \text{diam}(G) - 1$, for $k+1 \leq p, q \leq m$ by taking the path through z. For $1 \leq p, q \leq k$, the shortest path between the vertices v_i^p and v_j^q will be the path through level 0 which contains e and hence $d_{\mu_m(G)}(v_i^p, v_j^q) \leq \text{diam}(G) - 1$ in this case also. Take a pair of diametral vertices $(v_{i'}^0, v_{j'}^0)$in G. Then $d_{\mu_m(G)}(v_{i'}^0, v_{j'}^0) = \text{diam}(G) - 1$ by the path through level 0. Hence it follows that $D^{-1}(\mu_m(G)) = 1$.

Conversely suppose that $D^{-1}(\mu_m(G)) = 1$. Then clearly $\text{diam}(\mu_m(G)) = \text{diam}(G)$. If possible let $m > k + \frac{\text{diam}(G)-1}{2}$. Consider the pair$(v_{i'}^{k+1}, v_{j'}^{k+1})$ which are the dual vertices in the $k+1^{th}$ level of the diametral vertices $v_{i'}$ and $v_{j'}$ in G and let the edge e be added in $\mu_m(G)$. Then, Clearly the shortest path between these vertices is through the level 0 given by $v_{i'}^{k+1} - v_1^k - v_2^{k-1} \cdots - v_{k+1}^0 - v_{k+2}^0 - \cdots - v_{\text{diam}(G)-2(k+1)}^0 - v_{\text{diam}(G)-2k-1}^1 - \cdots - v_{\text{diam}(G)-1}^k - v_j'^{k+1}$, where $v_i^0 - v_1^0 - v_2^0 \cdots - v_{\text{diam}(G)-1}^0 - v_{j'}^0$ is the shortest path between $v_{i'}^0$ and $v_{j'}^0$ in G. Thus by the definition of k, we have $d(v_{i'}^{k+1}, v_{j'}^{k+1}) = \text{diam}(G)$ in $\mu_m(G) + e$.

Now, if we add any other edge in $\mu_m(G)$, then the distance $d(v_{i'}^0, v_{j'}^0) = \text{diam}(G)$. Thus we get a contradiction to the fact that $D^{-1}(\mu_m(G)) = 1$.

Similarly we can prove the case, when $\text{diam}(G)$ is even.

6 Diameter Minimality of the Generalized Mycielskian

A graph G is diameter minimal if $d(G-e) > d(G)$ for any $e \in G$ [3]. In this section, we have obtained a characterization for the generalized Mycielskian of a graph to be diameter minimal. Through out this section, we denote $d_{\mu_m(G)}(v_i, v_j)$ as $d(v_i, v_j)$ for the sake of convenience.

Theorem 6. *Let G be any connected graph. Then $\mu_m(G), m \geq 1$ is diameter minimal if and only if G is $K_{1,n}$.*

Proof. Let G be $K_{1,n}$, with $d(v_i) = n$. Then, $diam(\mu_m(G)) = m+1$. To prove $\mu_m(G)$ is diameter minimal, we consider the following possible cases of edge deletions in $\mu_m(G)$.

Case 1. Let an edge $v_i^0 - v_j^0$ be deleted.

First, suppose that m is even. Consider the pair of vertices (v_j^0, v_i^m) in $\mu_m(G)$. When the edge $v_j^0 - v_i^0$ is deleted, $d(v_j^0, v_i^m) = m+2$ by the path $v_j^0 - v_i^1 - v_j^2 - v_i^3 - \cdots - v_j^m - z - v_i^m$, where $d(v_j^0, v_j^m) = m$ and $d(v_j^m, v_i^m) = 2$. The distance between any other pair of vertices is not affected by the removal of this edge. When m is odd, $d(v_j^0, v_j^m) = m+2$ by the path $v_j^0 - v_i^1 - v_j^2 - v - i^3 - \cdots - v_i^m - z - v_j^m$, where $d(v_j^0, v_i^m) = m$ and $d(v_i^m, v_j^m) = 2$ and no other distance is affected by this deletion.

Case 2a. Let an edge $z - v_j^m$ be deleted.

First, take m is even. Then, $d(z, v_j^m) = 3$ by the path $z - v_x^m - v_y^{m-1} - v_j^m$, where $v_y^0 \in N(v_j^0)$ and $v_x^0 \in N(v_y^0)$. $d(v_j^m, v_j^1) = m+2$ by the path $v_j^m - v_i^{m-1} - v_j^{m-2} - v_i^{m-3} - \cdots - v_i^1 - v_j^0 - v_i^0 - v_j^1$. No other distance is affected by the removal of this edge. Next, assume that m is odd. Then, $d(z, v_j^m) = 3$ by the path $z - v_x^m - v_i^{m-1} - v_j^m$ and $d(v_j^m, v_i^1) = m+2$ by the path $v_j^m - v_i^{m-1} - v_j^{m-2} - v_i^{m-3} - \cdots - v_j^1 - v_i^0 - v_j^0 - v_i^1$. Other distances are unaffected by this deletion.

Case 2b. Let an edge $z - v_i^m$ be deleted.

Then, $d(z, v_i^m) = 2m$ by the path $v_i^m - v_j^{m-1} - v_i^{m-2} - \cdots - v_i^1 - v_j^0 - v_i^0 - v_j^1 - v_i^2 - \cdots - v_j^m - z$. The distance between any other pair of vertices is less than or equal to $2m$.

Case 3. Let an edge $v_i^k - v_j^{k+1}$ be deleted.

Then, $d(v_i^k, v_j^{k+1}) = 3$ by the path $v_i^k - v_x^{k+1} - v_i^{k+2} - v_j^{k+1}$ for $k < m-1$ and by the path $v_i^k - v_x^{k+1} - z - v_j^{k+1}$ for $k = m-1$. If m is even, consider the pair of vertices (v_j^k, v_i^{m-k}) in $\mu_m(G)$. Then $d(v_j^k, v_i^{m-k}) = m+2$ by the path $v_j^k - v_i^{k+1} - v_j^{k+2} - v - i^{k+3} - \cdots - v_i^m - z - v_j^m - v_i^{m-1} - \cdots - v_i^{m-k}$, where $d(v_j^k, z) = (m+1) - k$ and $d(z, v_j^{m-k}) = k+1$. The distance between any two other vertices is at most $m+2$. If m is odd, then $d(v_j^i, v_j^{m-k}) = m+2$ by the path $v_j^k - v_i^{k+1} - v - j^{k+2} - v_i^{k+3} - \cdots - v_j^m - z - v_i^m - v_j^{m-1} - \cdots - v_j^{m-k}$, where $d(v_j^k, z) = (m+1) - k$ and $d(z, v_j^{m-k}) = k+1$. The distance between other pairs of vertices is at most $m+2$.

Case 4. Let an edge $v_i^k - v_j^{k-1}$ be deleted.

In this case, if m is even, $d(v_j^{k-1}, v_j^{m+1-k}) = m+2$ by the path $v_j^{k-1} - v_i^{i-2} - \cdots - v_i^0 - v_j^0 - v_j^1 - v_i^2 - \cdots - v_i^{m-k} - v_j^{m+1-k}$. If m is odd, $d(v_j^{k-1}, v_i^{m+1-k}) = m+2$ by the path $v_j^{k-1} - v_i^{k-2} - \cdots - v_i^0 - v_j^0 - v_j^1 - v_i^2 \cdots - v_i^{m+1-k}$. All the other distances are at most $m+2$. Hence it follows that $\mu_m(G)$ is diameter minimal.

Conversely, assume that $\mu_m(G)$ is diameter minimal and if possible let $\delta(G) \geq 2$. Consider the following cases.

Case 1. $\text{diam}(\mu_m(G)) = \text{diam}(G)$.

Let an edge $v_i^0 - v_j^0$, be deleted. Then, for any $v_x^0, v_y^0 \in V^0$, $d(v_x^0, v_y^0) \leq diam(G)$ by the path, $v_x^0 - v_{x+1}^0 - v_{x+2}^0 - \cdots - v_i^0 - v_j^1 - \cdots - v_y^0$. The distance between any other pair of vertices remains same by the removal of this edge, since $\delta(G) \geq 2$.

Case 2: $\text{diam}(\mu_m(G)) = m+1$.

When an edge $z - v_i^m$ is deleted, $d(z, v_i^m) = 3$ by the path, $z - v_a^m - v_j^{m-1} - v_i^m$. Since, $\delta(G) \geq 2$, $d(z, v_a^{m-1}) = 2$, (see Fig. 4). Hence, $d(z, v_a^0) = m+1$, $\forall v_a^0 \in V^0$.

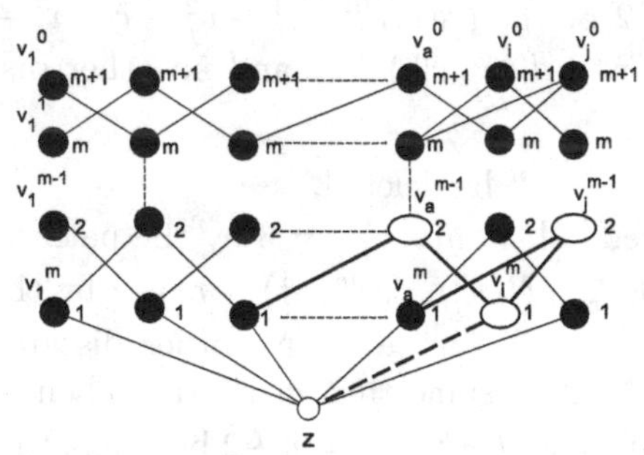

Fig. 4.

Case 3. $\text{diam}(\mu_m(G)) = 2(m+1)$.

Let an edge $v_i^0 - v_j^0$ be deleted. Then, $d(v_x^0, v_y^0) \leq 2(m+1)$ by the path, $v_x^0 - v_y^1 - \cdots - z - v_k^m - \cdots - v_y^0$. Also, the distance between any other pair is not affected by the removal of this edge, since $\delta(G) \geq 2$.

Thus the above arguments, show that $\mu_m(G)$ can not be diameter minimal. Therefore G must be a connected graph with at least one pendant edge.

Now, from the proof of Theorem 6, it is clear that deletion of an edge increases the diameter of the generalized Mycielskian if and only if it is a pendant edge and hence G must be $K_{1,n}$.

7 Some Bounds for $D^{-k}(\mu_m(G))$

In this section, let G be a connected graph and $m \geq 1$ be an integer.

Lemma 1. *If diam(G) $< m+1$, then $D^{-k}(\mu_m(G)) \leq n, 1 \leq k \leq \min\{\frac{m}{2}+1, diam(G)\}$.*

Proof. Add the edges $v_i^0 - z$ in $\mu_m(G)$. Then $d(v_i^j, z) \leq \frac{m}{2}+1$ and hence the bound.

Lemma 2. *Let d' be the diameter of G after adding $D^{-k}(G)$ edges to G and let $m+1 \leq diam(G) < 2(m+1)$. Then $D^{-k}(\mu_m(G)) \leq (m-|E|)D^{-k}(G)$ where $E = \{i|2i \leq d'\}$.*

Proof. Let $v_i^0 - v_{i'}^0, 1 \leq i \leq D^{-k}(G)$ be the edges added in G to reduce the diameter by at least k. Now, add $v_i^j - v_{i'}^j, j = 0, 1, \cdots, |E|$ in $\mu_m(G)$. This will clearly reduce the diameter of $\mu_m(G)$ by at least k.

Lemma 3. *Let $l = D^{-(2(m+1)-k)}(G)$ and diam(G) $\geq 2(m+1)$. Then $D^{-k}(\mu_m(G)) \leq l, k < 2(m+1)$.*

Proof. If $\text{diam}(G) \geq 2(m+1)$, then $\text{diam}(\mu_m(G)) = 2(m+1)$ and hence all shortest paths are the paths through z. Add those l edges which are used to get $D^{-(2(m+1)-k)}(G)$ in the 0th level of $\mu_m(G)$. Then the distance reduces by at least $2(m+1)-k$ in the level 0 and hence the diameter of $\mu_m(G)$ reduces by at least $k, k < 2(m+1)$.

Theorem 7

$$D^{-k}(\mu_m(G)) \leq \begin{cases} n & if \ \ diam(G) < m+1, \\ & \quad\ 1 \leq k \leq \min\{\frac{m}{2}+1, diam(G)\}, \\ (m-|E|)D^{-k}(G) & if \ \ m+1 \leq diam(G) < 2(m+1), \\ D^{-(2(m+1)-k)}(G) & if \ \ diam(G) \geq 2(m+1). \end{cases}$$

where $E = \{i|2i \leq d'\}, d'$ the diameter of G after adding $D^{-k}(G)$ edges to G.

Proof. The proof follows from the previous lemmas.

8 Conclusion

In this paper, we have obtained the diameter of the generalized Mycielskian of a graph. We have also considered the diameter variability problem arising from the changes of edges for the generalized Mycielskian. Moreover, we could characterize those generalized Mycielskian which are diameter minimal. The results obtained in this paper also point towards a future study of other network topological notions in the generalized Mycielskian of a graph.

References

1. Balakrishnan, R., Francis Raj, S.: Connectivity of the Mycielskian of a graph. Discrete Math. **308**, 2607–2610 (2008)
2. Bouabdallah, A., Delorme, C., Djelloul, S.: Edge deletion preserving the diameter of the hypercube. Discrete Appl. Math. **63**, 91–95 (1995)
3. Buckley, F., Harary, F.: Distance in graphs. Addison-Wesley, Redwood City (1990)
4. Chithra, M.R.: Studies on some topics in product graphs. Ph.D. thesis, Cochin University of Science and Technology, India (2013)
5. Chithra, M.R.: Changing and unchanging the diameter of Mycielski graphs. Ars Comb. (To appear)
6. Fisher, D.C., McKenna, P.A., Boyer, E.D.: Hamiltonicity, diameter, domination, packing and biclique partitions of Mycielski's graphs. Discrete Appl. Math. **84**, 93–105 (1998)
7. Francis Raj, S.: Connectivity of the generalised Mycielskian of digraphs. Graphs Comb. **29**, 893–900 (2012)
8. Graham, N., Harary, F.: Changing and unchanging the diameter of a hypercube. Discrete Appl. Math. **37**(38), 265–274 (1992)
9. Guo, L., Guo, X.: Connevtivity of the Mycielskian of a graph. Appl. Math. Lett. **22**, 1622–1625 (2009)
10. Guo, L., Liu, R., Guo, X.: Super connectivity and super edge connectivity of the Mycielskian of a graph. Graphs Comb. **28**, 143–147 (2012)
11. Lin, W., Wu, J., Lam, P.C.B., Gu, G.: Several parameters of generalized Mycielskians. Discrete Appl. Math. **154**, 1173–1182 (2006)
12. Savitha, K.S., Vijayakumar, A.: Some network topological notions of the Mycielskian of a graph. AKCE Int. J. Graphs Comb. **13**, 31–37 (2016)
13. Wang, J.J., Ho, T.Y., Ferrero, D., Sung, T.Y.: Diameter variability of cycles and tori. Inform. Sci. **178**, 2960–2967 (2008)
14. Wang, J.J., Ho, T.Y., Sung, T.Y., Ju, M.Y.: Changing the diameter in a diagonal mesh network. J. Inf. Sci. Eng. **29**, 193–208 (2013)

On Minimization of Deterministic Automaton with Rough Output

B.K. Sharma[1], S. Sharan[2(✉)], and S.P. Tiwari[3]

[1] Department of Computer Science and Engineering, Birla Institute of Technology, Ranchi 834001, India
bksharma1@gmail.com

[2] Department of Mathematics, School of Advanced Sciences VIT University, Vellore 632014, India
shambhupuremaths@gmail.com

[3] Department of Applied Mathematics, Indian Institute of Technology (Indian School of Mines), Dhanbad 826004, India
sptiwarimaths@gmail.com

Abstract. The purpose of the present work is to introduce and study the concept of minimal deterministic automaton with rough output which recognizes the given rough languages. Specifically, we use two concepts for such construction, one is based on Myhill-Nerode's theory and the other is on the basis of derivatives of the given rough language. Lastly, we discuss monoid representations of the given rough languages.

Keywords: Deterministic automaton · Rough language · Homomorphism · Monoid representation

1 Introduction and Preliminaries

Rough set theory, firstly proposed by Pawlak [8] in 1982, is a mathematical approach for the study of intelligent systems having insufficient and incomplete information. Rough set theory has vide applications in both mathematics and computer sciences e.g., artificial intelligence, cognitive sciences, particularly in the areas of knowledge acquisition, decision analysis and information systems. In the other direction, after the introduction of rough set theory by Pawlak, Kierczak [4] in 1985 introduced the concept of grammar theory based on rough set theory in terms of best lower and best upper approximation and discussed its relationship with languages. Subsequently, to reduce the gap between formal languages and natural languages, Paun et al. [7] introduced various kind of rough languages. They also investigated that the lower and upper approximations of context-free languages are regular languages. From the perspective of rough set, recently Basu [1] in 2005 introduced the notion of *rough* finite-state automaton which can recognize rough sets, considered as a generalization of nondeterministic finite-state automaton. This finite-state automaton differs from its crisp and fuzzy versions only in terms of return of transition map; specifically, in case of rough

S. Arumugam et al. (Eds.): ICTCSDM 2016, LNCS 10398, pp. 383–391, 2017.
DOI: 10.1007/978-3-319-64419-6_49

finite-state automaton, the transition map returns a rough set of states (rather than a single state or a subset of states or a fuzzy (sub)set of states). In order to enrich the algebraic study of rough finite-state automaton Tiwari and Sharan have made great effort in their series of papers (cf., [9,10,12,13]). During the recent years, the researchers paying attention on Myhill-Nerode's theory, which is a another significant branch of algebraic theory of languages and automata. In this view, there are numerous works have been done by many authors in many forms (cf., [2,5,6]), especially the recent one by Ignjatović et al. [3]. After that, Tiwari et al. ([11,14]) discussed the notions of Myhill-Nerode's type theory for the construction of minimal deterministic fuzzy automaton and minimal monoid for representation of fuzzy languages in terms of category theory.

Motivated by the work of Ignjatović et al. [3], in this paper we introduce and study the concept of minimal deterministic automaton with rough output which recognize rough language. One of such minimization is based on Myhill-Nerode's theory while another is based on derivatives of given rough languages. Lastly, we discuss monoid representation of the given rough languages.

Now, we collect some concepts associated with rough set theory, which are useful in the next sections. We start from the following concept of approximation space.

Definition 1 [8]. *An* ***approximation space*** *is a pair* (X, R)*, where* X *is a nonempty set and* R *is an equivalence relation on* X.

If R is an equivalence relation on a nonempty set X and $x \in X$, then let $[x] = \{y \in X : (x, y) \in R\}$, called an *equivalence class* or a *block* under R and $X/R = \{[x] : x \in X\}$.

Definition 2 [8]. *Given an approximation space* (X, R) *and* $A \subseteq X$*, the* ***lower approximation*** $\underline{A}$ *of* A *and the* ***upper approximation*** $\overline{A}$ *of* A *are defined as:*

$$\underline{A} = \bigcup\{[x] \in X/R : [x] \subseteq A\}, \text{and}$$

$$\overline{A} = \bigcup\{[x] \in X/R : [x] \cap A \neq \phi\}.$$

The pair $(\underline{A}, \overline{A})$ is called a **rough set**.
For an approximation space (X, R), let $A \subseteq X$, then $\underline{A}$ and $\overline{A}$ are interpreted as the collection of those elements of X that *definitely* and *possibly* belongs to A, respectively. Further, A is called **definable** (or **exact**) in (X, R) iff $\underline{A} = \overline{A}$. Equivalently, a definable set is a union of blocks under R. For any $A \subseteq X$, $\underline{A}$ and $\overline{A}$ are definable sets in (X, R).

Let X^* be the set of all 'words' on X (i.e., finite strings of elements of X), including the empty word (which we shall denoted by e). Then X^* forms a monoid under the 'concatenation' operation (with e as identity).
We close this section by introducing the following concept of rough language.

Definition 3. *Let* (X^*, R) *be an approximation space. A* ***rough language*** *in* X^* *is a map* $f : X^* \to D \times D$*, where* D *is a definable set.*

2 Minimal deterministic automaton of a rough language

In this section, we introduce and study the concept of deterministic automaton with rough output which recognizes the rough languages. Interestingly, we show that this automaton is minimal for a given rough language.
We begin by introducing the following concept of deterministic automaton with rough output.

Definition 4. *A* **deterministic automaton** *with rough output is a 6-tuples $N = (Q, R, X, \delta, q_0, \beta)$, where*

1. *Q is a finite non-empty* **set of states,**
2. *R is an* **equivalence relation** *on Q,*
3. *X is the* **set of input variables,**
4. *$\delta : Q \times X \to Q$ is a map called* ***transition map.***
5. *$q_0 \in Q$ is a fixed state called the* **initial state.**
6. *$\beta : Q \to D \times D$ is a map called* ***rough output map.***

Proposition 1. *For a given rough languages $f : X^* \to D \times D$ there exists a deterministic automaton with rough output which recognizes f.*

Proof. Define a relation R on X^* by $(u, v) \in R$ iff $\underline{f(uw)} = \underline{f(vw)}$ and $\overline{f(uw)} = \overline{f(vw)}, \forall w \in X^*$. Then R is an equivalence relation on X^*. For an equivalence relation R on X^*, let $Q_f = X^*/R = \{[u]_R : u \in X^*\}$, where $[u]_R = \{v \in X^* : (u, v) \in R\}$ denotes an equivalence class of R determined by u. It is clear that $R_f = Q/R = \{[p]_R : p \in Q\}$, where $[p]_R = \{q \in Q : (p, q) \in R\}$ is an equivalence relation on Q_f. Now, we define the maps δ_f and β_f as follows:

$$\begin{aligned} \delta_f : Q_f \times X^* \to Q_f \text{ such that } \delta_f([u]_R, v) = [uv]_R, \text{ and} \\ \beta_f : Q_f \to D \times D \text{ such that } \underline{\beta_f([u]_R)} = \underline{f(u)} \text{ and } \overline{\beta_f([u]_R)} = \overline{f(u)}. \end{aligned} \tag{1}$$

Both the maps are well-defined as shown below:
Let $u, v \in X^*$ such that $[u]_R = [v]_R$. Then

$$\begin{aligned} (u, v) \in R &\Rightarrow f(uw) = f(vw), \forall w \in X^* \\ &\Rightarrow (\underline{f(uw)}, \overline{f(uw)}) = (\underline{f(vw)}, \overline{f(vw)}) \\ &\Rightarrow \underline{f(uw)} = \underline{f(vw)} \text{ or } \overline{f(uw)} = \overline{f(vw)} \\ &\Rightarrow \underline{\beta_f([u]_R)} = \underline{f(u)} \text{ and } \overline{\beta_f([u]_R)} = \overline{f(u)}, \end{aligned}$$

whereby β_f is well-defined.
Again, let $u, v \in X^*$ such that $[u]_R = [v]_R$. Then

$$\begin{aligned} (u, v) \in R &\Rightarrow f(uw) = f(vw), \forall w \in X^* \\ &\Rightarrow (\underline{f(uw)}, \overline{f(uw)}) = (\underline{f(vw)}, \overline{f(vw)}) \\ &\Rightarrow (\underline{f(ux)}, \overline{f(ux)}) = (\underline{f(vx)}, \overline{f(vx)}), \forall x \in X \\ &\Rightarrow \underline{f(ux)} = \underline{f(vx)} \text{ or } \overline{f(ux)} = \overline{f(vx)} \\ &\Rightarrow (ux, vx) \in R \\ &\Rightarrow [ux]_R = [vx]_R \\ &\Rightarrow \delta_f([u]_R, x) = \delta_f([v]_R, x), \end{aligned}$$

whereby δ_f is well-defined. Thus $N_f = (Q_f, R_f, X, \delta_f, [e]_R, \beta_f)$ is a deterministic automaton with rough output. Also, by induction it is easy to verify that δ_f can be extended to $\delta_f^* : Q_f \times X^* \rightarrow Q_f$ such that $\delta_f^*([u]_R, v) = [uv]_R$, $\forall v \in X^*$. Finally, for all $u \in X^*$,

$$\begin{aligned} &\underline{f_{N_f}(u)} = \underline{\beta_f(\delta_f([e]_R, u))} \text{ and } \overline{f_{N_f}(u)} = \overline{\beta_f(\delta_f([e]_R, u))} \\ \Rightarrow &\underline{f_{N_f}(u)} = \underline{\beta_f([u]_R)} \text{ and } \overline{f_{N_f}(u)} = \overline{\beta_f([u]_R)} \\ \Rightarrow &\underline{f_{N_f}(u)} = \underline{f(u)} \text{ and } \overline{f_{N_f}(u)} = \overline{f(u)} \end{aligned}$$

Hence N_f recognizes f.

Definition 5. *A deterministic automaton with rough output is called* **reachable**, *if for all $q \in Q$, there exists $u \in X^*$ such that $\delta(q_0, u) = q$.*

Definition 6. *A rough language $f : X^* \rightarrow D \times D$ is said to be* **accepted** *by a reachable deterministic automaton $N = (Q, R, X, \delta, q_0, \beta)$ with rough outputs β, if for all $u \in X^*$*

1. *if $(p, q) \in R$ for $(u, v) \in R$, then $\delta(q_0, u) = p$ and $\delta(q_0, v) = q$, and*
2. *$\underline{\beta(\delta^*(q_0, u)} = \underline{f(u)}$ and $\overline{\beta(\delta^*(q_0, u)} = \overline{f(u)}$.*

Proposition 2. *The $N_f = (Q_f, R_f, X, \delta_f, [e]_R, \beta_f)$ is a reachable deterministic automaton with rough output β_f.*

Proof. For all $u \in X^*$, we have $\delta([e]_R, u) = [u]_R$. Hence N_f is reachable deterministic automaton with rough output β_f.

Proposition 3. *The reachable deterministic automaton with rough output recognizes the rough language.*

Proof. Let $N_f = (Q_f, R_f, X, \delta_f, [e]_R, \beta_f)$ be a reachable deterministic automaton with rough output β_f. Then $\forall u, v \in X^*$, if $([p], [q]) \in R_f$ for $(p, q) \in R$, $\delta_f([e]_R, u) = [p]$ and $\delta_f([e]_R, v) = [q]$. Also, for all $u \in X^*$, $\underline{\beta_f(\delta^*([q_0], u))} = \underline{f(u)}$, and $\overline{\beta_f(\delta^*([q_0], u))} = \overline{f(u)}$. Thus N_f recognizes a rough language f.

Analogous to the concept of homomorphism defined in [10], we now introduce the same between deterministic automata with rough output.

Definition 7. *Let $N = (Q, R, X, \delta, q_0, \beta)$ and $N' = (Q', R', X, \delta', q_0', \beta')$ be deterministic automata with rough output β and β'. A* **homomorphism** *from N to N' is a map $\Psi : Q \rightarrow Q'$ such that*

1. *$(p, q) \in R \Leftrightarrow (\Psi(p), \Psi(q)) \in R', \forall p, q \in Q$,*
2. *$(\underline{\Psi(\delta(p, w))}, \overline{\Psi(\delta(p, w))}) \subseteq (\underline{\delta'(\Psi(p), w)}, \overline{\delta'(\Psi(p), w)}), \forall p \in Q$ and $w \in X$, and*
3. *$\beta' \circ \Psi = \beta \Leftrightarrow (\underline{\beta'(\Psi(p))}, \overline{\beta'(\Psi(p))}) = (\underline{\beta(p)}, \overline{\beta(p)}), \quad \forall p \in Q.$*

Also, if Ψ is onto, N' is a **homomorphic** *image of N.*

Proposition 4. *For a rough language $f : X^* \to D \times D$, the deterministic automaton is a homomorphic image of any reachable deterministic automaton with rough output which recognizes f.*

Proof. Let $N = (Q, R, X, \delta, q_0, \beta)$ be a reachable deterministic automaton which recognizes rough language f. Then for all $q \in Q$ there exists $u \in X^*$ such that $\delta^*(q_0, u) = q$. We define a map $\Psi : M \to M_f$, such that

$$\Psi(p) = [u]_R, \text{ iff } \delta^*(q_0, u) = p,$$

$\forall p \in Q$ and $u \in X^*$. Then Ψ is well defined as for $p, q \in Q$ such that $p = q$, there exists $u, v \in X^*$ such that $\delta^*(q_0, u) = p, \delta^*(q_0, v) = q$. Now, $p = q \Rightarrow [p] = [q] \Rightarrow (p, q) \in R \Rightarrow (u, v) \in R \Rightarrow [u] = [v]$. Thus Ψ is well defined. Now, $p = q \Rightarrow \delta^*(q_0, u) = \delta^*(q_0, v)$. Hence for any $w \in X^*$, we have $\underline{f(uw)} = \underline{\beta(\delta^*(q_0, uw))} = \underline{\beta(\delta^*(p, w))} = \underline{\beta(\delta^*(q, w))} = \underline{\delta^*(q_0, vw)} = \underline{f(vw)}$, and $\overline{f(uw)} = \overline{\beta(\delta^*(q_0, uw))} = \overline{\beta(\delta^*(p, w))} = \overline{\beta(\delta^*(q, w))} = \overline{\delta^*(q_0, vw)} = \overline{f(vw)}$.

Now, $\forall [u] \in Q_f, u \in X^*, \exists\, p \in Q$ such that $\delta(q_0, u) = p$ and $\Psi(p) = [u]$, showing that Ψ is surjective.

Now, in order to show that Ψ is a homomorphism from N to N_f, we define a relation R_f on Q_f by $(p, q) \in R$ iff $([p], [q]) \in R_f$ iff $(\Psi(p), \Psi(q)) \in R_f$. Then $\forall w \in X^*$, $\underline{\Psi(\delta^*(p, w))} = \underline{\Psi(\delta^*(q_0, uw))}, = \underline{[uw]} = \underline{\delta_f^*([u], w)} = \underline{\delta_f^*(\Psi(p), w)}$, and $\overline{\Psi(\delta^*(p, w))} = \overline{\Psi(\delta^*(q_0, uw))} = \overline{[uw]} = \overline{\delta_f^*([u], w)} = \overline{\delta_f^*(\Psi(p), w)}$.

Finally, we need to show that $\beta_f \circ \Psi = \beta$. Therefore $\underline{\beta_f(\Psi(p))} = \underline{\beta_f([u])} = \underline{f(u)} = \underline{\beta(\delta^*(q_0, u))} = \underline{\beta(p)}$, and $\overline{\beta_f(\Psi(p))} = \overline{\beta_f([u])} = \overline{f(u)} = \overline{\beta(\delta^*(q_0, u))} = \overline{\beta(p)}$, showing that $\beta_f \circ \Psi = \beta$. Thus N_f is homomorphic image of N.

Now we define minimal deterministic automaton with rough output as follows:

Definition 8. *A deterministic automaton with rough output $N = (Q, R, X, \delta, q_0, \beta)$ is said to be a* ***minimal deterministic automaton with rough output*** *β of a rough language $f : X^* \to D \times D$, if it recognize f and $|N| \leq |N'|$, for any deterministic automaton $N' = (Q', R', X, \delta', q_0, \beta')$ with rough output β' which recognize f.*

Now, we have following result.

Proposition 5. *For a given rough language $f : X^* \to D \times D$, the deterministic automaton $N_f = (Q_f, R_f, X, \delta_f, [e], \beta_f)$ with rough output β_f is a minimal deterministic automaton.*

Proof. Let $N = (Q, R, X, \delta, q_0, \beta)$ be a deterministic automaton with rough output $\beta : Q \to D \times D$, which recognize f, and let $N' = (Q', R', X, \delta', q_0, \beta')$ be another deterministic automaton such that $Q' \subseteq Q$, where $Q' = \{p : \delta^*(q_0, u) = p, u \in X^*\}$, $\delta' = \delta/_{Q' \times X}$ and $\beta = \beta'/_{Q'}$. Now, from Proposition 4, the deterministic automaton N_f with rough output β_f is a homomorphic image of N, whereby $|N_f| \leq |N'| \leq |N|$. Hence the deterministic automaton N_f with rough output β_f of a rough language f is minimal.

3 The Derivative Automaton of a Rough Language

In this section, we introduce and study the concept of minimal derivative automaton with rough output which recognizes the rough languages.

Definition 9. *Let (X^*, R) be an approximation space. A rough language $f_u : X^* \to D \times D$ is called a* **derivative of rough language** $f : X^* \to D \times D$ *with respect to u if $f_u = (\underline{f}_u, \overline{f}_u)$ such that*

$$\underline{f_u}(v) = \underline{f(uv)} \text{ and } \overline{f_u}(v) = \overline{f(uv)}, \forall v \in X^* \text{ and} u \in X.$$

Proposition 6. *Let (X^*, R) be an approximation space and $f : X^* \to D \times D$ be a given rough language, then there exists a reachable deterministic automaton M_f which recognize f.*

Proof. Let $Q_f = \{f_u : u \in X^*\}$ be the set of **all derivatives** of a rough language f. Define the maps

$$\delta_f : Q_f \times X \to Q_f \text{ such that } \delta_f(\overline{f_u}, x) = \overline{f_{ux}}, \forall \overline{f_u} \in Q_f, x \in X.$$

and

$$\beta_f : Q_f \to D \times D \text{ such that } \underline{\beta_f(\overline{f}_u)} = \underline{\overline{f}_u(e)} \text{ and } \overline{\beta_f(\overline{f}_u)} = \overline{\overline{f}_u(e)}.$$

Also, define a relation R_f on Q_f as $(\overline{f}_u, \overline{f}_v) \in R_f$ iff $(u, v) \in R$. Then $M_f = (Q_f, R_f, X, \delta_f, \overline{f}_e, \beta_f)$ is a derivative automaton with rough output β_f. Since $\forall x \in X, \delta(\overline{f}_e, x) = \overline{f}(ex) = \overline{f}_x(e) = \overline{f}_x \in Q_f$, whereby, M_f is reachable derivative automaton with rough output β_f. Now, we have

$$\underline{\beta_f(\delta_f(\overline{f}_e, u))} = \underline{\beta_f(\overline{f}_{eu})} = \underline{\beta_f(\overline{f}_u)} = \underline{(\overline{f}_u(e))} = \underline{\overline{f}(u)},$$

and

$$\overline{\beta_f(\delta_f(\overline{f}_e, u))} = \overline{\beta_f(\overline{f}(eu)} = \overline{\beta_f(\overline{f}_u)} = \overline{(\overline{f}_u(e))} = \overline{\overline{f}(u)}.$$

Hence M_f recognize a rough language f.

Definition 10. *A* ***homomorphism*** *from a deterministic automaton $M_f = (Q_f, R_f, X, \delta_f, q_0, \beta_f)$ with rough output β_f to a derivative automaton $M'_f = (Q'_f, R'_f, X, \delta'_f, \overline{f}_e, \beta'_f)$ with rough output β'_f is a map $\Phi : M_f \to M'_f$, i.e., $\Phi : Q_f \to Q'_f$ such that*

1. $(\overline{f}_u, \overline{f}_v) \in R'_f \Leftrightarrow ((u, v) \in R \Leftrightarrow ([u], [v]) \in R_f$,
2. $(\underline{\Phi(\delta_f(\overline{f}_u, w))}, \overline{\Phi(\delta_f(\overline{f}_u, w))}) \subseteq (\underline{\delta'_f(\Phi(\overline{f}_u), w)}, \overline{\delta'_f(\Phi(\overline{f}_u), w)}), \forall \overline{f}_u \in Q_f$ *and* $w \in X$, *and*
3. $\beta'_f \circ \Phi = \beta_f \Leftrightarrow (\underline{\beta'_f(\Phi(\overline{f}_u))}, \overline{\beta'_f(\Phi(\overline{f}_u))}) = (\underline{\beta_f(\overline{f}_u)}, \overline{\beta_f(\overline{f}_u)}), \quad \forall \overline{f}_u \in Q_f$.

A bijective homomorphism Φ from a deterministic automaton M_f with rough output to a derivative automaton M'_f with rough output is called an isomorphism. If there is an **isomorphism** from M_f to M'_f. Then M_f is said to be isomorphic to M_f and is denoted by $M_f \cong M'_f$.
Now, we have following result.

Proposition 7. *For any rough language* $f_u : X^* \rightarrow D \times D$, *the derivative automaton* $M'_f = (Q'_f, R'_f, X, \overline{f_e}, \delta'_f, \beta'_f)$ *with rough output* β'_f *is a minimal deterministic automaton with rough output which recognize* f_u.

Proof. To prove M'_f is a minimal deterministic automaton with rough output β'_f, it is enough to show that M'_f is isomorphic to N_f. Let Φ be a map from $M'_f \rightarrow N_f$ i.e., $\Phi : Q'_f \rightarrow Q_f$, defined by $\Phi(\overline{f_u}) = [u]$, for all $u \in X^*$, i.e., $\delta'_f(\overline{f}_e, x) = (\overline{f}_e, x) = \overline{f}_x \in Q'_f$. Now we have $(\overline{f}_u, \overline{f}_v) \in R'_f \Rightarrow ((u,v) \in R \Rightarrow ([u],[v]) \in R_f$. Then $\forall w \in X$, $\underline{\Phi(\delta'^*_f(\overline{f}_u, w))} = \underline{\Phi(\delta'^*_f(\overline{f}_u, w))} = \underline{\Phi(\overline{f}_{uw})} = [uw] = \underline{\delta^*_f([u], w)} = \underline{\delta^*_f(\Phi(\overline{f}_u, w)}$, and $\overline{\Phi(\delta'^*_f(q, w))} = \overline{\Phi(\delta'^*_f(\overline{f}_u, w))} = \overline{\Phi(\overline{f}_{uw})} = [uw] = \overline{\delta^*_f([u], w)} = \overline{\delta^*_f(\Phi(\overline{f}_u, w))}$. Now, we need to show $\beta_f \circ \Phi = \beta'_f$. Therefore $\underline{\beta_f(\Phi(\overline{f}_u)} = \underline{\beta_f([u])} = \underline{\overline{f}_u} = \underline{\beta'_f(\delta'_f(\overline{f}_e, u))} = \underline{\beta'_f(\overline{f}_u)}$, and $\overline{\beta_f(\Phi(\overline{f}_u))} = \overline{\beta_f([u])} = \overline{\overline{f}_u} = \overline{\beta'_f(\delta'_f(\overline{f}_e, u))} = \overline{\beta'_f(\overline{f}_u)}$.

Thus Φ is a homomorphism of M'_f onto N_f. Now let $\overline{f}_u, \overline{f}_v \in Q'_f, \forall u, v \in X^*$, such that

$$\begin{aligned} \overline{f}_u = \overline{f}_v &\Leftrightarrow \overline{f}_u(e) = \overline{f}_v(e) \\ &\Leftrightarrow \overline{f}(eu) = \overline{f}(ev) \\ &\Leftrightarrow \delta'^*_f(\overline{f}_e(u)) = \delta'^*_f(\overline{f}_e(v) \\ &\Leftrightarrow [u] = [v] \\ &\Leftrightarrow \Phi(\overline{f_u}) = \Phi(\overline{f_v}), \end{aligned}$$

whereby Φ is well defined and one-one. As for each $[u] \in Q_f, \exists\ \overline{f}_u \in Q'_f$, such that $\Phi(\overline{f}_u) = [u]$, whereby Φ is onto. Hence Φ is well defined bijective homomorphism from M'_f to N_f, i.e., M'_f is isomorphic to N_f.

4 Recognition of Rough Language by Monoid

In this section, we continue our study regarding to minimization and try to develop a general theory of recognition of rough language by monoid.

We begin by introducing the following concept of monoid representation of rough language.

Definition 11. *A* ***monoid representation of rough language*** $f : X^* \rightarrow D \times D$ *is fourtuple* $N = (S, R, X^*, \Phi, \alpha)$, *where*

1. X^* *and* S *are the monoid for rough language* f, *such that*

$$\underline{f}(u) = \underline{\alpha}(\Phi(u)) \text{ and, } \overline{f}(u) = \overline{\alpha}(\Phi(u))$$

2. R is an equivalence relation on S,
3. $\Phi : X^* \to S$ is a map called **transition map,** *and*
4. $\alpha : S \to D \times D$ is a map called **rough output map.**

Definition 12. *A rough language $f : X^* \to D \times D$ is said to be* **recognized** *by a monoid S if there exist a homomorphism $\Phi : X^* \to S$ and $\alpha : S \to D \times D$ such that $\forall u \in X^*$*

$$\underline{f} = \underline{\alpha} \circ \Phi \text{ and } \overline{f} = \overline{\alpha} \circ \Phi,$$
$$\text{i.e., } \underline{f}(u) = \underline{\alpha}(\Phi(u)) \text{ and, } \overline{f}(u) = \overline{\alpha}(\Phi(u)).$$

Definition 13. *For any rough language $f : X^* \to D \times D$, we define a relation π_f on X^*, such that $\forall u, v, x, y \in X^*$,*

$$(u, v) \in \pi_f \Leftrightarrow \underline{f(xuy)} = \underline{f(xvy)} \text{ and } \overline{f(xuy)} = \overline{f(xvy)}$$

This relation π_f is called **syntactic congruence** *of rough language f. The factor monoid $N_f = (S_f, R_f, X^*, \Phi, \alpha)$, where $S_f = X^*/R = \{[u] : u \in X^*\}$, is called* **syntactic monoid** *of rough language and is denoted by $Syn(f)$.*

Definition 14. *The monoid representation $N_f = (S_f, R_f, X^*, \Phi, \alpha)$ with rough output α of a rough language f is* **minimal,** *if Φ is onto.*

Proposition 8. *For any rough language $f : X^* \to D \times D$, the syntactic monoid $Syn(f)$ is a minimal monoid of a rough language f.*

Proof. Let $Syn(f) = (S_f, R_f, X^*, \Phi, \alpha)$ be a syntactic monoid. We define a mapping $\Phi : X^* \to M$ by $\Phi(u) = [u]$, $\forall u \in X^*$. Then for each $[u] \in S_f, \exists u \in X^*$, such that $\Phi(u) = [u]$, showing that Φ is onto. Thus the monoid representation $N_f = (S_f, X^*, \Phi, \alpha)$ with rough output α of a given rough language f is minimal.

Remark 1. The syntactic monoid $Syn(f)$ of a rough language f is the smallest monoid which recognize the rough language f.

Definition 15. *The* **transition monoid** *of a derivative automaton denoted by $T(M_f)$ and define as*

$$(u, v) \in \pi_f \Leftrightarrow \delta(q, u) = \delta(q, v), \forall q \in Q_f.$$

Then π_f is equivalence relation and $T(M_f) = X^*/\pi_f$ and $T(M_f) = \{(\underline{\delta_u}, \overline{\delta_u}) : u \in X^*\}$ is a monoid with composition $\circ$ define by $\delta_u \circ \delta_v = \delta_{uv}, \forall u, v \in X^*$.

Proposition 9. *For any rough language $f : X^* \to D \times D$, the syntactic monoid $Syn(f)$ is isomorphic to the transition monoid of the derivative automaton for a rough language f.*

Proof. Let $M_f = (Q_f, R_f, X, \overline{f_e}, \delta_f, \beta_f)$ the derivative automaton with rough output β_f, and let the transition function of M_f obtained by u, is represented as $(\underline{\delta_u}, \overline{\delta_u}), \forall u \in X^*$. We know that a mapping $\Phi : X^* \to T(M_f)$, such that

$$\Phi(u) = (\underline{\delta_u}, \overline{\delta_u})$$
$$\text{i.e., } \Phi(u, v) = \delta_{uv} = \delta_u \circ \delta_v = \Phi(u) \circ \Phi(v)$$

is a homomorphism of X^* onto $T(M_f)$, thus $T(M_f) \cong X^*/\pi_f$.

On the other side, for any $u, v \in X^*$, consider any arbitrary $r \in Q_f$ then $r = (\underline{f}_u, \overline{f}_u)\ \forall u \in X^*$. Now,

$$\begin{aligned}
\delta_u = \delta_v &\Leftrightarrow \delta_u(r) = \delta_v(r), \forall r \in Q_f \\
&\Leftrightarrow r_u = r_v,\ \forall r \in Q_f \\
&\Leftrightarrow (\underline{f}_{xu}, \overline{f}_{xu}) = (\underline{f}_{xv}, \overline{f}_{xv}),\ \forall x \in X^* \\
&\Leftrightarrow (\underline{f}_{xu}(y), \overline{f}_{xu}(y)) = (\underline{f}_{xu}(y), \overline{f}_{xu}(y)),\ \forall x, y \in X^* \\
&\Leftrightarrow (\underline{f(xuy)}, \overline{f(xuy)}) = (\underline{f(xuy)}, \overline{f(xuy)}),\ \forall x, y \in X^* \\
&\Leftrightarrow (u, v) \in \pi_f
\end{aligned}$$

Therefore $T(M_f) \cong X^*/\pi_f = Syn(f)$. Hence the syntactic monoid $Syn(f)$ is isomorphic to the transition monoid of the derivative automaton for rough language f.

References

1. Basu, S.: Rough finite-state machine. Cybern. Syst. **36**, 107–124 (2005)
2. Borchardt, B.: The myhill-nerode theorem for recognizable tree series. In: Ésik, Z., Fülöp, Z. (eds.) DLT 2003. LNCS, vol. 2710, pp. 146–158. Springer, Heidelberg (2003). doi:10.1007/3-540-45007-6_11
3. Ignjatović, J., Ćiric, M., Bogdanović, S., Petković, T.: Myhill Nerode Type Theory for fuzzy languages and automata. Fuzzy Sets Syst. **161**, 1288–1324 (2010)
4. Kierczak, J.: Rough grammars. Fund. Inform. **VIII**(1), 73–81 (1985)
5. Kozen, D.: On the Myhill - Nerode theorem for trees. Bull. EATCS **47**, 170–173 (1992)
6. Maletti, A.: Myhill-Nerode theorem for recognizable tree series revisited. In: Laber, E.S., Bornstein, C., Nogueira, L.T., Faria, L. (eds.) LATIN 2008. LNCS, vol. 4957, pp. 106–120. Springer, Heidelberg (2008). doi:10.1007/978-3-540-78773-0_10
7. Paun, G., Polkowski, L., Skowron, A.: Rough set approximations of languages. Fund. Inform. **32**, 149–162 (1997)
8. Pawlak, Z.: Rough sets. Int. J. Comput. Inform. Sci. **11**, 341–356 (1982)
9. Sharan, S., Srivastava, A.K., Tiwari, S.P.: Characterizations of rough finite-state automata. Int. J. Mach. Learn. Cybern. (2015). doi:10.1007/s13042-015-0372-3
10. Tiwari, S.P., Sharan, S.: On coverings of rough transformation semigroups. In: Kuznetsov, S.O., Ślęzak, D., Hepting, D.H., Mirkin, B.G. (eds.) RSFDGrC 2011. LNCS (LNAI), vol. 6743, pp. 79–86. Springer, Heidelberg (2011). doi:10.1007/978-3-642-21881 1_14
11. Tiwari, S.P., Singh, A.K.: On minimal realization of fuzzy behaviour and associated categories. J. Appl. Math. Comput. **45**, 223–234 (2014)
12. Tiwari, S.P., Sharan, S.: Products of rough finite-state machines. J. Multiple-Valued Logic Soft Comput. **25**(4–5), 339–356 (2015)
13. Tiwari, S.P., Sharan, S., Singh, A.K.: On coverings of products of rough transformation semigroups. Int. J. Found. Comput. Sci. **24**, 375–391 (2013)
14. Tiwari, S.P., Yadav, V.K., Singh, A.K.: Constrution of a minimal realizatioin and monoid for a fuzzy languages: a categorical approach. J. Appl. Math. Comput. **47**, 401–416 (2015)

Statistical Approach to Trace the Source of Attack Based on the Variability in Data Flows

T. Subburaj, K. Suthendran(✉), and S. Arumugam

Kalasalingam University, Anand Nagar, Krishnankoil 626126, Tamilnadu, India
shubhurajo@gmail.com, {k.suthendran,s.arumugam}@klu.in

Abstract. Cyber attack is the most threatening factor of today's digital world and this is virtually doubled year by year. During February 2016, Bangladesh central bank was attacked by hackers through 35 illegal transactions in which five transactions resulted in loss of $81 Millions. However, the bank saved $850 Millions by reviewing the remaining thirty transactions. Later, Amazon.com was attacked by hackers which resulted in leakage of 80,000 login credentials. Distributed Denial of Service (DDoS) attack is one of the common forms of cyber attacks which have grown in size, become sophisticated, dangerous and also hard to detect. Tracing the source IP address of such an attack enables us to control the Internet crimes. In this work, a statistical approach based on metrics is presented to find the source of the attack. Six sigma approaches are used to set the threshold value based on which attack sources are predicted.

Keywords: DDoS · Statistical · Variation · Cyber security

1 Introduction

A Distributed Denial of Service (DDoS) attack is a malicious trial to create an internet service inaccessible by overwhelming it with traffic from multiple sources. Their target is important resources, banks, news websites etc. and presents a major challenge. The first major DDoS attack occurred at International Research Center of University of Minnesota in 1999 and affected the server of the University and was unusable for several days. Such attacks are happening more frequently across the world and it is hard to trace back since the source IP address in a packet can be spoofed when an attacker wants to hide himself from tracing [1]. The existing trace back methods by packet marking [20–23] are not suitable for identifying the source of attack in the network which does not use the router with the corresponding configuration. In this paper, a novel statistical approach is proposed to trace the source of attack by monitoring the variations in the data flow.

2 Related Works

Recent trends exhibit that Soft computing approaches are heavily used for DDoS attack detection. These methods are well suited for known attacks. However,

S. Arumugam et al. (Eds.): ICTCSDM 2016, LNCS 10398, pp. 392–400, 2017.
DOI: 10.1007/978-3-319-64419-6_50

unavailability of suitable training data set often turns out to be a major bottleneck for these methods. Knowledge based approaches are also used to identify the source of attack and employed in the victim side source network as well as intermediate network. Likewise such methods [7–11] depend on packet as well as information flow. Several security researchers have tried to handle DDoS attack using other data mining and machine learning approaches [12–19]. Conventional trace back methods like Link Testing which includes Input Debugging and Control Flooding [20], have emerged a decade ago. The packet marking era came in the picture with Node append, Node sampling, Edge sampling marking methods etc.

Based on literature survey it is observed that statistical methods can detect DDoS attack with high accuracy but it is not cost effective. Such methods are purely based on user input parameters which are essential from the system over and above user point of view. Knowledge-based methods [2,4] perform satisfactorily and the detection accuracy and real-time performance are reasonable. Data mining and machine learning methods fail to perform in real time. However it detects the DDoS attacks with high accuracy and generally these methods can be both supervised as well as unsupervised in nature. Statistical methods are useful to detect attacks from normal traffic.

Existing trace back mechanisms have crushing system overload, high calculation overhead and furthermore substantial number of false positives. In this paper we propose a new strategy of utilizing a measurable approach and our strategies moderate the calculation overhead, decrease the system overload and minimize false positives [24–26].

3 Trace Back Using Statistical Approach

Trace back using statistical approach method is superior to the Packet marking methods, since it does not require update on routing software.

3.1 Variation

Different types of data flows are used to trace on attack. The normal flow and attacking flow differentiation are known as Entropy Variation. Entropy variations are calculated based on statistical approach with metrics. The process is as follows:

- Difference between the non-attack and attack period is called Entropy Variation.
- The number of attack packets should not be less than the normal flow.
- Only one Distributed attack is continuing at a given time.
- The data flow for a certain router is same for both the attack cases and non-attack cases.

For example, Normal user sends 2 or 3 packets consecutively. After that, they wait for a while to receive the reply from receiver and then send the other packets.

But generally, the attackers send packets continuously because they won't stay for such amount of time.

Entropy is a [3] thought perceived by Shannon. It plays an important role in the detection methods of cyber attacks. It is particularly crucial in monitoring these unusual behaviors. Let $\{x_1, x_2, \ldots, x_n\}$ be the set of all goals of the packets traveling in the network. Let x be a random variable with

$$P(X = x_i) = \frac{number\ of\ packets\ going\ towards\ the\ goal\ x_i}{total\ number\ packets\ traveling\ in\ the\ network\ towards\ various\ goals}$$

$$Let\ P(X = x_i) = p_i,\ so\ that\ 0 \leq p_i \leq 1. \tag{1}$$

Discrete random variable for the entropy is defined by

$$H(X) = -\sum_{i=1}^{n} p_i * (log_2 p(x_i)) \tag{2}$$

Entropy is used to find the variations in the activity features and can be utilized to assess the irregularity of streams on a given router. Standardized entropy estimation is given by $\frac{H(x)}{log_2 n_0}$, where n_0 is the number of source hub.

The special flow monitoring algorithm is opted during the normal data flow to calculate the Mean and the standard deviation. From the calculation the threshold value is obtained. Threshold value is compared to the entropy value. If the entropy value is less than the threshold value, then the flow is identified as attack flow, otherwise it is normal flow.

3.2 Trace Back

The router stores [5,6] the upstream and downstream routers information, and destination address of the flow packets. Once an attack is detected based on the variations in the data flow, the router starts the trace back process and then it defines the data flow in descending order. The router finds the upstream router from where maximum data flow has occurred and passes this information to that router. There again the same process is carried out until the attack source is determined.

4 Simulation and Results

The environment setup is made with 3 sources, one intermediate router and one destination node. The bandwidth of the genuine traffic is set as constant and the attack is generated randomly from different sources. The packet flow variations at the router is monitored and compared with the threshold values, which leads to the decision as to whether the flow is attack flow or normal flow.

4.1 Detection Example

Consider the example given in Fig. 1, where there are three flows F1, F2 and F3 from three sources A, B and C respectively. R is the router and D is the destination.

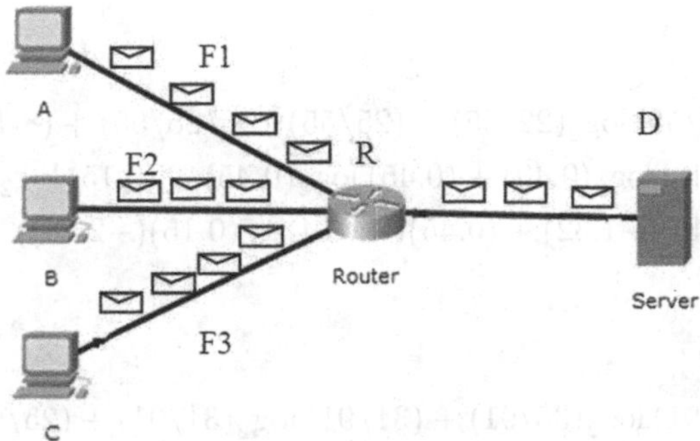

Fig. 1. Different packet flows

The attacker creates over traffic from the flows F2 and F3 on B and C after 2.0 s. The testing goes on for 2 s. We check out the number of packets got in each 0.5 s [27] and Table 1 summarizes the traced data. The Flow of information is demonstrated as follows: Flows are characterized as the number of packets originating from the source.

The details of computation of entropy are given below:

$$Entropy : H(X) = -\sum_{i=1}^{n} p(x_i) * (log_2 P(x_i)) \quad (3)$$

Table 1. Sample packet flows

Time interval	Normal flows			Entropy	Time interval	Attack flows			Entropy
	A	B	C			A	B	C	
0.0–0.5	45	26	41	1.55	2.0–2.5	26	323	144	1.14
0.5–1.0	22	25	8	1.45	2.5–3.0	23	196	420	1.09
1.0–1.5	35	31	25	1.57	3.0–3.5	15	132	140	1.24
1.5–2.0	51	32	25	1.52	3.5–4.0	6	114	145	1.12

Time: 0.0 to 0.5

$$\begin{aligned} H(X1) &= -((45/112)\log_2(45/112) + (26/112)\log_2(26/112) \\ &\quad + (41/116)\log_2(41/116)) \\ &= -((0.40)log_2(0.40) + (0.23)\log_2(0.23) + (0.37)\log_2(0.37)) \\ &= -((0.40)(-1.32) + (0.23)(-2.11) + (0.37)(-1.45)) \\ &= 1.55 \end{aligned}$$

Time: 0.5 to 1.0

$$\begin{aligned} H(X1) &= -((22/55)\log_2(22/55) + (25/55)\log_2(25/55) + (8/55)\log_2(8/55)) \\ &= -((0.40)\log_2(0.40) + (0.45)\log_2(0.45) + (0.15)\log_2(0.15)) \\ &= -((0.40)(-1.32) + (0.45)(-1.14) + (0.15)(-2.78)) \\ &= 1.45 \end{aligned}$$

Time: 1.0 to 1.5

$$\begin{aligned} H(X1) &= -((35/91)\log_2(35/91) + (31/91)\log_2(31/91) + (25/91)\log_2(25/91)) \\ &= -((0.38)\log_2(0.38) + (0.34)\log_2(0.34) + (0.27)\log_2(0.27)) \\ &= -((0.38)(-1.38) + (0.34)(-1.55) + (0.27)(-1.86)) \\ &= 1.57 \end{aligned}$$

Time: 1.5 to 2.0

$$\begin{aligned} H(X1) &= -((51/108)log_2(51/108) + (32/108)\log_2(32/108) \\ &\quad + (25/108)\log_2(25/108)) \\ &= -((0.47)\log_2(0.47) + (0.30)\log_2(0.30) + (0.23)\log_2(0.23)) \\ &= -((0.47)(-1.08) + (0.30)(-1.75) + (0.23)(-2.11)) \\ &= 1.52 \end{aligned}$$

Time: 2.0 to 2.5

$$\begin{aligned} H(X1) &= -((26/493)\log_2(26/493) + (323/493)\log_2(323/493) \\ &\quad + (144/493)\log_2(144/493)) \\ &= -((0.05)\log_2(0.05) + (0.66)\log_2(0.66) + (0.29)\log_2(0.29)) \\ &= -((0.05)(-4.25) + (0.66)(-0.61) + (0.29)(-1.78)) \\ &= 1.14 \end{aligned}$$

Time: 2.5 to 3.0

$$\begin{aligned} H(X1) &= -((23/639)\log_2(23/639) + (196/639)\log_2(323/639) \\ &\quad + (420/639)\log_2(420/639)) \\ &= -((0.04)\log_2(0.04) + (0.31)\log_2(0.31) + (0.66)\log_2(0.66)) \\ &= -((0.04)(-4.80) + (0.31)(-1.70) + (0.66)(-0.61)) \\ &= 1.09 \end{aligned}$$

Time: 3.0 to 3.5

$$\begin{aligned} H(X1) &= -(((15/287)\log_2(15/287) + (132/287)\log_2(132/287) \\ &\quad + (140/287)\log_2(140/287))) \\ &= -((0.05)\log_2(0.05) + (0.46)\log_2(0.46) + (0.49)\log_2(0.49)) \\ &= -((0.05)(-4.26) + (0.46)(-1.12) + (0.49)(-1.04)) \\ &= 1.24 \end{aligned}$$

Time: 3.5 to 4.0

$$\begin{aligned} H(X1) &= -((6/265)\log_2(6/265) + (114/265)log_2(114/265) \\ &\quad + (145/265)\log_2(145/265)) \\ &= -((0.02)\log_2(0.02) + (0.43)\log_2(0.43) + (0.55)\log_2(0.55)) \\ &= -((0.02)(-5.46) + (0.43)(-1.22) + (0.55)(-0.87)) \\ &= 1.12 \end{aligned}$$

The significant test in the discovery approach is to settle on a choice of the threshold value of entropy and entropy rate. Six-sigma approach is utilized to determine the threshold value of the entropy and entropy rate. To discover Six-sigma, figure sigma or standard deviation, duplicate by 6, and add or subtract the outcome to the ascertained mean [27]. Six-Sigma approach is ascertained utilizing the accompanying strategy:

$$\text{Average Entropy } (\beta) = \frac{1}{n}\sum_{i=1}^{n} H(X_i) \tag{4}$$

$$\begin{aligned} \text{Standarddeviation\&} &= \sigma \\ \text{Six} - \text{Sigmavalue } (\eta) &= 6\sigma \\ \text{Thresholdvalueoftheentropy\&} &= \beta \pm \eta \\ \text{AverageEntropyofthenormalflows } (\beta) &= \frac{1}{4}(1.55 + 1.45 + 1.57 + 1.52) \\ &= 1.52 \\ \text{Standard deviation } (\sigma) &= 0.053 \\ \text{Six} - \text{Sigma value } (\eta) &= 6\sigma = 0.315 \\ \text{Threshold value of the entropy } &= \beta \pm \eta \\ &= 1.52 - 0.315 \\ \textbf{Threshold value } &= \textbf{1.20} \end{aligned}$$

4.2 Flow Diagram

Figure 2 shows the entropy variations of flows at router R. The entropy value decreases significantly after the attack traffic to the router R that is, after 2.0 s.

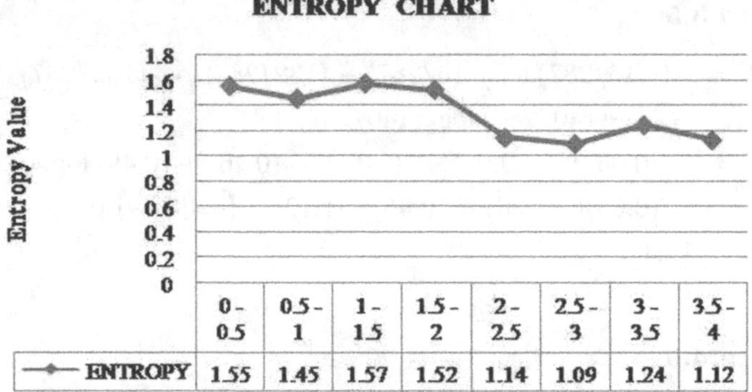

Fig. 2. Entropy variations of flows at router

From Fig. 2, it is observed that at time duration 1 to 1.5, the Entropy value is 1.57 which is greater than the threshold value 1.20, and so it is marked as normal flow. The following Entropy values are less than the threshold value and these flows are decided as attack flows. After the identification of the attack flows the source of packets is identified. This process is called trace back.

Once the attack has been detected, trace back is initiated by the router R which has 3 incoming flows F1, F2 and F3. Entropy values are used to find the maximum packets of the above flows and finally it is decided that F2 and F3 are Attack flows and the router blocks the incoming packets from F2 and F3.

5 Conclusion and Scope

Distributed denial of service (DDoS) attacks against the enterprises in the world continues as strong as ever. Based on latest data, 2016 has witnessed an increase of DDoS attacks in frequency and size, particularly as attackers are increasingly using DNS and DNSSEC to amplify attacks for greater impact against their victims using fewer botnet resources. Tracing out the source IP address of such attack enables us to control the Internet crimes. In this work, a statistical approach based on metrics was studied with suitable example to determine the source of attack. Six sigma approach is used to set the threshold value based on which attack sources are predicted. From the results, this method does not involve computational overhead on the switches. In addition, there is no stamping overhead for switches in trace back. In any case, the approach is not ready to recognize and trace back the digital attacks which are isotropic in nature. The reason is that the entropy diminishes just when one stream rules over different streams in the system. This strategy is used to identify the fundamental assaults and does not distinguish the low rate assault and high rate assaults. Trace back the source IP address in low rate and high rate attacks is still an open issue. When the network detects high rate attack, then that high rate is compared with flash crowd. If it is flash crowd then the flow is normal; otherwise it is high rate attack and after identifying its source the same can be blocked. When the network detects low rate attack, then after identifying the source the same can be blocked.

Acknowledgment. The first author is thankful to the management of Kalasalingam University for providing fellowship.

References

1. Peng, T., Leckie, C., Ramamohanarao, K.: Survey of network based defense mechanism countering the DoS and DDoS problems. Comput. J. ACM Comput. Surv. **39**, 123–128 (2007)
2. Nguyen, H.V., Choi, Y.: Proactive detection of DDoS attacks utilizing k-NN classifier in an anti DDoS framework. Int. J. Electr. Comput. Syst. Eng. **4**, 537–542 (2010)
3. Shanon, C.E.: A mathematical theory of communication. Bell Syst. Techn. J. **27**, 623–656 (1948)
4. Gavrilis, D., Dermatas, E.: Real-time detection of distributed denial-of-service attacks using RBF networks and statistical features. Comput. Netw. **48**, 235–245 (2005)
5. Wu, Y.C., Tseng, H.R., Yang, W., Jan, R.H.: DDoS detection and trace back with decision tree and grey relational analysis. Int. J. Ad-Hoc Ubiquit. Computing. **7**, 121–136 (2011)
6. Karimazad, R., Faraahi, A.: An anomaly based method for DDoS attacks detection using RBF neural networks. In: Proceedings of the International Conference on Network and Electronics Engineering, vol. 11, pp. 44–48 (2011)
7. Jeyanthi, N., Iyengar, N.C.S.N.: An entropy based approach to detect and disinuish DDoS aatacks from ash crowds in VoIP networks. Int. J. Netw. Secur. **14**, 257–269 (2012)
8. Thomas, R., Mark, B., Johnson, T., Croall, J.: NetBouncer: Client-legitimacy-based high performance filtering. In: Proceedings of the 3rd DARPA Information Survivability Conference and Exposition, p. 111. IEEE Explore (2003)
9. Limwiwatkul, L., Rungsawang, A.: Distributed denial of service detection using TCP/IP header and traffic measurement analysis. In: Proceedings of the IEEE International Symposium Communications and Information Technology, pp. 605–610. IEEE Explore (2004)
10. Zhang, G., Parashar, M.: Cooperative defense against DDoS attacks. J. Res. Pract. Inf. Technol. **38**, 69–84 (2006)
11. Wang, J., Phan, R.C.W., Whitely, J.N., Parish, D.J.: Augmented attack tree modeling of distributed denial of services and tree based attack detection method. In: Proceedings of the 10th IEEE International Conference on Computer and Information Technology, 1009–1014. IEEE Explore (2010)
12. Hwang, K., Dave, P., Tanachaiwiwat, S.: NetShield: Protocol anomaly detection with data-mining against DDoS attacks. In: Proceedings of the 6th International Symposium on Recent Advances in Intrusion Detection, pp. 1–20 (2003)
13. Li, L., Lee, G.: DDoS attack detection and wavelets. Telecommun. Syst. **28**, 435–451 (2005)
14. Sekar, V., Duffield, N., Spatscheck, O., van der Merwe, J., Zhang, H.: LADS: large-scale automated DDoS detection system. In: Proceedings of the Annual Conference on USENIX Annual Technical Conference, p. 16 (2006)

15. Gelenbe, E., Loukas, G.A.: Self-aware approach to denial of service defense. Comput. Netw. **51**, 1299–1314 (2007)
16. Lee, K., Kim, J., Kwon, K.H., Han, Y., Kim, S.: DDoS attack detection method using cluster analysis. Expert Syst. Appl. **34**, 1659–1665 (2008)
17. Li, M., Li, M.: A new approach for detecting DDoS attacks based on wavelet analysis. In: Proceedings of the 2nd International Congress on Image and Signal Processing, pp. 1–5. IEE Explore (2009)
18. Dainotti, A., Pescapé, A., Ventre, G.: A cascade architecture for DoS attacks detection based on the wavelet transform. J. Comput. Secur. **17**, 945–968 (2009)
19. Xia, Z., Lu, S., Li, J.: Enhancing DDoS flood attack detection via intelligent fuzzy logic. Informatica. **34**, 497–507 (2010)
20. Savage, S., Wetherall, D., Karlin, A.R., Anderson, T.E.: Network support for IP traceback. IEEE/ACM Trans. Netw. **9**, 226–237 (2001)
21. Lu, N., Wang, Y., Su, S., Yang, F.: A novel path-based approach for single-packet IP traceback. Secur. Commun. Netw. **7**, 309–321 (2014)
22. Song, D.X., Perrig, A.: Advanced and authenticated marking schemes for IP traceback. In: Proceedings of the IEEE INFOCOM 2001, pp. 878–886. IEE Explore (2001)
23. Seongjun, S., Lee, S., Kim, H., Kim, S.: Advanced probabilistic approach for network intrusion forecasting and detection. Expert Syst. Appl. **40**, 315–322 (2013)
24. Yu, S., Zhou, W., Doss, R., Jia, W.: Trace back of DDoS attacks using entropy variations. IEEE Trans. Parallel Distrib. Syst. **22**, 412–425 (2011)
25. Oshima, S., Takuo, N., Toshinori, S.: DDoS detection technique using statistical analysis to generate quick response time. In: Proceedings of 2010 International Conference on Broadband, Wireless Computing, Communication and Applications, pp. 672–677. IEEE explore (2010)
26. Bhandari, A., Sangal, A.L., Kumar, K.: Performance metrics for defense framework against distributed denial of service attacks. Int. J. Netw. Secur. **6**, 38–47 (2014)
27. Gupta, B.B., Misra, M., Joshi, R.C.: An ISP level solution to combat DDoS attacks using combined statistical based approach (2012). arXiv preprint arXiv: 1203.2400

Independent Resolving Number of Convex Polytopes

B. Suganya(✉) and S. Arumugam

National Centre for Advanced Research in Discrete Mathematics (n-CARDMATH), Kalasalingam University, Anand Nagar, Krishnankoil 626126, Tamil Nadu, India
suganyaptj@gmail.com, s.arumugam.klu@gmail.com

Abstract. Let $G = (V, E)$ be a connected graph. Let $W = \{w_1, w_2, \ldots, w_k\}$ be a subset of V with an order imposed on it. Then W is called a resolving set for G if for every two distinct vertices $x, y \in V(G)$, there is a vertex $w_i \in W$ such that $d(x, w_i) \neq d(y, w_i)$. The minimum cardinality of a resolving set of G is called the metric dimension of G and is denoted by $dim(G)$. A subset W is called an independent resolving set for G if W is both independent and resolving. The minimum cardinality of an independent resolving set in G is called the independent resolving number of G and is denoted by $ir(G)$. In this paper we determine the independent resolving number $ir(G)$ for three classes of convex polytopes.

Keywords: Resolving set · Metric dimension · Independent resolving set · Independent resolving number

1 Introduction

By a graph $G = (V, E)$ we mean a finite, connected and undirected graph with neither loops nor multiple edges. For graph theoretic terminology we refer to Chartrand and Lesniak [1].

The distance $d(u, v)$ between two vertices u and v in a connected graph G is the length of a shortest path between u and v. Let $W = \{w_1, w_2, \ldots, w_k\}$ be a subset of V with an order imposed on it and let $v \in V$. The representation $r(v \mid W)$ of v with respect to W is the $k-$ tuple $(d(v, w_1), d(v, w_2), \ldots, d(v, w_k))$. The set W is called a resolving set for G, if for any two distinct vertices $u, v \in V$, we have $r(u \mid W) \neq r(v \mid W)$. A resolving set of minimum cardinality is called a basis for G. The number of vertices in a basis for G is called the metric dimension of G and is denoted by $dim(G)$. Slater [11] introduced these ideas and used the term locating set and location number instead of resolving set and metric dimension. Harary and Melter [5] independently introduced these concepts and used the term metric dimension. The concept of metric dimension has applications in different areas including coin weighing problem, drug discovery, robot navigation, network discovery and verification, connected joins in graphs and strategies for mastermind game. For a survey of results in metric dimension we refer to Chartrand and Zhang [3].

Several types of resolving sets have been investigated by imposing conditions on the subgraph induced by a resolving set. Saenpholphat and Zhang [7–9]

S. Arumugam et al. (Eds.): ICTCSDM 2016, LNCS 10398, pp. 401–408, 2017.
DOI: 10.1007/978-3-319-64419-6_51

introduced the concept of connected resolvability [10]. Chitra and Arumugam [4] introduced the concept of resolving sets without isolated vertices. Imran et al. [6] considered three families of convex polytopes and proved that they have metric dimension 3.

Chartrand et al. [2] introduced the concept of independent resolving number of a graph.

Definition 1.1 [2]. *A subset $W \subseteq V$ in a connected graph G which is both a resolving set and an independent set is called an independent resolving set of G. The minimum cardinality of an independent resolving set of G is called the independent resolving number of G and is denoted by $ir(G)$.*

It has been observed in [2] that there are several families of graphs such as $K_n, n \geq 3, K_{m,n}$, where $m, n \geq 2$ and C_4 which do not have independent resolving sets. They have also determined the value of $ir(G)$ for several families of graphs.

Imran et al. [6] considered three families of convex polytopes S_n, T_n and U_n proved that they have metric dimension 3.

Definition 1.2 [6]. *The graph of convex polytope S_n consists of 2n 3-sided faces, 2n 4-sided faces and a pair of n-sided faces. We have $V(S_n) = \{a_i; b_i; c_i; d_i : 1 \leq i \leq n\}$ and $E(S_n) = \{a_ia_{i+1}; b_ib_{i+1}; c_ic_{i+1}; d_id_{i+1} : 1 \leq i \leq n\} \cup \{a_ib_{i+1}; a_ib_i; b_ic_i; c_id_i : 1 \leq i \leq n\}$ where the suffix is taken modulo n.*

The cycle induced by $\{a_i : 1 \leq i \leq n\}$ is the inner cycle, the cycle induced by $\{b_i : 1 \leq i \leq n\}$ is the interior cycle, the cycle induced by $\{c_i : 1 \leq i \leq n\}$ is the exterior cycle and the cycle induced by $\{d_i : 1 \leq i \leq n\}$ is the outer cycle.

The graph of the convex polytope S_n is given in Fig. 1.

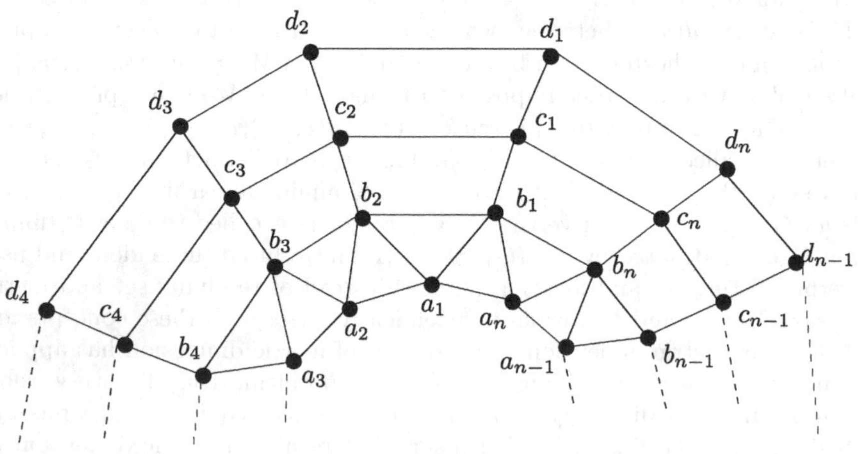

Fig. 1. The graph of convex polytope S_n

Theorem 1.3 [6]. *For $n \geq 6$, let the graph of convex polytope be S_n, then $dim(S_n) = 3$.*

Definition 1.4 [6]. *The graph of the convex polytope T_n consists of 4n 3-sided faces, n 4-sided faces and a pair of n-sided faces. We have $V(T_n) = \{a_i; b_i; c_i; d_i : 1 \leq i \leq n\}$ and $E(T_n) = \{a_ia_{i+1}; b_ib_{i+1}; c_ic_{i+1}; d_id_{i+1} : 1 \leq i \leq n\} \cup \{a_{i+1}b_i; a_ib_i; b_ic_i; c_id_i; c_{i+1}d_i : 1 \leq i \leq n\}$ where the suffix is taken modulo n.*

The cycle induced by $\{a_i : 1 \leq i \leq n\}$ is the inner cycle, cycle induced by $\{b_i : 1 \leq i \leq n\}$ is the interior cycle, cycle induced by $\{c_i : 1 \leq i \leq n\}$ is the exterior cycle, cycle induced by $\{d_i : 1 \leq i \leq n\}$ is the outer cycle.

The graph of the convex polytope T_n is given in Fig. 2.

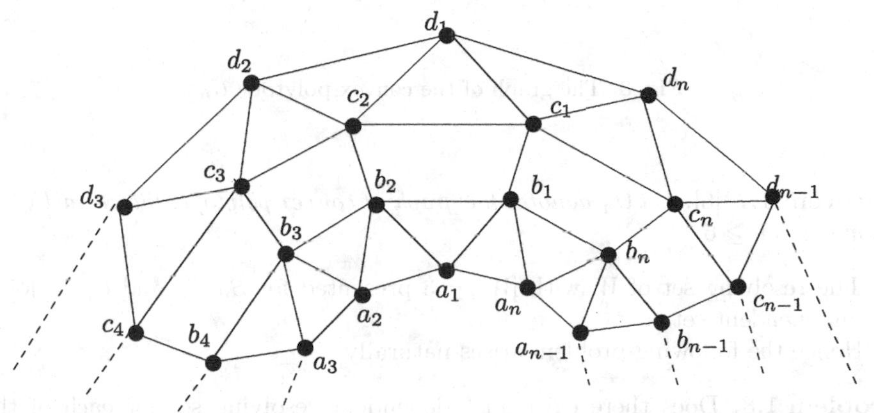

Fig. 2. The graph of convex polytope T_n

Theorem 1.5 [6]. *Let T_n denotes the graph of convex polytope, then $dim(T_n) = 3$ for every $n \geq 6$.*

Definition 1.6 [6]. *The graph of the convex polytope U_n consists of n 4-sided faces, 2n 5-sided faces and a pair of n-sided faces. We have $V(U_n) = \{a_i; b_i; c_i; d_i; e_i : 1 \leq i \leq n\}$ and $E(U_n) = \{a_ia_{i+1}; b_ib_{i+1}; e_ie_{i+1} : 1 \leq i \leq n\} \cup \{a_ib_i; b_ic_i; c_id_i; d_ie_i; c_{i+1}d_i : 1 \leq i \leq n\}$ where the suffix is taken modulo n.*

The cycle induced by $\{a_i : 1 \leq i \leq n\}$ is the inner cycle, the cycle induced by $\{b_i : 1 \leq i \leq n\}$ is the interior cycle, the cycle induced by $\{c_i : 1 \leq i \leq n\} \cup \{d_i : 1 \leq i \leq n\}$ is the exterior cycle and the cycle induced by $\{e_i : 1 \leq i \leq n\}$ is the outer cycle.

The graph of the convex polytope U_n is given in Fig. 3.

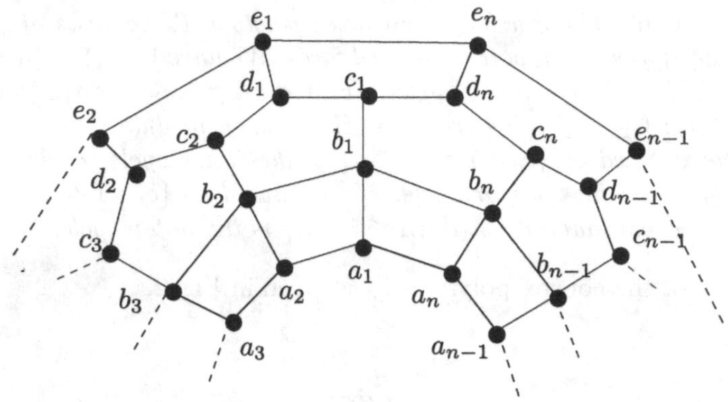

Fig. 3. The graph of the convex polytope U_n

Theorem 1.7 [6]. *Let U_n denotes the graph of convex polytope, then $dim(U_n) = 3$ for every $n \geq 6$.*

The resolving set of W with $|W| = 3$ presented for S_n, T_n and U_n in [6] is not independent set.

Hence the following problem arises naturally

Problem 1.8. Does there exist an independent resolving set for each of the convex polytopes S_n, T_n and U_n?

In this paper we investigate the above problem.

2 Main Results

Theorem 2.1. *The independent resolving number of the convex polytope S_n is 3 for all $n \geq 7$.*

Proof. Let $k = \lfloor \frac{n}{2} \rfloor$ and let $W = \{a_1, a_{k+1}, a_{n-1}\}$. We claim that W is an independent resolving set for S_n. The metric representations for the vertices of S_n are given below.

$$r(a_i \mid W) = \begin{cases} (i-1, k+1-i, i+1), & \text{if } n \text{ is odd for } 1 \leq i \leq k-1 \\ (i-1, k-i, i+1), & \text{if } n \text{ is even for } 1 \leq i \leq k-1 \\ (i-1, k+1-i, n-1-i), & \text{if } n \text{ is odd for } k \leq i \leq k+1 \\ (i-1, i-k, n-1-i), & \text{if } n \text{ is even for } k \leq i \leq k+1 \end{cases}$$

$$r(a_n \mid W) = \begin{cases} (1, k, 1), & \text{if } n \text{ is odd} \\ (1, k-1, 1), & \text{if } n \text{ is even} \end{cases}$$

$$r(b_1 \mid W) = \begin{cases} (1, k+1, 2), & \text{if } n \text{ is odd} \\ (1, k, 2), & \text{if } n \text{ is even} \end{cases}$$

$$r(b_i \mid W) = \begin{cases} (i-1, k+2-i, i+1), & \text{if } n \text{ is odd for } 2 \leq i \leq k-1 \\ (i-1, k+1-i, k+1), & \text{if } n \text{ is even for } 2 \leq i \leq k-1 \end{cases}$$

$$r(b_k \mid W) = \begin{cases} (i-1, k+1-i, k), & \text{if } n \text{ is even} \\ (i-1, k+2-i, k+1), & \text{if } n \text{ is odd} \end{cases}$$

$$r(b_{k+1} \mid W) = \begin{cases} (k, 1, k), & \text{if } n \text{ is odd} \\ (i-1, k+2-i, n-i), & \text{if } n \text{ is even} \end{cases}$$

$$r(b_i \mid W) = \begin{cases} (n+2-i, i-k-1, n-i), & \text{if } n \text{ is odd for } k+2 \leq i \leq n-1 \\ (n+2-i, i-k, n-i), & \text{if } n \text{ is even for } k+2 \leq i \leq n-1 \end{cases}$$

$$r(b_n \mid W) = (2, k, 1)$$

$$r(c_1 \mid W) = \begin{cases} (2, k+2, 3), & \text{if } n \text{ is odd} \\ (2, k+1, 3), & \text{if } n \text{ is even} \end{cases}$$

$$r(c_i \mid W) = \begin{cases} (i, k+3-i, i+2), & \text{if } n \text{ is odd for } 2 \leq i \leq k-1 \\ (i, k+2-i, i+2), & \text{if } n \text{ is even for } 2 \leq i \leq k-1 \end{cases}$$

$$r(c_k \mid W) = \begin{cases} (i, k+2-i, k+1), & \text{if } n \text{ is even} \\ (i, k+3-i, k+2), & \text{if } n \text{ is odd} \end{cases}$$

$$r(c_{k+1} \mid W) = \begin{cases} (k+1, 2, k+1), & \text{if } n \text{ is odd} \\ (i, k+3-i, n-i+1), & \text{if } n \text{ is even} \end{cases}$$

$$r(c_i \mid W) = \begin{cases} (n+3-i, i-k, n+1-i), & \text{if } n \text{ is odd for } k+2 \leq i \leq n-1 \\ (n+3-i, i-k+1, n-i+1), & \text{if } n \text{ is even for } k+2 \leq i \leq n-1 \end{cases}$$

$$r(c_n \mid W) = (3, k+1, 2)$$

$$r(d_1 \mid W) = \begin{cases} (3, k+3, 4), & \text{if } n \text{ is odd} \\ (3, k+2, 4), & \text{if } n \text{ is even} \end{cases}$$

$$r(d_i \mid W) = (i+1, k+4-i, i+3) \qquad \text{if } 2 \leq i \leq k-1$$

$$r(d_k \mid W) = \begin{cases} (i+1, k+3-i, k+2), & \text{if } n \text{ is even} \\ (i+1, k+4-i, k+3), & \text{if } n \text{ is odd} \end{cases}$$

$$r(d_{k+1} \mid W) = \begin{cases} (k+2, 3, k+2), & \text{if } n \text{ is odd} \\ (i+1, k+4-i, n-i+2), & \text{if } n \text{ is even} \end{cases}$$

$$r(d_i \mid W) = \begin{cases} (n+4-i, i-k+1, n+2-i), & \text{if } n \text{ is odd for } k+2 \leq i \leq n-1 \\ (n+4-i, i-k+2, n-i+2), & \text{if } n \text{ is even for } k+2 \leq i \leq n-1 \end{cases}$$

and $r(d_n \mid W) = (4, k+2, 3)$.

It can be easily verified that no two vertices of S_n have the same metric representation. Hence $ir(S_n) \leq 3$. Also by Theorem 1.3, $ir(S_n) \geq dim(S_n) = 3$. Thus $ir(S_n) = 3$.

Theorem 2.2. *The independent resolving number of the convex polytope T_n is 3 for all $n \geq 7$.*

Proof. Let $k = \lfloor \frac{n}{2} \rfloor$ and let $W = \{a_1, a_{k+1}, a_{n-1}\}$. We claim that W is an independent resolving set for T_n. The metric representations for the vertices of

T_n are given below.

$$r(a_i \mid W) = (i-1, k-i+1, i+1) \text{ if } 1 \le i \le k-1$$

$$r(a_k \mid W) = \begin{cases} (k-1, 1, k-1), & \text{if } n \text{ is even} \\ (k-1, 1, k), & \text{if } n \text{ is odd} \end{cases}$$

$$r(a_{k+1} \mid W) = \begin{cases} (k, 0, k-2), & \text{if } n \text{ is even} \\ (k, 0, k-1), & \text{if } n \text{ is odd} \end{cases}$$

$$r(a_i \mid W) = \begin{cases} (n+1-i, i-k-1, n-1-i), & \text{if } n \text{ is even for } k+2 \le i \le n-1 \\ (n+1-i, i-k, n-1-i), & \text{if } n \text{ is odd for } k+2 \le i \le n-1 \end{cases}$$

$$r(a_n \mid W) = \begin{cases} (1, k-1, 1), & \text{if } n \text{ is even} \\ (1, k, 1), & \text{if } n \text{ is odd} \end{cases}$$

$$r(b_1 \mid W) = \begin{cases} (1, k, 2), & \text{if } n \text{ is even} \\ (1, k+1, 2), & \text{if } n \text{ is odd} \end{cases}$$

$$r(b_i \mid W) = \begin{cases} (i-1, k-i+2, i+1), & \text{if } 2 \le i \le k-1 \\ (i-1, k-i+2, n-i), & \text{if } k \le i \le k+1 \\ (n+2-i, i-k-1, n-i), & \text{if } k+2 \le i \le n-1 \end{cases}$$

$$r(b_n \mid W) = \begin{cases} (2, k-1, 1), & \text{if } n \text{ is even} \\ (2, k, 1), & \text{if } n \text{ is odd} \end{cases}$$

$$r(c_1 \mid W) = \begin{cases} (2, k+1, 3), & \text{if } n \text{ is even} \\ (2, k+2, 3), & \text{if } n \text{ is odd} \end{cases}$$

$$r(c_i \mid W) = \begin{cases} (i, k-i+3, i+2), & \text{if } 2 \le i \le k-1 \\ (i, k-i+3, n-i+1), & \text{if } k \le i \le k+1 \\ (n+3-i, i-k, n-i+1), & \text{if } k+2 \le i \le n-1 \end{cases}$$

$$r(c_n \mid W) = \begin{cases} (3, k, 2), & \text{if } n \text{ is even} \\ (3, k+1, 2), & \text{if } n \text{ is odd} \end{cases}$$

$$r(d_i \mid W) = \begin{cases} (3, k+2, 4), & \text{if } i = 1 \\ (i+1, k-i+3, i+3), & \text{if } 2 \le i \le k-1 \\ (i+1, 3, n+1-i), & \text{if } k \le i \le k+1 \\ (n+3-i, i-k+1, n+1-i), & \text{if } k+2 \le i \le n-2 \\ (n+3-i, i-k+1, 3), & \text{if } n-1 \le i \le n \end{cases}$$

It can be easily verified that no two vertices of T_n have the same metric representation. Hence $ir(T_n) \le 3$. Also By Theorem 1.5 $ir(T_n) \ge dim(T_n) = 3$. Thus $ir(T_n) = 3$.

Theorem 2.3. *The independent resolving number of the convex polytope U_n is 3 for all $n \ge 7$.*

Proof. Let $k = \lfloor \frac{n}{2} \rfloor$ and let $W = \{a_1, a_{k+1}, a_{n-1}\}$. We claim that W is an independent resolving set for S_n. The metric representations for the vertices of U_n are given below.

$$r(a_i \mid W) = (i-1, k+1-i, i+1) \text{ if } 1 \le i \le k-1$$

$$r(a_k \mid W) = \begin{cases} (i-1, k+1-i, n-i-1), & \text{if } n \text{ is odd} \\ (k-1, 1, k-1), & \text{if } n \text{ is even} \end{cases}$$

$$r(a_{k+1} \mid W) = \begin{cases} (i-1, k+1-i, n-1-i), & \text{if } n \text{ is odd} \\ (n-i+1, i-k-1, n-i-1), & \text{if } n \text{ is even} \end{cases}$$

$$r(a_i \mid W) = \begin{cases} (n-i+1, i-k-1, n-i-1), & \text{if } k+2 \le i \le n-1 \\ (1, k, 1), & \text{if } i = n, n \text{ is odd} \\ (1, k-1, 1), & \text{if } i = n, n \text{ is even} \end{cases}$$

$$r(b_i \mid W) = (i, k+2-i, i+2) \text{ if } 1 \le i \le k-1$$

$$r(b_k \mid W) = \begin{cases} (i, k+2-i, n-i), & \text{if } n \text{ is odd} \\ (k, 2, k), & \text{if } n \text{ is even} \end{cases}$$

$$r(b_{k+1} \mid W) = \begin{cases} (i, k+2-i, n-i), & \text{if } n \text{ is odd} \\ (n-i+3, i-k+1, n-i+1), & \text{if } n \text{ is even} \end{cases}$$

$$r(b_i \mid W) = \begin{cases} (n+2-i, i-k, n-i), & \text{if } k+2 \le i \le n-1 \\ (2, k+1, 2), & \text{if } i = n, n \text{ is odd} \\ (2, k, 2), & \text{if } i = n, n \text{ is even} \end{cases}$$

$$r(c_i \mid W) = (i+1, k+3-i, i+3) \text{ if } 1 \le i \le k-1$$

$$r(c_k \mid W) = \begin{cases} (i+1, k-i+3, n+1-i), & \text{if } n \text{ is odd} \\ (k+1, 3, k+1), & \text{if } n \text{ is even} \end{cases}$$

$$r(c_{k+1} \mid W) = \begin{cases} (i+1, k+3-i, n+1-i), & \text{if } n \text{ is odd} \\ (n+4-i, i-k+2, n-i+2), & \text{if } n \text{ is even} \end{cases}$$

$$r(c_i \mid W) = \begin{cases} (n+3-i, i-k+1, n+1-i), & \text{if } k+2 \le i \le n-1 \\ (3, k+2, 3), & \text{if } i = n, n \text{ is odd} \\ (3, k+1, 3), & \text{if } i = n, n \text{ is even} \end{cases}$$

$$r(d_i \mid W) = (i+2, k+3-i, i+4) \text{ if } 1 \le i \le k-2$$

$$r(d_{k-1} \mid W) = \begin{cases} (i+2, k-i+3, i+4), & \text{if } n \text{ is odd} \\ (i+2, k-i+3, n-i+1), & \text{if } n \text{ is even} \end{cases}$$

$$r(d_k \mid W) = \begin{cases} (i+2, 3, n+1-i), & \text{if } n \text{ is odd} \\ (i+2, k-i+3, n+1-i), & \text{if } n \text{ is even} \end{cases}$$

$$r(d_{k+1} \mid W) = \begin{cases} (i+2, 3, n+1-i), & \text{if } n \text{ is odd} \\ (n-i+3, i-k+2, n-i+1), & \text{if } n \text{ is even} \end{cases}$$

$$r(d_i \mid W) = (n+3-i, i-k+2, n+1-i) \text{ if } k+2 \le i \le n-2$$

$$r(d_{n-1} \mid W) = \begin{cases} (4, k+2, 3), & \text{if } n \text{ is odd} \\ (4, k+1, 3), & \text{if } n \text{ is even} \end{cases}$$

$$r(d_n \mid W) = \begin{cases} (3, k+3, 4), & \text{if } n \text{ is odd} \\ (3, k+2, 4), & \text{if } n \text{ is even} \end{cases}$$

$$r(e_i \mid W) = (i+3, k-i+4, i+5) \text{ if } 1 \le i \le k-2$$

$$r(e_{k-1} \mid W) = \begin{cases} (i+3, k-i+4, i+5), & \text{if } n \text{ is odd} \\ (i+3, k-i+4, n-i+2), & \text{if } n \text{ is even} \end{cases}$$

$$r(e_k \mid W) = \begin{cases} (i+3, 4, n+2-i), & \text{if } n \text{ is odd} \\ (i+3, k-i+4, n-i+2), & \text{if } n \text{ is even} \end{cases}$$

$$r(e_{k+1} \mid W) = \begin{cases} (i+3, 4, n+2-i), & \text{if } n \text{ is odd} \\ (n+4-i, i-k+3, n-i+2), & \text{if } n \text{ is even} \end{cases}$$

$$r(e_i \mid W) = (n+4-i, i-k+3, n+2-i) \text{ if } k+2 \le i \le n-2$$

$$r(e_{n-1} \mid W) = \begin{cases} (5, k+3, 4), & \text{if } n \text{ is odd} \\ (5, k+2, 4), & \text{if } n \text{ is even} \end{cases}$$

$$r(e_n \mid W) = \begin{cases} (4, k+4, 5), & \text{if } n \text{ is odd} \\ (4, k+3, 5), & \text{if } n \text{ is even} \end{cases}$$

It can be easily verified that no two vertices of U_n have the same metric representation. Hence $ir(U_n) \leq 3$. Also by Theorem 1.7 $ir(U_n) \geq dim(U_n) = 3$. Thus $ir(U_n) = 3$.

2.1 Conclusion and Scope

We have proved that for three families of convex polytopes, the independent resolving number is a constant, namely 3. Characterizing the class of graphs which admit an independent resolving set is an open problem.

Acknowledgments. The first author is thankful to the management of Kalasalingam University for providing fellowship.

References

1. Chartrand, G., Lesniak, L.: Graphs and Digraphs, 5th edn. CRC Press, Boca Raton (2005)
2. Chartrand, G., Saenpholphat, V., Zhang, P.: The independent resolving number of a graph. Math. Bohemica **128**, 379–393 (2003)
3. Chartrand, G., Zhang, P.: The theory and application of resolvability in graphs: a survey. Congr. Numer. **160**, 47–68 (2003)
4. Chitra, P.J.B., Arumugam, S.: Resolving sets without isolated vertices. Procedia Comput. Sci. **74**, 38–42 (2015)
5. Harary, F., Melter, R.A.: On the metric dimension of a graph. Ars. Combin. **318**, 191–195 (1976)
6. Imran, M., Bokhary, S.A.U.H., Baig, A.Q.: On families of convex polytopes with constant metric dimension. Comput. Math. Appl. **60**, 2629–2638 (2010)
7. Saenpholphat, V., Zhang, P.: Some results on connected resolvability in graphs. Congr. Number **158**, 5–19 (2002)
8. Saenpholphat, V., Zhang, P.: On connected resolvability of graphs. Australas. J. Comb. **28**, 25–37 (2003)
9. Saenpholphat, V., Zhang, P.: Conditional resolvability in graphs: a survey. Int. J. Math. Math. Sci. **2004**, 1997–2017 (2004)
10. Saenpholphat, V., Zhang, P.: Connected resolvability of graphs. Czechoslov. Math. J. **53**, 827–840 (2003)
11. Slater, P.J.: Leaves of trees. Congr. Number **14**, 549–559 (1975)

Signed Cycle Domination in Planar Graphs

M. Sundarakannan[1(✉)] and S. Arumugam[2]

[1] Department of Mathematics, SSN College of Engineering, Kalavakkam, Chennai 603 110, Tamil Nadu, India
m.sundarakannan@gmail.com
[2] Kalasalingam University, Anand Nagar, Krishnankoil 626126, Tamil Nadu, India
s.arumugam.klu@gmail.com

Abstract. Let $G = (V, E)$ be a graph. A function $f : E \rightarrow \{-1, 1\}$ is called a signed cycle dominating function (SCDF) if $\sum_{e \in E(C)} f(e) \geq 1$ for every induced cycle C in G. The signed cycle domination number $\sigma(G)$ is defined as $\sigma(G) = \min \left\{ \sum_{e \in E} f(e) : f \text{ is an SCDF of } G \right\}$. In this paper, we prove that for any positive integer ℓ with $n - 2 \leq \ell \leq 2n - 6$, there exists a maximal planar graph G of order n such that $\sigma(G) = \ell$. We also prove that the problem of determining the signed cycle domination number is NP-complete.

Keywords: Signed cycle dominating function · Planar graph

1 Introduction

By a graph $G = (V, E)$ we mean a finite, undirected and connected graph with neither loops nor multiple edges. The order and size of G are denoted by n and m respectively. For graph theoretic terminology we refer to [2].

Let $v \in V$. The *open neighborhood* $N(v)$ and the *closed neighborhood* $N[v]$ are defined by $N(v) = \{u \in V : uv \in E\}$ and $N[v] = N(v) \cup \{v\}$. For any subset S of V, the subgraph of G induced by S is denoted by $\langle S \rangle$. A cycle C of G is an *induced cycle* if $\langle V(C) \rangle = C$. A *planar graph* G is called a maximal planar graph if G is not a spanning subgraph of another planar graph. Clearly, every face of a maximal planar graph is a triangle and $m = 3n - 6$. A graph G is *outerplanar* if it has an embedding in the plane with every vertex on the boundary of the unbounded face. An *outerplanar* graph G is called a maximal outerplanar graph if G is not a proper spanning subgraph of an outerplanar graph. Every inner face of a maximal outerplanar graph is a triangle and $m = 2n - 3$.

In [3], Xu introduced the concept of signed cycle domination of graphs and obtained the following results for maximal planar graphs and maximal outerplanar graphs.

Definition 1 [3]. Let $G = (V, E)$ be a graph. A function $f : E \longrightarrow \{-1, +1\}$ is called a *signed cycle dominating function* (SCDF) of G if $\sum_{e \in E(C)} f(e) \geq 1$

S. Arumugam et al. (Eds.): ICTCSDM 2016, LNCS 10398, pp. 409–414, 2017.
DOI: 10.1007/978-3-319-64419-6_52

holds for every induced cycle C of G. The *signed cycle domination number* of G is defined as $\sigma(G) = \min\left\{\sum_{e\in E} f(e) : f \text{ is a SCDF of } G\right\}$.

Lemma 1 [3].

(i) Let G be a maximal outerplanar graph G of order $n \geq 3$. Then $\sigma(G) \geq 1$.
(ii) Let G be a maximal planar graph G of order $n \geq 3$. Then $\sigma(G) \geq n-2$.

In [3] Xu also proposed the following conjecture.

Conjecture 1 If G is a maximal planar graph of order $n \geq 3$, then $\sigma(G) = n-2$.

Guan et al. [1] proved the following theorem.

Theorem 1 [1]. *Let G be a maximal planar graph of order $n \geq 3$. Then $\sigma(G) = n-2$ if and only if each induced cycle in G is a triangle.*

Using Theorem 1 the following counterexample to Conjecture 1 was given in [1]. The graph G given in Fig. 1. is a maximal planar graph and (x, y, z, a, x) is an induced cycle of length 4. Hence by Theorem 1 $\sigma(G) \neq n-2$.

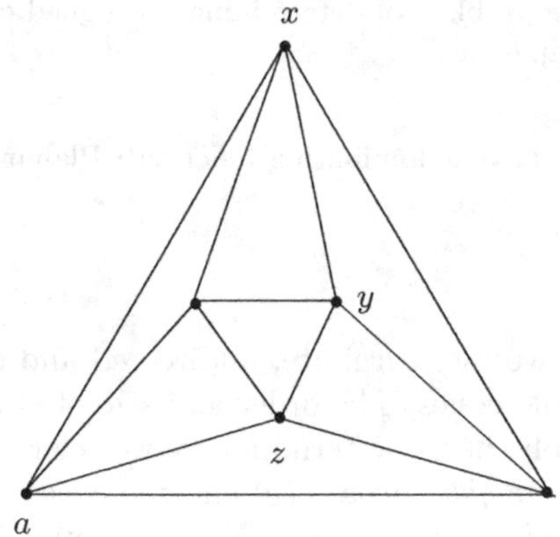

Fig. 1. Counterexample to Conjecture 1

In this paper we prove that for any two positive integers n and k with $n \geq 4$ and $n-2 \leq k \leq 2n-6$, there exists a maximal planar graph G of order n with $\sigma(G) = k$, which shows that the Conjecture 1 is not true. We also prove that the problem of determining the signed cycle domination number is NP-complete.

2 Main Results

Theorem 2. *Let G be a maximal outerplanar graph of order $n \geq 4$. Let α denote the number of vertices of degree 2. Then $\alpha - 1 \leq \sigma(G) \leq 2\alpha - 3$.*

Proof. Let $S = \{v \in V : deg\, v = 2\}$. Clearly S is an independent set and $|S| = \alpha$. Let $g : E \rightarrow \{1, -1\}$ be an SCDF of G. For each of the $n-2$ triangular faces ψ of G, we have $\sum\limits_{e \in E(\psi)} g(e) \geq 1$. Further for any $v \in S$, for at most one edge e_1 incident at v, $g(e_1) = -1$ and for the other edge e_2 incident at v, $g(e_2) = 1$. Hence $2 \sum\limits_{e \in E} g(e) \geq (n-2) + \alpha - (n - \alpha) = 2(\alpha - 1)$.

Thus $|g| \geq \alpha - 1$ for every SCDF g of G and hence $\sigma(G) \geq \alpha - 1$. Now let $H = \langle V - S \rangle$.

Define $f : E \rightarrow \{1, -1\}$ by

$$f(e) = \begin{cases} -1 & \text{if } e \text{ lies on the outerface of } H \\ 1 & \text{otherwise.} \end{cases}$$

Clearly f is an SCDF of G and $|f| = (2n-3) - 2(n-\alpha) = 2\alpha - 3$.

Hence $\sigma(G) \leq 2\alpha - 3$.

Corollary 1. *Let G be a maximal outerplanar graph G of order $n \geq 4$. Then $\sigma(G) = 1$ if and only if G has exactly two vertices of degree 2.*

Lemma 2. *Let G be a maximal planar graph of order $n \geq 4$. If G has a vertex $v \in V(G)$ with $\deg(v) = 3$, then $\sigma(G) = \sigma(G - v) + 1$.*

Proof. Let $v \in V(G)$ with $\deg(v) = 3$. Clearly $G - v$ is a maximal planar graph. Let $g : E(G-v) \longrightarrow \{+1, -1\}$ be a minimum SCDF of $G - v$, so that $\sigma(G-v) = \sum_{e \in E(G-v)} g(e)$. Since G is a maximal planar graph, $\langle N(v) \rangle = \langle \{a, b, c\} \rangle$ is a triangle and $\sigma(G) \geq \sigma(G - v) + 1$. Now, define $f : E(G) \longrightarrow \{-1, +1\}$ by

$$f(e) = \begin{cases} +1 & \text{if } e = va \text{ or } e = vb, \\ -1 & \text{if } e = vc, \\ g(e) & \text{otherwise.} \end{cases}$$

Clearly f is a SCDF of G and $\sum\limits_{e \in E} f(e) = \sigma(G-v)+1$. Hence $\sigma(G) = \sigma(G-v)+1$.

Corollary 2. *Let G be a maximal planar graph of order $n \geq 4$. If every maximal planar subgraph of G has minimum degree 3, then $\sigma(G) = n - 2$.*

Theorem 3. *Let n and k be positive integers such that $n \geq 4$ and $n - 1 \leq k \leq 2n - 6$. Then there exists a maximal planar graph of order n with $\sigma(G) = k$.*

Proof. Let $\ell = 6 + k - n$. Since $k \geq n - 1$, it follows that $\ell = 5$. Let $H = C_{\ell-2} + 2K_1$ where $C_{\ell-2} = (v_1, v_2, \ldots, v_{\ell-2}, v_1)$ and $V(2K_1) = \{a, b\}$. We claim that $\sigma(H) = 2\ell - 6$. Let $r = \left\lceil \frac{\ell-2}{2} \right\rceil$.

Define $f : E(H) \rightarrow \{1, -1\}$ by

$$f(v_i v_{i+1}) = \begin{cases} -1 & \text{if } 1 \leq i \leq r-1 \\ 1 & \text{otherwise} \end{cases}$$

and $f(av_r) = -1, f(bv_{r+1}) = -1$ and for all other edges e incident with a or b, $f(e) = 1$.

The possible induced cycles in H are $C_{\ell-2}$, triangles and cycles of length 4. Further no induced cycle of length 4 contains both the edge av_r and bv_{r+1}. Hence it follows that f is an SCDF of G and $|f| = 3\ell - 6 - (2r+2) = 2\ell - 6$. Thus $\sigma(H) \leq 2\ell - 6$.

Now, let g be any SCDF of H. If $f(e) = -1$ for two distinct edges av_i and av_j incident with a.

$$\text{Let } C = \begin{cases} (a, v_i, v_j, a) & \text{if } v_i \text{ and } v_j \text{ are adjacent} \\ (a, v_i, b, v_j, a) & \text{otherwise.} \end{cases}$$

Clearly $\sum\limits_{e \in E(C)} f(e) \leq 0$, which is a contradiction. Hence $f(e) = -1$ for at most one edge incident with a. Similarly $f(e) = -1$ for at most one edge incident with b. Further $\sum\limits_{e \in E(C_{e-2})} f(e) \geq 1$ or 2 according as $\ell - 2$ is odd or $\ell - 2$ is even. Hence $|f| \geq 2\ell - 6$, so that $\sigma(H) \geq 2\ell - 6$. Thus $\sigma(H) = 2\ell - 6$.

Since $k \leq 2n - 6$, it follows that $\ell \leq n$. If $\ell = n$, let $G_1 = H$. If $\ell < n$, add a vertex v and join v to every vertex of the outer face of H. Clearly G_1 is maximal planar and $\sigma(G_1) = \sigma(H) + 1$. Repeating this process $n - \ell$ times, we get a maximal planar graph G of order n such that $\sigma(G) = \sigma(H) + (n - \ell) = k$.

We now proceed to prove that the decision problem SIGNED CYCLE DOMINATION NUMBER is NP-complete. The reduction is from 3-SAT.

3-SAT

INSTANCE: A set $X = \{x_1, x_2, \ldots, x_r\}$ of boolean variables and a set $\mathcal{C} = \{C_1, C_2, \ldots, C_s\}$ of 3-element sets called clauses, where each clause C_i contains three distinct occurrences of either a variable x_i or its complement x'_i.

QUESTION: Does $\mathcal{C}$ have a satisfying truth assignment?

SIGNED CYCLE DOMINATION NUMBER(SCDN)

INSTANCE. A graph G and a positive integer k.

QUESTION. Does G have a signed cycle dominating function g with $\sum\limits_{e \in E} g(e) \geq k$?

Theorem 4. *SCDN is NP-complete.*

Proof. The proof is by reduction from 3-SAT. Clearly SCDN is in NP. Let H_i be the graph with its vertices labeled as in Fig. 2. There are three induced cycles $(x_{ia}, a_i, b_i, x'_{ia}, x_{ia})$, $(x_{ia}, x_{ib}, x'_{ib}, x'_{ia}, x_{ia})$ and $(x_{ib}, d_i, x'_{ib}, x_{ib})$ in H_i.

Define $f : E(H_i) \to \{-1, 1\}$ by

$$f(e) = \begin{cases} -1 & \text{if } e \in \{x_{ia}x_{ib}, a_ib_i, x_{ib}d_i\} \\ 1 & \text{otherwise.} \end{cases}$$

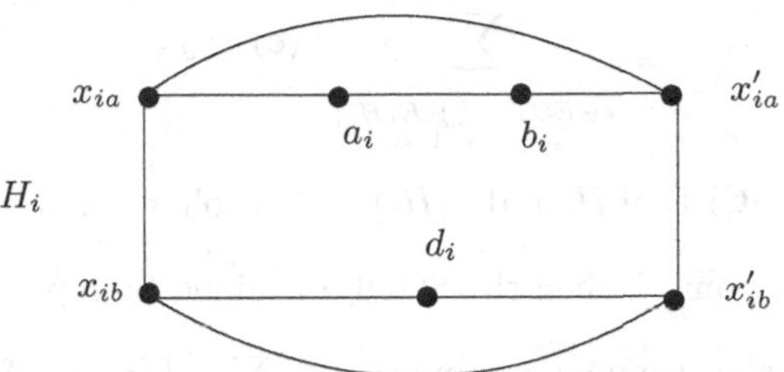

Fig. 2. Graph H with its SCDF

Clearly f is an SCDF of H_i and hence $\sigma(H_i) \leq \sum_{e \in E(H_i)} f(e) = 3$. Now any SCDF g of H_i contains at most three negative edges and hence $\sum_{e \in E(H_i)} g(e) \geq 3$. Thus $\sigma(H_i) \geq 3$ and so $\sigma(H_i) = 3$.

Now let I an instance of 3-SAT given by $X = \{x_1, x_2, \ldots, x_n\}$ and $\mathcal{C} = \{C_1, C_2, \ldots, C_j\}$. We construct an instance $g(I)$ of SCDN as follows. For each literal x_i we take a copy of H_i with its vertices labeled as in Fig. 1. For each clause $C_r = \{x_k, x_e, x_m\}$ we add an edge $c_j = v_r v'_r$ and join v_r and v'_r with $x_{kb}, x_{\ell b}$ and x_{mb}. Let G be the resulting graph and let $k = 3n + j$. Now suppose I has a satisfying truth assignment.

We define $h : E(G) \to \{1, -1\}$ as follows:

$$h(v_i v'_i) = 1 \text{ for all } i, 1 \leq i \leq j$$

$$h(x_{ia}x_{ib}) = \begin{cases} 1 & \text{if } x_i \text{ is assigned TRUE} \\ -1 & \text{otherwise} \end{cases}$$

$$h(x'_{ia}x'_{ib}) = \begin{cases} 1 & \text{if } x'_i \text{ is assigned TRUE} \\ -1 & \text{otherwise.} \end{cases}$$

If $C_i = \{x_k, x_\ell, x_m\}$, then $h(v_i x_{kb}) = h(v_i x_{\ell b}) = h(v_i x_{mb}) = 1$ and $h(v'_i x_{kb}) = h(v'_i x_{\ell b}) = h(v'_i x_{mb}) = -1$. If $C_i = \{x_k, x'_k, x_\ell\}$, then $h(v_i x_{kb}) = h(v'_i x'_{kb}) = h(v_i x_{\ell b}) = 1$ and $h(v_i x'_{kb}) = h(v'_i x_{kb}) = h(v'_i x_{\ell b}) = -1$. For all other edges e in H_i, let $h(e) = f(e)$. Clearly h is an SCDF of G and $\sum_{e \in E(H_i)} h(e) = \sigma(H_i) = 3$ for all $i, 1 \leq i \leq k$. Hence it follows that $\sum_{e \in E(G)} h(e) = 3k + j$. Hence the instance $g(I)$ of SCDN has YES answer.

Conversely suppose $g(I)$ has YES answer. Then there exists an SCDF $h : E(G) \to \{1, -1\}$ with

$$\sum_{e \in E(G)} h(e) \geq 3n + j. \tag{1}$$

For each r with $1 \leq r \leq j$, there exist seven edges in G incident with v_r, v'_r and these edges lie on three triangles. Hence $h(e) = -1$ for at most three of these seven edges, so that $\sum h(e) \geq 1$ where the summation is taken over the above seven edges. Thus

$$\sum_{e \in E(G) - \bigcup_{i=1}^{n} E(H_i)} h(e) \geq j \tag{2}$$

Since $h|v(H_i)$ is an SCFD of H_i and $\sigma(H_i) = 3$, it follows that $\sum_{e \in E(H_i)} h(e) \geq 3$. If $\sum_{e \in E(H_i)} h(e) > 3$ for some i, then the SCDF obtained by replacing $h|V(H_i)$ by f also satisfies (1). Hence we may assume that $\sum_{e \in E(H_i)} h(e) = 3$ for all $i, 1 \leq i \leq n$.

Since $(x_{ia}, x'_{ia}, x'_{ib}, x_{ib}, x_{ia})$ is an induced cycle of length 4, it follows that for one of the edges $x_{ia}x_{ib}, x'_{ia}x'_{ib}$ receive 1 under h. We now define

$$g(x_i) = \begin{cases} T & \text{if } h(x_{ia}x_{ib}) = 1 \\ F & \text{if } h(x_{ia}x_{ib}) = -1. \end{cases}$$

Then g is a satistiable truth assignment for the instance I of 3-SAT.

Acknowledgement. The first author is thankful to the management of SSN College of Engineering, Chennai for its support.

References

1. Guan, J., Liu, X., Lu, C., Miao, Z.: Three conjecture on the signed cycle domination in graphs. J. Comb. Optim. **25**, 639–645 (2013)
2. Chartrand, G., Lesniak, L.: Graphs & Digraphs. Chapman & Hall/CRC, Boca Raton (2005)
3. Xu, B.: On signed cycle domination in graphs. Discrete Math. **309**, 1007–1012 (2009)

An Optimization Algorithm for Task Scheduling in Cloud Computing Based on Multi-Purpose Cuckoo Seek Algorithm

R. Sundarrajan(✉) and V. Vasudevan

Department of Information Technology, Kalasalingam University, Anand Nagar, Krishnankoil 626126, Tamil Nadu, India
sunrsss@yahoo.co.in, vasudevan_klu@yahoo.co.in

Abstract. Cloud computing is used for providing as well as handling the services over the internet. The scheduling issue in cloud surroundings is NP-hard problem. The existing Cuckoo search provides some arranging criteria which details the major difficulties of process scheduling in the cloud. In this paper, we suggested a wizard known as Multi-Purpose Cuckoo Seek Algorithm (MPCSA) to schedule the tasks in Cloud processing. The main Purpose of MPCSA is to reduce the make span (overall completion time) period and also improve the source usage. Sources are assigned effectively in the cloud processing. We conduct experiments with Cloud sim (with surpass) to evaluate MPCSA. Simulation results show that our MPCSA algorithm can work efficiently compared with PSO and CSA.

Keywords: Cloud computing · Cuckoo search · Multipurpose cuckoo seek · NP-hard · Scheduling

1 Introduction

Cloud computing is internet based computing. If we put the data in the cloud we can access the data from anywhere, at anytime, and from any device through internet. Usually, Cloud Computing services will be tagged into 3 varieties:

1. PaaS-provider offers flat assisting floor for constructing applications
2. SaaS-provider provides licenses, packages to customers as a provider
3. IaaS-company gives uncooked calculating, garage, and network.

Cloud computing provides implementation ability, locality freelance, reserve combining, large net entrance irresponsibleness and measurability, snap and easy maintenance. Work could be a (computer-based) activity created by (more than 2, however, not plenty of) tasks that would need completely different processing skills and will have different resource necessities and restrictions, sometimes expressed among the task description. Every job could have varied parameters like needed information, desired completion time, typically referred to as the

S. Arumugam et al. (Eds.): ICTCSDM 2016, LNCS 10398, pp. 415–424, 2017.
DOI: 10.1007/978-3-319-64419-6_53

point in time, expected execution time, job priority etc. Job is taken into account as a private planning algorithmic program, which is typically associated in nursing abstract model that doesn't specify the physical location of the tasks where it were completed (executed) [2,4]. The population improvement algorithms such as Genetic algorithms (GA), Particle Swarm Optimization (PSO), microorganism search (searching) Optimization (MSO) and Bat algorithmic program (BA) use a population of individuals to unravel the issue [1] (Fig. 1).

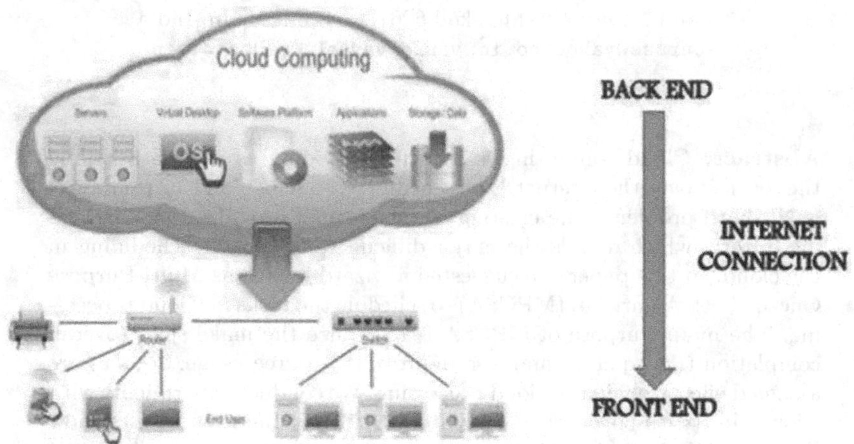

Fig. 1. Cloud architecture

2 Task Scheduling

Process arrangement performs a key role to improve the versatility and stability of systems. The primary purpose behind arranging the projects to the sources depending on the time bound includes finding out a complete and best schedule of various projects which can be implemented to give the best and acceptable result to the user [5]. While handling the source in any form, such as firewall program, network is always dynamically allocated according to the series and the requirements of the tasks and subtasks. This leads to the reasoning for the power problem that indicates no earlier described series may be useful during handling of task. The disadvantages of the existing method are such as the flow of task is unclear, performance routes are also unclear and at the same time sources available are also unclear because there are a numerous number of projects discussing them at the same time [2]. To overcome these disadvantages, MPCSA algorithm is proposed.

In this method, the arrangement of projects in reasoning indicates choose the best suitable resource available for performance of projects or to spend PC devices to projects in such a manner that the finalization time is reduced as much as possible. In arranging methods, the list of jobs is created based on various factors and the jobs are then selected according to their main concerns and are allocated to available processor chips and PC devices which satisfy a predetermined objective [2].

3 Algorithm Description

3.1 Cuckoo Search Algorithm

The Current suggested criteria known as Cuckoo Search Algorithm (CSA) routine the projects within Cloud computing. Each cuckoo gives single egg at an instance, and places it within a home that is arbitrarily selected, as well as the variety is able to find out an unfamiliar egg through a Pa [0, 1] possibility. The rate as well as protection of the criteria becomes extremely great if the significance of Pa is lower. Cuckoo Search (CS) criteria have been used to get over the regional optima issue by amalgamating with the PSO criteria. CSA criteria are in accordance with obliging family parasites actions of several cuckoo varieties along by the Levy flight an action of a few birds as well as fruits go [3,4].

Disadvantage of this algorithm:

- Randomly selects the Particular Single Object and Process On it.
- Not Suitable for Multiple Objective Optimization.
- It will be more complex when executing a large number of task optimization in cloud platform (Fig. 2).

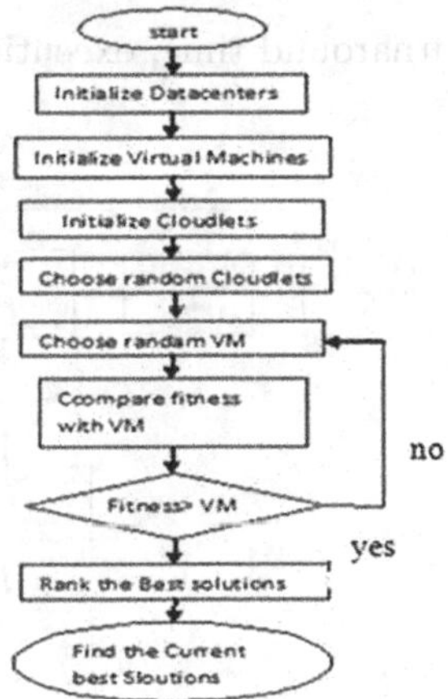

Fig. 2. Cuckoo Search Algorithm

3.2 Purpose Cuckoo Seek Algorithm: Proposed Work

The system suggested new transformative requirements which are known as several purpose cuckoos, which look for requirements to schedule the projects in Cloud Computing. A Multi-Purpose Cuckoo Search (MPCS) method has been developed to deal with multi-criteria marketing problems. This approach uses unique loads to merge several goals to a single purpose. As the loads differ arbitrarily, Pareto methodologies can be found and the points can be allocated differently over the methodologies (Fig. 3).

Advantage of this algorithm:

- Appropriate completion of the tasks into cloud.
- Resources utilization will be proper.
- Failure of the system is less.
- Load is minimized on different processors.
- Utilizing of resources from multiple centres.
- More Suitable when executing a large number of task optimization in cloud platform.

Algorithm:
Input: population of n host nest.
Output: best solution (nest with quality solution).
Step 1: Initialization.
Step 2: Initialize a populace of n host nests xi (I =1, 2, n);
Step 3: Calculate the Individual fitness value for each task;
Step 4: Find the total fitness value for all task;
Step 5: Evaluate the distance from Maximum Fitness value;
Rank the best solution.
Step 6: Calculate the total turnaround time, execution time, speed.

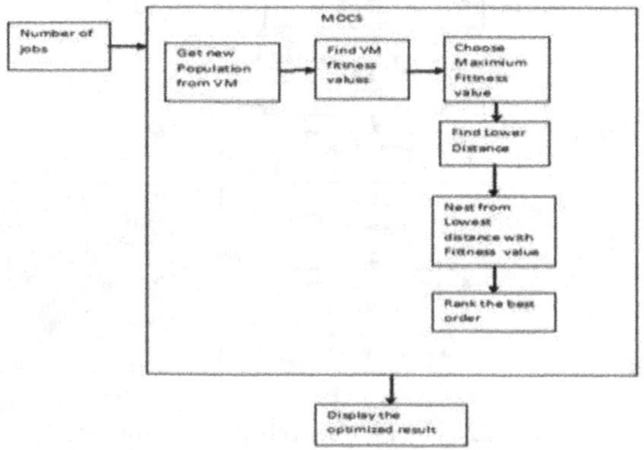

Fig. 3. Multi-purpose cuckoo seek algorithm

4 Working Methodology

As mentioned in the past area inhabitants, variety nests, fitness and place of the nodes, finding possibility, and impose flight tickets must be described for any MPCS. The first step in any MPCS is interpreting inhabitants and information reflection. The inhabitants are initialized by a vector, in which the duration of vector indicates the number of sources.

The successive actions of a cuckoo basically follow a power-law step-length submission with a larger end type of a unique move procedure. For convenience, we assume that every egg in a home symbolizes the remedy; whereas the cuckoo egg also symbolizes a new way. The objective is to utilize the latest way and possibly improved alternatives (cuckoos) towards substituting within the nests. It is not so excellent remedy. So, here we use the easiest strategy where there will be only one egg at each nest. Please always cancel any superfluous definitions that are not actually used in your text. If you do not, these may conflict with the definitions of the macro package, causing changes in the structure of the text and leading to numerous mistakes in the proofs.

Fitness value. Health and fitness value reveals the solution how it fits, i.e. whether it will be adjust in the direction of environment. For maximizing the issue, the fitness value of a remedy can be proportionate to the values of a purpose operate. For convenience, we assume that each and every egg in a home symbolizes a remedy, and also the cuckoo egg symbolizes latest remedy.

$$S = \sum M_n \tag{1}$$

$$I = \sum \frac{VM_{1\ldots n}}{M_n} \tag{2}$$

$$D = M_n - I(MIPS) \tag{3}$$

where, S = Total fitness value

M_n = Sum of VM RAM

I = Individual Fitness value

$VM_{1\ldots.n}$ = Individual Fitness value

$I(MIPS)1\ldots n$ = Individual (million instruction per second)

D = distance.

4.1 Choose the Virtual Machine and Compare Fitness

The task, while processing is a well-known NP-hard problem. The complex and challenging issue will be still more, where the fertilized groups are use to perform a huge quantity of tasks in the reasoning processing system. Here, many heuristics have been suggested, from the lower stage performance of projects in the numerous processor chips toward the advanced stage performance of projects in cloud computing.

4.2 Ranking Best Solution and Find Current

MPCSA is based on the oblige family parasitic actions on some cuckoo varieties along by the Impose journey actions of some parrots and fruit goes. Ecological features and the migrants of cultures of cuckoos hopefully lead them to meet and find the best environment for duplication. For the next creation levy journey operate is conducted and decided new home. Then fitness operate is conducted and the best home is selected.

5 Data Analysis

5.1 Experimental Evaluation

In this phase, we present consequences to assess the performance of the proposed technique with the use of cloudsim (Table 1).

We mention the Comparison of CS and MPCSA. From that, execution time of MPCS algorithm is less than the CSA (Tables 2, 3 and 4, Figs. 4 and 5).

Table 1. Comparison chart based on execution time

Number of virtual machine	Number of cloud lets	C.S.A execution time (m.s)	M.P.C.S.A execution time (m.s)
20	25	959	633
25	100	2959	2763
30	150	4499	4167

Table 2. Comparsion chart based on make span

Number of jobs	P.S.O	C.S.A	M.P.C.S.A
50	28	25	20
100	61	57	47
150	98	82	76

Table 3. Comparison chart based on speed

Number of virtual machine	Number of cloud lets	C.S.A speed (m.s)	M.P.C.S.A execution speed
20	25	511	717
25	100	277	477
30	150	130	330

Table 4. Comparsion chart based on turnaround time

Number of jobs	P.S.O	C.S.O	C.S.A	M.P.C.S.A
50	567	38	30	20
100	112	79	61	42
150	160	100	90	65

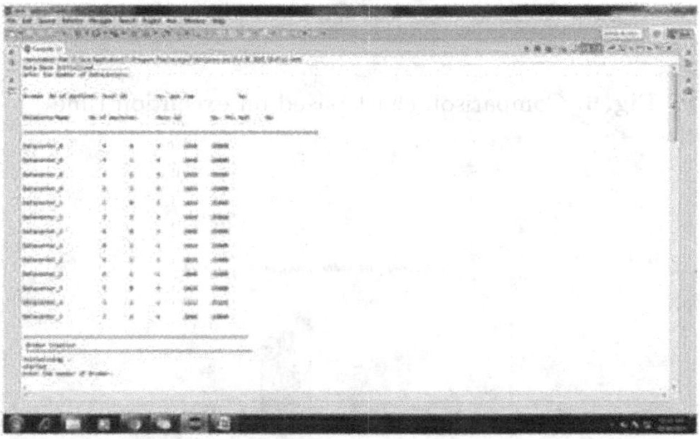

Fig. 4. Initialize numbers of data centers, virtual machines and cloudlets

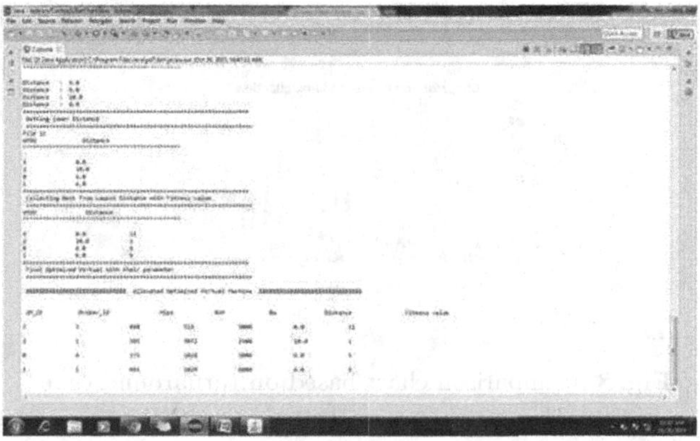

Fig. 5. Optimized results

5.2 Comparison Chart

From the above chart, we mention the Comparison of PSO, CS and MPCS. From that, make span of MPCS algorithm is lesser than the PSO, CSA (Figs. 6, 7, 8 and 9).

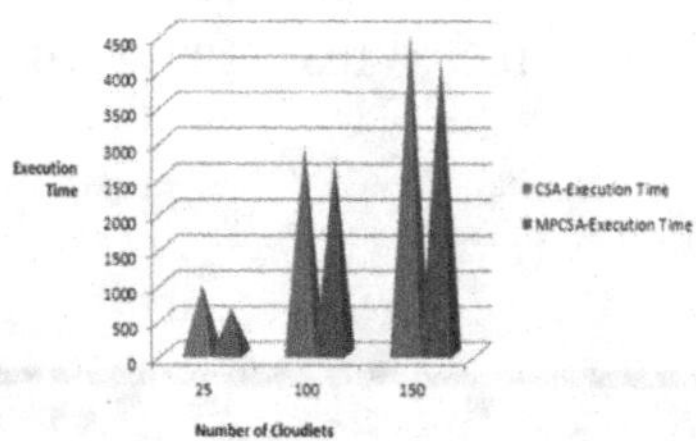

Fig. 6. Comparison chart based on execution time

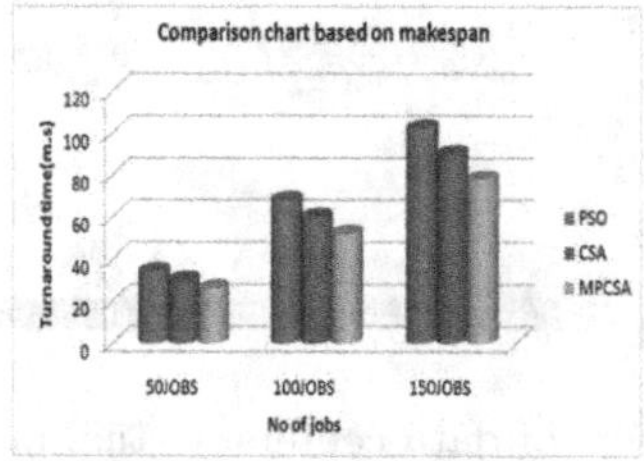

Fig. 7. Comparison chart based on makespan

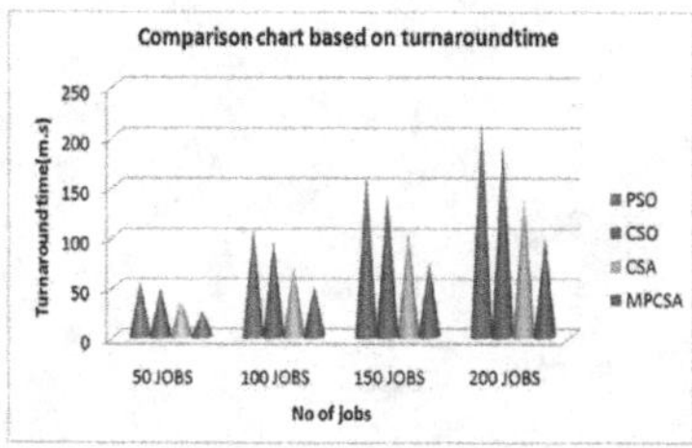

Fig. 8. Comparison chart based on turnaround time

While comparing from the above chart, the Speed of the Multi Objective cuckoo search algorithm will be very high. So the Coverage also will be very higher to the cuckoo search algorithm.

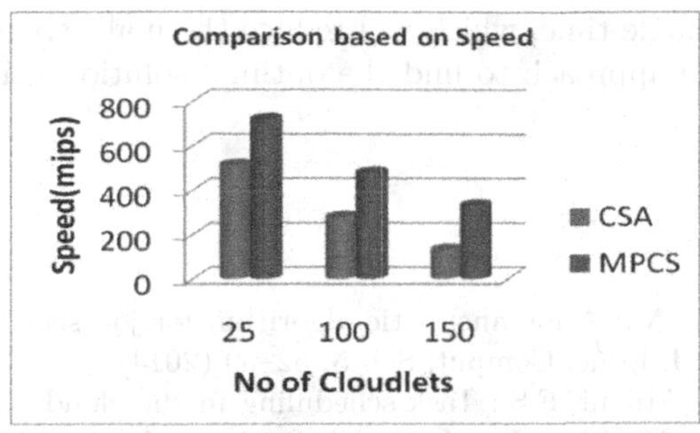

Fig. 9. Comparison chart based on speed

6 Explanation of Result

In this paper, we tend towards the tasks algorithmic rule to resolve task scheduling drawback in Cloud computing. Our main objective of this project is to reduce the makespan. Task arranging in reasoning processing is a very complicated problem. In order to effectively and price efficiently schedule the projects and data of programs onto these cloud computing surroundings, program schedulers have different guidelines that differ according to the purpose, function to reduce complete performance time, to balance the fill on resources used and so forth. Using the cuckoo search criteria for finding the maximum mixture is a very simple approach. This algorithm is the superior resolution for scheduling in Cloud sim console. As compared to the cuckoo search algorithm the speed and coverage of this algorithm is very high. Here in this algorithm we use the cuckoo bird behaviour with a multi objective job at a single time, which will reduce the makespan. In this scheduling it is very simple approach to find the optimal solution using multipurpose cuckoo search algorithms.

7 Future Work

By using the hybrid concept we can reduce the execution time and turnaround time. Resource Utilization can be improved. For example, we can combine some of the algorithms that are PSO, CAT, BAT, FIREFLIES. We are going to use the Fireflies algorithm with MPCS (Multi-purpose Cuckoo Seek) to avoid the wastage of resource and we can also allocate large numbers of flies.

8 Conclusion

We proposed a new evolutionary algorithm named MPCS algorithm which will reduce the total execution time, and minimize the turnaround time. As compared to the existing algorithm, the speed and coverage of this algorithm become very high. Here in this algorithm we use the cuckoo bird behaviour with a multi objective job at a single time, which will reduce the make span. In this scheduling it is very simple approach to find the optimal solution using cuckoo search algorithms.

References

1. Sajedi, H., Rabiee, M.: A metaheuristic algorithm for job scheduling in grid computing. Int. J. Mod. Educ. Comput. Sci. **5**, 52–59 (2014)
2. Navimipour, N.J., Milani, F.S.: Task scheduling in the cloud computing based on the cuckoo search algorithm. Int. J. Model. Optim. **5**(1), 44–47 (2015)
3. Prakash, M., Saranya, R., Rukmani Jothi, K., Vigneshwaran, A.: An optimal job scheduling in grid using cuckoo algorithm. Int. J. Comput. Sci. Telecommun. **3**(2), 65–69 (2012)
4. Kamat, S., Karegowda, A.G.: A brief survey of cuckoo search application. Int. J. Innov. Comput. Commun. Eng. **2**(2), 7–14 (2014)
5. Xue, S., Li, M., Xu, X., Chen, J.: An ant colony optimization-load balancing algorithm for task scheduling in cloud computing. J. Softw. **9**(2), 466–473 (2014)

P_3-Factorization of Triangulated Cartesian Product of Complete Graph of Odd Order

A. Tamil Elakkiya[1(✉)] and A. Muthusamy[2]

[1] Department of Humanities and Sciences,
Sri Venkateshwara College of Engineering and Technology,
Chittoor, Andra Pradesh, India
elakki.1@gmail.com

[2] Department of Mathematics, Periyar University, Salem, Tamil Nadu, India
ambdu@yahoo.com

Abstract. Partition of G into edge-disjoint H-factors is called an *H-factorization* of G. In this paper, we show that the necessary conditions $mn \equiv 0 (mod\ 3)$ and $3(mn+m+n-3) \equiv 0 (mod\ 8)$ for the existence of a P_3-factorization of $K_m \boxslash K_n$ are sufficient, where m and n are odd and $\boxslash$ denotes triangulated Cartesian product of graphs.

Keywords: Triangulated cartesian product · Cartesian product · Path factorization

1 Introduction

A *latin square* of order n is an $n \times n$ array, such that each row and each column of the array contains each of the symbols from $\{1, 2, \ldots, n\}$ exactly once. Two latin squares L_1 and L_2 of order n are said to be *orthogonal* if for each $(x, y) \in \{1, 2, \ldots, n\} \times \{1, 2, \ldots, n\}$ there is exactly one cell (i, j), in which L_1 contains the symbol x and L_2 contains the symbol y. In other words, if L_1 and L_2 are superimposed, the resulting set of n^2 ordered pairs are distinct. The latin squares $L_1, L_2, \ldots, L_t$ of order n are said to be *mutually orthogonal latin squares* ($MOLS(n)$) if for $1 \le a \ne b \le t$, L_a and L_b are orthogonal. $N(n)$ denotes the maximum number of $MOLS(n)$.

Partition of G into subgraphs $G_1, G_2, \ldots, G_r$ such that $E(G_i) \cap E(G_j) = \emptyset$ for $i \ne j \in \{1, 2, \ldots, r\}$ and $E(G) = \cup_{i=1}^{r} E(G_i)$ is called a *decomposition* of G; In this case we write G as $G = G_1 \oplus G_2 \oplus \ldots \oplus G_r$, where $\oplus$ denotes an edge-disjoint sum of subgraphs. In particular, if each G_i, $1 \le i \le r$, is isomorphic to some graph H then it is called an *H-decomposition* of G and is denoted as $H|G$. Let P_k, C_k, K_k and I_k respectively denote a path, cycle, complete graph and complement of complete graph (or independent set) on k vertices. A k-regular spanning subgraph of G is called a *k-factor* of G. A spanning subgraph of G is called a *P_3-factor* of G if each component of it is a P_3. We say that G has a *P_3-factorization* if G can be partitioned into edge-disjoint P_3-factors; existence of such path factorization is denoted by $P_3||G$. A *C_k-factor* of G is a 2-factor in which each component is a C_k. Decomposition of G into C_k-factors is called a *C_k-factorization* of G. A cycle containing all the vertices of G is called a *Hamilton cycle*.

S. Arumugam et al. (Eds.): ICTCSDM 2016, LNCS 10398, pp. 425–434, 2017.
DOI: 10.1007/978-3-319-64419-6_54

We say that G has a *Hamilton cycle decomposition* if it can be partitioned into edge-disjoint Hamilton cycles. For an integer λ, λG and $G(\lambda)$ respectively denote λ copies of G and a multigraph G with uniform edge-multiplicity λ. The *Cartesian product* $G \Box H$, the *wreath product* $G \otimes H$ and the *strong product* $G \boxtimes H$ of two graphs G and H are defined as follows: $V(G \Box H) = V(G \otimes H) = V(G \boxtimes H) = \{(u, v) \mid u \in V(G), v \in V(H)\}$. $E(G \Box H) = \{(u, v)(x, y) \mid u = x \text{ and } vy \in E(H) \text{ or } v = y \text{ and } ux \in E(G)\}$, $E(G \otimes H) = \{(u, v)(x, y) \mid u = x \text{ and } vy \in E(H) \text{ or } ux \in E(G)\}$ and $E(G \boxtimes H) = \{(u, v)(x, y) \mid u = x$ and $vy \in E(H)$ or $v = y$ and $ux \in E(G)$ or $ux \in E(G)$ and $vy \in E(H)\}$. It is well known that the Cartesian product is commutative. A graph G having partite sets $V_1, V_2, \ldots, V_m$ with $|V_i| = n$, $1 \leq i \leq m$ and $E(G)=\{uv \mid u \in V_i \text{ and } v \in V_j, \forall i \neq j\}$ is called *complete m-partite graph* and is denoted by $K_{m;n}$. Note that $K_{m;n}$ is isomorphic to $K_m \otimes I_n$.

Let $\vec{G} = (V, A)$ denote a digraph with vertex set V and arc set A. An arc of $\vec{G}$ with head at v and tail at u is denoted as $\vec{uv}$. The triangulated Cartesian Product of two digraphs $\vec{G}$ and $\vec{H}$, denoted as $\vec{G} \boxbslash \vec{H}$, is defined as follows: $V(\vec{G} \boxbslash \vec{H}) = \{(u, v) \mid u \in V(\vec{G}) \text{ and } v \in V(\vec{H})\}$. $A(\vec{G} \boxbslash \vec{H}) = \{\overrightarrow{(u, v)(x, y)} \mid u = x \text{ and } \vec{vy} \in A(\vec{H}) \text{ or } v = y \text{ and } \vec{ux} \in A(\vec{G}) \text{ or } \vec{ux} \in A(G) \text{ and } \vec{vy} \in A(H)\}$. The triangulated Cartesian product of two graphs G and H is isomorphic to the underlying graph of $(\vec{G} \boxbslash \vec{H})$, where $\vec{G}$ and $\vec{H}$ are oriented graphs of G and H, for example see Fig. 1. The triangulated Cartesian product of two regular graphs G and H is the underlying graph of $(\vec{G} \boxbslash \vec{H})$, where $\vec{G}$ and $\vec{H}$ are oriented graphs of G and H such that $| d^-(u) - d^+(u) | \leq 1$ for all $u \in V(\vec{G})$ and $V(\vec{H})$.

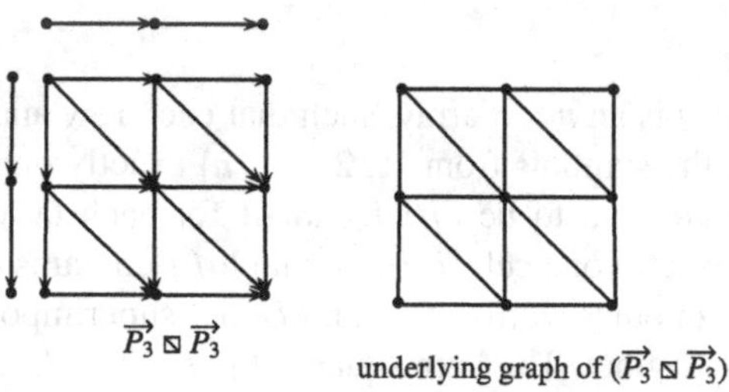

Fig. 1. $\vec{P}_3 \boxbslash \vec{P}_3$ and its underlying graph

The underlying graph of $(\vec{G} \boxbslash \vec{H})$ can be viewed as an edge-disjoint sum of $G \Box H$ and $d_E(G \boxbslash H)$, where $d_E(G \boxbslash H)$ represents the diagonal edges of $G \boxbslash H$ (see Fig. 2). We mean the triangulated cartesian product of two complete graphs of odd order is the same as the underlying graph of triangulated cartesian product of two diregular complete digraphs of odd order.

Study on path factorization of graphs is not new. Many Mathematicians have worked on path factorization of complete graphs, bipartite graphs and complete multipartite graphs and Proved its existence in [5,8–11,13,14].

Typical finite-element grids are neither Cartesian nor strong products, but it resembles a product structure like triangulated Cartesian product of paths, for example see

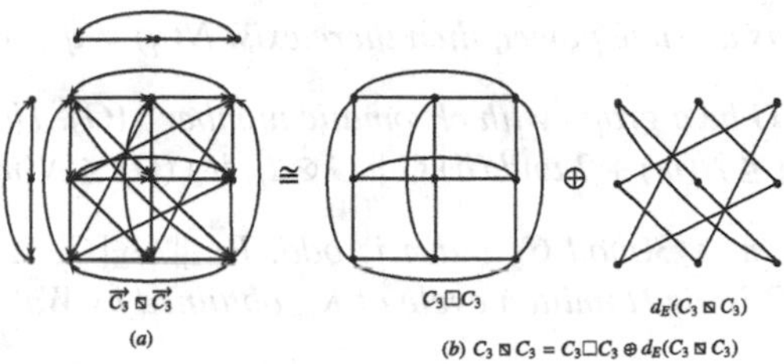

Fig. 2. $C_3 \boxtimes C_3$ and its edge-disjoint sum

Fig. 3 and reference [1]. Baglivo and Graver [4] have described the applicability of triangulated grid (triangulated Cartesian product of paths) for bracing an architectural structure. Recently, Shehzad Afzal and Clemens Brand [1] have factorized graphs into factors which are isomorphic to triangulated Cartesian product of two subgraphs. The above facts motivate the authors [12] to consider the existence of a P_3-factorization in triangulated Cartesian product of complete graphs and obtained few results.

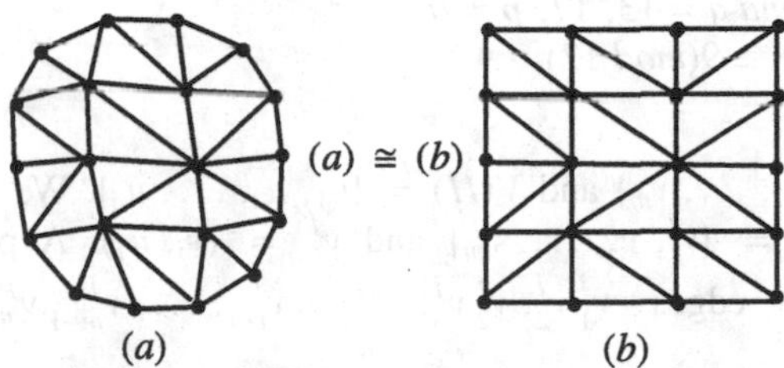

Fig. 3. (a) A simple finite-element grid; (b) Isomorphic structure of (a)

In this paper, we show that the necessary conditions $mn \equiv 0\,(mod\ 3)$ and $3(mn + m + n - 3) \equiv 0\,(mod\ 8)$ for the existence of a P_3-factorization of $K_m \boxtimes K_n$, m and n are odd are sufficient. In fact, the results of this paper together with the earlier results of the authors [12] completely settle the problem of existence of a P_3-factorization of $K_m \boxtimes K_n$, when m and n are odd.

We require the following to prove our main results.

2 Preliminary Results

Theorem 1 [14]. *The complete multigraph $K_r(\lambda)$ has a P_k-factorization if and only if $r \equiv 0\,(mod\ k)$ and $\lambda(r-1)k \equiv 0\,(mod\ 2(k-1))$.*

Theorem 2 (Walecki's Construction [2]). *For every positive integer m (m is odd), the graph K_m is Hamilton decomposable.*

Theorem 3 [7]. *There exists a pair of orthogonal latin squares of order n for every $n \neq 2, 6$.*

Theorem 4 [3]. *If q is a prime power, then there exist $N(q) = q - 1$.*

Theorem 5 [6]. *Let G be a graph with chromatic number $\chi(G)$. Then (i) $G \mid G \otimes I_m$ if $\chi(G) \leq N(m) + 2$ and (ii) $G \parallel G \otimes I_m$ if $\chi(G) \leq N(m) + 1$.*

Lemma 1 [12]. *For $m \equiv 3(mod\ 6)$ and n is odd, $P_3 \parallel d_E[C_m^i \boxtimes (C_n^j \oplus C_n^k)]$, where $C_m^i, 1 \leq i \leq (m-1)/2$ is the Hamilton cycle of K_m obtained by Walecki's construction.*

Lemma 2 [12]. *For $m \equiv 9\,(mod\ 12)$ and $|V(G)| = 5$, $P_3 \parallel d_E(C_m \boxtimes G) \oplus mG$.*

Lemma 3 [12]. *For $m \equiv 9\,(mod\ 12)$ and $|V(G)| = 7$, $P_3 \parallel d_E(C_m \boxtimes G) \oplus mG$.*

Lemma 4 [12]. *For $m \equiv 9\,(mod\ 12)$ and $|V(G)| = 11$, $P_3 \parallel d_E(C_m \boxtimes G) \oplus mG$.*

Lemma 5 [12]. *For $m \equiv 9\,(mod\ 12)$ and $|V(G)| = 13$, $P_3 \parallel d_E(C_m \boxtimes G) \oplus mG$.*

Theorem 6 [12]. *P_3-factorization of $K_m \boxtimes K_n$ exists if one of the following holds:*

(i) $m \equiv 9(mod\ 12)$, $n = 5, 13$ *and* 17,
(ii) $m \equiv 9(mod\ 12)$, $n^s, s > 1$, $n = 5, 13$ *and* 17,
(iii) $m \equiv 9(mod\ 12)$, $n = p^s q^t$ *for all* $s, t \geq 1$,
where $p = 5, 13$ *and* $q = 13, 17$, $p \neq q$
(iv) $m \equiv 9(mod\ 12)$, $n \equiv 9(mod\ 12)$.

Notation:
1. Let $V(G) = \{v_1, v_2, \ldots, v_m\}$ and $V(H) = \{u_1, u_2, \ldots, u_n\}$. We write $V(G \boxtimes H) = \bigcup_{j=1}^{n} U_j$, where $U_j = \{v_1^j, v_2^j, \ldots, v_m^j\}$ and $v_i^j = (v_i, u_j)$. A path on m vertices $v_1^j, v_2^j, \ldots, v_m^j$ with edges $v_1^j v_2^j, v_2^j v_3^j, \ldots, v_i^j v_{i+1}^j, \ldots, v_{m-1}^j v_m^j$ is denoted as $v_1^j v_2^j \ldots v_m^j$.

Note 1. Let $\overrightarrow{G} = \overrightarrow{G_1} \oplus \overrightarrow{G_2}$ and $\overrightarrow{H} = \overrightarrow{G_1} \oplus \overrightarrow{G_2} \oplus \overrightarrow{G_3}$, where $\overrightarrow{G_1} = v_1\, v_2\, v_3\, v_4\, v_5 \ldots v_{p-1}\, v_p\, v_1$, $\overrightarrow{G_2}$
$= v_1\, v_3\, v_5 \ldots v_p\, v_2\, v_4\, v_6 \ldots v_{p-3}\, v_{p-1}\, v_1$ and $\overrightarrow{G_3} = v_1\, v_5\, v_9 \ldots v_{p-1}\, v_3\, v_7\, v_{11} \ldots v_{p-3}\, v_1$ be three dicycles of length p. Throughout this paper, we always mean $G = G_1 \oplus G_2$ and $H = G_1 \oplus G_2 \oplus G_3$ are isomorphic to the underlying graphs of $\overrightarrow{G}$ and $\overrightarrow{H}$. The graphs G and H are shown in Figs. 4 and 5.

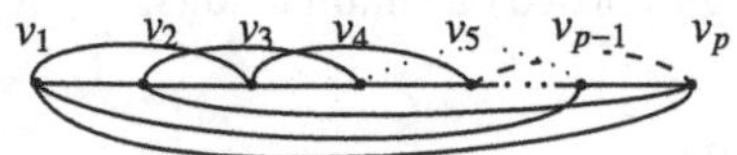

Fig. 4. $G = G1 \oplus G2$.

Remark 1. The Walecki's construction [2], gives the Hamilton cycle decomposition $\{C_m^1, C_m^2, \ldots, C_m^{(m-1)/2}\}$ of K_m (m is odd), where each $C_m^i = v_0\, v_i\, v_{i+1}\, v_{i-1}\, v_{i+2}\, v_{i-2} \ldots v_{[(m-1)/2]+i-1}\, v_{[(m-1)/2]+i+1}\, v_{[(m-1)/2]+i}\, v_0$, $i = 1, 2, \ldots, (m-1)/2$ is a Hamilton cycle of K_m and we write $K_m = C_m^1 \oplus C_m^2 \oplus \ldots \oplus C_m^{(m-1)/2}$.

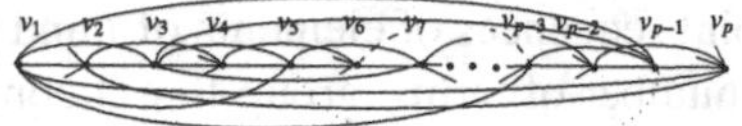

Fig. 5. $H = G1 \oplus G2 \oplus G3$.

Remark 2 [12]. Let $V(K_p)=\{v_1, v_2, \ldots, v_p\}$, p is a prime. For $1 \leq i \leq (p-1)/2$, let $H_i = v_1\, v_{2+(i-1)}\, v_{3+[2(i-1)]}\, v_{4+[3(i-1)]}\, v_{5+[4(i-1)]} \ldots v_{p+[(p-1)(i-1)]}\, v_1$, where the subscripts are taken modulo p with residues 1, 2, 3, ..., p. Note that H_i is a Hamilton cycle of K_p and $\{H_1, H_2, \ldots, H_{(p-1)/2}\}$ gives a Hamilton cycle decomposition of K_p, p is a prime.

Note: Whenever we consider the Hamilton cycle decomposition of K_m, we use the Walecki's construction when m is odd and non prime and we use the construction of Remark 2 when m is a prime.

Remark 3. **Construction of G and H from K_p, p is a prime.**

Let $I = \{1, 2, \ldots, (p-1)/2\}$ and $J = 2I$. We define $f : I \to J$ by $f(i) = 2i$ for all $i \in I$. We write I as the union of disjoint sequences as follows: we start the sequence with least available $i \in I$. If $p - i \in J$, then include $f^{-1}(p-i)$ as a second term of the sequence, otherwise if $f(i) \in I$, then include $f(i)$ as a second term of the sequence. If $p - i \notin J$, and $f(i) \notin I$, then complete the sequence and start the new sequence with the least element of I not in the sequences constructed earlier. Continue this process until all the elements of I are included in some sequence. If two sequences end with a same element of I, then join either with other in reverse order to form a single sequence.

As the elements of I are correspond to the Hamilton cycles of K_p, we group the Hamilton cycles as follows: If the sequence contains odd number of terms, take first 3 in one group and the remaining consecutive pairs as other groups. If the sequence contains even number of terms, take the consecutive pairs as groups. Note that, the subgraphs of K_p induced by each group of Hamilton cycles will be isomorphic to either G or H.

Example 1. By the Remark 2, each H_i, $i \in I = \{1, 2, \ldots, 9\}$, is a Hamilton cycle of K_{19} and we write K_{19} as $K_{19}=H_1 \oplus H_2 \oplus \ldots \oplus H_9$. Now we group the Hamilton cycles of K_{19} to form subgraphs which are isomorphic to the graphs G or H. By applying Remark 3, we get the following sequences:

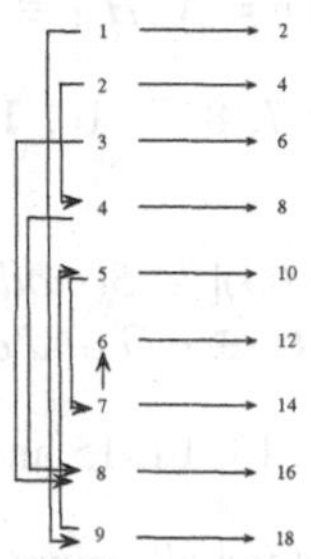

That is, S_1=(1, 9, 5, 7, 6), S_2= (2, 4, 8) and S_3=(3, 8). Here S_2 and S_3 end with the same element 8, so join together to get either S_1'=(2, 4, 8, 3) or S_1'=(3, 8, 4, 2).

Clearly S_1 and S'_1 are disjoint sequences of elements of I and their union is I. Since the sequence S_1 contains odd number of terms, group first 3 consecutive terms and group the remaining consecutive pairs. That is $1 \oplus 9 \oplus 5$ and $7 \oplus 6$. Since the sequence S'_1 contains even number of terms, group the consecutive pairs, (i.e.) $2 \oplus 4$ and $8 \oplus 3$. As the elements of I are correspond to the Hamilton cycles of K_p, we group the Hamilton cycles as follows: $H_1 \oplus H_9 \oplus H_5$, $H_7 \oplus H_6$, $H_2 \oplus H_4$ and $H_8 \oplus H_3$. Clearly the grouped Hamilton cycles of K_{19} form subgraphs of K_{19} which are isomorphic to either G or H.

Note that the triangulated Cartesian product of two complete graphs of odd order is same as the underlying graph of triangulated Cartesian product of two diregular complete digraphs of odd order. Also note that the triangulated Cartesian product of two dicycles can be written as follows: $\overrightarrow{C_m} \boxtimes \overrightarrow{C_n} \cong C_m \boxtimes C_n = C_m \Box C_n \oplus d_E(C_m \boxtimes C_n)$, where $C_m \boxtimes C_n$ is the underlying graph of $\overrightarrow{C_m} \boxtimes \overrightarrow{C_n}$.

3 P_3–Factorization of $K_m \boxtimes K_n$

Lemma 6. *For $m \equiv 9\,(mod\ 12)$ and $n = |V(G)| \equiv 17(mod\ 6)$, $P_3 \parallel d_E(C_m \boxtimes G) \oplus mG$.*

Lemma 7. *For $m \equiv 9\,(mod\ 12)$ and $n = |V(G)| \equiv 19(mod\ 6)$, $P_3 \parallel d_E(C_m \boxtimes G) \oplus mG$.*

Lemma 8. *For $m \equiv 9\,(mod\ 12)$ and $|V(H)| = 7$, $P_3 \parallel d_E(C_m \boxtimes H) \oplus mH$.*

Lemma 9. *For $m \equiv 9\,(mod\ 12)$ and $|V(H)| = 13$, $P_3 \parallel d_E(C_m \boxtimes H) \oplus mH$.*

Lemma 10. *For $m \equiv 9\,(mod\ 12)$ and $|V(H)| = 19$, $P_3 \parallel d_E(C_m \boxtimes H) \oplus mH$.*

Lemma 11. *For $m \equiv 9\,(mod\ 12)$ and $n = |V(H)| \equiv 25(mod\ 6)$, $P_3 \parallel d_E(C_m \boxtimes H) \oplus mH$.*

Lemma 12. *For $m \equiv 9\,(mod\ 12)$ and $|V(G)| = |V(H)| \equiv 1(mod\ 6)$, there exists a P_3-factorization of $d_E(C_m \boxtimes G) \oplus mG$ and $d_E(C_m \boxtimes H) \oplus mH$.*

Lemma 13. *For $m \equiv 9\,(mod\ 12)$ and $|V(H)| = 11$, $P_3 \parallel d_E(C_m \boxtimes H) \oplus mH$.*

Lemma 14. *For $m \equiv 9\,(mod\ 12)$ and $n = |V(H)| \equiv 23(mod\ 12)$, $P_3 \parallel d_E(C_m \boxtimes H) \oplus mH$.*

Lemma 15. *For $m \equiv 9\,(mod\ 12)$ and $|V(H)| = 17$, $P_3 \parallel d_E(C_m \boxtimes H) \oplus mH$.*

Lemma 16. *For $m \equiv 9\,(mod\ 12)$ and $n = |V(H)| \equiv 29(mod\ 12)$, $P_3 \parallel d_E(C_m \boxtimes H) \oplus mH$.*

The proofs of the above Lemmas 6, 7, 8, 9, 10, 11, 12, 13, 14, 15 are constructive. Therefore we omitted.

Lemma 17. *For $m \equiv 9\,(mod\ 12)$, $|V(G)| \equiv 5(mod\ 6)$ and $|V(H)| \equiv 11(mod\ 6)$, there exists a P_3-factorization of $d_E(C_m \boxtimes G) \oplus mG$ and $d_E(C_m \boxtimes H) \oplus mH$.*

Proof. Follows from Lemmas 2, 4, 6, 13, 14, 15 and 16.

Lemma 18. *For $m \equiv 9\,(mod\ 12)$ and for all prime p, there exists a P_3-factorization of $mK_p \oplus d_E(C_m \boxtimes K_p)$.*

Proof. Let $m = 12r + 9$, $r \geq 0$. By Remark 3, we have a factorization of K_p into the graphs isomorphic to G and H. So we can write $mK_p \oplus d_E(C_m \boxtimes K_p) = m[(G_1 \oplus G_2 \oplus \ldots \oplus G_s) \oplus (H_1 \oplus H_2 \oplus \ldots \oplus H_s)] \oplus d_E[C_m \boxtimes (G_1 \oplus G_2 \oplus \ldots \oplus G_s) \oplus (H_1 \oplus H_2 \oplus \ldots \oplus H_s)]$, where G_i's and H_i's are isomorphic to G and H. Then $mK_p \oplus d_E(C_m \boxtimes K_p) = [mG_1 \oplus d_E(C_m \boxtimes G_1)] \oplus [mG_2 \oplus d_E(C_m \boxtimes G_2)] \oplus \ldots \oplus [mG_s \oplus d_E(C_m \boxtimes G_s)] \oplus [mH_1 \oplus d_E(C_m \boxtimes H_1)] \oplus [mH_2 \oplus d_E(C_m \boxtimes H_2)] \oplus \ldots \oplus [mH_s \oplus d_E(C_m \boxtimes H_s)]$. By Lemmas 12 and 17 we have a P_3-factorization of $[mG_1 \oplus d_E(C_m \boxtimes G_1)]$ and $[mH_1 \oplus d_E(C_m \boxtimes H_1)]$. Since the remaining subgraphs of $mK_p \oplus d_E(C_m \boxtimes K_p)$ are isomorphic to either $[mG_1 \oplus d_E(C_m \boxtimes G_1)]$ or $[mH_1 \oplus d_E(C_m \boxtimes H_1)]$, we have a P_3-factorization of $mK_p \oplus d_E(C_m \boxtimes K_p)$. Hence we have a P_3-factorization of $mK_p \oplus d_E(C_m \boxtimes K_p)$, for all prime p.

Lemma 19. *For all prime p and $m \equiv 9\,(mod\, 12)$, there exists a P_3-factorization of $mK_{p^s} \oplus d_E(C_m \boxtimes K_{p^s})$, $s > 1$.*

Proof. Let $m = 12r + 9$, $r \geq 0$. Now $K_{p^s} = pK_{p^{s-1}} \oplus K_{p;\,p^{s-1}}$, $s > 1$. Then $mK_{p^s} \oplus d_E(C_m \boxtimes K_{p^s}) = m(pK_{p^{s-1}} \oplus K_{p;\,p^{s-1}}) \oplus d_E[C_m \boxtimes (pK_{p^{s-1}} \oplus K_{p;\,p^{s-1}})] = (pmK_{p^{s-1}} \oplus mK_{p;\,p^{s-1}}) \oplus p[d_E(C_m \boxtimes K_{p^{s-1}})] \oplus d_E[C_m \boxtimes K_{p;\,p^{s-1}}] = p[mK_{p^{s-1}} \oplus d_E(C_m \boxtimes K_{p^{s-1}})] \oplus [mK_{p;\,p^{s-1}} \oplus d_E(C_m \boxtimes K_{p;\,p^{s-1}})]$.

Case 1. Consider, $mK_{p;\,p^{s-1}} \oplus d_E(C_m \boxtimes K_{p;\,p^{s-1}})$, $s > 1$. By Theorem 5, we have a K_p-factorization of $K_{p;\,p^{s-1}}$. Corresponding to each K_p-factor of $K_{p;\,p^{s-1}}$, we have a $mK_p \oplus d_E(C_m \boxtimes K_p)$- factor of $mK_{p;\,p^{s-1}} \oplus d_E(C_m \boxtimes K_{p;\,p^{s-1}})$. Hence a K_p-factorization of $K_{p;\,p^{s-1}}$ implies a $mK_p \oplus d_E(C_m \boxtimes K_p)$- factorization of $mK_{p;\,p^{s-1}} \oplus d_E(C_m \boxtimes K_{p;\,p^{s-1}})$. By Lemma 18, we have a P_3-factorization of $mK_p \oplus d_E(C_m \boxtimes K_p)$. Thus we have a P_3-factorization of $mK_{p;\,p^{s-1}} \oplus d_E(C_m \boxtimes K_{p;\,p^{s-1}})$, for all prime p.

Case 2. Consider, $mK_{p^s} \oplus d_E(C_m \boxtimes K_{p^s}) = p[mK_{p^{s-1}} \oplus d_E(C_m \boxtimes K_{p^{s-1}})] \oplus [mK_{p;\,p^{s-1}} \oplus d_E(C_m \boxtimes K_{p;\,p^{s-1}})]$, $s > 1$.

(*a*) For $s = 2$, $mK_{p^2} \oplus d_E(C_m \boxtimes K_{p^2}) = p[mK_p \oplus d_E(C_m \boxtimes K_p)] \oplus [mK_{p;\,p} \oplus d_E(C_m \boxtimes K_{p;\,p})]$. Now the existence of a P_3-factorization of $p[mK_p \oplus d_E(C_m \boxtimes K_p)]$ and $mK_{p;\,p} \oplus d_E(C_m \boxtimes K_{p;\,p})$ follows from Lemma 18 and Case 1 respectively.

(*b*) For $s = 3$, $mK_{p^3} \oplus d_E(C_m \boxtimes K_{p^3}) = p[mK_{p^2} \oplus d_E(C_m \boxtimes K_{p^2})] \oplus [mK_{p;\,p^2} \oplus d_E(C_m \boxtimes K_{p;\,p^2})]$. Now the existence of a P_3-factorization of $p[mK_{p^2} \oplus d_E(C_m \boxtimes K_{p^2})]$ and $mK_{p;\,p^2} \oplus d_E(C_m \boxtimes K_{p;\,p^2})$ follows from Case 2(*a*) and Case 1 respectively.

By the induction hypothesis on s, we have a P_3-factorization of $p[mK_{p^{s-1}} \oplus d_E(C_m \boxtimes K_{p^{s-1}})]$. Hence combining all the above we have a P_3-factorization of $mK_{p^s} \oplus d_E(C_m \boxtimes K_{p^s})$, $s > 1$.

Lemma 20. *For $m \equiv 9\,(mod\ 12)$ and a prime $p \equiv 1, 5(mod\ 12)$, there exists a P_3-factorization of $mK_p \oplus d_E(K_m \boxtimes K_p)$.*

Proof. Let $m = 12r + 9$, $r \geq 0$ and $p \in \{(12t + 1) \text{ or } (12t + 5) \mid t \geq 1\}$. Now $mK_p \oplus d_E(K_m \boxtimes K_p) = mK_p \oplus d_E[(C_m^1 \oplus C_m^2 \oplus \ldots \oplus C_m^{(m-1)/2}) \boxtimes K_p] = mK_p \oplus d_E[(C_m^1 \boxtimes K_p) \oplus (C_m^2 \boxtimes K_p) \oplus \ldots \oplus (C_m^{(m-1)/2} \boxtimes K_p)] = mK_p \oplus d_E(C_m^1 \boxtimes K_p) \oplus d_E(C_m^2 \boxtimes K_p) \oplus \ldots \oplus d_E(C_m^{(m-1)/2} \boxtimes K_p)$. By Lemma 18, we have a P_3-factorization of $mK_p \oplus d_E(C_m^1 \boxtimes K_p)$. Further, $d_E(C_m^2 \boxtimes K_p) = d_E[C_m^2 \boxtimes \underbrace{(C_p^1 \oplus C_p^2 \oplus \ldots \oplus C_p^{(p-1)/2})}_{even}] = d_E[C_m^2 \boxtimes (C_p^1 \oplus C_p^2)] \oplus$

$d_E[C_m^2 \boxtimes (C_p^3 \oplus C_p^4)] \oplus \ldots \oplus d_E[C_m^2 \boxtimes (C_p^{(p-3)/2} \oplus C_p^{(p-1)/2})]$. By Lemma 1, we have a P_3-factorization of $d_E[C_m^2 \boxtimes (C_p^1 \oplus C_p^2)]$ and the remaining subgraphs of $d_E(C_m^2 \boxtimes K_p)$. Also we have a P_3-factorization of all the subgraphs isomorphic to $d_E(C_m^2 \boxtimes K_p)$. Thus combining all these we have a P_3-factorization of $mK_p \oplus d_E(K_m \boxtimes K_p)$.

Lemma 21. *For $m \equiv 9\,(mod\ 12)$ and a prime $p \equiv 1, 5(mod\ 12)$, $p \neq 5, 13, 17$, there exists a P_3-factorization of $K_m \boxtimes K_p$.*

Proof. Let $m = 12r + 9$, $r \geq 0$ and $p \in \{(12t + 1)$ or $(12t + 5) \mid t > 1\}$. Then $K_m \boxtimes K_p = (K_m \Box K_p) \oplus d_E(K_m \boxtimes K_p) = mK_p \oplus pK_m \oplus d_E(K_m \boxtimes K_p)$. Now the existence of a P_3-factorization of pK_m and $mK_p \oplus d_E(K_m \boxtimes K_p)$ follows from Theorem 1 and Lemma 20 respectively. Hence we have a P_3-factorization of $K_m \boxtimes K_p$.

Lemma 22. *For $m \equiv 9\,(mod\ 12)$ and $p^s \equiv 1, 5(mod\ 12)$, $p \neq 5, 13, 17$ and p is a prime, there exists a P_3-factorization of $K_m \boxtimes K_{p^s}$, $s > 1$.*

Proof. Let $m = 12r + 9$, $r \geq 0$ and for $s > 1$, $p^s \in \{(12t + 1)$ or $(12t + 5) \mid t \geq 1\}$. Then $K_m \boxtimes K_{p^s} = K_m \Box K_{p^s} \oplus d_E[K_m \boxtimes K_{p^s}] = p^s K_m \oplus mK_{p^s} \oplus d_E[K_m \boxtimes K_{p^s}]$. By Theorem 1, we have a P_3-factorization of $p^s K_m$.

Now, $mK_{p^s} \oplus d_E[K_m \boxtimes K_{p^s}] = mK_{p^s} \oplus d_E[(C_m^1 \oplus C_m^2 \oplus \ldots \oplus C_m^{(m-1)/2}) \boxtimes K_{p^s}] = mK_{p^s} \oplus d_E[(C_m^1 \boxtimes K_{p^s}) \oplus (C_m^2 \boxtimes K_{p^s}) \oplus \ldots \oplus (C_m^{(m-1)/2} \boxtimes K_{p^s})] = mK_{p^s} \oplus d_E(C_m^1 \boxtimes K_{p^s}) \oplus d_E(C_m^2 \boxtimes K_{p^s}) \oplus \ldots \oplus d_E(C_m^{(m-1)/2} \boxtimes K_{p^s})$. By Lemma 19, we have a P_3-factorization of $mK_{p^s} \oplus d_E(C_m^1 \boxtimes K_{p^s})$. Now for odd $n = p^s$, $d_E(C_m^2 \boxtimes K_n) = d_E[C_m^2 \boxtimes \underbrace{(C_n^1 \oplus C_n^2 \oplus \ldots \oplus C_n^{(n-1)/2})}_{even}]$ $= d_E[C_m^2 \boxtimes (C_n^1 \oplus C_n^2)] \oplus d_E[C_m^2 \boxtimes (C_n^3 \oplus C_n^4)] \oplus \ldots \oplus d_E[C_m^2 \boxtimes (C_n^{(n-3)/2} \oplus C_n^{(n-1)/2})]$. By Lemma 1, we have a P_3-factorization of $d_E[C_m^i \boxtimes (C_n^j \oplus C_n^k)]$. Thus we have a P_3-factorization of $d_E(C_m^2 \boxtimes K_{p^s})$. Also we have a P_3-factorization of all the subgraphs isomorphic to $d_E(C_m^2 \boxtimes K_{p^s})$. Thus we have a P_3-factorization of $mK_{p^s} \oplus d_E[K_m \boxtimes K_{p^s}]$. Hence combining all the above we have a P_3-factorization of $K_m \boxtimes K_{p^s}$, $s > 1$.

Lemma 23. *For $m \equiv 9\,(mod\ 12)$, there exists a P_3-factorization of $mK_{p^s.q^t} \oplus d_E(C_m \boxtimes K_{p^s.q^t})$, $s, t \geq 1$, p, q are primes and $p < q$.*

Proof. Let $m = 12r + 9$, $r \geq 0$. Now we write, $K_{p^s.q^t} = pK_{p^{s-1}.q^t} \oplus K_{p;(p^{s-1}.q^t)}$, $s, t \geq 1$. Then, $mK_{p^s.q^t} \oplus d_E(C_m \boxtimes K_{p^s.q^t}) = m[pK_{p^{s-1}.q^t} \oplus K_{p;(p^{s-1}.q^t)}] \oplus d_E[C_m \boxtimes (pK_{p^{s-1}.q^t} \oplus K_{p;(p^{s-1}.q^t)})] = p[mK_{p^{s-1}.q^t} \oplus d_E(C_m \boxtimes K_{p^{s-1}.q^t})] \oplus [mK_{p;(p^{s-1}.q^t)} \oplus d_E(C_m \boxtimes K_{p;(p^{s-1}.q^t)})]$.

Case 1. Consider, $mK_{p;(p^{s-1}.q^t)} \oplus d_E[C_m \boxtimes K_{p;(p^{s-1}.q^t)}]$, $s, t \geq 1$. By Theorem 5, we have a K_p-factorization of $K_{p;(p^{s-1}.q^t)}$. Corresponding to each K_p-factor of $K_{p;(p^{s-1}.q^t)}$, we have a $mK_p \oplus d_E(C_m \boxtimes K_p)$- factor of $mK_{p;(p^{s-1}.q^t)} \oplus d_E[C_m \boxtimes K_{p;(p^{s-1}.q^t)}]$. Hence a K_p-factorization of $K_{p;(p^{s-1}.q^t)}$ implies a $mK_p \oplus d_E(C_m \boxtimes K_p)$- factorization of $mK_{p;(p^{s-1}.q^t)} \oplus d_E[C_m \boxtimes K_{p;(p^{s-1}.q^t)}]$. By Lemma 18, we have a P_3-factorization of $mK_p \oplus d_E(C_m \boxtimes K_p)$. Thus we have a P_3-factorization of $mK_{p;(p^{s-1}.q^t)} \oplus d_E[C_m \boxtimes K_{p;(p^{s-1}.q^t)}]$, for all primes p, q.

Case 2. Consider, $mK_{p^s.q^t} \oplus d_E(C_m \boxtimes K_{p^s.q^t}) = p[mK_{p^{s-1}.q^t} \oplus d_E(C_m \boxtimes K_{p^{s-1}.q^t})] \oplus [mK_{p;(p^{s-1}.q^t)} \oplus d_E(C_m \boxtimes K_{p;(p^{s-1}.q^t)})]$, $s, t \geq 1$.

(a) For $s = 1, t = 1$, $mK_{p\cdot q} \oplus d_E(C_m \boxtimes K_{p\cdot q}) = p[mK_q \oplus d_E(C_m \boxtimes K_q)] \oplus [mK_{p;q} \oplus d_E(C_m \boxtimes K_{p;q})]$. Now the existence of a P_3-factorization of $p[mK_q \oplus d_E(C_m \boxtimes K_q)]$ and $mK_{p;q} \oplus d_E(C_m \boxtimes K_{p;q})$ follows from Lemma 18 and Case 1 respectively.

(b) For $s = 1, t = 2$, $mK_{p\cdot q^2} \oplus d_E(C_m \boxtimes K_{p\cdot q^2}) = p[mK_{q^2} \oplus d_E(C_m \boxtimes K_{q^2})] \oplus [mK_{p;q^2} \oplus d_E(C_m \boxtimes K_{p;q^2})]$. Now the existence of a P_3-factorization of $p[mK_{q^2} \oplus d_E(C_m \boxtimes K_{q^2})]$ and $mK_{p;q^2} \oplus d_E(C_m \boxtimes K_{p;q^2})$ follows from Lemma 19 and Case 1 respectively.

(c) For $s = 1, t \geq 3$, $mK_{p\cdot q^t} \oplus d_E(C_m \boxtimes K_{p\cdot q^t}) = p[mK_{q^t} \oplus d_E(C_m \boxtimes K_{q^t})] \oplus [mK_{p;q^t} \oplus d_E(C_m \boxtimes K_{p;q^t})]$. Now the existence of a P_3-factorization of $p[mK_{q^t} \oplus d_E(C_m \boxtimes K_{q^t})]$ and $mK_{p;q^t} \oplus d_E(C_m \boxtimes K_{p;q^t})$ follows from Lemma 19 and Case 1 respectively.

Case 3.(a) For $s = 2, t = 1$, $mK_{p^2\cdot q} \oplus d_E(C_m \boxtimes K_{p^2\cdot q}) = p[mK_{p\cdot q} \oplus d_E(C_m \boxtimes K_{p\cdot q})] \oplus [mK_{p;(p\cdot q)} \oplus d_E(C_m \boxtimes K_{p;(p\cdot q)})]$. Now the existence of a P_3-factorization of $p[mK_{p\cdot q} \oplus d_E(C_m \boxtimes K_{p\cdot q})]$ and $mK_{p;(p\cdot q)} \oplus d_E[C_m \boxtimes K_{p;(p\cdot q)}]$ follows from Case 2(a) and Case 1 respectively.

(b) For $s = 2, t = 2$, $mK_{p^2\cdot q^2} \oplus d_E(C_m \boxtimes K_{p^2\cdot q^2}) = p[mK_{p\cdot q^2} \oplus d_E(C_m \boxtimes K_{p\cdot q^2})] \oplus [mK_{p;(p\cdot q^2)} \oplus d_E(C_m \boxtimes K_{p;(p\cdot q^2)})]$. Now the existence of a P_3-factorization of $p[mK_{p\cdot q^2} \oplus d_E(C_m \boxtimes K_{p\cdot q^2})]$ and $mK_{p;(p\cdot q^2)} \oplus d_E(C_m \boxtimes K_{p;(p\cdot q^2)})$ follows from Case 2(b) and Case 1 respectively.

(c) For $s = 2, t \geq 3$, $mK_{p^2\cdot q^t} \oplus d_E(C_m \boxtimes K_{p^2\cdot q^t}) = p[mK_{p\cdot q^t} \oplus d_E(C_m \boxtimes K_{p\cdot q^t})] \oplus [mK_{p;(p\cdot q^t)} \oplus d_E(C_m \boxtimes K_{p;(p\cdot q^t)})]$. Now the existence of a P_3-factorization of $p[mK_{p\cdot q^t} \oplus d_E(C_m \boxtimes K_{p\cdot q^t})]$ and $mK_{p;(p\cdot q^t)} \oplus d_E(C_m \boxtimes K_{p;(p\cdot q^t)})$ follows from Case 2(c) and Case 1 respectively.

By the induction hypothesis on s, we have a P_3-factorization of $p[mK_{p^{s-1}\cdot q^t} \oplus d_E(C_m \boxtimes K_{p^{s-1}\cdot q^t})]$. Hence combining all the above we have a P_3-factorization of $mK_{p^s\cdot q^t} \oplus d_E(C_m \boxtimes K_{p^s\cdot q^t})$, $s, t \geq 1$.

Lemma 24. *For $m \equiv 9 \,(mod\ 12)$ and $p^s.q^t \equiv 1, 5(mod\ 12)$ (except $p = 5, 13$ and $q = 13, 17$, $p \neq q$) where p, q are primes, there exists a P_3-factorization of $K_m \boxtimes K_{p^s\cdot q^t}$, for all $s, t \geq 1$ and $p < q$.*

Proof. Let $m = 12r + 9$, $r \geq 0$ and for $s, t \geq 1$, $p^s.q^t \in \{(12v + 1) \text{ or } (12v + 5) \mid v \geq 1\}$. Then for $s, t \geq 1$, $K_m \boxtimes K_{p^s\cdot q^t} = K_m \square K_{p^s\cdot q^t} \oplus d_E[K_m \boxtimes K_{p^s\cdot q^t}] = (p^s \cdot q^t)K_m \oplus mK_{p^s\cdot q^t} \oplus d_E[K_m \boxtimes K_{p^s\cdot q^t}]$. By Theorem 1, we have a P_3-factorization of $(p^s \cdot q^t)K_m$.

Now, $mK_{p^s\cdot q^t} \oplus d_E[K_m \boxtimes K_{p^s\cdot q^t}] = mK_{p^s\cdot q^t} \oplus d_E[(C_m^1 \oplus C_m^2 \oplus \ldots \oplus C_m^{(m-1)/2}) \boxtimes K_{p^s\cdot q^t}] = mK_{p^s\cdot q^t} \oplus d_E(C_m^1 \boxtimes K_{p^s\cdot q^t}) \oplus d_E(C_m^2 \boxtimes K_{p^s\cdot q^t}) \oplus \ldots \oplus d_E(C_m^{(m-1)/2} \boxtimes K_{p^s\cdot q^t})$. By Lemma 23, we have a P_3-factorization of $mK_{p^s\cdot q^t} \oplus d_E(C_m^1 \boxtimes K_{p^s\cdot q^t})$.

Now for odd $n = (p^s \cdot q^t)$, $d_E(C_m^2 \boxtimes K_n) = d_E[C_m^2 \boxtimes \underbrace{(C_n^1 \oplus C_n^2 \oplus \ldots \oplus C_n^{(n-1)/2})}_{even}] = d_E[C_m^2 \boxtimes (C_n^1 \oplus C_n^2)] \oplus d_E[C_m^2 \boxtimes (C_n^3 \oplus C_n^4)] \oplus \ldots \oplus d_E[C_m^2 \boxtimes (C_n^{(n-3)/2} \oplus C_n^{(n-1)/2})]$. By Lemma 1, we have a P_3-factorization of $d_E[C_m^i \boxtimes (C_n^j \oplus C_n^k)]$. Thus we have a P_3-factorization of $d_E(C_m^2 \boxtimes K_{p^s\cdot q^t})$. Also we have a P_3-factorization of all the subgraphs isomorphic to $d_E(C_m^2 \boxtimes K_{p^s\cdot q^t})$. Thus we have a P_3-factorization of $mK_{p^s\cdot q^t} \oplus d_E[K_m \boxtimes K_{p^s\cdot q^t}]$.

Hence combining all the above we have a P_3-factorization of $K_m \boxtimes K_{p^s\cdot q^t}$, for all $s, t \geq 1$.

Theorem 7. *There exists a P_3-factorization of $K_m \boxtimes K_n$, m and n are odd if and only if $mn \equiv 0 \,(mod\ 3)$ and $3(mn + m + n - 3) \equiv 0 \,(mod\ 8)$.*

Proof. **Necessity.** Follows by counting the number of edges and vertices of the graph $K_m \boxtimes K_n$, m and n are odd.

Sufficiency. Follows from Theorems 6 and Lemmas 21, 22 and 24.

Conclusion: The results of Section 3 partially answer the problem of existence of a P_3-factorization of $K_m \boxtimes K_n$ when m and n are odd. Results of this paper together with the earlier results of the authors [12] completely solve the problem of existence of a P_3-factorization of $K_m \boxtimes K_n$ when m and n are odd. Existence of a P_3-factorization of $K_m \boxtimes K_n$ for the other parity of m and n is unknown.

Acknowledgement. The second author thank DST-SERB, New Delhi, for its financial support through Grant No. SR/S4/MS: 828/13.

References

1. Afzal, S., Brand, C.: Recognizing triangulated cartesian graph products. Discrete Math. **312**, 188–193 (2012)
2. Alspach, B., Bermond, J.C., Sotteau, D.: Decomposition into cycles I: Hamilton decompositions. In: Hahn, G. et al. (ed.) Cycles and Rays, pp. 9–18. Kluwer Academic Publisher, Dordrecht (1990)
3. Anderson, I.: Combinatorial Designs and Tournaments. Oxford University Press, Newyork (1997)
4. Baglivo, J.A., Graver, J.E.: Incidence and Symmetry in Design and Architecture. Cambridge University Press, New York (1983)
5. Chitra, V., Muthusamy, A.: P_3-factorization of cartesian product of complete graphs. Bull. ICA **63**, 60–72 (2011)
6. Haggkvist, R.: Decompositions of complete bipartite graphs. London Math. Soc. Lecture Note Ser. **141**, 115–147 (1989)
7. Lindner, C.C., Rodger, C.A.: Design Theory. Chapman & Hall/CRC Press, Taylor & Francis Group, Boca Raton (2009)
8. Muthusamy, A., Paulraja, P.: Path factorization of product graphs. In: Proceedings of the National Workshop on Graph Theory and its Applications, pp. 77–88. Manonmaniam Sundaranar University, Tirunelveli (1996)
9. Muthusamy, A., Paulraja, P.: Path factorizations of complete multipartite graphs. Discrete Math. **195**, 181–201 (1999)
10. Sampathkumar, R., Kandan, P.: P_4-factorization of cartesian product of complete Graphs. J. Combin. Math. Combin. Comput. **84**, 127–135 (2013)
11. Tamil Elakkiya, A., Muthusamy, A.: P_5-factorization of cartesian product of graphs. Discrete Math. Algorithms Appl. 6(2), 19 p. (2014). 1450019
12. Tamil Elakkiya, A., Muthusamy, A.: P_3-factorization of triangulated cartesian product of complete graphs. Discrete Math. Algorithms Appl. 7(1), 28 p. (2015). 1450066
13. Ushio, K.: P_3-factorization of complete bipartite graphs. Discrete Math. **72**, 361–366 (1988)
14. Yu, M.L.: On path factorization of complete multipartite graphs. Discrete Math. **122**, 325–333 (1993)

Further Progress on the Heredity of the Game Domination Number

Tijo James[1,4(✉)], Paul Dorbec[2,3], and Ambat Vijayakumar[1]

[1] Department of Mathematics, Cochin University of Science and Technology, Kochi 682022, India
tijojames@gmail.com, vambat@gmail.com
[2] Univ. Bordeaux, LaBRI, UMR 5800, 33400 Talence, France
dorbec@labri.fr
[3] CNRS, LaBRI, UMR 5800, 33400 Talence, France
[4] Department of Mathematics, Pavanatma College, Murickassery, Idukki 685604, India

Abstract. The domination game is a two-player game played on a finite, undirected graph G. During the game, the players alternately choose a vertex of G such that each chosen vertex dominates at least one previously undominated vertex. One player, called Dominator, tries to finish the game within few moves, while the second player, Staller, tries to make it last for as long as possible. The game domination number $\gamma_g(G)$ is the total number of moves in the game when Dominator starts and both players play optimally. The Staller start game domination number $\gamma_g'(G)$ is defined similarly when Staller starts the game. The behaviour of the game domination number on the removal of a vertex and an edge so as that no heredity is possible, in contrast with what is happening for domination. In this paper we consider the special case of *no-minus*-graphs.

Keywords: Domination game · Game domination number · No-minus graphs

1 Introduction

The game domination number was introduced by Brešar et al. [4]. This parameter is to domination what the game chromatic number is to graph colourings (see e.g. [1]).

Recall that a vertex is said to dominate itself and its neighbours. In the domination game, two players, named Dominator and Staller, alternate turns choosing a vertex in a finite, undirected graph G, and adding it to a set of vertices S. Whenever a player chooses a vertex to add to S, the vertex must dominate at least one vertex not yet dominated by the vertices of S. The game ends when no move is possible, that is when S is a dominating set of the graph. The total number of chosen vertices is called the *score* of the game. The two

S. Arumugam et al. (Eds.): ICTCSDM 2016, LNCS 10398, pp. 435–444, 2017.
DOI: 10.1007/978-3-319-64419-6_55

players have opposite goals, Dominator tries to minimize the final score while Staller tries to maximize it.

Two graph parameters relative to this game were introduced in [4]. Assuming both players play optimally, the game domination number $\gamma_g(G)$ is the score of the game on G when Dominator starts (we say in Game 1), and the Staller start game domination number $\gamma_g'(G)$ is the score when Staller starts (in Game 2). Both parameters are studied in parallel since many results hold for both of them. We thus may refer to them on the game domination numbers.

The first and the most natural question is to try to find bounds for the game domination number of a graph. In terms of the order n of the graph, Kinnersley et al. [8] conjectured that $\gamma_g(G)$ is bounded above by $\frac{3n}{5}$. Early results on this question for trees can be found in [3,6].

A natural technique to find bounds for the game domination number would be to find some heredity property: find a graph operation that involves a monotonous behaviour of the game domination number of the graphs. A first natural way of finding heredity is to consider the game within its course, and to have some vertices partially dominated. Given a subset S of vertices in a graph G, we denote by $G|S$ the graph where the vertices of S are considered already dominated. Kinnersley et al. [8] observed that whatever the set S of already dominated vertices, the game last no longer on $G|S$ than on G. More generally,

Theorem 1 (Continuation principle [8]). *Let G be a graph and $A, B \subseteq V(G)$. If $B \subseteq A$ then $\gamma_g(G|A) \leq \gamma_g(G|B)$ and $\gamma_g'(G|A) \leq \gamma_g'(G|B)$.*

This result together with earlier observations [4] on the problem allowed to deduce that the Staller start and the Dominator start game domination number may differ by at most one:

Theorem 2. [4,8] *For any graph G and subset S of vertices, $|\gamma_g(G|S) - \gamma_g'(G|S)| \leq 1$*

Another early consideration of heredity for the game domination numbers was made in [5], where the authors proved that the ratio of the game domination number of a graph and of a spanning subgraph could not be bounded. Then, the consequences of vertex and edge removal in a graph were considered in [2] and it is proved that in both cases, the game domination number can either increase or decrease.

Another main track of research on this topic is to compare the behaviour of the Staller start and Dominator start game domination numbers. As mentioned earlier, it is known that the difference is at most one, and that it can occur in both directions. Naturally, it comes that Staller can have the game last longer when she start the game, e.g. on a star; but it may also happen that she makes the game finish earlier when starting, as e.g. on the 5-cycle. This second behaviour is more surprising, and seems to happen on fewer graphs. It may not happen for example on trees. In [7], a special family of graphs was introduced, called *no-minus*: a graph G is no-minus if for any subset of vertices $S \subseteq V$, $\gamma_g(G|S) \leq \gamma_g'(G|S)$. In that cases, it is never interesting for Staller to pass a move. It is known already that forests [8], tri-split and dually chordal graphs [7] are no-minus graphs. In the following, we consider the case of no-minus graphs for earlier studied parameters, such as edge and vertex deletion.

2 Edge and Vertex Removal in No-Minus Graphs

2.1 Edge Removal

Here, we prove that removing an edge from a no-minus graph can either increase or decrease its game domination number by at most 1.

Theorem 3. *If G is a no-minus graph and $e \in E(G)$, then* $\left|\gamma_g(G) - \gamma_g(G-e)\right| \leq 1$ *and* $\left|\gamma_g'(G) - \gamma_g'(G-e)\right| \leq 1$.

Proof. First we prove that $\gamma_g(G) \leq \gamma_g(G-e)+1$. It is enough to show that Dominator has a strategy on G such that at most $\gamma_g(G-e)+1$ moves will be played. Both the players play a dominator start game on G, at the same time Dominator imagines a dominator start game played on $G-e$ with at most $\gamma_g(G-e)$ steps. Dominator's strategy on G is as follows. He copies every move of Staller in the real game to the imaginary game and responds optimally in $G-e$. Each response in the imagined game is then copied back to the real game in G. Let $e = uv$ and if every move of Dominator and Staller are legal, then the real game ends by at most $\gamma_g(G-e)$ steps. Suppose at the kth step Dominator chooses a vertex in the imagined game that is not a legal move in the real game. This is possible only if Dominator chooses a vertex that dominates either u or v itself and all other neighbours of that vertex are already dominated. Suppose that it dominates v only which is already dominated in G. After this move, the set of vertices dominated in both the graphs are same. At this stage the number of moves in the real game is $k-1$ and the next turn is that of Dominator. Therefore $\gamma_g(G) \leq k-1+\gamma_g(G|D)$ where D denotes the set of vertices already dominated in G. But in the imagined game, the next turn is that of Staller and the number of moves at this stage is k. Since Staller did not play optimally in $G-e$, $k+\gamma_g'(G-e|D) \leq \gamma_g(G-e)$. So,

$$\begin{aligned}
\gamma_g(G) &\leq k-1+\gamma_g(G|D)\\
&= k-1+\gamma_g(G-e|D)\\
&\leq k+\gamma_g(G-e|D)\\
&\leq k+\gamma_g'(G-e|D)\\
&\leq \gamma_g(G-e).
\end{aligned}$$

Hence, in this case the real game ends in at most $\gamma_g(G-e)$ steps.

Suppose, at the kth step Staller chooses a vertex in the real game and this is not a legal move in the imagined game. This is possible only if Staller chooses one of the end vertices of e and the other end vertex is the only vertex which is newly dominated. Let v denote the newly dominated vertex. Let D denote the set of vertices dominated in the real game after the kth move, at this stage the set of vertices dominated in the imagined game is $D - v$. In the real game, k vertices are already selected by both the players and the next is Dominator's turn. Therefore, $\gamma_g(G) \leq k + \gamma_g(G|D)$. But in the imagined game both the players selected $k-1$ vertices and the next turn is that of Staller. Therefore $k - 1 + \gamma_g'(G - e|D - v) \leq \gamma_g(G - e)$. So,

$$\begin{aligned}\gamma_g(G) &\leq k + \gamma_g(G|D)\\ &\leq k + \gamma_g(G - e|D)\\ &\leq k + \gamma_g(G - e|D - v)\\ &\leq k + \gamma_g'(G - e|D - v)\\ &= k - 1 + \gamma_g'(G - e|D - v) + 1\\ &\leq \gamma_g(G - e) + 1.\end{aligned}$$

Hence $\gamma_g(G) \leq \gamma_g(G - e) + 1$.

Now we prove that $\gamma_g(G - e) - 1 \leq \gamma_g(G)$. This proof is analogous to the proof of $\gamma_g(G - e) \leq \gamma_g(G) + 2$ in [2] but we substitute the condition $\gamma_g(G|D) \leq \gamma_g'(G|D)$ instead of $\gamma_g(G|D) \leq 1 + \gamma_g'(G|D)$. In both the cases the proof is independent of who moves the first. Hence this proof works for $\gamma_g'(G)$ also.

2.2 Vertex Removal

If a vertex from a graph G is removed, its game domination number either increases arbitrary large or decreases by at most two [2]. However, if G is a no-minus graph and v is a pendant vertex, we have the following lemma.

Lemma 1. *If G is a no-minus graph and v is a pendant vertex, then*

$$\gamma_g(G) - 1 \leq \gamma_g(G - v) \leq \gamma_g(G)$$

$$\gamma_g'(G) - 1 \leq \gamma_g'(G - v) \leq \gamma_g'(G).$$

Proof. First we prove that $\gamma_g(G - v) \leq \gamma_g(G|v)$. For that we need to show that Dominator has a strategy on $G - v$ that at most $\gamma_g(G|v)$ moves will be played. The strategy is as follows. Dominator and Staller play an ordinary Dominator start game played on $G - v$ and at the same time Dominator imagines another game played on $G|v$. He copies every move of Staller in the real game to the imagined game and respond optimally in the imagined game. He then copies back every optimal response in the imagined game to the real game. Every move of Staller in the real game is a legal move in the imagined game. Dominator never chooses v in the imagined game, so every move of Dominator in the imagined game is a legal move in the real game. Hence, the real game ends by at most $\gamma_g(G|v)$ steps. That is $\gamma_g(G - v) \leq \gamma_g(G|v)$. By the continuation principle $\gamma_g(G|v) \leq \gamma_g(G)$. Hence $\gamma_g(G - v) \leq \gamma_g(G)$.

Now, we prove that $\gamma_g(G) \leq \gamma_g(G-v)+1$. It is enough to show that Dominator has a strategy on G such that at most $\gamma_g(G-v)+1$ moves will be played. Dominator imagines a dominator start game played on $G - v$ simultaneously with the game played on G. He is copying every move of Staller in the real game to the imaginary game and respond optimally in it. Every optimal response in the imagined game is then copied back to the real game. If all the moves are legal then $\gamma_g(G) \leq \gamma_g(G-v)$. Suppose at the kth step Staller chooses a vertex that is not a legal move in $G - v$. This is possible only if Staller chooses a vertex whose neighbours are already dominated except v. Let D denote the set of vertices dominated in the real game after the kth step. But in the imagined game both the players played $k-1$ moves and the next move is that of Staller. Therefore $k-1+\gamma_g'(G-v|D-v) \leq \gamma_g(G-v)$ and hence

$$\begin{aligned}\gamma_g(G) &\leq k+\gamma_g(G|D)\\ &\leq k+\gamma_g(G-v|D-v)\\ &\leq k+\gamma_g'(G-v|D-v)\\ &\leq k-1+\gamma_g'(G-v|D-v)+1\\ &\leq \gamma_g(G-v)+1.\end{aligned}$$

Hence, $\gamma_g(G)-1 \leq \gamma_g(G-v) \leq \gamma_g(G)$. The proof is independent of who moves the first. So $\gamma_g'(G)-1 \leq \gamma_g'(G-v) \leq \gamma_g'(G)$ also holds.

3 Examples of No-Minus Graphs Attaining Possible Values

3.1 Edge Removal

Trees are no-minus graphs [8]. So, by Theorem 3, $|\gamma_g(T)-\gamma_g(T-e)| \leq 1$ and $\left|\gamma_g'(T)-\gamma_g'(T-e)\right| \leq 1$, for a tree T. Here, we show that all is possible except for $k = 1, 2$.

Case 1. $\gamma_g(T)-\gamma_g(T-e)=1$.

For $k=1$, 2, there is no tree T with $\gamma_g(T)=k$ and $\gamma_g(T-e)=k-1$.

For $k=3$, let T be the graph obtained from P_4 by attaching two vertices at one of the end vertex of P_4 and let e denote the middle edge of P_4. Clearly $\gamma_g(T)=3$ and $\gamma_g(T-e)=2$.

For $k=4$, let T be the graph obtained from P_3 by attaching two vertices at both end vertices of P_3 and subdivide one of the added edge. If e is the edge of P_3 incident to the vertex attached to the subdivided edge. Clearly $\gamma_g(T)=4$ and $\gamma_g(T-e)=3$.

For $k \geq 5$, let e be an edge of the star $K_{1,k-2}$ and attaching two vertices at the pendant vertex incident to e. Let T be the graph obtained by subdividing each edge incident to the center. Clearly $\gamma_g(T)=k$ and $\gamma_g(T-e)=k-1$.

Case 2. $\gamma_g(T) - \gamma_g(T-e) = -1$.

For $k = 1$, let T be the star $K_{1,t}$. Clearly $\gamma_g(T) = 1$ and $\gamma_g(T-e) = 2$.

For $k = 2$, let T be the graph obtained from P_3 by attaching two vertices at one end vertex. Let e be the newly added edge of T. Clearly $\gamma_g(T) = 2$ and $\gamma_g(T-e) = 3$.

For $k \geq 3$, let T be the graph obtained from the star $K_{1,k}$ by subdividing each edge except one. Let e be the edge that is not subdivided in the star. Clearly $\gamma_g(T) = k$ and $\gamma_g(T-e) = k+1$.

Case 3. $\gamma_g(T) - \gamma_g(T-e) = 0$

There is no tree T with an edge e such that $\gamma_g(T) = 1$ and $\gamma_g(T-e) = 1$. Any tree with $\gamma_g(T) = 1$ is of the form $K_{1,k}$ and $\gamma_g(K_{1,k} - e) = 2$.

Let e denotes the middle edge of P_4 and $\gamma_g(P_4) = \gamma_g(P_4 - e) = 2$.

For $k \geq 3$, Let T be the graph obtained from $K_{1,k-1}$ by subdividing each edge and e is any pendant edge of T. Clearly $\gamma_g(T) = \gamma_g(T-e) = k$.

Case 4. $\gamma_g^{'}(T) - \gamma_g^{'}(T-e) = 1$

For $k = 4$, let T be the graph obtained from P_4 by attaching three vertices at one of the end points of P_4 and let e be the middle edge of P_4. Clearly $\gamma_g^{'}(T) = 4$ and $\gamma_g^{'}(T-e) = 3$.

For $k = 5$, let T be the graph obtained from P_4 and $K_{1,t}$ by connecting them with an edge e in such a way that one end vertex of e is the degree two vertex of P_4 and the other end vertex is any pendant vertex of the star $K_{1,k}$. Clearly $\gamma_g^{'}(T) = 5$ and $\gamma_g^{'}(T-e) = 4$.

For $k \geq 6$, let e be an edge of the star $K_{1,k-3}$ and let T be the graph obtained from the star $K_{1,k-3}$ by subdividing each of its edge and attach three vertices to pendant vertex of the subdivided edge. Clearly $\gamma_g^{'}(T) = k$ and $\gamma_g^{'}(T-e) = k-1$.

Case 5. $\gamma_g^{'}(T) - \gamma_g^{'}(T-e) = -1$.

Let T be the graph obtained from K_2 by removing its edge. Clearly $\gamma_g^{'}(T-e) = 2$.

For $k = 2$, P_4 is the graph with $\gamma_g^{'}(P_4) = 2$ and let T be the graph obtained from P_4 by removing its pendant edge. Clearly $\gamma_g^{'}(T-e) = 3$.

For $k = 3$, let T be the graph obtained from P_3 by attaching three vertices at one of the end points of P_3. Clearly $\gamma_g^{'}(T) = 3$. If e is any pendant edge incident to the highest degree vertex of T then $\gamma_g^{'}(T-e) = 4$.

For $k \geq 4$, let T be the graph obtained from a star $K_{1,k-2}$ by subdividing each edge except one and attaching three vertices at the end vertex of the edge which is not subdivided. In this case $\gamma_g^{'}(T) = k$. Ife is the edge incident to one of the new vertices attached. Clearly $\gamma_g^{'}(T-e) = k+1$.

Case 6: $\gamma_g^{'}(T) - \gamma_g^{'}(T-e) = 0$

For $k \geq 2$, let T be the graph obtained from $K_{1,k-1}$ by subdividing each edge. Clearly $\gamma_g^{'}(T) = k$. If e is any pendant edge then $\gamma_g^{'}(T-e) = k+1$.

For $k = 1$, there is no tree with $\gamma_g^{'}(T) = 1$ and $\gamma_g^{'}(T) - \gamma_g^{'}(T-e) = 0$.

3.2 Vertex Removal

Here, we consider the effect of vertex removal in trees. It may be noted that, there are trees T whose game domination number becomes arbitrarily large after removing a vertex from T. It is proved [2] that there is no graph G with $\gamma_g(G) = k$ and $\gamma_g(G - v) = k - 2$ for $k \leq 4$. We give examples of trees with $\gamma_g(T) = k$ and $\gamma_g(T - v) = k - t$ for any $t \in \{0, 1, 2\}$ and any integer $k \geq 5$.

Proposition 1. *For any $k \geq 5$ there exists a tree T with a vertex v such that $\gamma_g(T) = k$ and $\gamma_g(T - v) = k - 2$.*

Proof. Let T be a tree obtained from $K_{1,k-2}$ with v as its center in which each edge is subdivided and two vertices are attached at an end vertex u of one subdivided edge as in Fig. 1.

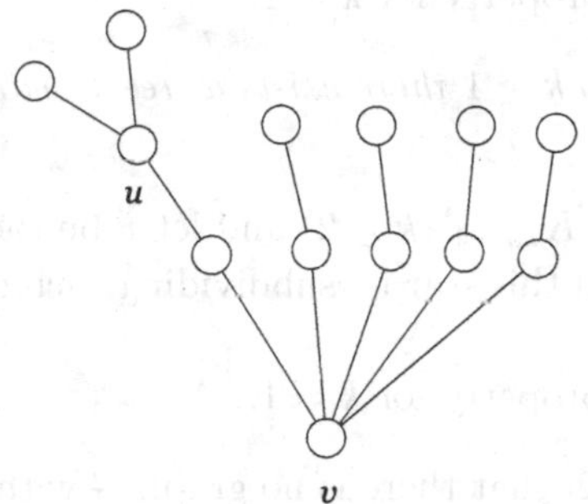

Fig. 1. A tree T with $\gamma_g(T) = 7$ and $\gamma_g(T - v) = 5$

Dominator first chooses the vertex v. So $\gamma_g(T) \leq 1 + \gamma_g'(T - v|N(v))$ and $\gamma_g'(T - v|N(v)) = 2 + k - 3$. Therefore $\gamma_g(T) \leq k$. Dominator never chooses a pendant vertex in T. Suppose that Dominator first chooses a vertex other than v and if it is u then Staller chooses the vertex adjacent to both u and v in T. In this case the game ends with atleast k moves. Suppose that Dominator's first turn is neither u nor v in T. In this case, Staller chooses a pendant vertex adjacent to u. If second move of Dominator is u then the game ends with atleast k moves. If second move of Dominator is a vertex other than u, then Staller chooses the other pendant vertex adjacent to u. In this case, the game ends with atleast k moves and hence the game domination number of T is k. Dominator first chooses the vertex u in $T - v$, after that Staller and Dominator alternately chooses a vertex from each component. So, the game on $T - v$ has $k - 2$ steps and hence $\gamma_g(T - v) = k - 2$.

Proposition 2. *For any $k \geq 1$ there exists a tree T with $\gamma_g(T) = k$ and $\gamma_g(T - v) = k - 1$ for some vertex $v \in V(T)$.*

Choose $T = P_n$, $n \geq 1$. This satisfies the above proposition, as mentioned in [2].

Proposition 3. *For any $k \geq 1$ there exists a tree T with $\gamma_g(T) = k$ and $\gamma_g(T - v) = k$ for some vertex $v \in V(T)$.*

Proof. Let k be a positive integer and let T' be an arbitrary tree with $\gamma_g(T') = k$ [4]. Let x be a first optimal move of Dominator in T'. Let T be the tree obtained from T' by attaching a vertex u to x [2]. In that case, T and $T - u$ have the same game domination number.

Proposition 4. *For any $k \geq 1$ there exists a tree T with $\gamma_g^{'}(T) = k$ and $\gamma_g^{'}(T - v) = k$ for some vertex $v \in V(T)$.*

Proof. Let T be a tree obtained from the star $K_{1,k-1}$ $(k \geq 2)$ by subdividing each edge. Let v be an end vertex of any subdivided edge. Then $\gamma_g^{'}(T) = k$ and $\gamma_g^{'}(T - v) = k$.
K_2 satisfies the desired property for $k = 1$.

Proposition 5. *For any $k \geq 1$ there exists a tree T with $\gamma_g^{'}(T) = k$ and $\gamma_g^{'}(T - v) = k - 1$*

Proof. Consider the star $K_{1,k-1}$ $(k \geq 2)$ and let v be the center of the star. Let T be a tree obtained from this star by subdividing each edge. Clearly, $\gamma_g^{'}(T) = k$ and $\gamma_g^{'}(T - v) = k - 1$.
K_1 satisfies the desired property for $k = 1$.

Remark 1. It is proved [2] that there is no graph G with $\gamma_g^{'}(G) = k$ and $\gamma_g^{'}(G - v) = k-2$ for $k < 4$ and there exist graphs G with $\gamma_g^{'}(G) = k$ and $\gamma_g^{'}(G-v) = k-2$ for $k \geq 4$.

Proposition 6. *There is no tree T with $\gamma_g^{'}(T) = 4$ and $\gamma_g^{'}(T - v) = 2$ for any vertex $v \in T$.*

Proof. Assume the contradiction. Let T be a tree with a vertex v such that $\gamma_g^{'}(T) = 4$ and $\gamma_g^{'}(T - v) = 2$. First, consider the case that v is a pendant vertex. Since T is a tree and tree is a no-minus graph [8], so $\gamma_g^{'}(T) - 1 \leq \gamma_g^{'}(T - v) \leq \gamma_g^{'}(T)$. Therefore $\gamma_g^{'}(T)$ is at most 3 and this contradicts $\gamma_g^{'}(T) = 4$. Now, consider the case that v is a cut vertex. In this case $T - v$ is disconnected with exactly two components. If $T - v$ has more than two components then Staller start game domination number of $T - v$ is at least 3. This is not possible. So clearly $T - v$ has exactly two components T_1 and T_2. Each component is either K_1 or K_2, otherwise it contradicts that $\gamma_g^{'}(T - v)$ is 2. Since T is a tree, v is adjacent to exactly one vertex in each component. In this case $\gamma_g^{'}(T)$ is at most 3. This contradicts that $\gamma_g^{'}(T) = 4$.

Proposition 7. *There is no tree with $\gamma_g^{'}(T) = 5$ and $\gamma_g^{'}(T - v) = 3$ for any vertex v in T.*

Proof. Assume the contradiction. Let T be a tree with a vertex v such that $\gamma_g^{'}(T) = 5$ and $\gamma_g^{'}(T-v) = 3$. Removal of a pendant vertex from a tree decreases its game domination number by at most 1. So clearly v is a cut vertex and $T-v$ has at most 3 components. Now, we prove that the vertex v is not an optimal first move of Staller in T. If possible let v be an optimal first move of Staller in T. Then

$$\begin{aligned}\gamma_g^{'}(T) &= 1 + \gamma_g(T|N[v]) \\ &\leq 1 + \gamma_g^{'}(T|N[v]) \\ &= 1 + \gamma_g^{'}(T - v|N(v)) \\ &\leq 1 + \gamma_g^{'}(T - v).\end{aligned}$$

Hence, Staller start game domination number of $T - v$ is decreased by at most 1. First, we consider the case that $T - v$ has 3 components. In this case each component is either K_1 or K_2. Staller first chooses a vertex from any of the three components and then Dominator chooses v. In this case the game is finished in at most 4 steps. This contradicts that $\gamma_g^{'}(T) = 5$.

Now consider the case that $T - v$ has exactly two components say T_1 and T_2. In this case one component say T_1 has $\gamma_g^{'}(T_1) = 1$ and the other component T_2 has $\gamma_g^{'}(T_2) = 2$. So T_1 is either K_1 or K_2 and there is a vertex in T_2 which is adjacent to all undominated vertices in T_2 after the first move of Staller. Consider a staller start game played on T and first optimal move of Staller is from either T_1 or T_2. If the first optimal move is from T_1 then Dominator chooses v after that Staller chooses a vertex from T_2 and Dominator chooses a vertex from T_2 that dominates all the undominated vertices in T. So the game on T is finished in at most 4 steps. This contradicts $\gamma_g^{'}(T) = 5$. If the first optimal move is from T_2 then Dominator chooses a vertex which is adjacent to all the undominated vertices in T_2. After that, Staller chooses a vertex from T and if with this move the game is not yet over, Dominator chooses a vertex in T_2 which is adjacent to

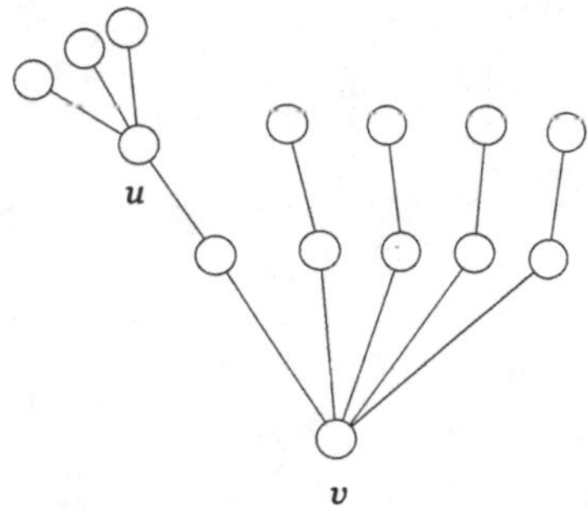

Fig. 2. A tree T with $\gamma_g^{'}(T) = 8$ and $\gamma_g^{'}(T - v) = 6$

v. So, the game is finished in at most 4 steps. This is a contradiction. So there is no tree with $\gamma_g'(T) = 5$ and $\gamma_g'(T - v) = 3$.

Proposition 8. *For any $k \geq 6$ there exists a tree T with a vertex v such that $\gamma_g'(T) = k$ and $\gamma_g'(T - v) = k - 2$.*

Proof. Let T be the tree obtained from a $K_{1,k-3}$ by subdividing each edge where v as the center of $K_{1,k-3}$ and attaching three vertices to one of the end points say u of a subdivided edge as in Fig. 2. Clearly, $\gamma_g'(T) = k$ and $\gamma_g'(T-v) = k-2$.

Acknowledgments. The authors thank the Erudite programme of the Kerala State Higher Education Council, Government of Kerala, India for funding the visit of the second author during March 2014.

References

1. Bartnicki, T., Grytczuk, J., Kierstead, H.A., Zhu, X.: The map-coloring game. Am. Math. Mon. **114**, 793–803 (2007)
2. Brešar, B., Dorbec, P., Klavžar, S., Košmrlj, G.: Domination game: effect of edge- and vertex-removal. Discrete Math. **330**, 1–10 (2014)
3. Brešar, B., Klavžar, S., Košmrlj, G., Rall, D.F.: Domination game: extremal families of graphs for the $\frac{3}{5}$- conjectures. Discrete Appl. Math. **161**, 1308–1316 (2013)
4. Brešar, B., Klavžar, S., Rall, D.F.: Domination game and imagination strategy. SIAM J. Discrete Math. **24**, 979–991 (2010)
5. Brešar, B., Klavžar, S., Rall, D.F.: Domination game played on trees and spanning subgraphs. Discrete Math. **313**, 915–923 (2013)
6. Bujtás, C.: Domination game on trees without leaves at distance four. In: Frank, A., Recski, A., Wiener, G. (eds.) Proceedings of the 8th Japanese-Hungarian Symposium on Discrete Mathematics and Its Applications, June 4–7, 2013, Veszprém, Hungary, pp. 73–78 (2013)
7. Dorbec, P., Košmrlj, G., Renault, G.: The domination game played on unions of graphs. Discrete Math. **338**, 71–79 (2015)
8. Kinnersley, W.B., West, D.B., Zamani, R.: Extremal problems for game domination number. SIAM J. Discrete Math. **27**, 2090–2107 (2013)

Modeling of Dissolved Oxygen Using Genetic Programming Approach

S. Vanitha(✉), C. Sivapragasam, and N.V.N. Nampoothiri

Department of Civil Engineering, Center for Water Technology, Kalasalingam University, Krishnankovil 626126, Tamil Nadu, India
svanithacivil@gmail.com, sivapragasam25@gmail.com

Abstract. Genetic Programming (GP) based modeling is suggested for modeling the variation of Dissolved Oxygen (DO) under controlled conditions in the presence and absence of toxicant. The results indicated that GP is able to evolve robust physically meaningful models even with small dataset by selecting the most relevant functions from the set of functions given for the modeling. It is interesting to note that the evolved models clearly reflect the underlying non-linearity of the process distinctly for both the case studies.

Keywords: Genetic Programming · Dissolved Oxygen · Mathematical Modeling

1 Introduction

Dissolved Oxygen (DO) in water bodies is essential for microorganisms and is a significant indicator of the state of aquatic ecosystem [7]. A certain minimum level of DO in water is required for aquatic life. DO can range between 0–18 parts per million (ppm) or milligram per litre (mg/l), but most natural water systems require 5–6 mg/l to support a diverse population [5]. DO can vary within a given ecosystem as a function of many interrelated complex parameters such as presence of organic pollutants, water body temperature, the biological activities of aquatic species etc. It is necessary, therefore, to develop models which can efficiently predict the DO variation. In the past, researchers have attempted both physical models as well as black box models for modeling DO. Radwan et al. [13] used MIKE 11 for DO modeling using climate, soil and crop data. AQUATOX is a commonly used simulation model for aquatic system and finds many applications by various researchers [4,8,9,12]. However, application of such models require many detailed information of the ecosystem being modeled which in many cases may be difficult to obtain or define accurately.

Of late, Artificial Neural Networks (ANN) based models have found increasing application in DO modeling due to its ability to model complex processes. For instance, Areerachakul et al. [5] used pH, Biochemical Oxygen Demand (BOD), Chemical Oxygen Demand (COD), Suspended Solids (SS), Total Kjeldhal Nitrogen (TKN), Ammonia Nitrogen(NH3N), Nitrite Nitrogen (NO2N),

S. Arumugam et al. (Eds.): ICTCSDM 2016, LNCS 10398, pp. 445–452, 2017.
DOI: 10.1007/978-3-319-64419-6_56

Nitrate Nitrogen (NO3N), Total Phosphorous(TP), Total Coliform as the inputs to model DO with 269 training data set and 115 test set. Chen et al. [7] used water temperature, pH, Electrical conductivity (EC), turbidity, Total Suspended Solids (TSS), Total Hardness (TH), Alkalinity and Ammonium nitrogen (NH4N) to model DO. A summary of ANN application for DO modeling by various investigators is given in Table 1. A variety of ANN training algorithm has been utilized such as GRNN, BPNN, RNN etc. The input data is seen to vary in time horizon from daily data to monthly data. It is reported that ANN performs well in the DO modeling.

It is important to note that there is no common consensus on the inputs being used with different researchers using different combination of inputs for modeling. However, many a times, it may be of interest for designers and engineers to know a quick estimate of DO variation in a given ecosystem without considering all the detailed parameters which affect its changes. It will be ideally more meaningful if such variations can be demonstrated through some simple mathematical models or equations. This study is an attempt in this direction. It is proposed to model the DO variation in a controlled ecosystem knowing only the initial DO concentration in the presence and absence of an externally induced toxicant as a function of time.

Genetic Programming (GP) is considered in this study in lieu of ANN. This is because, although ANN has a demonstrated potential to model complex non-linear processes, it doesn't reveal the nature of non-linearity [14]. GP, however, can evolve mathematical models clearly reflecting the non-linearity of the system [11,15,16]. To the authors' knowledge, this is the first attempt of GP in DO modeling of an ecosystem.

2 Experimental Setup and Tools Used

2.1 Experimental Setup

Two experimental tanks circular in size are created under controlled conditions (Figs. 1 and 2).

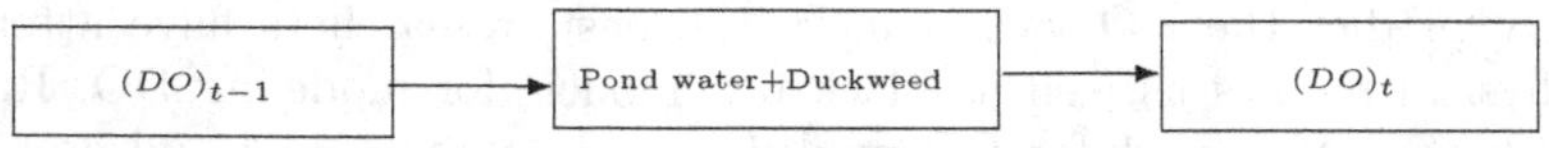

Fig. 1. Block diagram for Case Study 1

In first case, 15g duckweed is spread in 15 litre pond water. In second case, 0.5g glyphosate toxicant is added in 15 litre pond water containing 15 g duckweed. DO is measured daily in the two tanks using Winkler's method. The experimental case studies are run for 10 days without additional inflow loading.

Table 1. Review of Dissolved Oxygen model

S.No	Tools used	Input parameters	Time horizon	Data length	Training algorithm	References
1	ANN	pH, BOD, COD,TSS, TKN, NH3N, NO2N, NO3N, TP, Total Coliform	269 records training set 115 records test set	2006–2008 3 years	MLP was used with the Levenberg- Marquardt algorithm is used to train the ANN	Areerachakul et al. [5])
2	ANN	pH, EC, Ca2+,TA, TH,,Cl-, NO3N, NH3N	100-Training set, 32-Testing set, 32-Vaidation set	2003–2008	A three-layer back-propagation ANN was used with the Bayesian regularization training algorithm	Wen et al. [17]
3	ANN and MLR	T, pH, EC, Turbidity, SS, TH, TA, NH4N	280 -Training sets, 120- Testing sets	1993 to 2011	Two ANN models including BPNN and ANFIS approaches and MLR model	Chen et al. [7]
4	ANN	Water flow, pH, T, EC	monthly or semi-monthly	2004–2009	GRNN, BPNN and RNN	Antanasijevic et al. [2]
5	ANN	T, pH, HCO3-, SO42-, NO3-N, TH, Na, Cl-, EC, PO4, Ca, Mg, TH, TSS, CO2, K. BOD and TA	Each month at 17 different sites	9 years	GRNN. Uncertainty analysis of model results using the Monte Carlo Simulation (MCS) technique	Antanasijevic et al. [3]
6	ANN	Q, T, pH, BOD, DO	Monthly dataset	1990–1996	Feed Forward error back Propagation algorithm	Archana Sarkar & Prashant Pandey [1]
7	ANN, MLR	pH, Q, T, EC, DO	Daily dataset	6828 samples	MLP, RBNN, MLR	Murat Ay & Ozgur Kisi [6]

GRNN - General Regression Neural Network	Q - Discharge
MLP - Multi layer Perceptron	DO - Dissolved Oxygen
RBNN- Radial Based Neural Network	COD- Chemical Oxygen Demand
MLR-Multi Linear Regression	TSS-Total uspended Solids
RNN- Recurrent Neural Network	TKN-Total Kjeldahl Nitrogen
ANFIS-Adaptive neural-based fuzzy inference system	NH_3N-Ammonia Nitrogen
EC-Electrical Conductivity	NO_2N-Nitrite Nitrogen
Cl^-- Chloride	NO_3N-Nitrate Nitrogen
SO_4 -Sulphate	TP- Total Phosphorous
TH-Total Hardness	NH_4N-Ammonium Nitrogen
TA-Total Alkalinity	T-Water temperature

2.2 Genetic Programming

In this study, Genetic Programming (GP) is used to model DO. GP is very similar to Genetic Algorithm (GA), an evolutionary algorithm based on Darwinian theories of natural selection and survival of the fittest. However, GP operates on parse trees, rather than on bit strings as in a GA, to approximate the equation (in symbolic form) that best describes how the output relates to the input variables. The algorithm considers an initial population of randomly generated

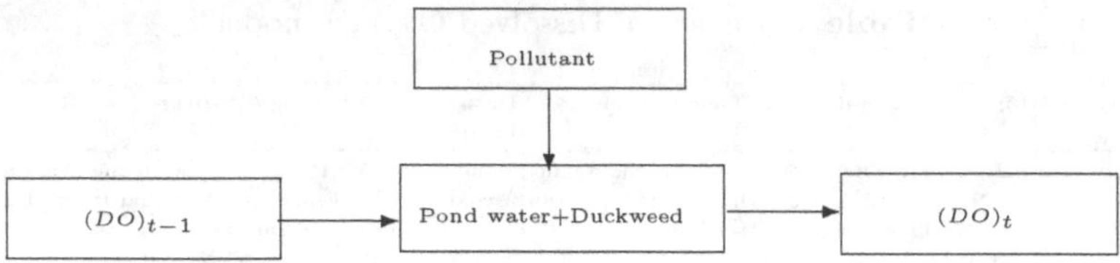

Fig. 2. Block diagram for Case Study 2

programs (equations), derived from the random combination of input variables, random numbers and functions. The functions can include arithmetic operators (plus, minus, multiply, divide), mathematical functions (sin, cos, exp, log), logical/comparison functions (OR/AND) etc., which have to be appropriately chosen based on some understanding of the process. This population of potential solutions is then subjected to an evolutionary process and the 'fitness' (a measure of how well they solve the problem) of the evolved programs are evaluated. Individual programs that best fit the data are then selected from the initial population. The programs that best fit are selected to exchange part of the information between them to produce better programs through 'crossover' and 'mutation', as used in GAs (to mimic the natural reproduction process). Here, exchanging the parts of best programs with each other is called crossover, copied exactly into the next generation s called reproduction and randomly changing programs to create new programs is called mutation [10]. The user must decide a number of GP parameters before applying the algorithm to model the data, such as population size, number of generations, crossover and mutation probability, etc. The programs that fitted the data less well are discarded. This evolution process is repeated over successive generations and is driven towards finding symbolic expressions describing the data, which can be scientifically interpreted to derive knowledge about the process being modeled [15]. GP is implemented in this study using Discipulus software.

3 Performance Measure

RMSE value is taken as the performance measure to check the performance of GP as shown in Eq (1).

$$RMSE = \sqrt{\frac{1}{n}\sum_{i=1}^{n}[(X_m)i - (X_s)i]^2} \tag{1}$$

where X is any variable that is being modeled; the subscripts m and s represent the observed and simulated values.

4 Results and Discussions

The results obtained from experimental studies and the development of mathematical model is described as below:

Training Data Set and Modeling Function. In order to evolve model which is more generic in nature, it is desired to model the percentage of residual DO at any time 't' (rather than actual value of DO) with residual DO in the previous time step. This can be functionally represented as

$$(DO_R)_t = f((DO_R)_{t-1}, t) \tag{2}$$

where$(DO_R)_t$ is the residual DO at time t (%);
$(DO_R)_{t-1}$ is the residual DO at previous time step $t-1$ (%);
$t=$ detention time (hours);

Table 2. Dissolved Oxygen (mg/l) variation as a function of time

Time 't' in (hours)	Case Study 1	Case Study 2
Initial	6.72	6.72
24	6.08	6.08
48	4.64	3.84
72	1.92	1.92
120	2	1.76
144	2.2	2.4
168	2.4	2.5
192	2.72	2.56
216	2.88	2.7
240	3.04	3.2

GP is run with the optimal parameter values (after trial and error) of population size of 500, cross over frequency at 50% and mutation frequency at 95% for 100 generations. Since it is expected that DO might vary drastically in a short time span, it is preferred to include exponential function besides using the arithmetic function in the GP run. It is also preferred to include trigonometric function (though it does not bear any direct relationship with the physics of the process) so as to capture a small non-linearity which might be superimposed on a predominantly linear process. Experiments are conducted for 10 days. Actual DO value obtained during experimental period is shown in Table 2. Since DO variation is a time dependent process, it is desired to use the first 4 data as training set, the next 3 as testing set and the last 2 data set as validation set (Table 3).

Table 3. Training data used for Dissolved Oxygen model

Data	Case Study 1			Case Study 2		
	$(DO_R)_{t-1}$ (%)	Time 't' (Hours)	$(DO_R)_t$ (%)	$(DO_R)_{t-1}$ (%)	Time 't' (Hours)	$(DO_R)_t$ (%)
Training	100	24	90.5	100	24	90.5
	90.5	48	69	90.5	48	57.2
	69	72	28.5	57.2	72	28.6
	28.5	120	29.7	28.6	120	26.2
Testing	29.7	144	32.7	26.2	144	35.7
	32.7	168	35.7	35.7	168	37.2
	35.7	192	40.4	37.2	192	38.0
Validation	40.4	216	42.8	38.0	216	40.2
	42.8	240	45.2	40.2	240	47.6

Modeling of DO Using GP. The GP evolved mathematical model for DO is shown in Table 4. For both the case studies, the nature of DO variation before and after 3 days is different. While during the first 3 days, the variation is non-linear, after 3 days, it is found to linearly vary with the residual DO in the previous time step. A close observation also indicates that the nature of non-linearity in Case Study 1 is different from that for Case Study 2 with residual DO varying more sharply as a function of time in the Case Study 2. This can be explained as due to the introduction of the toxicant. It can be interpreted that the effect of toxicant is quickly neutralized by the ecosystem with duckweed which reflects in terms of residual DO in the system becoming almost constant. However, the residual DO gain in the tail end of the study (in Case Study 2) is not reflected in the GP model due to lack of sufficient data. Whereas, in the absence of the toxicant (in Case Study 1), the residual DO after 3 days is found to marginally increase at the rate of 8% of the residual DO at the previous time step.

Table 4. Dissolved Oxygen model

S.No	Experimental Case Study	$(DO_R)_t$ model
1	Case Study 1	$(DO_R)_t(\%) = \begin{cases} Abs\left\{\sqrt{\frac{4}{1.18(DO_R)_{t-1}-t}} - 1.08\right\} \text{ for } t \leq 3 \text{ days} \\ *(DO_R)_{t-1}(\%), \\ 1.08 * (DO_R)_{t-1}(\%), \quad \text{for } t > 3 \text{ days} \end{cases}$
2	Case Study 2	$(DO_R)_t(\%) = \begin{cases} 0.09\frac{(DO_R)^2_{t-1}}{t}(\%) + 0.452 \text{ for } t \leq 3 \text{ days} \\ *(DO_R)_{t-1}(\%), \\ (DO_R)_{t-1}(\%), \quad \text{for } t > 3 \text{ days} \end{cases}$

It is to be noted that although exponential and trigonometric functions were included in the GP run, these functions do not appear in the final model. Figures 3 and 4 shows the comparison of actual DO and that estimated by GP evolved model. It can be seen that GP quite accurately models the two ecosystem conditions.

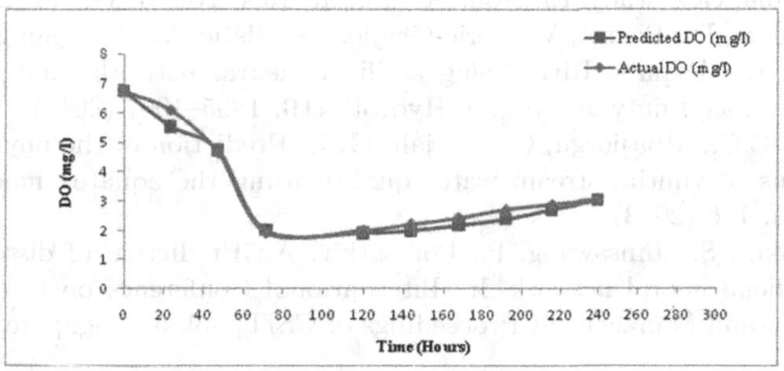

Fig. 3. Comparision of Predicted DO vs Observed DO for Case Study 1

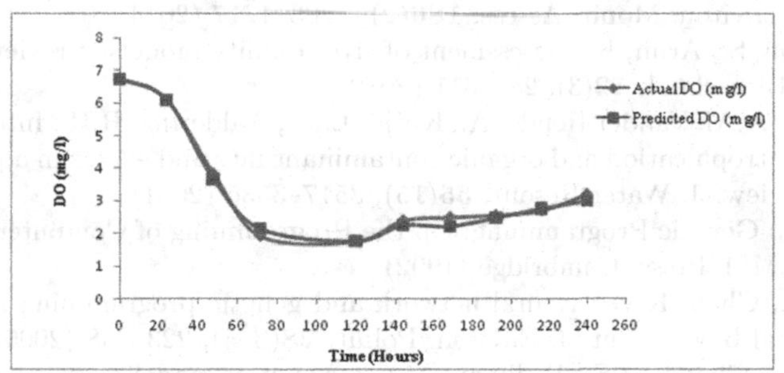

Fig. 4. Comparision of Predicted DO vs Observed DO for Case Study 2

5 Conclusions

Based on this study, the following conclusions have been arrived at:

(a) The nature of non-linearity is easily modeled in GP which is difficult in ANN.
(b) GP quite accurately models a natural ecosystem even with very less training data set, and hence can be used for modeling of aquatic systems.
(c) It is recommended to carry out more detailed research with additional parameters governing the ecosystem to have a deeper understanding of the physics of the ecosystem.

References

1. Archana, S., Prashant, P.: River water quality modelling using artificial neural network technique. Aquat. Procedia **4**, 1070–1077 (2015)
2. Antanasijevic, D., Pocajt, V., Povrenović, D., Perić-Grujić, A., Ristić, M.: Modelling of dissolved oxygen content using artificial neural networks: Danube River, North Serbia, case study. Environ. Sci. Pollut. Res. **20**, 9006–9013 (2013)
3. Antanasijevic, D., Pocajt, V., Peric-Grujic, A., Ristic, M.: Modelling of dissolved oxygen in the Danube River using artificial neural networks and Monte Carlo Simulation uncertainty analysis. J. Hydrol. **519**, 1895–1907 (2014)
4. Anyadike, C.C., Mbajiorgu, C.C., Ajah, G.N.: Prediction of the physico-chemical interactions of vimtim stream water quality using the aquatox model. IOSR J. Eng. **3**(10), 1–6 (2013)
5. Areerachakul, S., Junsawang, P., Pomsathit, A.: Prediction of dissolved oxygen using artificial neural network. In: International Conference on Computer Communication and Management Proceedings of CSIT, vol. 5, Singapore, pp. 524–528 (2011)
6. Ay, M., Kisi, O.: Modeling of dissolved oxygen concentration using different neural network techniques in foundation Creek, El Paso County, Colorado. J. Environ. Eng. **138**(6), 654–662 (2012)
7. Chen, W.B., Liu, W.C.: Artificial neural network modeling of dissolved oxygen in reservoir. Environ. Monit. Assess. **186**(2), 1203–1217 (2014)
8. Deepshikha, S., Arun, K.: Assessment of river quality models: a review. Rev. Environ. Sci. Biotechnol. **12**(3), 285–311 (2012)
9. Koelmans, A.A., Vander Heude, A., Knijff, L.M., Aalderink, R.H.: Integrated modelling of Eutrophication and organic contaminant fate and effects in aquatic ecosystems. a review. J. Water Resour. **35**(15), 3517–3536 (2001)
10. Koza, J.R.: Genetic Programming: On the Programming of Computers by Natural Selection. MIT Press, Cambridge (1992)
11. Muttil, N., Chau, K.W.: Neural network and genetic programming for modelling coastal algal blooms. Int. J. Environ. Pollut. **28**(3–4), 223–238 (2006)
12. Park, R.A., Clough, J.S., Wellman, M.C.: Aquatox modeling environmental fate and ecological effects in aquatic ecosystems. J. Environ. Model. **213**, 1–15 (2008)
13. Radwan, M., Willems, P., El-sadek, A., Berlamont, J.: Modelling of dissolved oxygen and biochemical oxygen demand in river water using a detailed and a simplified model. Intl. J. River Basin Manage. **1**(2), 97–103 (2003)
14. Schmid, B.H., Koskiaho, J.: Artificial neural network modeling of dissolved oxygen in a wetland pond: the case of Hovi, Finland. J. Hydrol. Eng. **11**(2), 188–192 (2006)
15. Sivapragasam, C., Muttil, N., Jeselia, M.C., Visweshwaran, S.: Infilling of rainfall information using genetic programming. Aquat. Procedia **4**, 1016–1022 (2015)
16. Sivapragasam, C., Mutti, N., Muthukumar, S., Arun, V.M.: Prediction of algal blooms using genetic programming. Mar. Pollut. Bull. **60**, 1849–1855 (2010)
17. Wen, X., Fang, J., Diao, M., Zhang, C.: Artificial neural network modeling of dissolved oxygen in the Heihe River, Northwestern China. Environ. Monit. Assess. **185**(5), 4361–4371 (2013)

A Stream Cipher for Real Time Applications

M. Venkatesulu and M. Ravi(✉)

Kalasalingam University, Anand Nagar, Krishnankoil 626126, Tamilnadu, India
venkatesulum2000@gmail.com, m.ravi@klu.ac.in

Abstract. Cryptography plays an important role in securing data against malicious attacks. Cryptography can be applied in designing and developing block ciphers and stream ciphers to protect data from intruders and unauthorized users. Based on usage of keys, cryptography is classified into symmetric cryptography and asymmetric cryptography. Asymmetric key based cryptosystems and block ciphers usually take more time to encrypt/decrypt data and hence found not suitable for real time applications. Symmetric key based cryptosystems and stream ciphers are generally recommended for securing data in real time. In this paper we propose a new stream cipher which can be used to encrypt/decrypt data in real time. As an application; we encrypt DVD content, playback the DVD content in real time by decrypting and integrating with a customized media player with no saving or copying possibilities.

1 Introduction

Every human activity is being digitized day in and day out. World is moving towards paperless arena. Be it audio, text, image, video all are digitized. Multimedia applications are entering into every sphere of activity. But with it carries lot of challenges, one such most important challenge is the Safety and Security of the data. We hear in every day report the data leakage, data theft and hacking of personal data, business data and governmental data. The reason being thieves and hackers are cleverer and hardworking than normal people. Protection of DATA is the biggest challenge of the present day digitized world. Cryptography is found to be the best tool to face the challenge of protecting data. It has given number of Block ciphers and Stream ciphers, some based on asymmetric keys and some based on symmetric keys. DES, 3-DES, AES, IDEA, RC2, RC5, Blowfish, Cast and Gost are some of the well known block ciphers and A5/I and RC4 are well known stream ciphers. Twofish, Serpant, AES, Blowfish, RC4, Grasshopper, 3-DES and IDEA are some of the well known symmetric key algorithms, and RSA, El Gamal and Elliptic curve cryptography are well known asymmetric key algorithms.

Arul Jothi and Venkatesulu [1–4] have presented a few schemes for data security. The first scheme uses only row and column XORing with the help of key, to increase confusion and diffusion in the resultant cipher text. Another scheme is presented which includes shifting and XORing with neighborhood bits and to further increase the efficiency of the encryption one more scheme is proposed which also performs rotation and substitution.

S. Arumugam et al. (Eds.): ICTCSDM 2016, LNCS 10398, pp. 453–456, 2017.
DOI: 10.1007/978-3-319-64419-6_57

Radha and Venkatesulu [5–9] have given the analysis of block ciphers AES, RC6 and IDEA in video streaming, also a block cipher algorithm which provides non-linear substitution and permutation components through the binary tree structure and random shifting row/column wise in a circular fashion in a square matrix, another block cipher algorithm which encrypts and decrypts a block of 512 bits based on primitive operation XOR, shuffler function and a chaotic map with non-linear transformation functions, and yet another block cipher algorithm using the shuffler operator and a binary tree which provides confusion and diffusion components for the round operations. They also present a block cipher design incorporating the 'bit-level' primitive GRP instruction and architecture for the processor instruction SHUFF to improve the performance of bit -level mappings.

Vidhya Saraswathi and Venkatesulu [10–12] present a few block ciphers; the first block cipher encryption for protecting multimedia data while transmitting through the network, the second encryption and decryption process in which chaotic maps are used for generating two keys, K_1 and K_2 of length 64 bits from the 128 bit master key, a third Image Encryption scheme for color images using chaotic maps and a fourth Block Cipher based on Boolean Matrices using Bit level Operations by generating lower triangular matrix as session key, performing GRP operation in encryption scheme, and UNGRP operation and inverse of lower triangular matrices in decryption scheme.

In this paper, we propose a new stream cipher to protect data in real time. As an application, the stream cipher is used to encrypt the DVD contents in off-line mode and to decrypt the encrypted DVD in real time to playback the contents. The decryption is integrated with the media player in such a way that no saving or copying is possible during playback mode.

2 Algorithm

Step 1: (Key Selection): Choose a binary string K of size $4n$, where $n =$ 4, 8, 16, 32, 64, ..., 512, etc.

Step 2: (Session Key generation): Choose a random permutation π of integers $\{1, 2, \ldots, n, \ldots, 2n, \ldots, 3n, \ldots, 4n\}$.
Let $\pi = \{p_1, p_2, \ldots, p_n, \ldots, p_{2n}, \ldots, p_{3n}, \ldots, p_{4n}\}$. Take a sub permutation P of π, say, $P = \{p_1, p_2, \ldots, p_n\}$.

Step 3: (Key stream generation): Arrange the elements of P in ascending order, say, $Q_1 < Q_2 < \cdots < Q_n$. Then $P_n = Q_s$ for some between **1** and **n**. Let $d_1 = Q_2 - Q_1$, $d_2 = Q_3 - Q_2, \ldots, d_{n-1} = Q_n - Q_{n-1}$. Let the initial Vector: $Q = \{Q_1, Q_2, \ldots, Q_n\}$ Define initial key stream bit $b_1 =$ Q_1 XOR Q_2 XOR Q_3,..., Q_{n-1} XOR Q_n. Subsequent key stream bits generation ($j > 1$):new Q_1= old Q_s +1. new Q_l =(new $Q_{l-1} + d_l$)mod 4n, for l = 2, ..., $n-1$, b_j =(new Q_1) XOR (new Q_2),...,Q_{n-1} XOR (new Q_n).

Step 4: (Encryption): Let **M** be the plain text converted into binary form, say, **M** =a_1, a_2,... a_j ...,a_n Encrypted text is given by C= (c_1,c_2... c_j ...,a_n), where $c_j = a_j$ XOR b_j.

Step 5: (Decryption): The plain text **M** can be obtained from the cipher text by XORing with the key bits stream, that is, c_j XOR $b_j = a_j$.

3 Application for DVD Content Protection

Initial Key K = 1001010101010101, user's random selection. Initial Permutation π is pseudo-random (system) generated and sub permutation P is the first four elements of π. The application was run on the i^5-64 bit processor @ 2.4 GHz speed with 8 GB RAM. The algorithm was implemented in C-Sharp using visual studio 2010 under windows 10 environment. The video file size is 710 MB and has taken just 1 min for encryption. The start-up delay for decryption and play-back is just 20 s and then on the play-back is continuous.

4 Conclusion

In this paper we proposed a new stream cipher for multimedia encryption in real time. As an application we encrypted DVD content and playback is done in real time by decrypting the encrypted DVD and integrating the decryption with customized media player with no saving and copying possibilities. In future, we wish to employ the cipher in different applications and also do the crypt analysis of the proposed cipher.

Acknowledgement. The second author is thankful to the management of Kalasalingam University for providing fellowship. The authors thank the referee for the valuable suggestions which resulted in the present form of the paper.

References

1. Arul Jothi, S., Venkatesulu, M.: Symmetric key cryptosystem based on randomized block cipher. In: 5th International Conference on Future Information Technology, 21–23 May 2010, Busan, South Korea, pp. 1–5. Future Tech (2010)
2. Jothi, S.A., Venkatesulu, M.: A generalized key scheme in a block cipher algorithm and its cryptanalysis. Int. J. Comput. Appl. **49**(9), 1–6 (2012)
3. Arul, J.S., Venkatesulu, M.: Encryption quality and performance analysis of GKSBC algorithm. J. Inf. Eng. Appl. **2**(10), 26–34 (2012)
4. Arul, J.S., Venkatesulu, M.: A new approach to randomized key in the symmetric block cipher encryption algorithm. Jokull J. **63**(7), 190–199 (2013)
5. Radha, N., Venkatesulu, M.: Performance of block ciphers in real-time multimedia applications. Eur. J. Sci. Res. **112**(4), 557–563 (2013)
6. Radha, N., Venkatesulu, M.: A complete binary tree structure block cipher for real-time multimedia. In: IEEE Technically Co-sponsored Science and Information Conference, pp. 346–352 (2013)
7. Radha, N., Venkatesulu, M.: A chaotic block cipher for real-time multimedia. J. Comput. Sci. **8**(6), 994–1000 (2012)
8. Radha, N., Venkatesulu, M.: Permutation induced shuffling based block cipher for real-time applications. Malays. J. Comput. Sci. **27**(1), 20–34 (2014)

9. Vidhya, S.P., Venkatesulu, M.: A block cipher for multimedia content protection with random substitution using binary tree traversal. J. Comput. Sci. **8**(9), 1541–1546 (2012)
10. Vidhya, S.P., Venkatesulu, M.: A block cipher for multimedia encryption using chaotic maps for key generation. In: Proceedings of International Conference, Advances in Information Technology and Mobile Computing (AIM-2013), pp. 277–282. Elsevier Science and Technology (2013)
11. Vidhya, S.P., Venkatesulu, M.: A secure image content transmission using discrete chaotic chaotic maps. Jokull J. **63**(9), 404–418 (2013)
12. Vidhya, S.P., Venkatesulu, M.: A block cipher based on boolean matrices using bit level operations. In: Proceedings of IEEE/ACIS International Conference on Computer and Information Science, pp. 59–63 (2014)

Erratum to: On Total Roman Domination in Graphs

P. Roushini Leely Pushpam[1] and S. Padmapriea[2(✉)]

[1] Department of Mathematics, D.B. Jain College, Chennai 600 097, India
roushinip@yahoo.com
[2] Department of Mathematics, Sri Sairam Engineering College,
Chennai 600 044, India
prieas@gmail.com

Erratum to: Chapter "On Total Roman Domination in Graphs" in: S. Arumugam et al. (Eds.), Theoretical Computer Science and Discrete Mathematics, LNCS 10398, https://doi.org/10.1007/978-3-319-64419-6_42

By mistake, the name of the second author was not spelled correctly. Instead of "Padmapriea Sampath" it shall read "S. Padmapriea".

The updated online version of this chapter can be found at
https://doi.org/10.1007/978-3-319-64419-6_42

S. Arumugam et al. (Eds.): ICTCSDM 2016, LNCS 10398, p. E1, 2017.
https://doi.org/10.1007/978-3-319-64419-6_58

Author Index

Printed in the United States
By Bookmasters